HANDBOOK ON TOURISM AND RURAL COMMUNITY DEVELOPMENT

RESEARCH HANDBOOKS IN TOURISM

Series Editor: Robin Nunkoo, *University of Mauritius*

This timely series brings together critical and thought-provoking contributions on key topics and issues in tourism and hospitality research from a range of management and social science perspectives. Comprising specially commissioned chapters from leading authors, these comprehensive *Research Handbooks* feature cutting-edge contributions and are written with a global readership in mind. Equally useful as reference tools or high-level introductions to specific topics, issues, methods and debates, these *Research Handbooks* will be an essential resource for academic researchers, practitioners, undergraduate and postgraduate students.

Titles in this series include:

Handbook of Social Tourism
Edited by Anya Diekmann and Scott McCabe

Handbook for Sustainable Tourism Practitioners
The Essential Toolbox
Edited by Anna Spenceley

Handbook on Tourism and Social Media
Edited by Dogan Gursoy and Rahul Pratap Singh Kaurav

Handbook on the Tourist Experience
Design, Marketing and Management
Edited by Dora Agapito, Manuel Alector Ribeiro and Kyle Maurice Woosnam

Handbook of Innovation for Sustainable Tourism
Edited by Irma Booyens and Patrick Brouder

Handbook of Niche Tourism
Edited by Marina Novelli, Joseph M. Cheer, Claudia Dolezal, Adam Jones and Claudio Milano

Handbook on Tourism, Public Health and Wellbeing
Edited by Robin Nunkoo, Jun Wen and Metin Kozak

Handbook on Tourism and Rural Community Development
Edited by Heather Mair

Handbook on Tourism and Rural Community Development

Edited by

Heather Mair

Department of Recreation and Leisure Studies, University of Waterloo, Canada

RESEARCH HANDBOOKS IN TOURISM

Edward Elgar PUBLISHING

Cheltenham, UK • Northampton, MA, USA

Published by
Edward Elgar Publishing Limited
The Lypiatts
15 Lansdown Road
Cheltenham
Glos GL50 2JA
UK

Edward Elgar Publishing, Inc.
William Pratt House
9 Dewey Court
Northampton
Massachusetts 01060
USA

A catalogue record for this book
is available from the British Library

Library of Congress Control Number: 2023931300

This book is available electronically in the **Elgar**online
Geography, Planning and Tourism subject collection
http://dx.doi.org/10.4337/9781800370067

Printed on elemental chlorine free (ECF)
recycled paper containing 30% Post-Consumer Waste

ISBN 978 1 80037 005 0 (cased)
ISBN 978 1 80037 006 7 (eBook)

Printed and bound in the USA

To Professor Keith Hollinshead.
A colleague, collaborator, champion, mentor, provocateur, and friend.
Wishing you blue skies and coloured paper.

Contents

Contributors

Roslizawati Che Aziz is a Senior Lecturer and has been affiliated to the Universiti Malaysia Kelantan (UMK), Malaysia since 2008. She received her PhD at the Universiti Putra Malaysia in 2016, a Master's in Business Administration (tourism and hospitality) in 2010 and a degree in tourism management in 2006. Her research interests are in sustainable rural tourism and community development, the Appreciative Inquiry approach, community-based rural tourism, recreation and outdoor management, and tourism impacts.

Karla A. Boluk is an Associate Professor in the Department of Recreation and Leisure Studies at the University of Waterloo, Canada. Guided by a social justice paradigm, the broad goals of her research programme are to examine the impacts of tourism social entrepreneurs. Her work critiques the United Nations Sustainability Goals (SDGs), as well as considers how the global framework could improve our understanding of how tourism may operate more responsibly.

Annæ Buchmann has an environmental science background with a PhD in social science. Her interdisciplinary research looks at long-term strategies for creating sustainable communities and organisations. Annæ is known for her contributions to film and media tourism, and researching semi-rural developments, and multi-use advocacy in decision-making processes of land management, as well as destination branding, marketing and emergency preparedness. Her research is disseminated in engaging presentations and publications, and Annæ contributes as an expert reviewer and advisory boards member.

Christine N. Buzinde is a Professor at Arizona State University, USA. Her research focuses on two areas: community development through tourism and the politics of tourism representations. Her work on representations regards tourism texts as cultural repositories through which inclusion/exclusion, North/South and core/periphery can be understood. Dr Buzinde's work on development adopts a grassroots approach and it aims to understand the relationship between community wellbeing and tourism development within marginalised communities.

Blanca Camargo is Professor of Tourism in the Business School at Universidad de Monterrey, Mexico. Originally from Colombia, she has Master's degrees in hospitality and tourism management from Purdue University and in international tourism management from AILUN in Italy. She holds a doctoral degree from Texas A&M University and her research is oriented towards sustainable tourism with a focus on ethics and justice in tourism development, marketing, and management. She is a member of the Mexican Academy of Tourism Research (AMIT) and Mexico's National System of Researchers (SNI, level I).

Doris A. Carson is an Assistant Professor at the Department of Geography at Umeå University in northern Sweden. Her research focuses on changing population mobilities and their contributions to new local development paths in sparsely populated areas, particularly those found in Sweden and Australia. Current work examines the implications of urbanisation for small 'hinterland' communities in sparsely populated areas. Doris has published widely on the topics of rural and remote area tourism, local innovation systems in peripheral areas, lifestyle migration, and demographic change in remote resource peripheries.

Katherine Dashper is Professor and Director of Research Degrees in the School of Events, Tourism and Hospitality Management at Leeds Beckett University, UK. She has published widely in international journals and books on issues related to human–animal relationships, gender, tourism, and events management. She has co-edited four books on tourism and sport, including *Rural Tourism: An International Perspective* (2014, Cambridge Scholars).

Tawsif Dowla is a PhD student in the Department of Recreation and Leisure Studies at the University of Waterloo, Canada. Led by a social justice and community-laden research paradigm, he aspires to harness tourism's power to foster 'mutual understanding and respect between peoples and societies' as articulated in its Global Code of Ethics. In his current research, Tawsif is invested in supporting resettled refugees' integration journey in their newly found homes by unpacking the transformative power of tourism.

Marco Eimermann is an Associate professor at the Department of Geography and is affiliated to the Arctic Research Centre (ARCUM), Umeå University, Sweden. His research interests relate to lifestyle migration, downshifting practices and sustainable entrepreneurship in rural, remote and northern sparsely populated areas. Eimermann is involved in the Lifestyle Migration Hub, a network of migration scholars studying social rather than economic reasons for voluntary mobility and migration across the globe. He has published peer-reviewed articles (e.g. in the *Journal of Rural Studies* and in *Population, Space and Place*), book chapters and books about transient and strategically switching populations in Europe.

Sinead Francis-Coan is a passionate Novocastrian with a background in tourism studies at the University of Newcastle, NSW, Australia, and a passion for the performing and creative arts. She believes both are powerful and transformative elements of our society. Sinead is an active member in our community, advocating for our environment, Indigenous rights, women's rights, workers' rights and refugee rights across a wide range of different community groups. Her postgraduate research focused on land use conflict in the Upper Hunter Valley, which required her to engage with a range of stakeholders to understand their perspectives.

Bryan S.R. Grimwood is a Settler Canadian of British ancestry, living and working in the homelands of Anishnaabeg, Haudenosaunee, and Attawandaron Peoples and in treaty territories that include the Huron Tract Purchase and the Haldimand Tract.

Bryan is an Associate Professor in the Department of Recreation and Leisure Studies at the University of Waterloo, Canada. Trained as a human geographer, and an engaged scholar, Bryan's research focuses on tourism and Indigenous–Settler relations, tourism ethics and responsibility, and northern landscapes. He is co-editor of two books as well as several special journal issues. Bryan sits on the editorial boards of *Leisure Sciences*, *Journal of Ecotourism*, *Tourism Geographies*, and *Annals of Tourism Research*, and co-founded the North American chapter of the Critical Tourism Studies network.

C. Michael Hall is Professor in Marketing and Tourism at the University of Canterbury, New Zealand; Guest Professor, Lund University, Campus Helsingborg, Sweden; Visiting Professor, Linneaus University, Kalmar, Sweden; and Docent and Visiting Professor, University of Oulu, Finland. His research interests include tourism, regional development, global environmental change, food, sustainability and World Heritage.

Keith Hollinshead was a critical analyst of long-time cultural inheritances, and the power of tourism and like declarative fields to project (and mis-project) the cosmologies and ontologies of peoples. Having worked in Wales, the US, and (mainly) Australia, his work examined the identificatory/representational systems through which particular 'visions' of place/space (and aspiration) are normalised (consciously or unconsciously). He was formerly a Distinguished Professor with the International Tourism Studies Association, founded at the University of Peking, Beijing, China in 2006. By inclination, Keith was a postdisciplinary researcher, and a posthumanist (postqualitative) researcher in spirit. He had both UK and Australian citizenship.

Tazim Jamal is Professor in the Department of Recreation, Park and Tourism Sciences at Texas A&M University (TAMU), USA. She is also a fellow of the International Academy for the Study of Tourism. Her research interests lie in tourism and sustainability, collaborative tourism planning, and ethical issues in travel and tourism. She is the author of *Justice and Ethics in Tourism* (2019, Routledge), co-editor of *The SAGE Handbook of Tourism Studies* (2009, SAGE Publications), and associate editor of the *Encyclopedia of Tourism* (2016, Springer) and the *Encyclopedia of Tourism Management and Marketing* (2022, Edward Elgar Publishing).

Sung Eun Jeon holds a BA in tourism development with two minors in human resources management and event management from the University of Waterloo, Canada. Sung Eun's research embracing a feminist ethic of care examines the roles of tourism social entrepreneurs contributing to regenerative tourism outcomes in Canada.

Carol Kline is a Professor and the Director of the Hospitality and Tourism Management programme at Appalachian State University, USA. Her teaching and research interests have historically focused broadly on tourism sustainability, including topics such as foodie segmentation, craft beverages, agritourism, tourism entrepreneurship, and tourism in developing economies. She is part of the Race, Ethnicity, and Social Equity in Tourism (RESET) initiative, which includes animals within the study of social equity. She is the founder of Fanimal Inc., a non-profit that helps individuals find animal-focused careers.

Bernard Lane is a consultant, writer and lecturer specialising in sustainable rural tourism, small towns, heritage and architectural conservation and rural development. He was the founder, and for 25 years was co-editor of the *Journal of Sustainable Tourism*, ranked in the top five of in the world ranking of over 200 tourism journals. He was also an Associate Editor of the *Journal of Ecotourism*, and has worked in many universities on all continents. He has over 170 publications, including six books, with key works translated into French, Spanish, Japanese, Korean, Finnish, Romanian, German and Polish.

Dominic Lapointe is a Professor in the Department of Urban and Tourism Studies at Université du Québec à Montréal, Canada. Trained in regional development at the Université du Québec à Rimouski (PhD 2011), he holds the Chaire de recherche sur les dynamiques touristiques et les relations socioterritoriales, and directs *Téoros*, the oldest French language tourism studies journal, while acting as the leader of the Groupe de recherche et d'intervention tourisme territoire et société (GRITTS) at UQAM. His work explores the production of tourism space, its role in the capitalist system expansion and its biopolitical dimensions.

Emma Lee is a *trawlwulwuy* woman of *tebrakunna* country, north-east Tasmania, Australia. She is a Professor at Federation University Australia, Victoria. Her research fields over the last 25 years have focused on Indigenous affairs, land and sea management, policy and governance of Australian regulatory environments. In 2021, she became the first Indigenous Australian editor of a Best Practice Guideline for the International Union for the Conservation of Nature (IUCN). Emma was nominated for 2022 Australian of the Year (Tasmania) in recognition of her rights work.

Linda Lundmark is an Associate Professor at the Department of Geography at Umeå University, Sweden. She has published extensively on the topics of regional development and tourism in rural areas in the north. Her research interests also include climate change, natural resources and mobility as part of contemporary and future economic development prospects in sparsely populated nations like Sweden. She is currently the Chair of the Centre of Regional Science Scientific Advisory Board, has received funding for several large research projects from Swedish research councils, and is part of several research networks on tourism and tourism in polar areas.

Michael Mackay is Senior Social Scientist at Aotearoa New Zealand's Crown Research Institute, AgResearch, and adjunct Senior Research Fellow at Lincoln University, New Zealand. His research interests include tourism-led rural regeneration, agri-tourism, design-led innovation, creative place-making, and the globalisation of rural areas. Mike is also a social impact assessment practitioner, specialising in the effects of land use change, including tourism development, on rural communities.

Heather Mair is a Professor in the Department of Recreation and Leisure Studies at the University of Waterloo in Ontario, Canada. She has authored, co-authored or co-edited six books and more than 50 journal articles. Her research interests include critical investigations of the role of sport, leisure and tourism in community building and development, particularly in rural areas. Heather was the President of the Canadian Association for Leisure Studies from 2014 to 2020.

David W. Marcouiller is Professor of Rural Development at the University of Wisconsin – Madison, USA. A resource economist by training, his work focuses on the linkages between natural resources and community economic development with interests in income distribution, multifunctional rural landscapes, amenity driven migration, and the compatibility of alternative land uses. He has published over 200 manuscripts in a variety of outlets that span tourism and forest economics, outdoor recreation planning, regional science, landscape and urban planning, and rural development.

Gianna Moscardo is a Professor in the College of Business, Law and Governance at James Cook University, Australia. Her research interests include customer and tourist experience design, heritage interpretation, sustainability communication and new approaches to tourism planning and development for sustainability.

Meghan Muldoon is an Assistant Professor of Sustainable Tourism at the University of Groningen's Campus Fryslân in Leeuwarden, the Netherlands, where she teaches tourism, culture and sustainability as well as qualitative methods. Her PhD is from the University of Waterloo, Canada, where she undertook a PhotoVoice research study with tourism stakeholders residing in the townships of Cape Town, South Africa. Her research interests include the intersections of tourism and poverty, postcolonialism, feminisms, digital discourses, representations of Indigeneity, and arts-based methodologies.

Tracy Nelson is Science Objective Leader (Sustainable Agri-Food Production Systems) at Aotearoa New Zealand's Crown Research Institute, AgResearch. She has an extensive research background in rural entrepreneurship, agri-tourism, and the geography of rural change.

Can-Seng Ooi is a sociologist and the Professor of Cultural and Heritage Tourism at the University of Tasmania, Australia. Often framed within critical tourism studies, he conducted research on many society and tourism issues, and in different countries, including Australia, Singapore, Denmark, Malaysia and China. More information on him is available at www.cansengooi.com.

Gaurav Panse is a PhD student in the Department of Recreation and Leisure Studies at the University of Waterloo, Canada. His research focuses on the confluence of tourism destination management and marketing, urban resilience, sustainable development, and socio-environmental justice.

Mery Angeles Perez is Adjunct Professor in the School of Science and the Environment at Memorial University, Grenfell Campus, Canada. She holds a PhD in rural studies from the University of Guelph with a focus on community capacity building and development through creativity. In her master's and doctoral programmes, her research focused on the role of the arts, particularly music, in capacity development, community building, and social and environmental change in Canada and in Latin America. Her research interests include the relationship between environmental governance and community development, and the role of Indigenous and rural communities in sustainability.

Harvey C. Perkins is Emeritus Professor of Planning at the University of Auckland and Director of People and Places Ltd, a research and consultancy firm. He was formerly Director of the University of Auckland's thematic research initiative, Transforming Cities: Innovations for Sustainable Futures, and Professor of Human Geography at Lincoln University in Canterbury, New Zealand. A past President and Fellow of the New Zealand Geographical Society, Harvey has an extensive research background in urban and rural change.

Girish Prayag is a Professor of Tourism Marketing at the University of Canterbury Business School in Christchurch, New Zealand. He has co-authored more than 100 articles and published two books, while co-editing several others. His research interests include place attachment, organisational resilience, disaster tourism and food/wine tourism.

Brady Reid holds a Master of Arts in Environmental Policy from Memorial University, Canada, during which time he worked collaboratively with the community of Ewipkek (Flat Bay) on Newfoundland's west coast. Currently, Brady is working towards a PhD in sustainability management at the University of Waterloo. His research interests intersect sustainable rural communities and decision-making processes within a resource management context. As a settler scholar, Brady also explores the roles and responsibilities inherent to engaged, cross-cultural research relationships between Indigenous and non-Indigenous collaborators – including critical self-reflection and positionality throughout the research process.

Donald G. Reid is University Professor Emeritus in the School of Environmental Design and Rural Development (SDRED) at the University of Guelph in Canada. He is an international scholar whose work focuses on sustainable development, social planning, poverty, community development, leisure, and tourism. His earlier books include *Work and Leisure in the 21st Century: From Production to Citizenship* (1995, Wall & Emerson); *Tourism, Globalization and Development: Responsible Tourism Planning* (2003, Pluto Press); *Social Policy and Planning in the 21st Century: In Search of the Next Great Social Transformation* (2016, Routledge); and *A New*

World-System: From Chaos to Sustainability (2020, Routledge). His latest book is *Social Progress and the Authoritarian Challenge to Democracy* (2022, Routledge).

Kyle Rich is an Associate Professor in the Department of Recreation and Leisure Studies at Brock University in St. Catharines, Ontario, Canada. Kyle's research looks at the impact of policy, community, and inclusion/exclusion on experiences in sport, recreation, and physical activity programming. This research has examined a variety of local, regional, and national programmes and how they are implemented in both urban and rural contexts. In particular, Kyle is interested in using community-based and participatory research methodologies in order to help community organisers understand and address local issues in efforts to improve community health.

Alexander Safonov is a doctoral student in the Department of Management, Marketing and Tourism at the University of Canterbury, New Zealand. With an extensive background in adventures, hospitality, and event management and marketing, his research examines the associational tourism economy and a broad range of tourism-related topics.

Becky Shelley studied political science and is Deputy Director at the Peter Underwood Centre for Educational Attainment at the University of Tasmania, Australia. Becky is interested in work at the intersections of policy, research and practice. The Children's University Tasmania is delivered by the Peter Underwood Centre.

Mark C.J. Stoddart is Professor in the Department of Sociology at Memorial University of Newfoundland, Canada, with research interests in environmental sociology, social movements, and communications and culture. He is author of the book *Industrial Development and Eco-Tourism: Can Oil Extraction and Nature Conservation Co-Exist?* (2020, Palgrave Macmillan). His work appears in a range of international and Canadian journals, including *Journal of Sustainable Tourism*, *Mobilities*, *Island Studies Journal*, *Sociological Quarterly*, and the *Journal of Rural and Community Development*.

Jennifer Sumner is the Coordinator of the Adult Education and Community Development Program at OISE/University of Toronto, Canada. She has a PhD in rural studies and her research interests include sustainable food systems, the social economy and rural tourism. She is the author of *Sustainability and the Civil Commons: Rural Communities in the Age of Globalization* (2005, University of Toronto Press), the editor of *Learning, Food and Sustainability: Sites for Resistance and Change* (2016, Palgrave Macmillan) and the co-editor of *Critical Perspectives in Food Studies* (3rd edition, 2021, OUP Canada) and *Leisure and Food* (2015, Routledge).

Nick Taylor has broad experience applying social impact assessment and research in New Zealand and internationally. He has worked on tourism projects, community development strategies and economic regeneration, as well as on land and water plans, irrigation developments, and a wide range of resource developments and infrastructure projects. Nick was a founding director of Taylor Baines & Associates and is now an independent researcher and consultant. He is a senior adjunct of Lincoln University and a past President of the International Association of Impact Assessment (IAIA).

Kelly Vodden is a Professor (Research) at the Environmental Policy Institute, Grenfell Campus, Memorial University, Canada. Dr Vodden has been actively involved in community and regional development research, policy and practice in Canadian rural communities since 1995. Her research examines community involvement in planning and natural resource governance, community and regional development, collaborative governance, and creating sustainable local economies. For more on Kelly's research and that of her students and collaborators see: ruralresilience.ca.

Michael Woods is Professor of Human Geography at Aberystwyth University in Wales, UK and Co-Director of the Wales Institute of Social and Economic Research and Data (WISERD). He studies changing rural communities, rural policy and politics, rural development, and the 'global countryside', including the impacts on rural communities of international tourism and amenity migration. He is the author of books including *Rural Geography* (2004, SAGE) and *Rural* (2011, Routledge) and is a former editor of the *Journal of Rural Studies*.

Xiaotao Yang is a Lecturer in the School of Tourism and Urban–Rural Planning, Zhejiang Gongshang University, China. Primarily interested in social justice, her interests lie in indigenous knowledge, poverty alleviation, participatory research, and knowledge production.

Yixi Yang is a PhD candidate at the Memorial University of Newfoundland, Canada. Her research interests include climate change communication, environmental sociology, the environmental movement in China, digital sociology, and computational social science. Her current research examines environmental discourses in China's digital media sphere, public participation in climate politics in China, and the role of social networks in shaping public perceptions of climate change.

Acknowledgments

This project, like all projects, is the work of many and not the few. For anyone who has faced the challenge of coordinating dozens of authors, ideas, formats, word limits, and time constraints, you know how exhilarating and exhausting this all can be. Doing it during a pandemic … well … enough said about that.

I want to take this opportunity to extend my thanks and deepest appreciation for every author who took such great care with their chapters, meeting deadlines (some many months in advance!) and responding to my suggestions and comments in timely, kind, and sometimes humorous ways. As I put the finishing touches to the book, I realised being a sole editor can leave one feeling lonely and more than a bit vulnerable, so I am truly grateful to all those who said 'yes' and came on board to complete this work.

Thanks to Katy Crossan and Stephanie Hartley at Edward Elgar Publishing for their supportive and timely responses to my many and frequent email messages.

This book is dedicated first to all those living in rural areas and contemplating their own tourism trajectory – every author in this collection wishes you well.

Finally, to George and Sophie who gave me all the love, treats, time and space I needed to complete this work.

1. Introduction: understanding rural community development and tourism in challenging times

Heather Mair

As I was beginning to put together the plan for this handbook in earnest, news of COVID-19 vaccine production and availability was, of course, welcome after months filled with news of death, economic ruin, and personal sacrifice. We all needed to be hopeful in the face of such dark times. Indeed, growing awareness of the social divisions and inequality, which were laid bare by the uneven impacts of the virus upon communities of colour and anyone living without adequate financial support, not to mention the looming threat of a climate catastrophe, made us all want to entertain a vision of a world that we could 'reset' or 'build back better' (Bakker & Elkington, 2021). Rural areas, particularly but not exclusively in North America, were beginning to reel from the mixed blessing of urban-to-rural migration, and increasingly concerned about rising house prices, economic decline, and a near complete collapse of health and social services under the weight of the pandemic.

Perhaps not surprisingly, it also seemed likely that travel and tourism development, positioned both as a response to the very human need for connection, learning, and mobility (not to mention an opportunity for much needed economic recovery), would play a central role in this return to normal. An update from the Organisation for Economic Co-operation and Development (OECD) in early 2021 predicted a global recovery was in sight (OECD, 2021). In my home country, Canada, news was filled with forecasts of a 'Roaring '20s Rebound' (Pittis, 2021); an era of potentially unrivalled economic growth, spurred on by pent-up demand, government investment, and a citizenry desperate to return to normal, perhaps even better, times. Stories about the potential growth of wilderness, luxury, and even 'revenge' travel seemed to be everywhere (see for example, Robins, 2021; Wallace, 2021), as prognosticators and travel writers attempted to predict what would happen next for the travel industry, which had been all but devastated by COVID-19.

Fast forward to the middle of 2022 and the news of COVID-19 seems to have subsided (at least for now), but instead of a grand reset and a roaring rebound, life is shaped by skyrocketing inflation, soaring gas and food prices (exacerbated by the invasion of Ukraine by Russia), travel delays, and airport chaos. And, of course, the climate emergency has not taken a holiday. While there are indications that some lessons from decades of over-tourism have been learned (Borghese et al., 2022), much of the discussion about tourism, especially away from the halls of academia, remains little different from business as usual.

The purpose of this handbook is to bring together scholars to contemplate issues of rural community development and to critically explore the role of tourism therein. For those who have long been skeptical of tourism's economic and social benefits (see for example, Higgins-Desbiolles et al., 2019; Fletcher, 2011; Mair, 2006), there has perhaps never been a more relevant time to take up these ideas and to prepare ourselves for responding to what is surely going to be primarily an economic argument for accepting some manifestation of travel and tourism (good and bad) as a way to address these interrelated crises.

As I secured commitments from authors both well known and relatively unknown to me, I worried about the impact of COVID-19 on progress. I was right to worry – the virus slowed down and exacerbated hardships for many authors; and, of course, members of the rural communities about whom this book is being written faced great challenges too. As I put the final pieces together and began to contemplate where we head next as the pandemic ebbs (or appears to be ebbing), the efforts of the scholars featured here to move beyond the dominance of economic issues related to tourism in rural areas and other essential ideas explored critically in this volume (e.g., ethics, resilience, power, regeneration, etc.) are more important than ever. If we are to 'recover', what are our options when it comes to tourism and rural community development?

TOURISM AND RURAL COMMUNITY DEVELOPMENT IN CHALLENGING TIMES

It is beyond dispute that more humans than ever are moving from rural to urban areas (United Nations, 2018) and yet rural areas remain essential, vital spaces on our planet. For the scholars whose work is featured in this collection, and others who have spent years (and even decades) researching and thinking about rural places and spaces, there is a strong commitment to continue to support these areas and especially the people who live there. And yet, the challenges they face are daunting.

As noted in their thought-provoking volume, *The Right to Be Rural*, Foster and Jarman (2022) present a vivid description of the impact of the forces of industrialization, deindustrialization, economic restructuring and urbanization on many rural communities:

> In rural communities, shuttered mines, fallowed or converted farmland, idle mills, empty manufacturing plants, and boarded-up main streets are no accidents of culture. They are signs that companies have moved on, but people have not. (p. 3)

Indeed, the complex and interrelated forces affecting rural life and their implications (e.g., out-migration, aging population, transportation/distance challenges, reduced or non-existent social and health supports, dwindling tax base, etc.) are well known to students of rural life. Add to this list other complicated and complicating factors, including climate change and related events and disasters, the rights of Indigenous

Peoples, the counter-forces of in-migration or counter urbanization, often driven by tourism (Mitchell, 2004; 2019; Mitchell & Shannon, 2017; 2018) and exacerbated by global events such as the COVID-19 pandemic, and one can easily appreciate the need to take up questions of tourism in an effort to both understand what is happening in rural communities and to identify not just opportunities for survival but for thriving in challenging times.

When I was invited by Edward Elgar to serve as editor for this handbook, the proposed title was *Handbook of Rural Tourism Development*. After thinking about the opportunity for some time, I agreed to take on the role on the condition that the title be edited to reflect what I thought should be its primary focus: *Tourism and Rural* Community *Development.*

In their recent, systematic review of the rural tourism literature, Rosalina, Dupre, and Wang (2021) define rural tourism in the following way:

> … a type of tourism located in areas within a destination that are characterised by rural functions (such as traditional, locally-based, authentic, remote, sparsely populated and mainly agricultural areas) where the tourists can physically, socially, or psychologically immerse themselves in this specific destination. (p. 137)

Of course, in a close reading of this definition, one can see the emphasis is again placed on the tourist and tourism and less so on the community. However, given what we have come to know about the negative impacts of tourism in rural areas, shifting the focus away from the tourist and tourism, and towards the local community, is warranted. Further, students of tourism – rural and otherwise – understand that tourism very rarely, if ever, holds the key to solving a community's problems. In fact, it is often the very source of those problems itself, especially in rural areas.

This emphasis on community is why this handbook is important. As rural communities around the world begin to contemplate their post-pandemic existence, gathering up-to-date, provocative, and inspiring work that puts their interests, needs, and potential for thriving at the centre is essential.

OUTLINE OF THE HANDBOOK

This collection includes chapters designed to introduce key topics and issues relevant for exploring and understanding the relationship between tourism and rural community development. The chapters include a wide range of topics and approaches, from discussions of relevant concepts and methodological approaches to reflections on projects located in rural places found in Australia, southern Canada, Arctic Sweden, New Zealand, and Tibet. Taken together, the chapters exemplify the complexity and heterogeneity of rurality and rural places, and yet each also aims, in its own way, to explore the complexity and heterogeneity of the communities of human and non-human residents therein. The collection is divided into five parts: conceptual foundations; epistemological foundations, paradigms, and worldviews; key conceptualizations; processes and practices; and emerging matters and new directions. While not meant to be a complete accounting of all scholarship relevant to tourism and rural community development, the work represented here is comprehensive and broad.

Conceptual Foundations

The first major section is devoted to fostering an understanding and appreciation of three undeniably complex notions: rural communities, rural tourism, and rural development. It features three chapters, which have been written by established, leading, and emerging scholars in these areas.

Michael Woods begins by tracing the evolution of the notion of community, particularly as a subject of academic inquiry. Moving on to describe the growing awareness of the complexity of community and how it has come to be understood with regard to tourism, Woods draws on relational theory as well as the work of Liepins (2000) to consider the ways community is enacted, and also its material context. The author engages with notions of mobility to highlight the multiplicity of rural communities and introduces four ways they have been understood or positioned in relation to tourism: communities as recipients, communities as resources, communities as commodities, and communities as assemblages. Readers will see evidence of these four manifestations of community through tourism throughout the rest of the collection.

The next chapter, by Donald G. Reid, centres our attention on the notion of rural tourism by providing a detailed discussion of its complexity, including aspects of the structure, planning and management of rural tourism but also through an engagement of relevant issues and challenges. While underscoring the economic argument that accompanies much writing about rural tourism, Reid also broadens the lens to talk more directly about the other dynamic factors shaping tourism in rural areas (for example, geographical or cultural uniqueness). The chapter pays special attention to issues related to the planning of tourism in rural areas and makes a strong case for the need for all planning efforts to be both grounded in the local context (history, geography, culture) and to be guided by rural community members themselves. The themes of community power and control reverberate throughout this contribution as they do in many of the chapters that follow this foundational analysis.

Kelly Vodden, Mery Angeles Perez, and Brady Reid complete this first part of the collection by tackling the complex and contested topic of rural development. The authors trace the evolution of the notions of rural and rurality and provide the reader with a solid understanding of these challenging terms by highlighting not just their social practice and discursive elements but also the powers that shape them. Next, they turn their attention to the concept of development, tracing its origins in primarily economic thinking to more recent considerations of different worldviews and perspectives that introduce notions of development as living a good and (ecologically) balanced life. They conclude by bringing these ideas together within the broader context of rural tourism, arguing for collaborative and participatory approaches to rural tourism development, which take into account these new ways of understanding rural, rurality, and development while balancing economic, social, cultural, and environmental factors.

Epistemological Foundations, Paradigms, and Worldviews

The next major section represents an effort to draw direct attention to broad knowledge-related issues and methodological questions, which are essential for understanding research related to tourism and rural community development. In short, how we know what we know about these issues plays a key role in how we develop a sense of pressing issues and, perhaps more importantly, how we begin to imagine what solutions to those issues might be.

Girish Prayag takes the reader through the long-standing history of post-positivist inquiry in tourism research – rural and non-rural. Prayag poses five important questions that the reader can use to guide their research decisions and provides high-level detail in this regard. Importantly, this chapter does not present a kind of 'recipe' for conducting post-positivist or mixed methods research (frequently oversimplified as an effort to merely undertake both quantitative and qualitative methods together into one project), but instead encourages the reader to challenge themselves and to think through not just their own assumptions regarding what is 'good' research but also to question the purpose of social research in tourism more broadly.

Meghan Muldoon captures the spirit of interpretivist work in rural tourism by beginning with a story about a community tourism/recreation project that was challenged due to assumptions about which methodological approach would suffice. Muldoon uses this story to open the reader's thinking about the possibilities of research in rural tourism by focusing on the opportunities afforded by a paradigm that is grounded in, and centres, meaning, stories, experiences, and shared histories. In this chapter, readers are treated to an overview of compelling examples of interpretive research, with a particular focus on the role(s) of emotion, values, and our embodied identities in research (something that can never be detangled from the social research experience, even in rural tourism).

Next, Christine N. Buzinde puts critical and transformative pedagogy at the centre of the discussion and challenges the reader to think carefully about the relationships that shape not just tourism development (for instance, colonialism) but also opportunities to reframe and reshape those relationships and, importantly for educators and those concerned with community development in and outside tourism, to create spaces for new ways of knowing (knowledge plurality). Introducing the notion of decoloniality and positing the potential for decolonial tools (such as critical consciousness and culture circles), Buzinde lights the way for rethinking community engagement, and she invites readers to contemplate how they might act as educators and community organizers and to envision and confront tourism learning in new ways.

Michael Mackay, Nick Taylor, Harvey C. Perkins, and Tracy Nelson describe social impact analysis (SIA), a relatively prevalent methodological approach in rural tourism research. The authors reflect on the utility of SIA for understanding tourism-induced social impacts and rural social change as a result of the development of the A20, the longest continuous trail in the Nga Haerenga (New Zealand Cycle Trail Project), which is an extensive network of publicly accessible and off-road bike trails. The authors carefully describe their use of SIA and demonstrate how it shaped their understanding of the complex and unevenly distributed impacts of the A20, including opportunities for socio-economic regeneration of rural areas near the trail, but also in regard to challenges that have resulted, such as issues related to carrying capacity and stress on local infrastructure.

Keith Hollinshead and I foreground opportunities for critical thinking and encourage readers to challenge the ways in which tourism's ways of knowing and judging have become normalized, as well as the implications of this normalization. We introduce the reader to scholarship outside tourism studies and provide a list of 12 key conceptual areas (culture gene banks, site experiencing, *darshana*, performativity, fragmentation, authorization, critical mix, critical mass, entry statement, tradition, transition, and traditionality), which are selected to help foster a broader understanding of the complexity of the relationship between tourism and rural community development. We also invite the reader to develop their *rapport à soi* (relation to oneself) as a way to foster not just critical thinking about the worldmaking power of tourism, but to become aware of the impact of their own judgements about tourism and the extent to which they have played a part in underscoring the taken-for-granted-ness of tourism in rural areas.

The last chapter in this section comes from *tebrakunna* country, Emma Lee and Bryan S.R. Grimwood, who model a new and provocative way of knowing about tourism. Self-described as a trawlwulwuy woman of *tebrakunna* country, Tasmania, Australia, and as a Settler Canadian, the authors use storywork to 'restory' tourism by presenting two personal narratives to illustrate what they refer to as respectful 'listening' to Indigenous elders. They also build in and on the notion of refusal, arguing, 'For Indigenous Peoples, refusal speaks to the persistent and ongoing struggles for decolonization, sovereignty, and resurgence' (p. 126). Their chapter serves to challenge the reader's understandings of place, of time, language, and of tourism more broadly, and stands as a compelling effort to centre Indigenous ways of knowing, teaching, and practices in tourism studies and to invite the reader to consider 'how we teach and learn not *for* Indigenous Peoples, but rather *from* and *with* Indigenous Peoples' (p. 135).

Key Conceptualizations

The third part is linked to the first foundational discussion but aims to widen the lens and to bring in scholarship related to other essential concepts. While some authors stay at the conceptual level, others draw in extant research and/or case studies in order to demonstrate how these key concepts are highly relevant to understanding tourism and rural community development.

Katherine Dashper leads us off by taking up the gendered, unequal structures shaping rural community life. Diving deep into both the forces affecting rural communities and how these impacts are gendered, the author then explores what this means for tourism development, especially given the complexities of geography, culture, history, and changing political economic landscapes, not to mention the increasingly blurred lines between rural and urban communities. Dashper describes the centrality of gender to achieving broad sustainability goals, including social, economic, and environmental, and argues for the adoption of a 'gender-aware lens' (p. 140), through which the gendered nature of rural tourism development, where women's knowledge, status, interests, and priorities can be marginalized, but can also be exposed, understood, and addressed.

Jennifer Sumner provides a discussion of food, an increasingly visible and valuable part of rural community development and tourism. She introduces broad conceptualizations of food, which serve to illustrate the ways it is political, multifaceted, and 'much more than merely sustenance' (p. 153). This approach allows for a nuanced way of thinking about the role(s) food in tourism and rural community development, and the chapter includes examples such as farmers' markets, 100 mile stores, food trails, food festivals, and farm-to-table restaurants to make the case. Sumner also presents this way of thinking about food and food tourism as an opportunity to (re)consider the role tourism can and should play in community (economic) development and illustrates her points drawing on the work of Loxley (2007) and concepts of linkages, leakages, and leveraging.

Bernard Lane describes the complex and sometimes contested relationships between heritage, heritage interpretation, and rural tourism. Drawing on the classic book first published in 1957 by Tilden (2007), Lane presents long-standing principles of interpretation and explores the relationship between interpretation and broader goals of rural tourism including understanding, appreciation, preservation, and protection. The chapter takes a practical turn, considering the tasks facing a rural heritage interpreter and the changing nature, forms, and practices of heritage and interpretation more broadly.

In their discussion of climate change and its relationship to tourism and rural community development, Mark C.J. Stoddart and Yixi Yang complicate our understanding of the role of tourism in exacerbating climate change (due primarily to its carbon intensity) by describing their recent work on the oil–tourism interface in the North Atlantic region and by identifying ways the tourism sector can adapt and respond to climate change (e.g., the potential for decarbonizing tourism and other initiatives). Their chapter sets the foundation for an essential, critical conversation about the extent to which tourism really can contribute to the UN Sustainable Development Goals and pinpoints the 'vexing challenge' (p. 186) of reducing its extra-local role in global warming, especially when much of rural tourism's lauded benefits are framed as much-needed, localized, short-term economic gains.

David W. Marcouiller's chapter follows on well as he introduces key and increasingly relevant conceptualizations of vulnerability and resiliency and relates them to the broader context of tourism and rural community development. Given how tourism is founded upon relationships that are frequently framed, first and foremost, as economic, Marcouiller points out that rural areas can be especially susceptible to the effects of, not just fluctuations in levels of tourism demand, but bigger problems such as political strife, weather events, and, of course, pandemics. Further, because rural regions tend to lack economic diversity (and often rely on tourism as a means to diversify economic support), resources for addressing vulnerability and building resiliency can be scarce. As Marcouiller shows us, high reliance on tourism often also means high vulnerability and relatively lower resiliency for communities and residents.

Alexander Safonov and C. Michael Hall take on the concept of degrowth, or *décroissance*, another increasingly prominent notion in tourism studies and deeply relevant to the context of rural community development. Using degrowth as a 'missile term' (Hickel, 2021) to support a critique of growthism and an opportunity to highlight the biophysical processes need to support tourism, the authors also describe the 'Janus-faced' nature of rural tourism development, making rural areas prone to both under- and over-tourism (p. 208). Safonov and Hall then seek to challenge the foundational economic and environmental premise of tourism development, by asking whether even a local, community-based approach to rural tourism would be sufficiently sustainable to address the growing list of negative impacts and challenges. The authors also reflect skeptically on the so-called opportunities presented by the COVID-19 pandemic to foster a domestic or local turn in tourism, and the related desire to 'right-size' or 're-set' tourism (rural and otherwise) in the face of growing awareness of its lasting climate effects.

Next, Sinead Francis-Coan and Annæ Buchmann present a complex view of rural communities as multifunctional, socially constructed landscapes. The authors present a case study of the contested processes of land use, planning, and management in the Upper Hunter Valley, Australia, where changing use, values, and meanings of the local landscape have engendered debate and opposition about 'appropriate' land use, including (but not exclusively), tourism. Francis-Coan and Buchmann investigate the potential of collaborative approaches for land use planning and management as a way to address and perhaps prevent these debates and conflicts. Their work encourages readers to embrace multifaceted conceptualizations of landscapes and to look critically at their contested role in rural tourism development.

Tawsif Dowla and Karla A. Boluk contribute to the debate about degrowth as they investigate the ways rural communities can work to challenge the overriding economic orientation of tourism. Based on research undertaken in St. Jacobs Village in southwestern Ontario, Canada, the authors introduce what they describe as a 'new emergent paradigm' of regenerative tourism (p. 236), which they situate in an ethic of care and creative tourism. Dowla and Boluk also draw on notions of cultural tourism and consider the growing appeal of 'authentic', interactive experiences and co-created tourism products. Their work in and with this rural community illustrates how an intentional and engaged approach to tourism development and planning can help to build a spirit of regeneration and creativity that puts members of local communities at the centre.

Processes and Practices

Much of the research in tourism and rural community development eventually gets to the question of 'how', or what I call processes and practices. That is, if scholars are concerned about the overriding economic focus of rural tourism, or the extent to which tourism's challenges can outweigh the benefits for rural communities, what is to be done? While many of the chapters in this collection touch on these vital conversations and debates in some way, each of the chapters in this section, albeit in different ways, directly takes up questions regarding what processes and practices should be in place for rural tourism to unfold in ways that address at least some of the concerns and issues raised above.

First, Tazim Jamal and Blanca Camargo build on conceptualizations of regenerative tourism but posit the need for further development of ethical priorities and principles of justice, which can be used to guide these new notions and approaches. Jamal and Camargo draw on fascinating case examples, including the Nyambu Ecological Tourism Village in rural Bali and San Basilio de Palenque, known as Palenque, a rural community in Colombia, to support the need for 'widening our circle of compassion' (p. 255) and to think relationally, far beyond the traditional economics of tourism. Further, they identify key principles and practices, such as local ownership and control, direct resident participation, inclusiveness, human dignity, and respect. The authors push the reader to consider how all of these principles and practices, when situated within a non-dualistic, posthumanist paradigm, could guide regenerative – even restorative – rural tourism development towards a future of human flourishing.

Roslizawati Che Aziz takes up the discussion of processes and practices in tourism and rural community development with a focus on appreciative inquiry (AI). Situated within a broadly critical consideration of the notion of sustainable development, Aziz argues the power to determine the direction of sustainable rural tourism development is in the hands of community members. Introducing the notion of AI, a relatively well-established approach to community development for more than a decade, the author first discusses its potential as an asset-based approach, which is grounded in local knowledge and engagement, and then considers the role it can and should play as a mechanism for sustainable tourism development in rural areas.

Gianna Moscardo concentrates on governance processes and practices and their role in the development of Destination Community Wellbeing (DCW) in order to challenge the ways tourism has previously been used as a strategy for development in rural communities. Moscardo argues for a better understanding of the forces and conditions that shape barriers and challenges facing effective community development planning. The author draws the reader's attention to how business-dominated governance approaches view rural communities as resources to be developed into successful tourism businesses (as opposed to seeing tourism as a resource for rural development and wellbeing) and then explores, by reviewing a series of case studies of story-based rural community engagement in tourism, the opportunities and implications of developing ways for community residents to be better empowered and engaged in the planning, management, and practice of tourism.

Dominic Lapointe also identifies the dominance of business and economics-oriented approaches to tourism development and situates the chapter within a consideration of the role of tourism in (uneven) regional development while delineating clear linkages between tourism and the broader notion of salvage capitalism (Tsing, 2015). Illustrating the extent to which tourism destinations can be conceptualized regionally, which allows us to see how they play a complex and coordinated role in the system of capitalist production/accumulation, Lapointe applies the notion of salvage capitalism to illustrate how tourism processes and practices are built on the translation – the production and consumption – of products and experiences both in and outside the capitalist system. This opens opportunities for understanding that rural communities are much more than passive destinations for tourists but are producers (in and out of the capitalist tourism system) and for which attention to issues such as quality of life must be paid.

Next, Bernard Lane offers some practical advice for academics interested in the processes and practices of tourism and rural community development. Tracing the appeal of tourism for rural communities, or what Lane describes as the 'love–hate' relationship (p. 315) that exists between many rural areas and their tourism offerings, as well as the evolution of the role of universities and academics in rural tourism development, the author looks back on a wealth of experience in the field. The chapter provides a series of pragmatic tips, techniques, and considerations that all readers will find engaging and enlightening.

Emerging Matters and New Directions

While arguably many of the chapters included in the previous sections also touch on new ideas and matters that are really just coming to the fore for students of rural community development and tourism (e.g., degrowth, regenerative tourism, etc.), this section seeks to spark the reader's attention in regard to exciting and rich areas of emerging scholarship.

Kyle Rich uses the lens of mobility to consider community tourism and sport events and to explore the social processes and impacts of rural–urban interdependencies as well as their role in rural community development. Focusing on events as opportunities for the movement of capital, people and ideas between rural and urban spaces, Rich discusses the potential of notions such as 'new regionalism' (Wheeler, 2002) for encouraging a way of thinking about communities as not discrete entities but instead as sets of economic, political, cultural, social, and environmental relationships, linkages, and interdependencies. In particular and somewhat provocatively, Rich considers the impact of these interdependencies on notions of identity while also setting out an agenda for rural community development and tourism research that takes sport and community events into account in more compelling and satisfying ways.

Carol Kline focuses on the use of animals, which is an area of growing interest in rural community development and tourism. While present in tourism offerings since the beginning of travel, the role of animals has been largely neglected in tourism studies research and Kline identifies animal use as one of 'tourism's blind spots' (Tribe, 2008, p. 251). The author uses a critical theory lens to centre animal use and welfare within the broader discussion of rural tourism and challenges the reader, through powerful case examples of milking cows and heritage hunting dogs, to consider how ideas of heritage and tradition in tourism can mask serious issues of animal abuse and welfare. Kline clearly illustrates how notions of responsibility and morality in tourism, as well as regenerative and transformative tourism, can take into account the more-than-human world.

Can-Seng Ooi and Becky Shelley provide an instructive case study of regenerative tourism, particularly in terms of intergenerational change through education, by exploring a programme called the Children's University of Tasmania, where tourism destinations and their cultural capital assets are developed as learning destinations. Bundled with more 'traditional' strategic policy goals related to the role of tourism in economic development for rural regions and communities, Ooi and Shelley advance our understanding by exploring the relatively understudied role tourism can play in service to communities where local people and their needs (specifically education attainment levels and cultural capital more broadly) are centred.

Karla A. Boluk, Gaurav Panse, and Sung Eun Jeon draw on notions of the feminist ethic of care to propel our understanding of tourism through a study of the regenerative practices of Canadian women tourism social entrepreneurs. The authors situate the chapter within the broader context of the Sustainable Development Goals and growing critiques and frustration with the notion of sustainability, in tourism studies and elsewhere. The chapter illustrates how the work of women tourism social entrepreneurs, a heretofore understudied and undervalued area in our field, was grounded in a 'web of care', where care for self, community, and even the ecosystem come together. Their chapter lights the way for a future of learning about regenerative tourism and rethinking 'traditional' tourism systems more broadly.

Marco Eimermann, Doris A. Carson, and Linda Lundmark expand our knowledge about the unique context and complexity of rural communities in northern, sparsely populated areas (SPAs) with their chapter on intra-European lifestyle migrants who are attracted to, and living in Gafsele, a dogsledding community in Arctic Sweden. As the authors make clear, SPA communities have long been dependent on external sources for development, and tourism is often heralded as a way to attract not just visitors but lifestyle migrants whose activities, including civic engagement, can create significant changes in these dynamic communities. Indeed, the implications of these changes are fodder for much future research, as Eimermann, Carson and Lundmark illustrate.

Xiaotao Yang's work takes us into the world of Tibet and challenges not just what we know about tourism, but whose voices we listen to when we conduct research and think about the relationship between travel and work. Building on an argument about the importance of the local context and domestic tourism, Yang uses narrative visual inquiry to take us deep into the lives of young Tibetans who, by virtue of their need to travel to work, take advantage of the opportunity to explore the rural places and communities around them. Yang's work encourages us to continue to both rethink the work/not-work dichotomy that frames so much of tourism studies and, using Tibetan language to make the case, to create room for other ways of knowing about tourism.

In the final, concluding chapter (Chapter 30), I briefly review the primary contributions of each chapter, which I organize around four key arguments or themes: (1) tourism and rural community development is about far more than the economy; (2) there is opportunity for rethinking the economic dimensions; (3) the potential of 'new' voices and (non-Western) knowledges; and (4) the centrality of community. After engaging these primary arguments in turn, I conclude the chapter by drawing on these and other ideas to consider what could (or should) be next for rural community development and tourism.

REFERENCES

Bakker, P., & Elkington, J. (2020). To build back better, we must reinvent capitalism. Here's how. World Economic Forum. https://www.weforum.org/agenda/2020/07/to-build-back-better-we-must-reinvent-capitalism-heres-how/, accessed March 9, 2020.

Borghese, L., Buckley, J., & O'Hare, M. (2022). Venice reveals details of its €10 tourist entry fee. CNN Travel. https://www.cnn.com/travel/article/venice-entrance-fee-launch-date/index.html, accessed July 5, 2022.

Fletcher, R. (2011). Sustaining tourism, sustaining capitalism? The tourism industry's role in global capitalist expansion. *Tourism Geographies*, *13*(3), 443–461.

Foster, K., & Jarman, J. (eds) (2021). *The Right to Be Rural*. Edmonton: University of Alberta Press.

Hickel, J. (2021). What does degrowth mean? A few points of clarification. *Globalizations*, *18*(7), 1105–1111. https://doi.org/10.1080/14747731.2020.1812222.

Higgins-Desbiolles, F., Carnicelli, S., Krolikowski, C., Wijesinghe, G., & Boluk, K. (2019). Degrowing tourism: Rethinking tourism. *Journal of Sustainable Tourism*. 10.1080/09669582.2019.1601732.

Liepins, R. (2000). New energies for an old idea: Reworking approaches to 'community' in contemporary rural studies. *Journal of Rural Studies*, *16*, 23–35.

Loxley, J. (2007). *Transforming Or Reforming Capitalism: Towards a Theory of Community Economic Development*. Halifax: Fernwood Publishing.

Mair, H. (2006). Global restructuring and local responses: Investigating rural tourism policy in two Canadian communities. *Current Issues in Tourism*, *9*(1), 1–45.

Mitchell, C.J. (2004). Making sense of counter-urbanization. *Journal of Rural Studies*, *20*(1), 15–34.

Mitchell, C.J. (2019). The patterns and places of counterurbanization: A 'macro' perspective from Newfoundland and Labrador, Canada. *Journal of Rural Studies*, *70*, 104–116.

Mitchell, C.J., & Shannon, M. (2017). Establishing the routes to rural in-migrant proprietorship in a Canadian tourism region: A mobilities perspective. *Population, Space and Place*, *24*(3), e2095.

Mitchell, C.J., & Shannon, M. (2018). Are in-migrant proprietors driving or enhancing cultural heritage tourism in transitioning resource-dependent communities? The case of Trinity, Newfoundland and Labrador. *The Canadian Geographer/Le Géographe Canadien*, *62*(3), 398–413.

OECD (2021). https://www.oecd.org/economic-outlook, accessed March 15, 2021.

Pittis, D. (2021). https://www.cbc.ca/news/business/canadian-economy-weath-recession-rebound-1.5933651, accessed March 15, 2021.

Robins, J. (2021). Revenge travel is coming after a year of pent-up demand. *Globe and Mail.* https://www.theglobeandmail.com/life/travel/article-revenge-travel-is-coming-are-you-ready/, accessed March 13, 2021.

Rosalina, P.D., Dupre, K., & Wang, Y. (2021). Rural tourism: A systematic literature review on definitions and challenges. *Journal of Hospitality and Tourism Management, 47,* 134–149.

Tilden, F. (2007). *Interpreting Our Heritage*, 4th edn. Chapel Hill: University of North Carolina Press.

Tribe, J. (2008). Tourism: A critical business. *Journal of Travel Research, 46*(3), 245–255.

Tsing, A.L. (2015). *The Mushroom at the End of the World.* Princeton: Princeton University Press.

United Nations (2018). The speed of urbanization around the world. *Population Facts.* United Nations Department of Economic and Social Affairs. Available from https://population.un.org/wup/Publications/Files/WUP2018-PopFacts_2018-1.pdf, accessed March 31, 2021.

Wallace, D. (2021). The future of travel is private. *Globe and Mail.* https://www.theglobeandmail.com/life/travel/article-the-future-of-travel-is-private/, accessed March 15, 2021.

Wheeler, S. (2002). The new regionalism: Key characteristics of an emerging movement. *Journal of the American Planning Association, 68*(3), 267–278.

PART I

CONCEPTUAL FOUNDATIONS

2. Understanding rural communities

Michael Woods

INTRODUCTION

In the summer of 2020, the partial easing of coronavirus restrictions in Britain prompted a short but intensive holiday season as visitors flocked to coastal and countryside areas. In communities such as Morfa Nefyn, perched on the Llŷn Peninsula in north-west Wales, the sudden influx of tourists provoked mixed feelings. The local economy was heavily dependent on tourism and things had been tough since a national lockdown had been imposed in March. Business owners were desperate to welcome back visitors and recoup some of their lost income. Others in Morfa Nefyn, however, faced the prospect with trepidation, fearful that tourists would bring the virus back into the community, and complained about overcrowding and parking problems as record numbers of holidaymakers arrived. Some residents told the media that they felt unable to go to the local shop because it was full of visitors, and tensions erupted within the community, requiring councillors to act as peacemakers.

This anecdote captures both dimensions through which rural communities are commonly approached in academic studies of tourism. On the one hand, tourism is widely recognized as a key component in the diversification of rural economies away from traditional primary industries or manufacturing, and especially in neo-endogenous rural development led by communities themselves. In this perspective, encapsulated in the terminology of 'community-based tourism' or 'community-oriented tourism', tourism becomes internalized to rural communities – part of their social fabric, identity, and material reality. On the other hand, studies have analysed the impacts of tourism on rural communities, whether through the dilution of social coherence and local cultures, pressures on housing or employment, population change from the arrival of amenity migrants and migrant workers, or the appropriation and commodification of local cultural practices and experiences for tourist consumption. In these representations, tourism is external to rural communities, as a potential threat or challenge, with tourists 'othered' and set apart from community members.

In both approaches the concept of 'community' frequently remains unpacked. Studies rarely dwell on questions such as: What does 'community' mean in these contexts? Who does it include? How is it defined and delimited? What are the attributes and dynamics of 'community' that can either make it a resource to be mobilized to develop a tourist economy, or expose its fragility to new conditions introduced by tourism? What actually happens when communities react, adapt and evolve in response to interactions with tourism?

This chapter takes a step back to examine these foundational questions about the conceptualization of 'community' and its deployment in research on tourism and rural development. It starts by revisiting the initial theorization of 'community' in early rural sociology and anthropology and its use in the corpus of rural community studies through the mid twentieth century, as well as the critiques directed at this work. The chapter then proceeds to discuss the revitalization of interest in community in rural studies at the start of this century and its reconceptualization influenced by new insights from social theory, notably from relational theory. Finally, the chapter considers how these conceptual understandings of community have been, or could be, applied in research on tourism and rural development, identifying four broad models: community as recipient, community as resource, community as commodity, and community as assemblage.

INTRODUCING COMMUNITY

'Community' has long been a core unit of analysis in rural sociology and anthropology since the emergence of these disciplines around the beginning of the twentieth century. The essential premise is that the rural population can be divided into largely self-defined and partly autonomous groups of individuals that are associated with demarcated territories and are characterized by a density of social relations and a collective sense of identity. Moreover, to describe these groups as 'communities' implies that they exhibit certain attributes and features that distinguish them from other forms of social organization. In particular, early studies in rural sociology were strongly influenced by the work of Ferdinand Tönnies, originally published in 1887, which contrasted the model of 'community', or *gemeinschaft*, with that of 'society' or *gesellschaft*. Whereas *gesellschaft* was defined by Tönnies as being based on individualism mediated by contractual relations, *gemeinschaft* or community is based on kinship, a shared habitat and collective endeavour (Tönnies, 1963). Significantly, although Tönnies himself did not conceptualize the distinction geographically, his discussion of *gesellschaft* as the dominant social form in the expanding cities of the late nineteenth century led by implication to an association of *gemeinschaft* with rural settlements that was subsequently embellished by geographers and sociologists (Panelli, 2006; Woods, 2011).

The adoption of Tönnies's concept of *gemeinschaft* was aided by its compatibility with the 'structuralist-functionalist' approach that dominated early studies of community. As Harper (1989) and Liepins (2000a) have documented, this concept perceived communities to be stable, discrete entities with observable structures and functions. If this description sketched out the parameters of 'community' as a research object, the idea of *gemeinschaft* was seized on to fill out the content, presenting a theorization of what structures and functions might be expected to be observed in rural communities. Studies thus sought to record the social structures and dynamics of rural communities and to measure them against the *gemeinschaft* ideal, in effect treating 'the notion of "community" as a term signifying a particular social arrangement, one generally fixed in terms of both the form and function it would take' (Liepins, 2000a, p. 25).

The popularity of community studies expanded after the Second World War, prompted in part by appreciation that rural communities in many parts of the global north were on the cusp of transformational social and economic change. Britain was a notable locus for community studies (see for example Frankenberg, 1957; Littlejohn, 1964; Rees, 1950; Williams, 1956; 1963) but the approach was replicated in other parts of Europe, as well as in North America and further afield (Arensberg & Kimball, 1948; Dempsey, 1990; Kolars, 1962; Poiner, 1990; Sanders & Lewis, 1976; Vidich & Bensman, 1957). The common format was to focus on a single, discrete, settlement – usually a village or a small town – and to document over a period of time the multiple aspects of its social relations, including kinship networks and generational dynamics, social stratification, working life, religion, politics, customs and culture, and the role of various institutions from schools to sports clubs. Increasingly, community studies employed an ethnographic methodology, which as Harper (1989) and Liepins (2000a) note, marked a departure from the 'structuralist-functionalist' paradigm by shifting attention from recording community structures and purposes to attempting to demonstrate the reality of 'community' through 'the careful recounting of "authentic" lived experiences and relationships' (Liepins, 2000a, p. 25). Yet, they continued to treat communities as relatively fixed and observable entities.

Although most community studies were undertaken in rural or small-town settings, a few focused on more urban communities, and some efforts were made to compare different studies in order to elucidate the differentiation of rural and urban communities and the granularities of the rural–urban transition (Frankenberg, 1966). Community studies hence served to position communities spatially, but they also did so temporally. Influenced by Tönnies and other writers such as Louis Wirth, the conceptualization of community employed in the studies intrinsically understood the social relations of rural communities as embedded in deep-rooted and enduring histories that contrasted with the dynamism and modernity of urban society. As such, rural communities, and indeed the very notion of 'community', faced threats not only from urbanization, but from modernization and globalization. Alwyn Rees's seminal study of a Welsh rural community, for instance, described kinship that was 'woven together "like a pig's entrails"' (Rees, 1950, p. 75) and lamented the encroachment of modern mobility that 'uproots the ablest members of the community and scatters them indiscriminately into official positions up and down the country' (ibid., p. 165), though as Crow (2008) observes, the hostile assessment at least acknowledged that the community was not static and enclosed.

These implicit assumptions about continuity and change in effect framed community studies as being of a time and contributed to them appearing increasingly anachronistic as urbanization and globalization advanced. Furthermore, community studies were subjected to methodological and theoretical critiques that criticized the parochialism and common lack of detail of methods in community studies that inhibited comparative analysis, as well as their largely descriptive and atheoretical nature. Although Crow (2008) refutes some aspects of these critiques, especially for later studies conducted in the 1980s and 1990s, the approach declined in popularity, supplanted in rural studies by political economy perspectives that emphasized broader social structures and favoured the concept of 'locality' over that of 'community' (Panelli, 2006; Woods, 2005).

RETHINKING COMMUNITY

Whilst the prominence of 'community' as a unit of analysis in rural studies faded, the notion of 'community' (and some of cultural associations linked to it) nonetheless persisted in popular and policy discourse. Indeed, the notion of 'community' arguably gained currency in rural development policy through the 1990s and early 2000s, as the modernization paradigm of rural development through state intervention and inward investment gave way to a new approach of neo-endogenous rural development (van der Ploeg et al., 2000; Woods, 2011). Neo-endogenous rural development emphasizes the valorization of a region's own internal resources, mobilized from the bottom up through the collective action of rural communities acting in place. In Europe, the new approach was encoded in the Cork Declaration in 1996, which established the framework for EU rural development policy over the next 25 years, including the methodology of community-led local development (CLLD) enacted in the LEADER programme (Woods, 2005; 2011). Renewed emphases on community action or community self-help could similarly be observed emerging in rural development policies in Australia, Canada, New Zealand, and the United States, as well as in some approaches to rural development in the global south (Bruce and Whitla, 1993; Cheshire, 2006; Shepherd, 1998). In liberal democracies, at least, this transition reflected a broader shift in the dominant mode of governmentality, from a concern with 'governing the social' in the managed liberalism of the mid twentieth century (exemplified by the welfare state), to a concern with 'governing through communities' in a new advanced liberal or neoliberal regime (Rose, 1996). The latter model was predicated on the rolling-back of the state through deregulation and privatization and the transferal of responsibility for governing societal needs to the private and voluntary sectors, and to individual citizens operating collectively within communities. As a governmental strategy, it was applied to both urban and rural societies, but was afforded additional purchase in rural contexts by its resonance with the embedded cultural associations of 'community' and rurality articulated in the earlier community studies (Murdoch, 1997; Woods & Goodwin, 2003).

The re-emergence of 'community' as a central idea in rural development and governance raised questions about the nature and dynamics of rural communities, which in turn attracted academic attention. However, unlike earlier community studies, the new wave of research was not concerned with approaching communities as fixed vessels in which to study social relations, or with exploring the essence of rural communities, but rather engaged with rural communities as *active entities* through which individual agency was translated into collective outcomes. One thread of this work drew on the concept of 'social capital', the collective resource created by ties and interactions between members of a community. Ideas of social capital had been implicit in community studies but were made explicit and popularized in the 1990s by the political scientist Robert Putnam, who distinguished between *bonding social capital* that holds a community together and *bridging social capital* that connects members of a community with other communities (Putnam, 2000). Putnam further echoed earlier community studies by arguing that bonding social capital in American communities had been eroded by the rise of individualism and modern media and consumer culture. Whilst rural researchers critiqued the normative dimensions of Putnam's thesis and questioned its ability to demonstrate causality in the connections it asserted, the idea of social capital as an important factor in community mobilization that was in need of strengthening in many rural communities where it had become weakened, was embraced in work that advocated community capacity-building initiatives as tools for inclusive rural development (Anderson & Bell, 2003; Árnason et al., 2009; Magnani & Struffi, 2009; Woods, 2011).

A second research concern focused on the question of how communities are defined, and issues of inclusion and exclusion in rural communities. The positioning of communities as delivery vehicles for rural development and governance politicized the drawing of community boundaries (socially as well as geographically), especially when linked to entitlements to vote, participate or benefit from community development schemes. As the implementation of land reforms in Scotland that introduced a 'community right to buy' demonstrated, the matter of determining community membership had become complicated by mobility and migration, with kinship networks extending beyond geographical locales and resident populations becoming increasingly heterogeneous (Mackenzie, 2006). Accordingly, rural research developed new lines of enquiry into practices of community bounding and into the power relations that included and excluded participants in rural community governance (see for example Little & Jones, 2000).

To inform these enquiries, rural researchers also turned to conceptual frameworks that followed from the 'cultural turn' in geography and sociology in the 1990s, especially theories of social constructivism and symbolic interactionism. In so doing, rural researchers reconnected with anthropologists, who had contributed to early community studies and who had retained a core concern with community as a concept as interest had faded in geography and sociology (Panelli, 2006). Anthropologists had responded to critiques of community studies by drawing on hermeneutic and interpretative theories to develop a critical conceptualization of community that emphasized its symbolic construction, as articulated by Anthony Cohen:

> the 'community' as experienced by its members does not consist in the social structure or in 'the doing' of social behaviour. It inheres, rather in 'the thinking' about it. It is in this sense that we can speak of the 'community' as a symbolic, rather than a structural, construct. (Cohen, 1985, p. 98)

Accordingly, from a social constructivist perspective, communities do not exist as a priori, fixed units; rather they are imagined into being through lay, media, and policy discourses, acts of mapping and naming, and processes of 'othering'. Othering establishes group identities by contrasting 'us' and 'them'; it allows communities to define who belongs by discursively articulating the differences between, say, the residents of neighbouring villages. Yet, as a flourish of critical research in rural geography in the mid and late 1990s demonstrated, it can also produce fissures within rural communities by 'othering' minorities or non-conforming lifestyle groups that do not fit with the imagined, idealized community identity (Cloke & Little, 1997; Philo, 1992).

Furthermore, engaging with the anthropological tradition highlighted the significance of performance and practices in symbolically constructing community as well as texts. Rituals, traditions, shows, and festivals are ways of performing community that can reinforce internal social norms and signal messages about community identity to external audiences. As such, they can frequently involve acts of othering, for example in the representation of stereotypical characters or the commemoration of divisive historical events, which can symbolically exclude certain groups from belonging to the community, but which when challenged can also become rallying points for defensive community mobilization (Smith, 1993). More routinely, community is also performed through everyday social interactions in pubs, churches, schools, and shops (Bell, 1994; Neal & Walters, 2008; Woods, 2011).

However, as the last observation about the spaces through which community is enacted indicates, whilst an emphasis on performativity allowed studies of community to escape from the trap of structuralism, it cannot be entirely divorced from materiality. Liepins (2000a) argues that the spaces of community include 'both the material sites filled by communal activities, and the symbolic and metaphoric spaces in which people connect "in community" even while existing in different physical or social locations' (p. 28). From this starting point, Liepins elaborated a model of community comprised of four intersecting components: people, practices, meanings, and spaces and structures.

People are placed at the centre of Liepins's model, recognizing that a community is essentially defined by social collectivity and connection, but the translation of a collection of individuals into a community is achieved through the mobilization of the elements in the outer ring. Communities are socially constructed through the articulation and circulation of shared 'meanings' that define a collective identity and values. Communities are performed through 'practices' of everyday social interactions, rituals, and communal gatherings. Finally, communities are embedded in place by sites and structures that include 'the physical sites of schools and halls and bridges where people gather in their practice of "community"' (Liepins, 2000a, p. 32) but also metaphorical spaces, such as community newsletters. These elements are interconnected and mutually constitutive. Meanings legitimate practices, which occur in spaces and through structures, which enable the mobilization of meanings, which are embodied in spaces and structures, which affect how practices can occur, which enables the circulation and challenging of meanings. Liepins's model is methodological as much as conceptual, with the dimensions of the meanings, practices, and spaces and structures of community representing different ways in which communities can be studied, or the 'objects and processes from which we can create readings of "community"' (ibid., p. 31). This was illustrated by Liepins in a follow-up paper that applied the model to research on three case study rural communities in Australia and New Zealand (Liepins, 2000b).

Significantly, Liepins's model also acknowledges the wider context of terrains of power and discourse in which communities are positioned and with which they continually interact. Indeed, Liepins noted that although she placed 'people' at the centre of the model, this should not imply that the only people that matter are those 'inside' a community. Rather, 'recognition must be given to the constitutive capacity of others external to any given "community" who may nevertheless name and construct notions about "community" which could either enable or constrain such collectives' (Liepins, 2000a, p. 30). As such, Liepins's understanding of community is fundamentally relational, part of a broader relational turn in social science that has foregrounded connectivity, interdependency, contingency and temporality (Heley & Jones, 2012).

The adoption of relational perspectives has allowed rural studies to move away from a concern with distilling the essence of rural spaces, societies and communities to exploring the rural as multifaceted, heterogeneous, ever-changing, globally engaged and contested, existing materially and discursively in relation to the urban (Heley & Jones, 2012; Murdoch, 2003; Woods, 2007). Within rural geography, the focus has tended to be on the relationality of rural *places*, yet Liepins's model is a reminder that *communities* too are relational.

Recognizing that communities are made up of *social relations* is of course not a new insight, but a relational viewpoint also highlights the *elasticity* and *multiplicity* of rural communities. Elasticity here can be understood as the capacity of a community to stretch over space, as its members move but maintain a collective affinity. Rural communities have predominantly been places of emigration, but in many cases kinship ties have been maintained. Migrants to cities from rural communities in many parts of Africa and Asia, for example, frequently form 'village associations' and continue to mix with one-time neighbours and to keep local customs, transposing aspects of rural life into the city and giving the community a translocal footprint (Brogger, 2019; Chen, 2020; Englund, 2002; Naumann & Greiner, 2017; Velayutham & Wise, 2005). Across the world, mobile phones and the internet have made it possible for out-migrants to continue to participate in the practices of community events even at a distance, and rural communities benefit from remittances, or from dispersed members who act as agents and intermediaries for business (Aguayo, 2008; Petrou & Connell, 2017). Equally, members of the extended community may be mobilized in internal community politics or conflicts around new developments or other changes.

At the same time, intensified global mobility brings new people into rural localities, unsettling and reconfiguring collective meanings and practices and contributing to the multiplicity of rural communities. These include transitory members, such as second-home owners, migrant workers or even regular visitors, who are only partially present in a rural place and simultaneously part of other communities elsewhere, but who nonetheless may identify with the rural community and participate in community practices and spaces. The emotional attachment to the community of such individuals may be as strong as or stronger than that of established permanent residents, but may be tied to different interests or values, creating potential for conflict (Stedman, 2006; Tuulentie & Kietavainen, 2020).

Where the integration of new residents into community structures and practices is limited, multiple parallel communities may co-exist in the same geographical space, as Nelson and Hiemstra (2008) documented for the resort town of Leadville, Colorado. Butler and Ben (2021) have more recently highlighted the racialized layers of community in settler colonial societies such as Australia, including the discursive and material separation of indigenous communities as distinct from the 'rural community'. Even in the seemingly most integrated and homogeneous community, relational theory suggests the 'shape' of the community may look very different from the different vantage points of its individual members, perceived through different cultural and political frames, especially by those individuals who may experience exclusion or marginalization by dominant community discourses around sexuality, religion, lifestyle, and so on. Indeed, Neal and Walters (2008) comment that the conviviality that is enacted through community practices as a glue for collective identity may often be fragile, such that the 'warmth of conviviality expressed by the inclusive, non-other defined community may conceal more difficult emotions such as anxiety, boredom and loneliness' (p. 291).

MOBILIZING COMMUNITY IN RURAL TOURISM RESEARCH

Research on rural tourism has engaged with each of the conceptual perspectives discussed above. Some studies have implicitly or explicitly adopted a structuralist-functionalist approach, whilst others have employed social constructivist or relational positions. With reference to Liepins's framework, tourism studies have variously approached rural communities through the entry points of meanings, practices, and structures and spaces. Together these multiple strategies have afforded a wide range of insights across different aspects of tourism processes and experiences as they intersect with rural communities. Indeed, in addition to the different conceptualizations of community that have been applied in rural tourism research, it is also possible to differentiate studies by the ways in which they position communities in relation to tourism and thus the aspects of community that are prioritized in the analysis. Four of the most significant are discussed in this section: communities as recipients, communities as resources, communities as commodities, and communities as assemblages.

The first approach positions communities as relatively passive recipients of tourism, with studies typically examining residents' perceptions of tourism (e.g., Boonzaaier & Grobler, 2012), attitudes towards new tourism developments (e.g., Bender & Kerstetter, 2020; Rocket & Ramsey, 2017), or perspectives on second homes (e.g., Rye, 2011), or seeking to measure the impact of tourism on community wellbeing (e.g., Moscardo & Murphy, 2016). Commonly, the definition of the 'local community' is not problematized, such that 'community' means little more than a group of individuals living in a specific place that provides the population for the study. Similarly, communities are often implicitly represented as homogeneous groups, although some studies, such as Ulrich-Schad's (2018) work in a US amenity destination, recognize the diversity and dynamism of the community through the categorization of different group responses to tourism.

In the second approach, communities are regarded more actively as resources that can be mobilized in support of tourism. This understanding of community is evident in the idea of 'community-based tourism', where collective action utilizes embedded social and cultural resources such as social capital to create a tourist offer. That community-based tourism is more commonly associated with rural communities than with urban communities and is especially promoted as a development strategy for rural communities in the global south, implicitly echoes conceptualizations that have portrayed rural settlements as retaining a truer essence of collectivity than cities (see for instance Hwang et al., 2016; Zielinski et al., 2020). Studies following this approach have discussed methods for community mobilization and engagement, such as community arts (Chatkaewnapanon & Kelly, 2019) or collaborative governance (Keyim, 2018), whilst more critical research has questioned the limits to community agency and the extent to which communities are empowered through tourism, for example in Ruiz-Ballesteros and Hernandez-Ramirez's (2010) analysis of community-based tourism in rural Ecuador.

The third approach can also encompass work on community-based tourism but differs from the above in seeing the community itself as a commodity that is in effect traded through tourism. In other words, endogenous rural tourism is based not only on the commodification of landscapes, recreational activities, or material objects, but also on the social interactions and cultural practices that constitute the community (Mannon & Glass-Coffin, 2019). Research accordingly explores the production and reproduction of staged encounters and experiences that enable visitors to temporarily share in a version of community life, for instance through cultural festivals (Lin & Bestor, 2020), immersive working or creative holidays (Blapp & Mitas, 2018), or home stays. Much of this work adopts a critical stance, identifying risks of cultural appropriation, devaluation, and disruption to established social structures (Kirtsogiou & Theodossopoulos, 2004; Mannon & Glass-Coffin, 2019; Mbaiwa, 2011).

The above three approaches may each involve engagement with all three dimensions of community in Liepins's model – meanings, practices, and sites and structures – but the final approach is the most explicitly relational, with a community understood as an assemblage of heterogeneous entities. From this perspective, tourism and tourists are entangled with the community and part of its dynamic constitution, introducing new people, ideas, and practices into the community through external flows of visitors, workers, migrants, commodities, capital, cultural influences and so on. As such, research in this final approach is focused on the processes and power relations through which tourism changes communities and communities seek to engage with tourism and its effects, with example studies investigating dynamics of amenity gentrification (Matarrita-Cascante & Stocks, 2013; Nelson & Hines, 2018; Rainer, 2016), the engagement and integration of second-home owners

(Farstad, 2016; Gallent, 2014; 2015; Kietavainen et al., 2016), and local policymaking (Lysgard, 2019).

CONCLUSION

Rural communities exist in relation to tourism in a number of ways. Communities share geographical spaces with tourists and tourism practices, but communities may also be directly engaged in promoting, supporting and managing tourism, and aspects of community life may be commodified as part of the tourism offer of a locality. As such, communities have the capacity both to affect and to be affected by rural tourism. Numerous studies have surveyed 'community' perceptions of tourists and attitudes towards tourism developments, but implicit assumptions in such work, that communities are bounded, homogeneous entities to which tourism is external, have been critiqued and countered by more critical research investigating the processes and dynamics through which communities are actively engaged and entangled in tourism.

The shift towards a more nuanced perspective has been supported by conceptual tools that have themselves evolved from the structural-functional models of early community studies to approaches informed by social constructivist and relational theories. In particular, the multidimensional character of communities, composed of meanings, practices, and sites and structures, has been highlighted, most notably by Liepins (2000a). Rural tourism has touched on each of these aspects, as research has documented – for example, residents' perceptions of tourism are framed through prisms of community meanings; certain community practices and rituals have been commodified as tourist experiences; and sites through which community is enacted, such as pubs and shops, have become places where residents encounter tourists. These points of intersection are also vectors of change for communities. Indeed, relational ideas of community emphasize the porosity, dynamism, and interconnectivity of rural communities and for many, tourism is a driver of connectivity and change. Second-home owners, migrant workers in hospitality industries, and regular tourists all contribute to a blurring of the boundaries of communities, disrupt settled meanings and networks, but also create new capacities for communities – including new capacities for tourism development. Far from being separate, rural communities and rural tourism are mutually constitutive, and developing an understanding of 'community' as a concept is important for enriching the potential of rural tourism research.

REFERENCES

Aguayo, B.E.C. (2008). Global villages and rural cosmopolitanism: Exploring global ruralities. *Globalizations*, *5*, 541–554.

Anderson, C., & Bell, M. (2003). The devil of social capital: A dilemma of American rural sociology. In P. Cloke (ed.), *Country Visions* (pp. 232–244). Harlow: Pearson.

Arensberg, C.M., & Kimball, S.T. (1948). *Family and Community in Ireland*. London: Peter Smith.

Árnason, A., Shucksmith, M., & Vergunst, J. (eds) (2009). *Comparing Rural Development*. Aldershot: Ashgate.

Bell, M.M. (1994). *Childerley: Nature and Morality in a Country Village*. Chicago: University of Chicago Press.

Bender, A.K., & Kersletter, D.L. (2020). Residents' attitudes towards tourism development in a rural community: A qualitative approach. *Journal of Park and Recreational Administration, 38*, 90–108.

Blapp, M., & Mitas, O. (2018). Creative tourism in Balinese rural communities. *Current Issues in Tourism, 21*, 1285–1311.

Boonzaaier, C., & Grobler, J. (2012). Community perceptions of tourism in the Tshivhase area of Limpopo Province, South Africa. *Anthropology Southern Africa, 35*, 60–70.

Brogger, D. (2019). Urban diaspora space: rural–urban migration and the production of unequal urban spaces. *Geoforum, 102*, 97–105.

Bruce, D., & Whitla, M. (eds) (1993). *Community-Based Approaches to Rural Development*. Sackville: Mount Allison University.

Butler, R., & Ben, J.H. (2021). Centring settler colonialism in rural Australian multicultures: race, place and local identities. *Journal of Ethnic and Migration Studies, 47*, 2179–2197.

Chatkaewnapanon, Y., & Kelly, J.M. (2019). Community arts as an inclusive methodology for sustainable tourism development. *Journal of Place Management and Development, 12*, 365–390.

Chen, N. (2020). Rural community unbound: trans-locality, rural-to-rural connections and the formation of inter-regional surname associations in China. *Journal of Rural Studies, 78*, 378–386.

Cheshire, L. (2006). *Governing Rural Development*. Aldershot: Ashgate.

Cloke, P., & Little, J. (eds) (1997). *Contested Countryside Cultures*. London: Routledge.

Cohen, A. (1985). *The Symbolic Construction of Community*. London: Tavistock.

Crow, G. (2008). Recent rural community studies. *International Journal of Social Research Methodology, 11*, 131–139.

Dempsey, K. (1990). *Smalltown*. Melbourne: Oxford University Press.

Englund, H. (2002). The village in the city, the city in the village: Migrants in Lilongwe. *Journal of Southern African Studies, 28*(1), 137–154.

Farstad, M. (2016). Worthy of recognition? How second home owners understand their own group's moral worth in rural host communities. *Sociologia Ruralis, 56*, 408–412.

Frankenberg, R. (1957). *Village on the Border*. London: Cohen and West.

Frankenberg, R. (1966). *Communities in Britain*. Harmondsworth: Pelican.

Gallent, N. (2014). The social value of second homes in rural communities. *Housing Theory and Society, 31*, 174–191.

Gallent, N. (2015). Bridging social capital and the resource potential of second homes: The case of Stintino, Sardinia. *Journal of Rural Studies, 38*, 99–108.

Harper, S. (1989). The British rural community: An overview of perspectives. *Journal of Rural Studies, 5*, 161–184.

Heley, J., & Jones, L. (2013). Relational rurals: Some thoughts on relating things and theory in rural studies. *Journal of Rural Studies, 28*, 208–217.

Hwang, D., Chi, S.H., & Lee, B. (2016). Collective action that influences tourism: Social structural approach to community involvement. *Journal of Hospitality and Tourism Research, 40*, 497–515.

Keyim, P. (2018). Tourism collaborative governance and rural community development in Finland: The case of Vuonislahti. *Journal of Travel Research, 57*, 483–494.

Kietavainen, A., Rinne, J., Paloniemi, R., & Tuulentie, S. (2016). Participation of second home owners and permanent residents in local decision making: The case of a rural village in Finland. *Fennia, 194*, 152–167.

Kirtsogiou, E., & Theodossopoulos, D. (2004). 'They are taking our culture away': Tourism and cultural commodification in the Garifuna community of Roatan. *Critique of Anthropology, 24*, 135–157.

Kolars, J.F. (1962). Community studies in rural Turkey. *Annals of the Association of American Geographers, 52*, 476–489.

Liepins, R. (2000a). New energies for an old idea: Reworking approaches to 'community' in contemporary rural studies. *Journal of Rural Studies, 16*, 23–35.

Liepins, R. (2000b). Exploring rurality through 'community': Discourses, practices and spaces shaping Australian and New Zealand rural 'communities'. *Journal of Rural Studies, 16*, 325–341.

Lin, Y. & Bestor, T. (2020). Embedding food in place and rural development: Insights from the Bluefin Tuna Cultural Festival in Donggang, Taiwan. *Journal of Rural Studies, 79*, 373–381.

Little, J., & Jones, O. (2000). Masculinity, gender and rural policy. *Rural Sociology, 65*, 621–639.

Littlejohn, J. (1964). *Westrigg: The Sociology of a Cheviot Parish*. London: Routledge and Kegan Paul.

Lysgard, H.K. (2019). The assemblage of culture-led policies in small towns and rural communities. *Geoforum, 101*, 10–17.

Mackenzie, A.F.D. (2006). 'S Leinn Fhèin am Fearann' (The land is ours): Reclaiming land, re-creating community, North Harris, Outer Hebrides, Scotland. *Environment and Planning D: Society and Space, 24*, 577–598.

Magnani, N., & Struffi, L. (2009). Translation sociology and social capital in rural development initiatives: A case study from the Italian Alps. *Journal of Rural Studies, 25*, 231–238.

Mannon, S.E., & Glass-Coffin, B. (2019). Will the real rural community please stand up? Staging rural community-based tourism in Costa Rica. *Journal of Rural and Community Development, 14*, 71–93.

Matarrita-Cascante, D., & Stocks, G. (2013). Amenity migration to the global south: Implications for community development. *Geoforum, 49*, 91–102.

Mbaiwa, J.E. (2011). Cultural commodification and tourism: The Goo-Moremi community, central Botswana. *Tijdschrift voor Economische en Sociale Geografie, 102*, 290–301.

Moscardo, G., & Murphy, L. (2016). Using destination community wellbeing to assess tourist markets: A case study of Magnetic Island, Australia. *Journal of Destination Marketing and Management, 5*, 55–64.

Murdoch, J. (1997). The shifting territory of government: Some insights from the Rural White Paper. *Area, 29*, 109–118.

Murdoch, J. (2003). Co-constructing the countryside: Hybrid networks and the extensive self. In P. Cloke (ed.), *Country Visions* (pp. 263–282). Harlow: Pearson.

Naumann, C., & Greiner, C. (2017). The translocal villagers: Mining, mobility and stratification in post-apartheid South Africa. *Mobilities, 12*, 875–889.

Neal, S., & Walters, S. (2008). Rural be/longing and rural social organizations: Conviviality and community-making in the English countryside. *Sociology, 42*(2), 279–297.

Nelson, L., & Hiemstra, N. (2008). Latino immigrants and the renegotiation of place and belonging in small town America. *Social and Cultural Geography, 9*, 319–342.

Nelson, P., & Hines, J.D. (2018). Rural gentrification and networks of capital accumulation – A case study of Jackson, Wyoming. *Environment and Planning A, 50*(7), 1473–1495.

Panelli, R. (2006). Rural society. In P. Cloke, T. Marsden, & P. Mooney (eds), *Handbook of Rural Studies* (pp. 63–90). London: Sage.

Petrou, K., & Connell, J. (2017). Food, morality and identity: Mobility, remittances and the translocal community in Paama, Vanuatu. *Australian Geographer, 48*, 219–234.
Philo, C. (1992). Neglected rural geographies: A review. *Journal of Rural Studies, 8*, 193–207.
Poiner, G. (1990). *The Good Old Rule: Gender and Other Power Relationships in a Rural 'Community'*. Sydney: Sydney University Press.
Putnam, R. (2000). *Bowling Alone: The Collapse and Renewal of American Community*. New York: Simon and Schuster.
Rainer, G. (2016). Constructing globalized spaces of tourism and leisure: Political ecologies of the Salta Wine Route (NW Argentina). *Journal of Rural Studies, 43*, 104–117.
Rees, A.D. (1950). *Life in a Welsh Countryside*. Cardiff: University of Wales Press.
Rockett, J., & Ramsey, D. (2017). Resident perceptions of rural tourism development: The case of Fogo Island and Change Islands, Newfoundland, Canada. *Journal of Tourism and Cultural Change, 15*, 299–318.
Rose, N. (1996). The death of the social? Refiguring the territory of government. *Economy and Society, 25*, 327–356.
Ruiz-Ballesteros, E., & Hernandez-Ramirez, M. (2010). Tourism that empowers? Commodification and appropriation in Ecuador's Turismo Comuntario. *Critique of Anthropology, 30*, 201–229.
Rye, J.F. (2011). Conflicts and contestations: Rural populations' perspectives on second homes phenomenon. *Journal of Rural Studies, 27*, 253–274.
Sanders, I.T., & Lewis, G.F. (1976). Rural community studies in the United States – A decade in review. *Annual Review of Sociology, 2*, 35–53.
Shepherd, A. (1998). *Sustainable Rural Development*. Basingstoke: Macmillan.
Smith, S.J. (1993). Bounding the Borders: Claiming space and making place in rural Scotland. *Transactions of the Institute of British Geographers, 18*, 291–308.
Stedman, R.C. (2006). Understanding place attachment among second home owners. *American Behavioral Scientist, 50*, 187–205.
Tönnies, F. (1963). *Community and Society*. New York: Harper and Row.
Tuulentie, S., & Kietavainen, A. (2020). New rural community? Narratives from second home owners about everyday life in a tourist region in Finnish Lapland. *Sociologia Ruralis, 60*, 357–374.
Ulrich-Schad, J.D. (2018). 'We didn't move here to move to Aspen': Community making and community development in an emerging rural amenity destination. *Journal of Rural and Community Development, 13*, 43–66.
Van der Ploeg, J.D., Renting, H., Brunori, G., Knickel, K., Mannion, J., Marsden, T., de Roest, K., Sevilla Guzmán, E., & Ventura, F. (2000). Rural development: From practice to policies to theory. *Sociologia Ruralis, 40*, 391–408.
Velayutham, S., & Wise, A. (2005). Moral economies of a translocal village: Obligation and shame among South Indian transnational migrants. *Global Networks, 5*, 27–47.
Vidich, A., & Bensman, J. (1957). *Small Town in Mass Society*. Urbana: University of Illinois Press.
Williams, W.M. (1956). *The Sociology of an English Village: Gosforth*. London: Routledge and Kegan Paul.
Williams, W.M. (1963). *A West Country Village: Ashworthy*. London: Routledge and Kegan Paul.
Woods, M. (2005). *Rural Geography*. London: Sage.
Woods, M. (2007). Engaging the global countryside: Globalization, hybridity and the reconstitution of rural place. *Progress in Human Geography, 31*, 485–507.
Woods, M. (2011). *Rural*. Abingdon: Routledge.
Woods, M., & Goodwin, M. (2003). Applying the rural. In P. Cloke (ed.), *Country Visions* (pp. 245–262). Harlow: Pearson.

Zielinski, S., Jeong, Y., Kim, S., & Milanes, C. (2020). Why community-based tourism and rural tourism in developing and developed nations are treated differently? A review. *Sustainability*, *12*, 5938.

3. Understanding rural tourism

Donald G. Reid

INTRODUCTION

This chapter introduces the reader to the structure, planning, and management of tourism in rural areas. The question driving this discussion is not only what is rural tourism, but more precisely, what makes rural tourism unique in comparison to tourism in other locations? More importantly, this chapter presents the challenges and issues facing rural tourism today. Although there are several potential answers to the question raised above, it must be addressed within the setting in which the specific tourism product is found. Context is everything and everything flows from context. Although we speak of rural tourism as if it is a single entity, it is not homogeneous in its makeup but multifaceted with differences as great as the rural–urban divide, with which we are so often concerned. Each landscape and community is unique and will need to be treated as such. That said, there are some unifying features that can provide a definition of rural tourism generally.

THE CONFIGURATION OF RURAL TOURISM

As laid out in many of the other chapters in this book, there is a unique way of seeing the world from the rural point of view. The countryside has traditionally provided a link to an authentic, some would say tranquil, time in history. For many, the rural lifestyle is thought to be an abrupt contrast to the complex and hurried world most of us live in today. This notion of the countryside as essentially simple and uncomplex may not be accurate but is held by many visitors to the countryside, nonetheless. This concept of rural includes an inimitable set of values and a socio-psychological paradigm that pervades throughout the countryside. Many people feel a special attachment to the idea of rural, and deep in their soul they understand it to be different than the urban setting beyond simply the landscape. Even though there is a recognizably distinct urban culture, most people understand that all human culture germinates from the rural environment, even though it may have lost its predominance in the everyday lives of urbanites and been driven to the subconscious. There is often a spiritual aspect to rural. All that being said, rural tourism is changing, hopefully without losing its authenticity and uniqueness. That is the challenge today.

Rural tourism is taking on many new challenges in the modern world. In the opening paragraph of their book, titled *Rural Tourism Development: Localism and Cultural Change*, George et al. sum up the challenges to rural tourism development in rural spaces succinctly yet vividly, when they state:

> Rural tourism represents a merging of perhaps two of the most influential yet contradictory features of modern life. Not only are the forces of economic, social, cultural, environmental and political change working to define rural spaces the world over, but broad global transformations in consumption and transportation patterns are reshaping leisure behaviour and travel. For those concerned with both the nature of change in rural areas and tourism development, the dynamics and impacts of integrating these two dramatic shifts are not well known but are becoming increasingly provocative discourses for study. (George, Mair, & Reid, 2009, p. 1)

Because the overwhelming majority of citizens in most countries live in cities, rural dwellers often feel neglected or taken advantage of by larger society. And it is quite possible their concern has merit. Given that most rural areas are thought of as limited to agricultural production and other extractive industries, most countries have developed an agricultural or resource use policy but not a comprehensive rural policy that encompasses a variety of economic and social issues including the place and role of tourism on the landscape and in rural communities. Land zoning in most countries treats tourism as any other industry, but tourism has several unique characteristics which demand special zoning laws to facilitate planning and mitigate tourism's externalities. Tourism around the world is becoming increasingly important economically and sociologically, as society transitions from an industrial to a highly digital technological economy, one that is based in services and finance. The tourism industry needs to begin, if it hasn't already started, to lobby for the development of rural policy that addresses this deficiency. Tourism is no longer just an add-on to the rural economy, but a major part of it. In some places it is even the driving force.

Rural tourism at its core is defined by people visiting an area that is lower in population than in the urban setting, that offers some type of unique natural environmental feature, such as agricultural land, forested areas, lakes, rivers, and valleys. Rural tourism also includes village and small-town visitation. It is distinguished by the natural features and the activities that traditionally go on in its unique environment. Rural tourism also embraces exceptional features such as archaeological sites that contain historical representations of history and prehistory. It is a preserver of human and natural history. It features arts and cultural attractions as well as outdoor recreation and adventure experiences. Tourism in rural areas pretty much offers something to everybody.

Rural tourism is often defined by governments, as areas that are a designated distance from a major urban population, but this is not a satisfying definition beyond its administrative simplicity. Kokt and Hattingh suggest that there are two fundamental types of rural tourism:

> The first type is 'pure' rural tourism that conjures up images of the unspoilt, idyllic countryside with activities related to enjoying the setting. The type of rural tourism typically targets a number of niche markets offering a range of traditional, peaceful rural experiences such as hiking, fishing, picnicking, observing wildlife and enjoying the countryside and its people. The second type of rural tourism is a large scale, high-impact phenomenon in which the rural location is less important. This type of tourism imposes urban values on the rural space and attracts large numbers of people. (Kokt & Hattingh, 2019, p. 126)

In addition to the foregoing definitions of rural tourism there are two additional ways of approaching the understanding of tourism in the rural setting. Given that tourism is first and foremost an economic activity, examining it from a supply and demand perspective is appropriate. From a supply perspective, a rural community may exploit a unique natural geological feature that exists in their locale. If, for example, a settlement is located at the base of a mountain then it is natural for that community to exploit the mountain for downhill skiing and other types of outdoor activities that take advantage of that type of feature. On the other hand, a community may analyse secondary sources of data and determine there is pent-up demand for an activity that may have nothing in common with the area but is projected to attract visitors because of the absence of that event or facility in the catchment area.

From the community's point of view, the fundamental goal of rural tourism is to enhance local, and in some cases, national economic development. Tourism in the countryside provides the rural community with substantial economic development by bringing visitors into the area from outside its jurisdiction. In my own country, Canada, Statistics Canada reports that, 'from January to September of 2018, tourism spending in Canada was $80.8 billion, an increase of 5.9% compared to the same period in 2017. Domestic revenues edged up 6.9% to almost $63.3 billion, and tourism revenues from international travellers increased 2.3% to $17.5 billion' (Innovation, Science and Economic Development Canada, 2019), and a fair share of those receipts found their way into rural Canada. In Ontario, Canada's most populous province, the gross receipts for tourism in 2019 were 36.1 billion Canadian dollars. Again, a goodly portion of this revenue flowed to rural areas. As a comparison, 'in 2018, tourism receipts for South Africa (SA) were 9,789 million US dollars' (World Data Atlas, 2018), that is approximately 11.5 billion Canadian dollars. South Africa's population is 59.5 million people whereas Ontario's is 14.73 million. Whether or not it is fair to compare South Africa to Canada is questionable, but the contrast does give you some idea of the importance of tourism to countries that have a large rural sector in both the developed world and in the developing part of the globe.

It is often thought that tourism is an environmentally friendly industry that naturally lends itself to the rural setting. That premise should not be automatically assumed. Successful tourism in rural spaces may also contribute to environmental destruction. Certainly, the overwhelming inundation of people in a confined space creating and dispensing waste into settings that are not equipped to handle that pressure is anti-environmental. Of course, air transportation of holidaymakers to far-flung destinations is a major part of air pollution and a serious contributor to global warming. On the positive side, the increase over the last few decades in the number of parks and protected areas for tourism purposes in many of the African countries and in other developing nations has increased the biodiversity in those parks for all humankind. Increasing the number of parks in the system disperses visitors to new areas, reducing the pressure on the previously established parks. Management for multipurpose use of these areas is critical but difficult to achieve. In the future, we may need to integrate wildlife and humans into rural spaces and not wall off the wildlife into museum-like enclaves. Sustainable management of the environment means sustainable management of tourism within an area where humans and wildlife live together.

THE FUTURE DEVELOPMENT OF RURAL TOURISM

Given the fast-paced change occurring in both the urban and rural setting, what niche can rural tourism fill in this fast-changing world? Tourism in the countryside has often been a lifeline when the primary industries have faltered or collapsed. In the agricultural sector, for example, tourism was initially seen as a secondary line of revenue to augment the traditional family farm income. As factory farming has become dominant in the rural landscape, this former auxiliary revenue generator has now become a major economic activity for many modern-day family farms. Tourism has also been a fallback industry to fishers when the fishing industry has undergone hard times. In the future, tourism may become an increasingly important economic generator, and augment those declining industries.

Given that global warming is now a major issue throughout the world, it will likely influence travel patterns in the future, possibly giving more emphasis to domestic tourism, particularly in rural areas. Rural tourism may have a significant role to play in environmental education. Locally based tourism may replace some international travel if the global warming crisis is to be adequately addressed. Like the carbon-based extractive industries, such as coal and oil production, the travel industry will need to examine its goals and objectives and bring them in line with environmentally acceptable practices. It could be that the value of spending time in a rural setting that provides nature-based leisure activity will heighten in value as time moves on and people become more environmentally conscious.

Unlike most manufacturing or extractive industries that ship their product out of the area to other locales for sale, with the proceeds being returned to the producing area, tourism brings people into the rural setting who carry environmental and social disruptions with them. This unique relationship increases the number of people in the tourism site for short or long periods of time, creating stresses that otherwise would not occur in places that don't import large numbers of people to consume their product. Stresses are placed on the infrastructure that often exceed the capacity of the area to absorb them. These stresses are not only taxing on the local facilities but become psychologically and socially stressing to the permanent residents as well. The influx of people beyond what the social and ecological environment can absorb is potentially damaging to the local community. Psychological stress multiplies as the distance between the cultures of the host and guest increases. In most countries, municipal zoning regulations treat tourism as a normal retail business and do not account for a mass influx of outsiders to the area. There is great need for tourism zoning regulations to account for the increased visitor load that tourism brings.

PLANNING AND MONITORING RURAL TOURISM DEVELOPMENT

In Canada, we have witnessed resistance to tourism development by local residents who have become disillusioned by the intrusion of large numbers of visitors into their community. These and other considerations must be addressed in the planning phase of tourism development in any rural region that undertakes to make tourism a substantial part of their economy. That said, many rural areas are so eager for any kind of economic development, including tourism, that they often overrate the community's capability to undertake this development. Planning that anticipates this overestimation is badly needed prior to development. One of the most critical administrative functions of the tourism committee in rural areas is the development of the tourism plan. The tourism plan outlines the concept of the destination and the theme(s) it will try to create and organize into a coherent destination package. It is important to create a destination theme that fits the natural surroundings or history of the area. Tourism themes that have no relation to the natural area are less likely to be successful and, in some cases, turn out to be nothing more than a blight on the landscape. There are exceptions to this general rule, of course, such as entertainment centres of various kinds.

Planning begins by setting aspirational goals that outline in broad terms what is intended to be achieved. Goals are followed up with a set of objectives that are measurable and, if achieved, will accomplish the overall goal of the plan. The objectives are often wide ranging, addressing such subjects as facilities, programmes, volunteer training and deployment, and the organizational structure by which the destination site will be administered, including the financing needed to achieve the objectives. Undertaking feasibility assessments for each proposed element in the plan is an important component of the planning process. Feasibility does not just mean economic feasibility, but takes into account social and environmental impacts as well.

It is important for the plan to be completely transparent and open to public scrutiny. In fact, planners should develop opportunities and schemes that will encourage the general public to not only review what is being proposed but play an active part in developing the proposal. The development of the plan should be as much about protecting the community and satisfying the recreational needs of the local residents as it is about the financial benefits to be achieved. If done correctly, the tourism enterprise will be enjoyed by the locals as much as visitors to the site. It often produces a significant quantity of leisure opportunity for local residents to enjoy.

An honest assessment of the capacity of the community to absorb what is being imagined is a critical step in the planning process. This assessment must be based on carefully conducted research and not simply on the aspirations of the project promoters. In too many cases in the past, tourism developed incrementally over time, and it is only in hindsight that its negative consequences were detected, making them more difficult to control or rectify retroactively. The impact of the tourism sector is often felt more so in rural settings than in urban areas simply because of population size. Large numbers of tourists are easier to absorb and blend into the city than in the rural context. All of this points to the need in rural areas for an in-depth and ongoing planning and monitoring process. The mitigation of the externalities produced by the tourism product before they become overly disruptive and contentious is highly desirable. All too often planning for tourism focuses solely on the product and not on the context in which it is situated. The goal of tourism planning must be for community protection and enhancement including, but not exclusively, economic development. Tourism development can also provide amenities to residents that other types of rural communities do not possess. Because of the revenue tourism can provide, other community services such as hospitals, education facilities, and recreation areas, among other amenities, can be realized. Often these facilities and programmes are not available to non-tourism communities because of the lack of resources.

Sustainability is often a consideration in the tourism literature and discourse, but one never really knows whether that sustainability interest focuses on the product or on the habitat in which the product is located. For example, in Africa and other developing regions, I often get the impression that the industry focuses on the tourism product when they speak about sustainability, not the environment in which it is embedded. All too often, the industry sees its facility as the product and not the surrounding natural feature or culture that needs and deserves protecting. What is important to consider is the relationship between the tourism infrastructure, including the number of visitors that can be sustainably supported in the area (carrying capacity), and the local culture and environment that provide the tourism product. It is important for the tourism product to not overwhelm the basic and vital uniqueness of the community.

An examination of the substantive aspects of rural tourism is fundamental to understanding the industry. The components and role of those components in rural tourism include local and regional history, product development, planning, signage, host/guest relations, accommodations including sewage and garbage disposal, mitigation procedures (if necessary), travel patterns within the area, community development and internal and external relations. In addition, it is important to analyse the overall organizational structure of the industry. There may be other issues unique to the area under review that also need to be included and preserved in the planning process. Many small-town tourism communities designate areas that can accept commercial tourism enterprises and other areas that are off-limits to development. These various matters must be addressed in the plan and with more detail on each subject supplied in the supporting documentation.

The rural tourism product is dynamic, not static. Constant monitoring of the development and the changes to the social and physical environment that the attraction produces is important to keep the resource fresh and attractive and the local residents positively engaged. We have seen the rise and fall of tourism communities and products because they were left to drift and were not monitored sufficiently or managed adroitly throughout their lifecycle. Butler (2006) has applied the product lifecycle model to tourism development to assist with the conceptualization and management of the changing nature and longevity of the rural tourism product in all its forms. Butler's model provides an excellent analytical framework that demonstrates the lifecycle stages of tourism development, and it can act as a guide for those commissioned to implement the planning and monitoring process over the life of the tourism product.

COMMUNITY INVOLVEMENT AND CONTROL

It is often the case that the tourism enterprise is funded by outside investors who have a different perspective on the tourism development project than the local community. The issue of power dynamics in tourism development at the local level needs great consideration and wide-ranging discussion prior to development. To date, much of the power in these relationships has been held by developers and project proponents, including national governments, but not the rural community. Many national politicians, in both the developing and developed world, view tourism development from a national economic perspective and not from a community point of view. While national balance of trade accounts and the generation of foreign currency are certainly considerations in the development discussion, they must not drive or control the process. This is not to suggest that national governments will not have an interest or stake in the product, but that the local community cannot be overlooked in the decision-making process. All the players in the system, particularly those that will be directly affected by the plan, need to be identified and their role and involvement in the control of the project determined a priori. The distribution of the eventual rewards must be completely spelled out and accepted, not only to the local officials but to the larger community itself. The role of local representation in the process needs to be comprehensively delineated prior to the development of a plan of action and long before the project planning and eventual implementation stage of the project begins.

Tourism products in the rural setting are not limited to those that fit into the natural environment, but there are times when attractions have no relation to the setting whatsoever. Whether this is a positive or negative feature depends on your point of view. The most outstanding example of this type of tourism are the theme attractions that have little relationship to the landscape or the people that live on it. Many rural villages and towns promote casinos and entertainment centres as their tourism venue. They are often treated like big box stores in the urban mall, an anchor industry to which other smaller retailers can attach themselves. Although these attractions produce income, they often have little relationship to the local area or businesses.

Many communities use their surrounding natural environment as the anchor for their tourism product. The natural flora and fauna of the area can provide the motivation for a visit to the rural setting. The buildings on the main street of villages are often constructed from local materials that are extracted from the immediate area giving them a unique and attractive facade. Many of these main streets have been turned into boutique shops where entrepreneurs sell themed products like Christmas ornaments or other specialty products.

People have been very inventive in turning their natural landscape into vacation lands. Skiing and rock climbing in the mountains, basking in the sun and surf on the coasts and inland lakes and waterways are prime examples of where visitors congregate. Just about any activity, whether authentic or not, can be introduced into the rural landscape and made into a tourism product. Ranches for horseback riding, waterways for fishing, boating, and swimming, mountains for skiing and paragliding, large tracts of forests for hiking and operating all-terrain vehicles, open lands for cross-country skiing and cycling, or simply using the landscape for spiritual contemplation, can be found in the rural countryside. The imagination is unlimited when it comes to using the rural landscape for tourism and recreation purposes. Tourism is not defined by activity but by intent and the personal meaning obtained from the activity. It is also defined by the income it provides to local residents.

Education and personal fulfilment are often a motivation for tourism. Many archaeological sites located in rural areas have become attractions. Some archaeological digs are now being worked by tourists, signifying that a good portion of rural tourism is not simply holidaymaking but educational as well. In addition to business conferences, many resorts include public seminars on all types of subjects as part of their calendar of events. Conference centres in rural areas also offer a tranquil setting for serious debate and dialogue and for firms to make business decisions.

History often plays a large role in rural tourism development. In addition to the themed main streets mentioned above, former working facilities such as mills, that were once used for grinding wheat into flour or for generating electricity, have been turned into major attractions in many rural settings. Often, these historic buildings have been converted into hotels and restaurants for use by the visiting public. Historic fortifications have been preserved and made into attractions in North America in order to preserve and portray the history of the area. Famous battles have also been preserved through the sanctification of the ground on which those battles were fought. Not only is history preserved, but in some cases these facilities deliver much-needed economic development to an area. Cultural tourism is a major attraction in many Indigenous communities. Culture is important to the tourism product in Africa and in other developing countries as well. What is critical to address in cultural tourism is the issue of authenticity. This is a sensitive area of development and proponents need to resist cultural appropriation. Working with the people whose culture is being produced for tourism is required. It takes great skill to display a unique culture authentically and without turning it into a garish theme park. Many of the cultural attractions in rural areas are a representation of a time past and need to be understood in those terms. On the African continent, visitors want to see the animals, particularly the big five, but they are also attracted to unique cultures that exist throughout the continent. In Africa and in other developing countries where culture is a significant tourist attraction, the culture needs to be experienced in combination with wildlife, and that culture must be authentic not an inauthentic McDonaldization of something that is precious to many people.

In Canada, there are many pioneer villages that depict the interplay between Indigenous people and settlers. These are often dressed up to portray a peaceful coming together of two or more distinct cultures. These depictions may not represent reality, however. The early encounters between Indigenous people and European settlers were often less peaceful than is now depicted, and tourism representations of that relationship are often riddled with erroneous notions about the noble savage and other such negative images of the Indigenous person. Residential schools are a product of the early interaction between Indigenous groups and Europeans and this deleterious encounter needs to be recorded in history and explained to today's citizens through educational tourism.

Theatre and music are becoming a mainstay in the rural countryside during the summer season. In my own country, Canada, Shakespeare has found a long-term home in Stratford, Ontario, a small community of approximately 32,000 people located in southern Ontario and within driving distance of Toronto and Detroit, Michigan. The Stratford Festival operates four theatres within the community boundaries. The theatre festival attracts a large number of visitors from around the world each summer. Artistic Director Antoni Cimolino and Executive Director Anita Gaffney announced

> the artistic and financial successes of the 2018 season – its longest on record. ... they announced a $1.9-million surplus, with a 10% increase in attendance to 502,605. Total revenue for the 2018 season was $65.8 million, up 8% from $60.7 million in 2017, when the Festival posted a surplus of $232,765. (Cimolino & Gaffney, 2018)

Summer and regional theatre has become prominent in the rural and small-town setting. The Stratford Shakespearian Festival not only enjoys local and regional acclaim but international recognition as well. This is a major accomplishment for a theatre company that operates from just April to October each year. Stratford is just one example of a small town that can be considered a fully functioning tourism community. There are at least four other major and several minor theatres in small-town southern Ontario in addition to Stratford. These attractions produce considerable spin-offs in the smaller adjacent communities that provide accommodation and restaurants to the theatre patrons. Independent music and art exhibitions are growing up around the theatres as well. Summer theatre is expanding in southern Ontario, Canada, as it is in other parts of the world. Thus, rural tourism should be viewed as a system and not only as a single event or venue.

THE TOURISM SYSTEM

When one examines the scope of the catchment area it is easy to recognize that Stratford is part of a larger tourism system that operates across the region. It is important for single tourism enterprises to be aware of the larger picture and how each component fits into the larger system. I raise the Stratford example not because it is unique but because there are many such configurations throughout the world that view their tourism operation from a systems perspective and not just from a single component point of view. The trail system in Nova Scotia, Canada, is another prime example of a well-functioning tourism system, in fact many systems. The systems perspective for tourism in the rural area is critical but not always functional due to the lack of coordination on the part of the tourism industry. In order to help with organizing the region's tourism product in Ontario, Canada, the Ontario Ministry of Heritage, Sport, Tourism and Cultural Heritage established 13 tourism regions within the province, to promote the development of a strong regional tourism organization. Although the day-to-day tourism decisions are created, controlled, and managed by the regional and local actors, support is provided by the Ontario government to encourage rural tourism development and its domestic and international marketing. Similar institutional infrastructure frameworks that promote tourism on behalf of their constituents exist throughout the world. Many of them also supply a logistical support service to rural and small-town tourism organizations. Most developing country governments recognize the importance of tourism to conservation and provide aid to local areas for promotion and organization of the tourism product. Nature-based tourism in many conservation areas provides funds for such things as flood control and other environmentally necessary projects that may not be tourism related. It takes a coordinated effort involving all players in the system to make a country's tourism offerings successful. This is particularly important for rural tourism given that small-town and rural communities do not usually have sufficient financial resources to launch an adequate tourism development and marketing campaign on their own.

In a rural tourism system, signage that provides direction from one tourism site to another is extremely important, particularly if the components are some distance apart. Maps and brochures describing the various points along the trail need to be meticulously developed. These types of activities are often the focus for a number of volunteer organizations that spring up in various rural tourism areas. Festivals, be they music, visual arts, agricultural fairs, or others, rely on volunteers. The training and deployment of those volunteers often fall in the purview of the local volunteer tourism committee or business association, such as the chamber of commerce. These groups are regularly supported by the local municipal council. Volunteer training, recruitment, deployment, and the development and implementation of a volunteer reward structure, in addition to marketing, form a large part of the tourism committee's mandate. If a local tourism area is lucky enough to have a paid professional to lead the efforts of a volunteer committee, this asset can be most valuable in making the tourism sector stable and successful over the long term.

It is important to note that the enthusiasm of the community often outstrips its ability and the resources available to realize their creativity. Often the most potentially lucrative project may not be feasible for many reasons, including maintaining the integrity of the rural space. Also, building the final product may need to be incremental and achieved in stages. Once the plan is completed and implemented, the evaluation of the outcome should be commenced. The plan will need continued monitoring and revision during its lifetime. Impacts of the project must be continually monitored once implementation of the plan begins. A warning has been issued by Gursoy and Nunkoo and needs to be stressed in tourism development, when they state that the

> development of tourism is also accompanied by several economic, social, cultural, and environmental costs that affect the lives of local residents. The tourism sector has been found to disturb, disrupt, and destroy local communities and bring changes that negatively affect residents' daily lives. The negative consequences of tourism development have led to growing concerns for the conservation and preservation of natural resources, human well-being, and the long-term economic prosperity of host communities. If the negative impacts are not managed, the local population can easily turn to open hostility toward the sector's development, eventually contributing to the destination's decline. Residents' negative perceptions toward tourism development also affect tourist satisfaction and the image of a destination. (Gursoy & Nunkoo, 2019)

What Gursoy and Nunkoo outline above does not need to become reality. There is nothing inherently 'evil' about tourism. Good planning and ongoing management of the destination site can avoid the negative outcomes about which Gursoy and Nunkoo caution us. Unfortunately, there are many rural communities that suffer these negativities because in their eagerness to gain economic advantage they do not take a methodical and transparent approach to the planning and development of the tourism project. There is no substitute for good planning and management to avoid such catastrophe.

A sound approach to community tourism planning and development will include a research and monitoring scheme. This scheme needs to be developed before the plan is implemented in order to develop a baseline against which the plan and its products can be evaluated over time. The research and monitoring protocol should include such areas as social characteristics, changing population and demographics and a continuing profiling of the population, cultural sensitivities and environmental characteristics, in addition to the usual financial data. The evaluation of the project should be continuous, and steps taken to ward off any of the negativities that Gursoy and Nunkoo warn us about before they become a major point of contention in the community.

The world in general, and rural and small-town communities specifically, are continuing to change dramatically as we pass through the early part of the 21st century. We continue to experience the consolidation of agriculture and the mechanization of the resource extraction industries, and the constant expansion of urban culture into the rural setting. These dynamics are having a profound effect on the rural landscape and culture. As a consequence, people, both rural and urban, are seeking to find solace from the relentless push of modernity into all facets of life. Rural tourism will need to be resilient in the face of vast changes occurring throughout society and particularly in the rural setting. The tourism industry must become aware of what those major changes will mean to the rural landscape, and adapt to them, if it is not only to survive but flourish over the short and long term.

BIBLIOGRAPHY

Butler, R. (2006). *The Tourism Area Lifecycle (Vol. 1)*. Clevedon: Channel View Publications.

Cimolino, A., & Gaffney, A. (2018). *Stage Door News: Stratford Festival celebrates record-setting season and announces a surplus of $1.9 million.* Retrieved from https://www.stage-door.com/3/news/2019-News/Entries/2019/3/3/stratford-stratford-festival-celebrates-record-setting-season-and-announces-a-surplus-of-1.9-million.html.

George, W.E., Mair, H., & Reid, D.G. (2009). *Rural Tourism Development: Localism and Cultural Change*. Bristol: Channel View Publications.

Gursoy, D., & Nunkoo, R. (2019). Introduction to tourism impacts. In D. Gursoy & R. Nunkoo, *The Routledge Handbook of Tourism Impacts: Theoretical and Applied Perspectives* (pp. 1–20). New York: Routledge.

Innovation, Science and Economic Development Canada. (January 11, 2019). *Statistics Canada reports that tourism spending is growing in Canada.* Retrieved from https://www.canada.ca/en/innovation-science-economic-development/news/2019/01/statistics-canada-reports-that-tourism-spending-is-growing-in-canada.html.

Kokt, D., & Hattingh, J. (2019). Local economic development, rural tourism and communities. In D. Gursoy & R. Nunkoo, *The Routledge Handbook of Tourism Impacts* (pp. 121–132). London: Routledge.

Reid, D.G. (2003). *Tourism, Globalization and Development: Responsible Tourism Planning.* London: Pluto Press.

Reid, D.G. (2017). *Social Policy and Planning for the 21st Century: In Search of the Next Great Social Transformation.* New York: Routledge.

World Data Atlas. (2018). *South Africa – international tourism receipts.* Retrieved from https://knoema.com/atlas/South-Africa/Tourism-receipts.

4. Understanding rural development

Kelly Vodden, Mery Angeles Perez and Brady Reid

INTRODUCTION

Tourism is increasingly presented as a promising avenue for rural development across the globe. This trend has become more prevalent as communities shift from single resource-based economies to more diversified economic systems that build upon local assets (Price, 2017; Quintana, 2021). Growing global wealth and travel accessibility have created a market for the environment and culture, which enabled the tourism industry to commodify the characteristics of rural regions by constructing rural "attractions" (Crouch, 2006). These attractions are a promising option for diversifying and strengthening local economies. However, we caution that tourism can also negatively impact rural livelihoods and lifeways in myriad ways. Adoption of tourism strategies in rural regions requires ongoing assessment and careful, participatory planning and implementation to mitigate negative consequences and ensure positive contributions toward community well-being (Khan et al., 2020). In this chapter we explore tourism as a promising, yet challenging, approach for rural development by unpacking the complex concepts that surround it. Both "rural" and "development" are context-specific, value-laden, and heavily debated terms. We examine the evolution and prevailing conceptualizations of rural/rurality and development, as well as outlining the intersection between these contested terms, to contextualize and illuminate potential implications for rural tourism and community development.

About 45 per cent of the global population lives in what can be defined as rural regions (De La O Campos et al., 2018). Despite accelerating trends of urbanization since the mid 20th century (Jedwab et al., 2017), rural regions remain important arenas for decision-making and policy regarding development. They often provide most of a country's sources of water, minerals, energy, fibre, food, and other ecological services for urban and rural consumers alike (Brown & Schafft, 2011). Rural areas can also be seen as "repositories of cultural values or even national identities" (Shucksmith & Brown, 2016, p. 2), while making significant contributions to national economies. Different definitions of rural and rurality have implications for the scale and reach of rural research, policy, and decision-making (Statistics Canada, 2015). Scott et al. (2007) note, for example, that "over 30 definitions of 'rural' are currently being used by different government departments in the UK" (p. 3).

The concept of development has also evolved over time and continues to be a source of debate (Mirakhor & Askari, 2010). Generally, development is understood as efforts that seek to build a "good life" for populations (Solo de Zaldivar, 2013). Guided by an economics-focused, neoliberal perspective, many argue that development projects have, however, often been detrimental to community well-being

(Mowforth, 2014). Shifting toward alternative approaches can help to ensure that ecological, social, and cultural dimensions and the voices of local and Indigenous communities are incorporated (Pike et al., 2007).

World Tourism Day 2020 celebrated the theme of "Tourism and Rural Development", with the UN World Tourism Organization (UNWTO) Secretary claiming that "tourism has proven to be a lifeline for many rural communities" (UNWTO, 2020a, para. 3). While the COVID-19 pandemic has created challenges, the UNWTO suggests that tourism as a tool for rural development is "more relevant than ever" (UNWTO, 2020b, p. 3). Despite this potential, there is much to consider in the pursuit of tourism-based rural development. Sharpley (2004) suggests, for example, that as demand for rural tourism grows, there is a need for "a new approach to the management and planning of rural tourism" (p. 374). Scholars and practitioners further warn that environmental and cultural sustainability must always be prioritized to avoid the pitfalls of mainstream development approaches of the past (Saufi et al., 2017). Before examining these implications further, however, we first interrogate the conceptual foundations of "rural" and "development".

DEFINING RURAL AND RURALITY

The Evolution of the Concept of Rurality

For decades, scholars have debated the concepts of rural and rurality as an object of analysis through which we can understand "what the rural is, represents, or accommodates in a given context, whether specific or broad" (Peng et al., 2018, p. 236). Understanding rurality and its changing nature is relevant when considering implications for rural tourism development as Sharpley (2004) notes that since "the fragmentation of rural systems in general, and the declining role of the agrarian economy in particular … tourism has assumed a more central role" (p. 374). In this section we first describe the evolution of the term rurality emerging in the 1980s as a "chaotic conception" (Pratt, 1996, p. 69). We then explore prevailing "concrete" and "abstract" conceptualizations of the rural (Rye, 2006) and discuss connections to the "rural idyll" (Halfacree, 2003), noting that differing conceptualizations have varied implications for rural people and rural community development (Bennett et al., 2019).

Rural sociology, focused on social processes and organization in rural areas (Warner, 1974), emerged as a new social scientific discipline near the turn of the 20th century (Newby, 1980). By the 1970s, prominent theorists in the field portrayed a "crisis in the subject" (Newby, 1980, p. 4), exemplified by a lack of understanding, strong conceptual foundations or clear paradigms concerning the rural despite a growing body of research (Warner, 1974; Cloke, 2006). While rurality remained "the everyday" for rural people, the academic debate subsided until the latter part of the 20th century. A revival of these discussions included assertions of the multiplicity of rurality as a subjective, socially constructed, and changing concept (Mingay, 1989;

Pratt, 1996). Instead of competing discourses that debated the right and wrong representations of rurality, theorists focused on understanding nuance and asked, "How do actors socially construct their rurality?" (Rye, 2006, p. 409).

Pratt (1996) considers the rural and rurality as objects of analysis that remain elusive through differing perspectives. The rural often represents some "quality of landscape or preponderance of social and economic practices" (Pratt, 1996, p. 70), while rurality is generally understood as a discourse that shapes a place through the development of "symbolic meaningful connections mutually affecting rural space and people" (Aquilino et al., 2021, p. 139). Further to what Rye (2006) describes as "concrete" and "abstract" conceptualizations of the rural and rurality, rural definitions have also been categorized as statistical/demographic or constructivist (Shucksmith & Brown, 2016).

Conceptualizing Rural: Statistical/Demographic Approaches

Scott et al. (2007) conceptualize the rural through five "dimensions" that focus on contrasts to the urban, demographics, land use, economic and social activity. The "demographic approach" (Shucksmith & Brown, 2016, p. 2) often encompasses some combination of population levels (census rural in Canada, for example, refers to populations less than 1000) (Statistics Canada, 2015) and population density (e.g., less than 150 persons per square kilometre (OECD, 2011)), which can differ depending on context (see the "hybrid geography" explored by Wandl et al., 2014). Douglas (2018) suggests that "rural regions function as distinct economic spaces in the increasingly open, global economy" (p. 10). Often, these conceptualizations have positioned rural areas as primarily natural resource-dependent, and/or primarily agricultural (Marsden, 1995; Sutherland, 2020). However, scholars recognize the diversity that exists in rural conditions, and describe a decline of natural resources as "a basis for rural work, family, and community life" (Deavers, 1992, p. 184). Indeed, different rurals can be conceptualized empirically based on characteristics such as distance to urban centres or economic dependence on natural resource sectors (Subedi et al., 2020) or through place-based asset analysis and mobilization (Halseth et al., 2010). Defining rural through demographic/statistical approaches is complicated by the diversity of "types" of rural, such as urban-adjacent/urban-fringe, or small villages and hamlets. These varying "degrees" of rural tend to confirm what Isserman (2005) argues is rural separation vs integration, the former considering urban *or* rural, which leads to homogeneous representations of the rural vs the latter considering urban *and* rural, which allow for heterogeneous representations of the rural. While distance from urban centres is one variable employed in statistical definitions, the OECD's new rural development paradigm for the 21st century includes urban–rural linkages as one of eight components for successful rural development (Dahlman 2016). Rural–urban relationships and interconnections are highly relevant for rural tourism development, which often depends on urban "markets". While urban residents value rural areas as a retreat from urban life, they often include expectations for urban-style amenities. Scholars caution against simplistic binary depictions of the rural as "non-urban". Cloke (2006), for example, notes that ongoing restructuring has led to a "blurring of conventional boundaries between country and city" (p. 18).

Statistical approaches to defining rural are also relevant for government policymakers, investors, and others who support rural tourism initiatives. Even from a demographic/statistical approach there are many ways that rural can be, and is, defined. Cromartie and Bucholtz (2008) note that in the United States, depending on the population threshold and boundary selection, the percentage of the population that could be classified as rural ranges from 7 per cent to 49 per cent – either extreme could result in significantly different policy decisions, with implications for programming, services, and ultimately rural lives (Goodwin, 1998; Isserman, 2005). Social constructivist or more abstract approaches similarly have direct implications for rural people and livelihoods. This alternative approach to defining rural and rurality is based on more explicitly subjective interpretations of the rural that are simultaneously perceived through hegemonic discourse and produced by rural people and other social actors.

Conceptualizing Rural: Social Constructivist/Relational Approaches

The "social constructivist framework" (Shucksmith & Brown, 2016, p. 2) conceptualizes rural and rurality through a more philosophical debate including an imaginative, relative space that is embedded in nature (Cloke, 2006). Sometimes described as the "rural idyll", rural is often associated with nostalgia (Shucksmith, 2018) and "romanticised views of nature, wilderness, tradition and authenticity … a healthier environment where the sense of belonging, community and attachment to nature can be nourished" (Cortes-Vazquez, 2017, p. 116).

Rural life is often also seen as connected to social relationships and notions of shared social and cultural identities, community traditions and interactions "that occur through behaviours, self-efficacy and local cultural and social experiences" (Aquilino et al., 2021, p. 138). Blanchette and LaFlamme (2019) add that definitions provided by rural residents themselves are accountable to their lived realities and therefore warrant particular emphasis. Local perspectives often highlight common ground but also the diversity of rural – both within and across communities. Without settling on any consistent definition of the rural or rurality, recent scholarship asserts that the diversity of the rural must be recognized and there are "many different ruralities that articulate distinctly across nation, region, occupation, gender, race, and class" (Blanchette & LaFlamme, 2019, p. 53).

Importantly, Frisvoll (2012) critiqued much of the debate on rurality around the turn of the 21st century as having been neglectful of the subject of power and the role of various social actors (including both individuals and groups) that create development directions and outcomes. They do so in part through their chosen representations, part of "a complex dynamic tangle of social relations, representations, networks, material features, humans and their actions" (p. 448). Frisvoll (2012) thus describes the making of rural in part through the chosen representations of various actors and highlights power, embedded in social practice, as key to rural studies.

In practice, the range of definitions or conceptualizations of both rural/rurality and development make the work of rural planners, policymakers and practitioners even more challenging (Johansen & Nielsen, 2012). When considering how various definitions of the rural implicate the "everyday life" of rural people and communities, we look to Dymitrow and Brauer's (2017) interpretation of the increased subjectivity in rural conceptualizations. They note that due to increased rural–urban blurring,

> there is an ever greater likelihood that current understandings of "rural" as used in formal contexts (legislation, administration, land use, funding and research) may misapprehend the societal phenomena this concept purports to explain and, by that, get in the way of making sound planning, policy and development decisions. (p. 28)

It becomes important for scholars engaged in rural research to consider the *context* of their inquiry. As Murdoch and Pratt (1993) note "any bid to impose a 'definitive' rural domain is itself an exercise in power" (p. 423), supporting Frisvoll's (2012) emphasis on the important role of power in constructions of rurality, and the need to question whose versions of rural identity are being told and adopted, for what reasons.

DEFINING DEVELOPMENT

The Evolution of the Concept of Development

The evolution of the concept of development has long been marked by paradigm shifts and debates, which have shaped trends of thought and practice across the globe (Tombazos & Miller, 2006). Diversity in cultural perceptions, contexts, and epistemologies has fuelled many of these debates (Redclift, 2005). Throughout history, development has predominantly carried economic connotations and its practice has tended to implement neoliberal and modernization approaches to address the problem of global poverty (Mowforth, 2014). In recent decades, perspectives and voices that have historically been dismissed are challenging these deeply ingrained trends, drawing attention to human and social aspects, as well as non-anthropocentric ways to relate to the world. In this section we discuss the evolution of the concept of development from this economic focus toward one that considers multiple community assets, and local and Indigenous worldviews.

The basic understanding of development as a field of policy and practice relates to the idea of improving, building up, and advancing the well-being and quality of life of people at various scales, including communities of place (Tombazos & Miller, 2006). While defining the concept of community is beyond the scope of this chapter, given its essential role in rural development, it is relevant here that community is a social construct that conveys ideas of sharing (Moseley, 2003), identity, belonging (Blackshaw, 2009), and civic order (Delanty, 2003). Communities of place have been described as including three aspects: the physical space that a group occupies, the society that emerges, and the collective actions that the group takes, forming a system of interactions that create common interests (Bridger & Alter, 2006).

Various perspectives on what it means to improve human life make defining development and its goals a challenge. Many authors since the 1980s have advocated for sustainable development, with the goal of attaining a good life for all humanity (Kates et al., 2005). But defining this "good life" is far from straightforward. Since the middle of the 17th century, humanity has upheld science and reason as the key tools to find solutions to its problems (Allan, 2019) and positioned economics as a guiding beacon for development. Neoclassical economics offered logical solutions to poverty, positing that economies are social systems with interacting components that strive toward equilibrium and maximum utility (Lawson, 2013). Capitalism, linear growth, and international trade were offered as paths toward prosperity, supported by scientific knowledge and technological advancement, and toward the eventual equalization of the incomes of rich and poor countries (Mirakhor & Askari, 2010). Following the "growth imperative", development was quantified using markers such as per capita income and gross domestic product (Allan, 2019).

The dominating discourse has focused on alleviating poverty through economic means, often at a cost to other aspects of well-being. Major global events such as the recessions of the 1980s and 1990s, the rise of neoliberal ideologies, and dismantling of the welfare state brought about an economic crisis globally, and in many rural communities. Their challenges were exacerbated by forces such as rising environmental concerns, consumer demands for more specialized yet low-cost goods, and growing competition from low-cost global competitors (Hayter, 2000). Criticisms of the neoliberal paradigm have intensified, however, as some suggest that decades of its implementation have failed to truly address issues of poverty and environmental sustainability (Pike et al., 2007; Ruben & Bastiaensen, 2000).

Human Needs and Sustainable Development

The evolution of how development is understood and implemented reflects a change in how we understand the relationships of people with each other and with the ecosystems they are part of. Development has largely been guided by an anthropocentric perspective. It assumed that nature's sole purpose was to serve humanity and that knowledge and technology could overcome problems such as natural disasters and resource depletion and ultimately control and dominate ecosystems (Hopwood et al., 2005). Examples of colonization and unsustainable patterns of resource extraction can be seen in rural areas across the globe, demonstrating negative ecological, social and health impacts that "development" can have on communities (Gonzales, 2018; Hanrahan, 2017). While also contributing to food, water, and/or health insecurity, economics-centred approaches to development have often failed to achieve the goal of economic prosperity. The result is economic insecurity as well, exemplified by resource depletion, fluctuating global commodity markets, and inequities in technology change and its impacts (Postero & Zamosc, 2006; Mowforth, 2014). For rural communities, often dependent on natural resources and related global commodity markets, these failures highlight a need to move away from reliance on resource extraction alone (Hipsher, 2020), but also to address vulnerabilities to climate change and threats from energy, food and water scarcity (Dahlman, 2016).

First introduced and defined by the Brundtland Report in 1987 (World Commission on Environment & Development, 1987) sustainable development (SD) "refers to a manner of human living, which can exist and reproduce on a long-term basis while providing good living conditions" (Borowy, 2014, p. 2). Despite debates and ambiguity, Kates et al. (2005) argue that SD entails a widely agreed-upon core set of principles and values, including recognition of ecosystem limits that must be respected if these "good living conditions" are to be achieved for current and future generations. The United Nations (UN) Sustainable Development Goals (SDGs), the Agenda for Sustainable Development and related Resolution to work toward full implementation by 2030 have promoted a more holistic understanding of development. While critics have pointed out that the concept of SD remains focused on the needs of humanity (Lee, 2000), these global commitments highlight that economic, social, and environmental dimensions of development must be balanced and integrated for lasting solutions to be implemented (Hopwood et al., 2005) and recognize that humanity does not exist in a compartmentalized relationship of dominance over nature, but rather is embedded within and interconnected with it. Additional fundamental principles of SD relate to equity and ensuring those affected by development are involved in decision-making processes (World Commission on Environment & Development, 1987; Kates et al., 2005). Thus, SD is a call for development that incorporates and respects the knowledge and perspectives of rural residents, including Indigenous peoples.

Local and Indigenous Knowledge and Perspectives

Historically, dominant discourses on development have tended to dismiss local and Indigenous knowledge systems as backward or irrelevant (Charnysh, 2017). However, the worldviews of many Indigenous peoples across the globe challenge assumptions that economic growth models uphold and offer alternative understandings. Concepts such as "Sumak Kawsay" or "Buen Vivir", and "mino-bimaadiziwin" or "the good life", which are central to the worldviews of Andean Indigenous peoples in Latin America and the Anishinaabe in Canada, are two examples. These perspectives offer key differences from the Western definition of development (Vanhulst & Beling, 2014). "Sumak Kawsay" upholds that the good life is one where we live in harmony with the Earth and its natural cycles (Thomson, 2011). For the Anishinaabe, the good life means balance, only achieved when human beings exercise their responsibility to take care of the land (Alvernaz, 2015).

In addition to Indigenous perspectives, the relevance of other local voices in the conversations on development cannot be dismissed. Not only are definitions of the good life shaped by local values, experiences, histories, and landscapes, but so are the specific challenges faced by communities (Pike et al., 2007). Because place is both the reason behind and the expression of a community's course of development,

development is, in essence, a process that has the element of place at its heart (Pike et al., 2007). Scholars underline the role of communities of place, groups of residents sharing a common physical location, in development since people and places are interconnected in a circular relationship of influences. Community residents, it is argued, best understand local issues such as environmental change or aging infrastructure (Roberts & Townsend, 2015), and can find ways to attain their common goals and interests through community development (Zhu et al., 2015). Additionally, local communities offer knowledge that is relevant at a regional scale to better inform strategies related to larger scale development issues, such as those related to environmental sustainability, or come together as groups of communities to address common opportunities or challenges through regional development approaches (Vodden et al., 2019).

An emerging trend in conceptualizations of development shifts the focus toward the capacities and relationships present in communities, while monitoring and considering influencing external conditions (McKnight & Kretzmann, 2012). Development is seen as beginning with community and being fuelled by local assets, including networks that allow for the flow of information and resources. From this perspective, development is no longer a top-down process that seeks to address a community's deficiencies (often emphasized in rural contexts), but a mobilization of existing assets or capitals.

Community Capitals and Asset-Based Approaches

As discussions on development continue to shift, there has been a trend in recent decades to inventory what a community has versus what it is lacking (Lowery et al., 2020). Resources are referred to as assets or "capitals" that can enhance community well-being. Four main aspects often emerge: social, natural, human, and economic (or financial) capitals (Dale, 2014). Additional categories such as built/manufactured, cultural and political capitals are also considered (Emery and Flora, 2006). Building on a community capitals framework, Lowery et al. (2020) suggest that a holistic vision is needed in rural development, and that storytelling can help us understand sustainability in rural settings more fully, in contrast to data-driven approaches that fail to fully consider local assets and priorities.

The relationships that exist within communities can influence and help or hinder development (Talen, 2000). Defined as the "norms and networks that enable people to act collectively" (Woolcock & Nayaran, 2000, p. 225) social capital can serve as a catalyst for mobilizing all other capitals for development (Butters et al., 2017). It can facilitate dialogue, negotiation and consensus-building, which are essential for sustainable community development (Dale, 2014). Social capital is also intimately linked to cultural capital, including various forms of heritage, traditions, cultural performance, and practice that are drawn upon in tourism development (Song & Yuan, 2021; Mitchell & Shannon, 2018), and to shared community identity (or identities).

Rural tourism scholars also point to the importance of natural capital as an asset, including amenities such as water bodies, scenic landscapes and/or favourable climate (Hunter et al., 2005). Brown and Schafft (2011) emphasize the role of human capital, citing challenges of labour force adjustment when transitioning from extractive resource sectors to tourism and the importance of training and workplace skill development, while Quintana (2021) highlights the need for financial and physical capital resources (e.g., airports, roads, etc.). Strategies for sustainable rural development must take all forms of capital, which are intricately interconnected, into consideration.

In summary, viewing development through a sustainability lens, we conceptualize rural community development as contextualized, purposeful, and systematic interventions that sustain and enhance multiple community assets and capitals. Such interventions may take the form of policy (or policies), programmes, projects, or practices intended to influence the development trajectory within one or more rural communities (Minnes & Vodden, 2019). This understanding of development moves away from linear, economic growth models that have ignored ecosystem functions and dynamics and excluded human needs such as health, education, and social justice, failing to help communities reach or maintain a "good life".

IMPLICATIONS FOR RURAL TOURISM DEVELOPMENT

Tourism is now a common strategy for rural development across the globe as rural regions face the decline of traditional industries, among other intersecting challenges, while market demand for experiences associated with nature and culture that many rural areas offer continues to grow (Mitchell & Shannon, 2018; Sizek, 2019). Tourism also offers a promising economic development alternative for regions with limited natural resources and labour supplies such as small islands (Armstrong & Read, 2021). Thus, the role of tourism and other forms of amenity-based development, and the cultural economy in rural communities are now widely recognized (Walsh, 2017; Dinis et al., 2019). Tourism can create jobs and income, both directly and indirectly through multiplier effects, help reduce out-migration (Fytopoulou et al., 2021; Song & Yuan, 2021) and encourage in-migration. In-migrants (both newcomers and returnees) can become important contributors to the local economy through their knowledge, skills and expertise, social networks and/or financial capital (Brown & Schafft, 2011; Cortes-Vazquez, 2017; Mitchell & Shannon, 2018). Access to other commercial and service offerings in a community can also increase, or be sustained, in turn (Kocabıyık & Loopmans, 2020).

Tourism activities have also contributed to environmental conservation (Panzer-Krause, 2020; Dinis et al., 2019; Christensen & Jones, 2020) and thriving cultures in rural areas. World Heritage sites, for example, have become tourist attractions while also delineating guidelines for environmental and community sustainability (Panzer-Krause, 2020; Christensen & Jones, 2020). By allowing residents to maintain livelihoods and relationships with their land, culture, and other residents, that they may have not been able to do otherwise, rural tourism can sustain and even strengthen community identity and cohesion (Dinis et al., 2019, Song & Yuan, 2021).

Tourism in rural communities also comes with costs, however. Serious concerns exist about ecological degradation, pressure on local infrastructure and services, loss of culture and introduction or exacerbation of inequities and social conflicts, for example. Amenity migrants may capture employment opportunities over long-time residents, and tourism enterprises may be externally owned and controlled, leaving limited and precarious sources of income for local residents from a sector that predominantly benefits outsiders (Mitchell & Shannon, 2018; Stotten et al., 2021). At the same time low-income and long-term residents often experience increased costs of land, housing, taxes, goods and services (Fytopoulou et al., 2021; Hunter et al., 2005). Similarly, some groups within the community, such as younger generations or those with assets such as land or businesses, may benefit more than others, contributing to internal divides. In some cases, increases in criminal activity, decreased cohesion and trust, and changes to identities and rhythms and practices of daily life occur and contribute to a sense of "un-homing", or affective and emotional displacement, among original residents (Fytopoulou et al., 2021; Kocabıyık & Loopmans, 2020).

Also of concern is how the rural idyll is constructed, packaged and marketed to tourists (Crouch, 2006). This idealization often constructs rural life as existing in harmony with nature and as a "backward, simple and innocent way of life characterized by traditional values and culture" (Panser-Kraus, 2020, p. 237). Often, these constructions dilute local or Indigenous sovereignty, perspectives, and realities. Rural tourism activities and their promotion may build upon a romanticized vision that may or may not accurately represent the values and experiences of local communities (Muller, 2004), thus running the risk of erasing "rural people and lived-experiences through abstraction" while positioning them as "the exotic 'other'" (Halfacree, 2003; D'Hauteserre, 2004). Conflicts and/or unwelcome changes to local identities and cultures can arise when the visions held by visitors are incongruent with local lived realities (Lovell & Bull, 2017; Panzer-Krause, 2020).

Given that rural communities are dynamic and diverse, we recognize that the impacts of tourism and development and appropriate strategies for rural tourism development must also be varied and place based. Previous experiences and research point to three key directions for good practice: (1) supporting other sectors in the community and regional economy; (2) ensuring tourism development pays attention to socio-cultural and ecological sustainability; and (3) recognizing the critical role of community participation and leadership. We conclude by pointing to collaborative governance models of sustainable rural tourism development as a way forward in these directions that prioritize sustainability and the knowledge and experiences of local and Indigenous people.

Supporting Other Sectors of the Community and Regional Economy

As rural communities consider tourism as an avenue to shift from natural resource dependency, the dangers associated with shifting from one single sector dependency to another must be considered. Instead, tourism development can help to develop complementary sectors and thus economic diversification, or "multifunctional rural transition" (Holmes, 2010) – with multiple functions extending beyond economic sectors to conservation and other non-economic values as well. The importance of economic diversity is highlighted by the often-seasonal nature of tourism and vulnerability to economic downturns or other disruptions such as the COVID-19 pandemic, requiring additional sources of income and employment to support a viable local economy.

The sale of local products is an important way that tourism has helped stimulate other aspects of rural economies, benefits which ideally extend beyond one community to the surrounding region (Panzer-Krause, 2020). The use of local products within the tourism industry helps to reduce leakage and increase local benefits while enhancing brand recognition for the community and region (Kastenholz et al., 2016). Drawing from evolutionary economic geography, Mitchell and Shannon (2018) discuss the potential for extending existing development paths through tourism. Stotten et al. (2021) provide the example of the symbiosis between subsistence farming and tourism in European alpine communities. Tourism provides new markets and revenues for farmers, while farmers are key actors in preserving the landscape that helps to draw visitors and enhance visitor experience. In other cases, tourism has incorporated and enhanced local arts and culture sectors (e.g., Vodden et al., 2016; Slawinski et al., 2019).

Novel recombination of assets can also create branching from existing development trajectories or even entirely new development paths (Mitchell & Shannon, 2018). Perkins et al. (2015) provide the cogent example of the Cromwell District in New Zealand, now globally recognized for its top-quality pinot noir. The district's economy shifted from reliance on gold mining to wool and fruit production, then to tourism and amenity-based in-migration. This led to growth in winemaking, including wine-related hospitality enterprises, driven largely by in-migrants' capital, labour and networks, and in turn to expansion in real estate, construction, and service industries. This "re-resourcing" transition has occurred in iterative development processes involving both local and global actors.

Attention to Socio-Cultural and Ecological Sustainability

Lasting sustainable development must balance economic, social and environmental aspects (Hopwood et al., 2005). In the case of tourism, strategies that neglect to consider these interconnections can lead toward unmanaged growth that reaches beyond the capacity of local social-ecological systems and leaves communities vulnerable to asset degradation (Brown & Schafft, 2011; Sizek, 2019). Tourism can help address the multifaceted problem of poverty but requires well-integrated strategies that support economic and social development *and* environmental sustainability (Dahlman, 2016). A community's values and experience of their local landscape are embedded in local systems that both affect and draw from various community assets (Talen, 2000). Focusing on increasing gains in one form of capital at the expense of another will inevitably affect the sustainability of the local tourism industry as a whole. The UN guidelines pertaining to the sustainable development goals (SDGs) highlight the multidimensional nature of sustainability. As Dahlman (2016, para. 4) suggests, "the SDGs and rural development are closely interconnected" and their implementation will require effective rural development policies, programmes, and initiatives to ensure that rural people and communities are not left behind.

A growing literature on sustainable community tourism offers further guidance on strategies for tourism development that pay attention to the intricate connections within socio-cultural and ecological systems, connections that are often recognized by local and Indigenous knowledge systems (Quintana, 2021; Waligo et al., 2013). These examples include destinations that draw on local culture and traditions to craft an appealing destination by celebrating and supporting what makes local cultures unique and authentic rather than tokenizing or degrading them. Additionally, tourism requires resources and skills, which are embedded into a region's human, social, and political capital. Therefore, skills and capacity building strategies must also be in place for local communities to meet industry needs (Dinis et al., 2019). This includes harnessing and developing the leadership required to ensure tourism planning and policies that stimulate the economy, ensuring economic viability and fair distribution of socio-economic benefits (UNEP and UNWTO, 2005), while promoting cultural and ecological sustainability and effectively incorporating diverse local voices.

Critical Role of Community Leadership and Participation

Community engagement and support are vital to any long-term, sustainable tourism strategy. Community participation is key to a sound sustainable tourism development strategy in the areas of knowledge, capacity building, and support. As previously discussed, it is now widely recognized that the knowledge and participation of local and Indigenous people is crucial in articulating and pursuing sustainable and locally appropriate development planning and implementation (Sirai & Ndlangamandla, 2021). Following principles of inclusion and social justice, this participation should include a wide range of local voices, including political leaders, businesses, non-government organizations and segments of the community who may often be underrepresented but can contribute and/or be affected (Quintana, 2021). This local engagement can help communities to better meet the needs of the industry by drawing upon local assets and capacities, such as a well-trained local labour force, local capacity for innovation and entrepreneurship, sustainable environmental practices, and adequate and appropriate local infrastructure (Dinis et al., 2019; Walmsley, 2019). By including community members from multiple sectors, such as the socio-cultural, environmental, human, and economic realms, a community-led strategy for sustainable tourism development can draw from and help foster collaboration and social capital in a realistic, respectful, and balanced manner (Douglas, 2006). Inventorying local assets, which may be otherwise taken for granted, can empower communities and provide opportunities to protect and celebrate them (Slawinski et al., 2019). Knowledge of where local assets must be further developed to successfully implement tourism strategies is also critical to enable community-based ownership and benefits.

Further, local engagement is recognized as key in fostering supportive attitudes toward tourism, thus increasing the likelihood for tourism to prosper as a local, sustainable industry over time (Yuan et al., 2019). Implementing a participatory approach can help prevent conflicts and setbacks. Several models have been articulated in recent decades to address power imbalances and disregard for local and Indigenous rights, assets, and identities in tourism development. These include community-based tourism planning, participatory community tourism, and others (Giampiccoli & Saayman, 2018; Grybovych, 2008; Reid et al., 2004). While examining each of these in depth is beyond the scope of this chapter, we note that key, shared components are multi-stakeholder collaboration and encouragement of local ownership (Jackson & Morpeth, 2000). Additionally, various authors suggest that for any model to bring about tangible outcomes, a well-articulated definition of the concept and values of community and formal mechanisms to gather the perspectives of local citizens are needed (Giampiccoli & Saayman, 2018; Reid et al., 2004). Discussions about how rural and community are conceptualized locally can help to create a common vision of the rural idyll based on local perspectives and lived experiences that, once incorporated into tourism products and marketing, will be reinforced and therefore perpetuated, and shaped, by tourism (Panzer-Krause, 2020).

CONCLUSION: SUSTAINABLE RURAL TOURISM DEVELOPMENT THROUGH COLLABORATIVE GOVERNANCE

The debates behind the definition of rural and rurality increasingly consider power relations and the implications of differing definitions for rural people. The concept of development also continues to be contentious, with its initial focus on economic strategies evolving to consider more holistic understandings of sustainability and engagement of local and Indigenous knowledge systems in community development processes. These varying conceptualizations are brought together to create an understanding of rural community development and, more specifically, rural community tourism development. We conclude that tourism development strategies are more effective and sustainable in the long term when they draw upon local understandings of the rural and acknowledge and enhance multiple community capitals, while mitigating potential negative implications.

Tourism development requires expertise and capital from a range of sources and while we have emphasized the critical role of community-level stakeholders, the examples provided also demonstrate the important role that diverse actors external to rural communities play, influencing the ability of communities to mobilize their assets and ultimately the outcomes of development processes. A "new governance paradigm" has emerged in recent decades due to failures of traditional modes of governing to address increasingly complex and dynamic public concerns (Salamon, 2002). The result has been calls for governance models based on collaboration rather than top-down government authority through multilevel processes of deliberation, consensus-building and shared decision-making among divergent interests, with an important role for local actors (Vodden, 2015). The potential for prioritizing the local within these multilevel governance structures and processes is key for rural tourism development given the arguments provided above that plans and policies that best safeguard the well-being or "good life" of local communities are those that favour local and Indigenous leadership and ownership (Presenza et al., 2013; Quintana, 2021).

This shift in public policy and administration has led to increased interest in collaborative, community-oriented tourism governance models that help address power dynamics by combining bottom-up and top-down approaches, and bringing together resources and capital from various stakeholders as they identify and work toward common goals to build a strong tourism sector. Yet there has been limited research on how collaborative governance might be applied in tourism (Bichler & Lösch, 2019). The need for further work in this area is, we suggest, a final implication of our review of how rural development principles and good practices can be applied in sustainable rural community tourism development.

These proposed future directions are not without limitations. Both theoretical and empirical evidence suggest that community-led collaborative strategies for sustainable rural tourism development, and rural development more generally, are difficult in practice and face many barriers. Espoused values are not always acted upon, for example, and sustainable tourism plans may be dominated by powerful actors and do little more than pay lip-service to genuine change in the face of entrenched paradigms (Hall, 2011; Novaczek, 2015), to name just two important challenges. We invite rural tourism scholars and professionals to reflect on community priorities, assets, and capacities in an effort to ensure that tourism activities uphold and foster local and Indigenous representations of the rural and appropriate intentions behind development, but also to further explore the barriers that exist to these aspirations and how they might be overcome.

REFERENCES

Allan, B. (2019). Paradigm and nexus: Neoclassical economics and the growth imperative in the World Bank, 1948–2000. *Review of International Political Economy: RIPE*, *26*(1), 183–206.

Alvernaz, G. (2015). Changing our goals from ownership to stewardship: Lessons from America's Indigenous people. In D. Nemeth (ed.), *Ecopsychology: Advances from the Intersection of Psychology and Environmental Protection* (Volume 1: *Science and Theory*, pp. 149–165). Santa Barbara, CA: Praeger.

Aquilino, L., Harris, J., & Wise, N. (2021). A sense of rurality: Events, placemaking and community participation in a small Welsh town. *Journal of Rural Studies*, *83*, 138–145. https://doi.org/10.1016/j.jrurstud.2021.02.013.

Armstrong, H.W., & Read, R. (2021). Tourism and sustainable growth in small (island) economies. In J.L. Roberts, S. Nath, S. Paul, & Y.N. Madhoo (eds), *Shaping the Future of Small Islands*. Singapore: Palgrave Macmillan. https://doi-org.qe2a-proxy.mun.ca/10.1007/978-981-15-4883-3_5.

Bennett, K.J., Borders, T.F., Holmes, G.M., Kozhimannil, K.B., & Ziller, E. (2019). What is rural? Challenges and implications of definitions that inadequately encompass rural people and places. *Health Affairs*, *38*(12), 1985–1992. https://doi.org/10.1377/hithaff.2019.00910.

Bichler, B.F., & Losch, M. (2019). Collaborative governance in tourism: Empirical insights into a community-oriented destination. *Sustainability*, *11*(23), 6673. https://doi.org/10.3390/su11236673.

Blackshaw, T. & Fielding-Lloyd, B. (2009). Community and identity. In T. Blackshaw (ed.), *Key Concepts in Community Studies* (pp. 112–117). London: SAGE UK.

Blanchette, A., & LaFlamme, M. (2019). Introduction: An Anthropological Almanac of Rural Americas. *Journal for the Anthropology of North America*, *22*(2), 52–62. https://doi.org/10.1002/nad.12119.

Borowy, I. (2014). Sustainable health: The need for new developmental models. *Bulletin of the World Health Organization*, *92*, 699. https://doi.org/10.2471/I/BLT.12.145219.

Bridger, J.C., & Alter T.R. (2006). The engaged university, community development, and public scholarship. *Journal of Higher Education Outreach and Engagement*, *11*(1), 163–178.

Brown, D.L., & Schafft, K.A. (2011). *Rural People and Communities in the 21st Century: Resilience and Transformation*. Cambridge: Polity Press.

Butters, L., Okusipe, O., Eledi, S., & Vodden, K. (2017). Engaging the past to create a new future: A comparative study of heritage-driven community development initiatives in the Great Northern Peninsula. *Journal of Rural and Community Development*, *12*(2/3), 186–209.

Charnysh, V. (2017). *Migration, diversity, and economic development: Post-WWII displacement in Poland.* [Doctoral dissertation, Harvard University].

Christensen, J., & Jones, R. (2020). World Heritage and local change: Conflict, transformation and scale at Shark Bay, Western Australia. *Journal of Rural Studies*, *74*, 235–243. https://doi.org/10.1016/j.jrurstud.2019.11.017.

Cloke, P. (2006). Conceptualizing rurality. In P. Cloke, T. Marsden, & P. Mooney (eds), *The Handbook of Rural Studies* (pp. 18–28). London: SAGE Publications Ltd. https://doi.org/10.4135/97818486080.16.n2.

Cortes-Vazquez, J. (2017). The end of the idyll? Post-crisis conservation and amenity migration in natural protected areas. *Journal of Rural Studies*, *51*, 115–124. https://doi.org/10.1016/j.jrurstud.2017.02.005.

Cromartie, J., & Bucholtz, S. (2008). Defining the "rural" in rural America. *Amber Waves*, *6*(3), 28–35.

Crouch, D. (2006). Tourism, consumption and rurality. In P. Cloke, T. Marsden, & P. Mooney (eds), *The Handbook of Rural Studies* (pp. 355–364). London: SAGE Publications Ltd.

Dahlman, C. (2016, March 31). A new paradigm for rural development. *OECD Insights* blog. http://wp.me/p2v6oD-2rg.

Dale, A. (2014). Social capital and sustainable community development: Is there a relationship? In A. Dale, & J. Onyx (2014). *A Dynamic Balance* (pp. 13–30). Vancouver: UBC Press.

D'Hauteserre, A.-M. (2004). Postcolonialism, colonialism, and tourism. In A.A. Lew, M. Hall, & A.M. Williams (eds), *A Companion to Tourism* (pp. 235–245). Oxford: Blackwell Publishing Ltd. https://doi.org/10.1002/9780470752272.ch19.

De La O Campos, A.O., Villani, C., Davis, B., & Takagi, M. (2018). *Ending extreme poverty in rural areas: Sustaining livelihoods to leave no one behind.* FAO: Rome. Licence: CC BY-NC-SA 3.0 IGO. Retrieved from: https://digitallibrary.un.org/record/3829378?ln=en.

Delanty, G. (2003). *Community*. London: Routledge. https://doi.org/10.4324/9781315011417.

Deavers, K. (1992). What is rural? *Policy Studies Journal*, *20*(2), 184–189.

Dinis, I., Simose, O., Cruz, C., & Teodoro, A. (2019). Understanding the impact of intentions in the adoption of local development practices by rural tourism hosts in Portugal. *Journal of Rural Studies*, *72*, 92–103. https://doi.org/10.1016/j.jrurstud.2019.10.002.

Douglas, C.H. (2006). Small island states and territories: Sustainable development issues and strategies – challenges for changing islands in a changing world. *Sustainable Development*, *14*(2), 75–80. https://doi.org/10.1002/sd.297.

Douglas, D. (2018). Governance in rural contexts: Toward the formulation of a conceptual framework. *EchoGéo*, *43*, 1–15. https://doi.org/10.4000/echogeo.15265.

Dymitrow, M., & Brauer, R. (2017). Performing rurality. But who? *Bulletin of Geography. Socio-economic series*, *38*, 27–45. https://doi.org/10.1515/bog-2017-0032.

Emery, M., & Flora, C. (2006). Spiraling-up: Mapping community transformation with community capitals framework. *Community Development*, *37*(1), 19–35.

Frisvoll, S. (2012). Power in the production of spaces transformed by rural tourism. *Journal of Rural Studies*, *28*, 447–457. https://doi.org/10.1016/j.jrurstud.2012.06.001.

Fytopoulou, E., Tampakis, S., Galatsidas, S., Karasmanaki, E., & Tsantopoulos, G. (2021). The role of events in local development: An analysis of residents' perspectives and visitor satisfaction. *Journal of Rural Studies*, *82*, 54–63.

Giampiccoli, A., & Saayman, M. (2018). Community-based tourism development model and community participation. *African Journal of Hospitality, Tourism and Leisure*, *7*(4), 1–27.

Gonzalez, A. (2018). Control, fear and suppression: A political ecology of voice study into resource extraction industry-community engagement in Peru's Loreto Region. *The Extractive Industries and Society*, *5*(3), 319–329.

Goodwin, M. (1998). The governance of rural areas: Some emerging research issues and agendas. *Journal of Rural Studies*, *14*(1), 5–12.

Grybovych, O. (2008). *Deliberative democratic practices in tourism planning: Towards a model of participatory community tourism planning*. [Doctoral dissertation, University of Northern Iowa].

Halfacree, K. (2003). Landscapes of rurality: rural others/other rurals. In I. Robertson and P. Richards (eds), *Studying Cultural Landscapes* (pp. 141–164). London: Arnold.

Hall, C.M. (2011). Policy learning and policy failure in sustainable tourism governance: From first- and second-order to third-order change? *Journal of Sustainable Tourism*, *19*(4–5), 649–671.

Halseth, G., Markey, S., Reimer, B., & Manson, D. (2010). Introduction: The next rural economies. In G. Halseth, S. Markey, & D. Bruce (eds), *The Next Rural Economies: Constructing Rural Place in Global Economies* (pp. 1–16). Wallingford: CABI.

Hanrahan, M. (2017). Water (in)security in Canada: National identity and the exclusion of Indigenous peoples/L'(in)sécurité de l'eau au Canada: L'identité nationale et l'exclusion des peuples indigènes. *British Journal of Canadian Studies*, *30*(1), 69–89.

Hayter, R. (2000). *Flexible Crossroads: The Restructuring of British Columbia's Forest Economy*. Vancouver, BC: UBC Press.

Hipsher, S. (2020). Global value chains: Tourism, agriculture, and resource extraction. In S. Hipsher (ed.), *Wealth Creation Approach to Reducing Global Poverty* (pp. 165–192). Singapore: Palgrave Macmillan. https://doi.org/10.1007/978-981-15-4116-2_6.

Holmes, J. (2010). The multi-functional transition in Australia's tropical savannas: The emergence of consumption, protection and indigenous value. *Geographical Research*, *48*, 265–280.

Hopwood, B., Mellor, M., & O'Brien, G. (2005). Sustainable development: Mapping different approaches. *Sustainable Development*, *13*(1), 38–52.

Hunter, L., Boardman, J., & Saint Onge, J.M. (2005). The association between natural amenities, rural population growth, and long-term residents' economic well-being. *Rural Sociology*, *70*(4), 452–469. https://doi.org/10.1526/0036011057775012714.

Isserman, A.M. (2005). In the national interest: Defining rural and urban correctly in research and public policy. *International Regional Science Review, 28*(4), 465–499. https://doi.org/10.1177/0160017605279000.

Jackson, G., & Morpeth, N. (2000). Reclaiming community ownership in tourism or stalled process? In D. Hall & G. Richards (eds), *Tourism and Sustainable Community Development*. (pp. 119–134). London: Routledge.

Jedwab, R., Christiaensen, L., & Gindelsky, M. (2017). Demography, urbanization and development: Rural push, urban pull and … urban push? *Journal of Urban Economics, 98*, 6–16. https://doi.org/10.1016/j.jue.2015.09.002.

Johansen, P.H., & Nielsen, N.C. (2012). Bridging between the regional degree and the community approaches to rurality: A suggestion for a definition of rurality for everyday use. *Land Use Policy, 29*, 781–788. https://doi.org/10.1016/j.landusepol.2011.12.003.

Kastenholz, E., Eusébio, C., & Carneiro, M.J. (2016). Purchase of local products within the rural tourist experience context. *Tourism Economics, 22*(4), 729–748.

Kates, R.W., Parris T.M., & Leiserowitz, A.A. (2005). What is sustainable development? Goals, indicators, values, and practice. *Environment: Science and Policy for Sustainable Development, 47*(3), 8–21.

Khan, A., Bibi, S., Ardito, L., Lyu, J., Hayat, H., & Anas, M.A. (2020). Revisiting the dynamics of tourism, economic growth, and environmental pollutants in the emerging economies—Sustainable tourism policy implications. *Sustainability, 12*(6), 2533. doi:http://dx.doi.org.qe2a-proxy.mun.ca/10.3390/su12062533.

Kocabiyik, C., & Loopmans, M. (2020). Seasonal gentrification and its (dis)contents: Exploring the temporalities of rural change in a Turkish small town. *Journal of Rural Studies, 87*(2), 482–493. https://doi.org/10.1016/j.jrurstud.2020.09.002.

Lawson, T. (2013). What is this "school" called neoclassical economics? *Cambridge Journal of Economics, 37*(5), 947–983.

Lee, K. (2000). Global sustainable development: Its intellectual and historical roots. In K. Lee, A. Holland, & D. McNeill (eds), *Global Sustainable Development in the 21st Century* (pp. 31–47). Edinburgh: Edinburgh University Press.

Lovell, J., & Bull, C. (2017). *Authentic and Inauthentic Places in Tourism*. Abingdon: Routledge. https://doi.org/10.4324/9781315676685.

Lowery, B., Dagevos, J., & Vodden, K. (2020). Goal-driven or data-driven? Inventory of sustainability indicator initiatives in rural Canada. *Sustainability, 12*(20), 8601.

Marsden, T. (1995). Beyond agriculture? Regulating the new rural spaces. *Journal of Rural Studies, 11*(3), 285–296. https://doi.org/10.1016/0743-0167(95)00027-K.

McKnight J., & Kretzmann, P. (2012). Mapping community capacity. In M. Minkler (ed.), *Community Organizing and Community Building for Health and Welfare*, 3rd edition (pp. 171–186). New Brunswick: Rutgers University Press.

Mingay, E. (1989). Introduction. In E. Mingay (ed.), *The Rural Idyll* (pp. 1–6). London: Routledge.

Minnes, S., & Vodden, K. (2019). Introduction. In K. Vodden, D. Douglas, S. Markey, S. Minnes, & B. Reimer (eds), *The Theory, Practice, and Potential of Regional Development: The Case of Canada* (pp. 1–11). Abingdon: Routledge.

Mirakhor, A., & Askari, H. (2010). The evolution of the Western concept of development. In A. Mirakhor & H. Askari (eds), *Islam and the Path to Human and Economic Development* (pp. 1–25). New York: Palgrave Macmillan.

Mitchell, C.J.A., & Shannon, M. (2018). Establishing the routes to rural in-migrant proprietorship in a Canadian tourism region: A mobilities perspective. *Population, Space and Place, 24*(3), e2095. https://doi.org/10.1002/psp.2095.

Moseley, M. (2003). *Rural Development: Principles and Practice*. London: SAGE Publications Ltd.

Mowforth, M. (2014). *The Violence of Development: Resource Depletion, Environmental Crises and Human Rights Abuses in Central America*. London: Pluto Press.
Muller, D.K. (2004). Mobility, tourism, and second homes. In A.A. Lew, C.M. Hall, & A.M. Williams (eds), *A Companion to Tourism* (pp. 387–398). Oxford: Blackwell Publishing Ltd.
Murdoch, Jonathan, & Pratt, Andy C. (1993). Rural studies: Modernism, postmodernism, and the "post-rural". *Journal of Rural Studies*, *9*(4), 411–427.
Newby, H. (1980). Trend report: Rural sociology. *Current Sociology*, *28*, 3–141.
Novaczek, N. (2015). We live differently than the others: Culture, islandness, and alternative economies on Chiloe and Prince Edward Island. In K. Vodden, R. Gibson, & G. Baldacchino (eds), *Place Peripheral: Place-Based Development in Rural, Island and Remote Regions* (pp. 127–155). St. John's, NL: ISER Books.
OECD (2011). *OECD Regions at a Glance 2011*. Paris: OECD Publishing. https://doi.org/10.1787/reg_glance-2011-en.
Panzer-Krause, S. (2020). The lost rural idyll? Tourists' attitudes towards sustainability and their influence on the production of rural space at a rural tourism hotspot in Northern Ireland. *Journal of Rural Studies*, *80*, 235–243. https://doi.org/10.1016/j.jrurstud.2020.09.013.
Peng, B., Liu, Z., Zhang, B., & Chen, X. (2018). Idyll or nightmare: What does rurality mean for farmers in a Chinese village undergoing commercialization? *Inter-Asia Cultural Studies*, *19*(2), 234–251. https://doi.org/10.1080/14649373.2018.1463072.
Perkins, H.C., Mackay, M.D., & Espiner, S. (2015). Putting pinot alongside merino in Cromwell District, Central Otago, New Zealand: Rural amenity and the making of the global countryside. *Journal of Rural Studies*, *39*, 85–98. https://doi.org/10.1016/j.jrurstud.2015.03.010.
Pike, A., Rodríguez-Pose, A., & Tomaney, J. (2007). What kind of local and regional development and for whom? *Regional Studies*, *41*(9), 1253–1269.
Postero, N.G., & Zamosc, L. (2006). *The Struggle for Indigenous Rights in Latin America*. Brighton: Sussex Academic Press.
Pratt, A. (1996). Discourses of rurality: Loose talk or social struggle? *Journal of Rural Studies*, *12*(1), 69–78. https://doi.org/10.1016/0743-0167(95)00046-1.
Price, R. (2017). *Ecotourism and Sustainable Tourism: Management, Opportunities and Challenges*. Hauppauge: Nova Science Publishers.
Presenza, A., Del Chiappa, G., & Sheehan, L. (2013). Residents' engagement and local tourism governance in maturing beach destinations. Evidence from an Italian case study. *Journal of Destination Marketing & Management*, *2*(1), 22–30.
Quintana, V.M. (2021). Eco-cultural tourism: Sustainable development and promotion of natural and cultural heritage. In S. Khan & Z. Yu (eds), *Tourism* (pp. 231–247). London: IntechOpen.
Redclift, M. (2005). Sustainable development (1987–2005): An oxymoron comes of age. *Sustainable Development*, *13*(4), 212–227. doi:10.1002/sd.281.
Reid, D.G, Mair, H., & George, W. (2004). Community tourism planning: A self-assessment instrument. *Annals of Tourism Research*, *31*(3), 623–639.
Roberts, E., & Townsend, L. (2015). The contribution of the creative economy to the resilience of rural communities: Exploring cultural and digital capital. *Sociologia Ruralis*, *56*(2), 197–218. https://doi.org/10.1111/soru.12075.
Ruben, R., & Bastiaensen, J. (2000). *Rural Development in Central America: Markets, Livelihoods and Local Governance*. New York: St. Martin's Press.
Rye, J.F. (2006). Rural youths' images of the rural. *Journal of Rural Studies*, *22*(4), 409–421.
Salamon, L. 2002. The new governance and the tools of public action: An introduction. In L. Salamon (ed.), *The Tools of Government: A Guide to the New Governance* (pp. 1–47). Oxford: Oxford University Press.

Saufi, A., Andilolo, I., Othman, N., & Lew, A. (2017). *Balancing Development and Sustainability in Tourism Destinations: Proceedings of the Tourism Outlook Conference 2015*. Singapore: SpringerLink.

Scott, A., Gilbert, A., & Gelan, A. (2007). *The Urban–Rural Divide: Myth or Reality*. Aberdeen: The Macaulay Land Use Research Institute.

Sharpley, R. (2004). Tourism and the countryside. In A.A. Lew, C.M. Hall, & A.M. Williams (eds), *A Companion to Tourism* (pp. 374–386). Oxford: Blackwell Publishing Ltd.

Shucksmith, M. (2018). Re-imagining the rural: from rural idyll to Good Countryside. *Journal of Rural Studies*, *59*, 163–172. https://doi.org/10.1016/j.jrurstud.2016.07.019.

Shucksmith, M., & Brown, D.L. (2016). Framing rural studies in the Global North. In M. Shucksmith and D.L. Brown (eds), *Routledge International Handbook of Rural Studies* (pp. 1–26). Abingdon: Routledge.

Sirayi, M., & Ndlangamandla, S. (2021). Using Southern theories in sustainable development: Indigenous knowledge, indigenous planning, and the land question. In M. Sirayi, M. Kanyane, & G. Verdini (eds), *Culture and Rural–Urban Revitalisation in South Africa: Indigenous Knowledge, Policies, and Planning* (pp. 54–71). Abingdon: Routledge.

Sizek, J. (2019). Playground. *Journal for the Anthropology of North America*, *22*(2), 125–127. https://doi.org/10.1002/nad.12104.

Slawinski, N., Winsor, B., Mazutis, D., Schouten, J.W., & Smith, W.K. (2019). Managing the paradoxes of place to foster regeneration. *Organization & Environment*, *34*(4), 595–618. doi:10.1177/1086026619837131.

Solo de Zaldivar, V.B. (2013). Etnicidad, desarrollo y "Buen Vivir": Reflexiones críticas en perspectiva histórica. *European Review of Latin American and Caribbean Studies*, *95*(95), 71–95.

Song, Y., & Yuan, M. (2021). Tourism and its impact on Dong traditional music and life in Xiaohuang. *Journal of Tourism and Cultural Change*, *19*(2), 200–215, doi:10.1080/14766825.2019.1707839.

Statistics Canada (2015). *Data and definitions* [Bulletin]. https://www150.statcan.gc.ca/n1/pub/21-006-x/2008008/section/s2-eng.htm.

Stotten, R., Schermer, M., & Wilson, G. (2021). Lock-ins and community resilience: Two contrasting development pathways in the Austrian Alps. *Journal of Rural Studies*, *84*, 124–133.

Subedi, R., Roshanafshar, S., & Greenberg, T.L. (2020). *Developing meaningful categories for distinguishing levels of remoteness in Canada*. Statistics Canada. Catalogue no. 11-633-X-No. 026.

Sutherland, L.-A. (2020). The "desk-chair countryside": Affect, authenticity and the rural idyll in a farming computer game. *Journal of Rural Studies*, *78*, 350–363. https://doi.org/10.1016/j.jrurstud.2020.05.002.

Talen, E. (2000). The problem with community in planning. *Journal of Planning Literature*, *15*(2), 171–183. https://doi.org/10.1177/08854120022092971.

Thomson, B. (2011). Pachakuti: Indigenous perspectives, buen vivir, sumaq kawsay and degrowth. *Development (Society for International Development)*, *54*(4), 448–454.

Tombazos, C.G., & Miller, S.M. (2006). Returning to the classical tradition: The Relevance and application of inframarginal analysis to development economics. In C.G. Tombazos & X. Yang (eds), *Inframarginal Contributions to Development Economics* (pp. 3–25). Singapore: World Scientific Publishing Pte.

UNEP and UNWTO (2005). *Making Tourism More Sustainable: A Guide for Policy Makers*. Nairobi: UNEP Publications.

UNWTO (2020a). *Tourism and rural development* [News release]. https://www.unwto.org/news/tourism-and-rural-development.

UNWTO (2020b). *UNWTO Recommendations on Tourism and Rural Development: A Guide to Making Tourism an Effective Tool for Rural Development*. Madrid: UNWTO. https://doi.org/10.18111/9789284422173.

Vanhulst, J., & Beling, A.E. (2014). Buen vivir: Emergent discourse within or beyond sustainable development? *Ecological Economics*, *101*, 54–63.

Vodden, K. (2015). Governing sustainable coastal development: The promise and challenge of collaborative governance in Canadian coastal watersheds. *The Canadian Geographer*, *59*(2), 167–180.

Vodden, K., Hunt, L., & Bell, R. (2016). Tourism and the arts: Arts, culture and First Nations-led tourism development, broadening conceptions of the social economy. In M. Roseland, M. Gismondi, S. Markey, M. Beckie, & S. Connelly (eds), *Scaling Up: The Convergence of Social Economy and Sustainability* (pp. 147–168). Edmonton: Athabasca University Press.

Vodden, K., Douglas, D., Markey, S., Minnes, S., & Reimer, B. (2019). *The Theory, Practice, and Potential of Regional Development: The Case of Canada*. London: Routledge.

Waligo, V.M., Clarke, J., & Hawkins, R. (2013). Implementing sustainable tourism: A multi-stakeholder involvement management framework. *Tourism Management*, *36*, 342–353.

Walmsley, A. (2019). *Entrepreneurship in Tourism*. Abingdon: Routledge.

Walsh, M. (2017). Strategies for building community resilience: The Duhallow experience. In L. Brinklow & R. Gibson (eds), *From Black Horses to White Steeds: Building Community Resilience* (pp. 33–61). Charlottetown, PEI: Island Studies Press.

Wandl, A., Nadin, V., Zonneveld, W., & Rooij, R. (2014). Beyond urban–rural classifications: Characterising and mapping territories-in-between across Europe. *Landscape and Urban Planning*, *130*, 50–63. https://doi.org/10.1016/j.landurbaplan.2014.06.010.

Warner, W.K. (1974). Rural society in a post-industrial age. *Rural Sociology*, *39*, 306–318.

Woolcock, M., & Narayan, D. (2000). Social capital: Implications for development theory, research, and policy. *The World Bank Research Observer*, *15*(2), 225–249.

World Commission on Environment & Development (1987). *Our Common Future*. Oxford: Oxford University Press.

Yuan, Q., Song, H., Chen, N. & Shang, W. (2019). Roles of tourism involvement and place attachment in determining residents' attitudes toward industrial heritage tourism in a resource-exhausted city in China. *Sustainability*, *11*(19), 5151.

Zhu, Y., Lan, H., Ness, D.A., Xing, K., Schneider, K., Lee, S., & Ge, J. (2015). *Transforming Rural Communities in China and Beyond*. Cham: Springer International Publishing.

PART II

EPISTEMOLOGICAL FOUNDATIONS, PARADIGMS, AND WORLDVIEWS

5. Post-positivism and mixed methods research in rural tourism

Girish Prayag

INTRODUCTION

All research is based on a set of assumptions that guides the achievement of the project's goals. Whether qualitative or quantitative research, or both, are undertaken, a firm belief in the underlying paradigm of the research allows the results to be understood in relation to the scientific phenomenon or 'reality' that the researcher initially sought to elucidate. The phrase 'paradigm shift' was first used by the American Philosopher Thomas Kuhn (1962) to describe not only a philosophical way of thinking but also the critics levelled against science and its pretension to understand reality. 'Paradigms' or 'world views' are understood to be the perspective, school of thought, and the set of beliefs that inform the research design and the interpretation of the results (Lincoln & Guba, 2000). Denzin and Lincoln (2000) define paradigms as human or social constructions, which deal with the beliefs upon which the research study is built, and which indicate where researchers are coming from and how they construct meaning from the data. Examples of different 'paradigms' include positivism, post-positivism, interpretivism, critical theory, feminism, post-modernism, and chaos theory (Jennings, 2001). A paradigm consists of four elements, namely epistemology, ontology, methodology and axiology (Lincoln & Guba, 1985). This chapter provides an overview of the main beliefs surrounding post-positivism, as a paradigm, and reviews its application in the field of rural tourism. Specifically, the chapter is organized around answering five questions related to this paradigm: (1) *When* should researchers use post-positivism and mixed methods? (2) *Why* should researchers use post-positivism and mixed methods? (3) *How* should researchers apply post-positivism and mixed methods? (4) *What* are its critical design elements? And (5) *How* should researchers analyse mixed methods data?

Post-positivism is based on a traditional scientific research approach, which has elements of reductionism, logical empiricism in data collection, a cause-and-effect orientation, and determinism based on a priori theories (Creswell, 2007). It is often associated with quantitative approaches (Creswell & Plano-Clark, 2007) and called the 'pragmatic' paradigm (Tashakkori & Teddlie, 2003; Patton, 1990). The central philosophy underlying this paradigm is that 'truth' cannot be understood solely from the use of a single scientific method and therefore a mono-paradigmatic approach to research does not uncover the whole truth (Tashakkori & Teddlie, 2003). Therefore, according to this approach, paradigms such as positivism and interpretivism are inherently limited in providing evidence on the whole truth. Post-positivism supports a *relational epistemology* (i.e., relationships in research are best determined by the context and specificities of the study), a *non-singular reality ontology* (i.e., there is no single reality and participants will each have their own and unique interpretations of reality), a *mixed methods methodology* (i.e., mixing of qualitative and quantitative research methods), and a *value-laden axiology* (i.e., conducting research that benefits people) (Kivunja & Kuyini, 2017). Yet mixed methods studies are not exclusively associated with post-positivism.

Nevertheless, followers of this paradigm are often adhering to mutually incompatible metaphysical assumptions about, for example, the nature of reality (ontology), the nature of knowledge (epistemology), and methodology, or the ways in which scientific knowledge can and should be acquired (Guba & Lincoln, 1994). Others believe there can be a common ground between the two traditions of qualitative and quantitative research on epistemological grounds (Holtz & Odağ, 2020; Tashakkori & Teddlie, 2003).

A recent search using the key words 'rural tourism' and 'mixed method*', limited to journal articles only in the database Scopus, up until the year 2020, revealed only 75 studies employing mixed methods, while a similar search using the key words 'rural community development' and 'mixed method*' revealed only four studies. These highlight the lacuna regarding the application of post-positivism, specifically mixed methods, as a way to understand rural tourism; and more so to understand the relationship between tourism and rural community development.

Unfortunately, in tourism studies and those focused on rural tourism (Khoo-Lattimore et al., 2019), researchers employing mixed methods fail to give clear justifications for their approach. For example, justifications for how the chosen methods in rural tourism studies are compatible with their stated paradigms are not clear. This is an issue which is replicated in the broader tourism literature (see Khoo-Lattimore et al., 2019). Yet Lincoln and Denzin (2003) argue that researchers can use multiple paradigms in, for example, qualitative research, that are compatible, but they also recognize that the borders and boundary lines separating the various paradigms and perspectives have blurred. Therefore, contradictions, tensions and oppositions between different paradigms reflect different ways of knowing about, and valuing, the social world (Creswell & Plano-Clark, 2007). Hence, this paradigm-method fit issue remains the most contested area in mixed methods research (Greene, 2008; Johnson et al., 2007). For example,

> critical rationalism could indeed serve as guiding epistemological theory for most researchers, independent of their quantitative versus qualitative standpoints, beyond a mere pragmatic utilization of whatever is useful to the researcher. Such a unifying epistemological approach could help to connect some of the loose ends existing in different methodological approaches. (Holtz & Odağ, 2020, p. 542)

At a time when the tourism industry is facing some of its biggest challenges globally, due to COVID-19, rural tourism can provide products and experiences that can be considered safer when compared to relatively high-density urban tourism experiences. With a rise in stress and anxiety in society, travellers are seeking connections to the natural environment, ways to improve health and social connections, and ways to disconnect from the increasingly complex home and work environments. To this end, some forms of rural tourism, through their focus on wellness, agritourism, and slow tourism products and experiences, have an unprecedented opportunity to capitalize on the negative impacts of the global pandemic. However, this will require use of multiple theoretical lenses and multiple methods to understand how rural tourism can be developed and marketed for greater individual and community well-being and resilience. Next, the chapter provides answers to the questions posed earlier.

WHEN SHOULD RESEARCHERS USE POST-POSITIVISM AND MIXED METHODS?

There are multiple reasons for choosing this paradigm. First, if researchers believe that there is more than a single reality and consider that reality can only be imperfectly understood, then gathering multiple perspectives from participants can provide a better understanding of these 'realities' (Creswell, 2007; Lincoln & Guba, 2000). Second, rather than relying on one data collection method or analytical method, employing multiple methods and levels of data analysis can improve rigour and understanding of a phenomenon. Therefore, another potential reason for using this paradigm is as a means to improve rigour or reliability/validity or trustworthiness of the findings. This often implies the use of triangulation of qualitative and quantitative research approaches (Jick, 1979). Tashakkori and Teddlie (2003) define triangulation as 'the combination of multiple data sources, data collection and analysis procedures, research methods, and/or inferences that occur at the end of the study' (p. 717). Third, if the nature of the phenomenon being studied requires the researcher to have some influence on the subject and object of the study, but objectivity and researcher–subject independence are still critical for achieving the research objectives (Ponterotto, 2005), then post-positivism is the ideal paradigm. Post-positivists advocate a modified dualism/objectivism, through its recognition of the researcher having some influence on what is being researched while using some level of objectivity in the research design process (Ponterotto, 2005). Fourth, mixed methods are particularly useful when researchers would like to use one method to further clarify, enhance or elaborate on the results of data gathered using a different method (complementarity) (Wisdom et al., 2012). Fifth, if theory falsification rather than verification is the main goal of the study, post-positivism is the right choice (Gephart, 2004). For example, in applying social exchange theory (SET) to understand residents' support for rural tourism development, researchers might want to show that residents may not always require social, economic and environmental benefits to outweigh costs to show support. In essence, the researcher may want to achieve theory falsification rather than verification. Using mixed methods would provide more nuances and opportunities to triangulate multiple perspectives to derive conclusions. Thus, mixing prevents the firing of a silver 'falsification bullet', given that there are multiple perspectives to consider before theory falsification.

In general, mixed methods is used when there is a belief that qualitative methods have something to add to the findings of quantitative data (Giddings & Grant, 2007) and vice versa. In a rural tourism study, Mordue et al. (2020) argue that they integrated qualitative and quantitative research allowing for very specific subjective opinions to be analysed alongside more representative survey findings, fitting the complementarity argument put forward by Wisdom et al. (2012). Triangulation, in particular, also emerges as an important justification for using this paradigm in rural tourism studies (Cheng et al., 2020; Su et al., 2019).

WHY SHOULD RESEARCHERS USE POST-POSITIVISM AND MIXED METHODS?

Although mixed methods are commonly used in tourism studies, Khoo-Lattimore et al. (2019), in their systematic review of such studies, found that 89% of studies do not explicitly declare their epistemological, ontological, and axiological stances. It seems that 'mixing' is often undertaken for the sake of 'mixing' rather than as a result of a deeper connection to the study's epistemological stance. The same trend follows in rural tourism studies, where the explicit epistemological stance is not discussed (see Cheng et al., 2020; Pratama & Wulandari, 2019; Su et al., 2019).

According to Greene (2006; 2008), there are four conceptually distinct domains that underpin any paradigm (see Table 5.1). The reasons why researchers should choose post-positivism in comparison to other paradigms are deeply rooted in these four criteria. For example, if reducing a complex phenomenon to a few key ideas or dimensions (reductionism) is a necessary criterion for improving understanding of the phenomenon, then clearly post-positivism is a good choice. If the inquiry logic requires that researchers use one method to discover paradoxes and contradictions in the findings from another method, which can then offer ideas for reframing the research question (Greene et al., 1989; Wisdom et al., 2012), then post-positivism should be adopted. If there is no measurement scale for a variable or existing measures are inadequate, the results from one method can be used in order to develop a new measure (Greene et al., 1989; Wisdom et al., 2012) as part of a scale development procedure to enhance validity and reliability of measurements.

The research inquiry logic, which includes, for example, the sample design and instruments used, is commonly referred to as the 'methodology' (Greene, 2006) and there are many issues associated with the design of mixed methods studies (Greene, 2008). While these are discussed later, the ability to give voice to multiple perspectives and the explicit consideration of, for example, stakeholders and competing perspectives (e.g., economic and social benefits for tourism developers versus residents) are the critical reasons to employ mixed methods. For example, giving voice and doing justice to research that is focused on indigenous and marginalized communities involved in rural tourism development might be better achieved through

Table 5.1 Post-positivism as a paradigm

Criteria	Key questions	Post-positivism
Philosophical stance and beliefs (epistemology and ontology)	– What are the key philosophical or epistemological assumptions of this paradigm?	– Determinism, reductionism and objectivism
Inquiry logics (methodology)	– What are the broad inquiry purposes and questions? – What logic underpins the study? – What are the procedures and designs, quality standards and writing and reporting forms that guide the researcher's logic?	– Single data source not sufficient to explain the phenomenon – Results need additional framing and explanation – Multiple phases and types of data needed to understand the phenomenon
Guidelines for research practice – The philosophical stance and inquiry logics translate into specific strategies and procedures that are used to conduct research	– How is the methodology put into practice?	Research design components: – How to mix different aspects of qualitative and quantitative research? – How to sample for each phase of the research? – How to conduct data analysis that 'mixes' the various research stages properly?
Socio-political commitments – The interests that are being served through the research	– Whose socio-political interests are being served? – What power relations surround the location or study context? – What value commitments are advanced in the study?	Researchers should give explicit consideration to audience, perspective, voice and advocacy. Researchers should use multiple ways of seeing and hearing participants and making sense of the social world

Source: Adapted from Greene (2008).

mixed methods by providing stakeholders with both quantitative and qualitative data. In a way, the need for explicit consideration of the tensions that socio-political interests can bring forth in designing and conducting research, is a strength of post-positivism (see Table 5.1). For example, residents' concerns with rural tourism development might not align with those of local government and tourism developers. The socio-political interests at play can be better captured through mixed methods by giving power, through the research design process, to residents' voices. Post-positivism can provide ways to navigate the 'value-free' neutrality inherent in the dominant, positivistic discourse, allowing for more reflexivity in research practices from conceptualization to execution (Ateljevic et al., 2005).

Together, Greene's (2008) four domains provide a comprehensive and cohesive framework, alongside the necessary practical guidelines that allow the application of mixed methods in social science research, including tourism studies. Although these domains have been more fully developed with respect to both positivism and interpretivism paradigms, this is not the case for mixed methods research (Onwuegbuzie et al., 2009). Indeed, not all of Greene's (2006) domains have been fully articulated and developed in the post-positivism paradigm (Greene, 2006; 2008; Johnson et al., 2007). There are still many unresolved issues for using mixed methods, providing reasons as to why researchers should pursue inquiries using this approach. For example, one cannot separate methods from paradigms and data cannot always be divided into a dichotomy of quantitative or qualitative data. Some qualitative data, for example, can be quantified. Research using mixed methods is not simply about conceptualizing how to use two different types of data or using two different data collection techniques (Gilbert, 2006), requiring researchers to reflect on why they are using this approach.

HOW SHOULD RESEARCHERS APPLY POST-POSITIVISM AND MIXED METHODS?

According to Collins et al. (2006) there are 13 distinct, interactive, and iterative steps for conducting mixed methods research:

- determining the goals of the study and which goals pertain to what aspect of the mixing;
- formulating the mixed research objective(s);
- determining the rationale(s) for mixing quantitative and qualitative approaches;
- determining the purpose(s) for mixing quantitative and qualitative approaches;
- determining the mixed research question(s);
- selecting the mixed sampling design;
- selecting the mixed research design;
- collecting quantitative and qualitative data;
- transforming and analysing the quantitative and qualitative data;
- legitimating the data sets and mixed research findings;
- interpreting the mixed research findings;
- writing the mixed research report;
- reformulating the mixed research question(s).

While tourism research does not often articulate 'mixed' research objectives, the rationale for mixing qualitative and quantitative research is underpinned with arguments of complementarity, with one approach adding or enhancing the other. In tourism, researchers have the option between the sequential implementation of the mixed method research design (i.e., qualitative study first followed by the quantitative study or vice versa) or the simultaneous study design (Molina-Azorin & Font, 2016), also known as the concurrent design, where both the qualitative and quantitative elements of the research are conducted at the same time. Researchers have privileged the sequential design in tourism studies (Molina-Azorin & Font, 2016). Concurrent designs can be seen in Table 5.2, where usually the timing of the qualitative and quantitative phases distinguishes concurrent from sequential designs. In concurrent designs, both the qualitative and quantitative phases happen at the same

Table 5.2 Research designs for mixed method studies

Research Design	Research Approach (primary data collection)	Description
Concurrent	QUAL + QUAN	The qual and quant parts of the study are given equal weight
	QUAL + quan	The qual part is given substantially more weight than the quant
	qual + QUAN	The quant is given substantially more weight than the qual
	qual + quan	Both have lower weight in comparison to, for example, secondary data analysis
Sequential	QUAL→QUAN	The qual stage precedes the quant stage but both have the same weight
	QUAL→quan	The qual stage precedes the quant stage with the former having more weight in the study
	qual→QUAN	The qual stage precedes the quant stage but the latter has more weight in the study
	QUAN→QUAL	The quant precedes the qual and both have equal weight in the study
	QUAN→qual	The quant precedes the qual and the quant has more weight in the study
	quan→QUAL	The quant precedes the qual but the qual has more weight in the study.

time. Creswell and Plano-Clark (2007) identify four different types of sequential design. The first type, *triangulation design*, evaluates the extent to which the quantitative and qualitative data converge, but also pinpoints if there is convergence at all. The second type, the *embedded design*, focuses on the qualitative component of the study informing the quantitative component. The qualitative data add to an understanding of the mechanisms embedded in the quantitative data. The third type, the *explanatory design*, has as its main purpose the use of qualitative data to help explain the quantitative results and the qualitative data serve as follow-up explanations. The last type, *exploratory design*, is used primarily for instrument or taxonomy development. The survey instrument is designed based on items or attributes identified from a qualitative phase. There are also guidelines for when quantitative data are collected first and when qualitative data are collected first (see Prayag, 2019).

In tourism studies, however, often the rationale(s) and purpose(s) of mixing have to be deduced from the findings of the study rather than the authors making explicit statements about these in their research design. The exception is in scale development studies, where researchers generally specify that the qualitative phase is used to generate ideas and items related to the theoretical domain of a concept, while the quantitative phase is used to develop and refine the proposed measurement scale (see Dedeoğlu et al., 2020; Kock et al., 2019).

The mixed sampling and research design tend to be more clearly articulated in recent tourism studies, as the field matures. However, as Onwuegbuzie and Collins

(2007) observed, researchers in general tend to stick to random sampling designs attributed to the quantitative paradigm and the non-random sampling, including convenience, purposive and quota sampling strategies generally attributed to the qualitative paradigm when using mixed methods. There is almost a strong hesitance to consider alternative ways of mixing the sampling design in tourism research in particular. Similarly, sample size recommendations for undertaking mixed methods research tend to be polarized around large samples being prescribed for quantitative research while smaller sample sizes tend to be recommended for qualitative research. In mixed methods research, there are at least 24 different sampling schemes that can be used by researchers (Onwuegbuzie & Collins, 2007). These different sampling schemes, combined with the 35 different mixed methods research designs proposed by Tashakkori and Teddlie (2003), offer tourism researchers and rural tourism studies a panoply of options, but many of them have rarely been considered beyond the sequential design and sampling using traditional qualitative and quantitative schemes. For example, in using concurrent designs where the qualitative and quantitative parts of the study have the same weight, researchers can use probability sampling schemes (Onwuegbuzie & Collins, 2007) that involve multistage random sampling, maximum variation, homogeneous and critical case beyond the traditional probability sampling schemes of simple, stratified, cluster and systematic sampling.

As an example, in a study of residents' support for rural tourism development where both qualitative and quantitative approaches are being adopted simultaneously (concurrent design) to understand the problem at hand, a researcher would not only sample residents but also other stakeholders (e.g., tourism operators, representatives from local government and other national policymakers). In the qualitative phase, the researcher would select stakeholders that are similar (homogeneous) if the goal is to identify congruence in thoughts and behaviours. However, if the goal is to capture diversity in thoughts and behaviours, a maximum variation approach would be better. This implies that the researcher is still looking for shared patterns in the data that cut across cases, but their significance is derived from a heterogeneous sample (Palinkas et al., 2015). In a critical case sampling approach, researchers would choose stakeholders in a way that permits logical generalization and maximum application of information because if it is true in one case, it must be true of all other cases (Palinkas et al., 2015). A more detailed view of these approaches can be seen in other studies (Hesse-Biber, 2010; Palinkas et al., 2015).

Samples for the different stages of the mixed method research can be identical, parallel, nested or multilevel (see Onwuegbuzie & Collins, 2007). For example, an identical relationship would imply that exactly the same sampling units participate in both the qualitative and quantitative phases of the study while a parallel relationship implies that the samples for the qualitative and quantitative stages of the research are different but are drawn from the same population of interest (see Onwuegbuzie & Collins, 2007).

WHAT ARE THE CRITICAL DESIGN ELEMENTS IN MIXED METHODS?

From the outset, and based on the previous discussion, it seems that a central feature of post-positivism is the consideration of which qualitative and quantitative approaches to mix and how to go about mixing them. In the COVID-19 pandemic, there are further additional issues related to methods to be considered, such as face-to-face interviews versus interviews conducted online through platforms (e.g., Microsoft Teams, Zoom, and Skype). The health and safety of both participants and interviewers for a qualitative phase need to be considered in more depth. A proliferation of quantitative approaches using secondary data and surveys is now emerging in the broader tourism literatures as a result of COVID-19. This will be to the detriment of mixed methods studies, which will be perceived as more difficult to undertake. Cross-country tourism studies, using mixed methods that include community engagement activities through participatory approaches, may also face setbacks due to lockdowns and quarantine regulations for researchers. The critical elements of mixed method design might indeed have to rely more on a 'pragmatic' approach.

Nevertheless, commonly used methodologies that use mixed methods include naturalist methodology, narrative inquiry, case study, phenomenology, ethnography, action research, experimental methodology, quasi-experimental methodology, and causal comparative methodology (Kivunja & Kuyini, 2017). For example, Xue and Kerstetter (2018) use ethnographic fieldwork and critical discourse analysis to understand implicit power relationships between local residents and a private organization on a range of tourism development issues. While action research is used in rural tourism studies (see Nyaupane & Poudel, 2012), its complementarity with other methods has yet to be explored in the field. The case study approach remains by far the most common method (see Ma et al., 2020; Mordue et al., 2020). Yet, longitudinal case studies may offer far better insights into rural tourism development than cross-sectional data (Xu et al., 2017). Experimental methods remain by far the least used by researchers in this field. Table 5.2 offers some options in terms of how tourism researchers can design studies using mixed methods. However, it must be noted that some studies can have multiple phases of qualitative and or quantitative research. A common practice in community-based tourism research is to use an initial qualitative design to uncover tourism-related problems, followed by a survey of residents and/or tourists, and a third stage where the findings are discussed among stakeholders through focus groups and other participatory approaches.

HOW SHOULD RESEARCHERS ANALYSE MIXED METHODS DATA?

The mixed analysis step is often considered to be the most difficult, as researchers have to use different analytical methods and integrate these methods to extract the most value from the data analysis. Typically, this step necessitates competence in conducting both quantitative and qualitative data analyses and depending on doctoral traditions in different parts of the world, some researchers might be exposed to only one type of analysis. This limits the potential for true 'mixing' of data analytical techniques. In this respect, the skills of the research team are critical in determining how the mixing will be achieved and using what techniques. Despite the growing use of mixed methods, there is still the need for a widely accepted framework or set of ideas that can be used to analyse findings from mixed method studies (Greene, 2008). In fact, tourism research shows very little mixing at the data analysis stage, with researchers primarily sticking to traditional methods of qualitative and quantitative data analysis. For example, Li et al. (2019) mix different data sources to offer an understanding of how toured places are imagined, presented and consumed as part of China's urban–rural transition in the process of development and change, but this does not necessarily imply mixing of data analysis.

Several authors (Onwuegbuzie et al., 2007; 2009; Onwuegbuzie & Combs, 2011) provide guidelines for mixing the data analytical procedures, but these approaches are neither widespread in tourism studies nor in rural tourism studies. For example, the analysis emphasis of interest can be used as a basis for mixing – comprising case-oriented analyses (i.e., analyses that focus primarily or exclusively on the selected case(s)), variable-oriented analyses (i.e., identifying relationships among entities that are conceived as variables) and/or process/experience-oriented analyses (i.e., evaluating processes or experiences pertaining to one or more cases within a specific context over time). Other recommendations by these previously mentioned studies include, for example, whether or not the analysis types associated with one tradition will be used to analyse data associated with a different tradition (i.e. cross-over mixed analysis vs non-cross-over mixed analysis) (see Onwuegbuzie et al., 2009; Onwuegbuzie & Combs, 2011).

Researchers need to also give more attention to the data analysis itself. For example, the time sequence refers to whether the quantitative and qualitative analysis components occur in a chronological order and how this sequence affects interpretation of the results (Creswell & Plano-Clark, 2007). For example, Greene (2007, p. 155) identified the following four phases of analysis: (a) data transformation, (b) data correlation and comparison, (c) analysis for inquiry conclusions and inferences, and (d) using aspects of the analytic framework of one methodological tradition within the analysis of data from another tradition. Researchers also have to give some thoughts to options such as qualitative analysis of quantitative data, quantitative analysis of qualitative data, quantitative analysis of both qualitative and quantitative data, and qualitative analysis of both qualitative and quantitative data (Onwuegbuzie & Combs, 2011).

Existing studies in tourism and rural tourism specifically show very little mixing in reporting the research findings with studies tending to report each phase (e.g., qual and quant) separately. For example, Robinson (2021) mixes secondary research with semi-structured interviews in studying rural food and wine tourism. The same approach is used by Mordue et al. (2020), where they report the survey findings first followed by the focus group findings, with little overt mixing of the qualitative and quantitative findings. A typical method that is particularly favoured by rural tourism researchers is the combination of analysing government documents on policies to support and/or develop rural tourism combined with primary data collection (either qualitative or quantitative) (see Liu et al., 2020). The Q-methodology, which involves some form of quantification combined with in-depth interviews, was used by Juschten et al. (2020) to understand urban–rural tourism mobility. Ma et al. (2020) use non-participant observation and interviews to understand residents' attitudes towards land expropriation for tourism development. The combination of more than two methods is rare in this field of study with the exception of, for example, the study by Arbogast et al. (2020). In the first stage of the project, the authors use multiple methods (qualitative and quantitative) such as key informant interviews, resident surveys, visitor preferences surveys and economic impact analysis to understand the research problem. This first stage highlights opportunities and challenges in using multiple methods in rural tourism. However, the use of multiple methods broadens Arbogast et al.'s (2020) understanding of the research problem. In the second stage, the authors use the findings from the first stage to inform activities such as asset mapping, landscape design/visualization and cultural identity design.

CONCLUSION

This chapter examined five key questions that underpin the use of post-positivism as a paradigm, and mixed methods in tourism research and more specifically in rural tourism studies. The chapter highlights the increasing use of mixed methods in conducting rural tourism studies with the sequential design leading the way in terms of application to the research phenomenon under study. While there are philosophical debates around the integration of qualitative and quantitative paradigms in the same study, a pragmatic solution is often the preferred stance of researchers. Both qualitative and quantitative research in the same study can improve understanding of the phenomenon in ways that would be difficult to achieve using either approach alone. As pointed out in the chapter, the method and sampling approaches when using mixed methods are not so well understood in terms of the options available to researchers. More importantly, mixing of the analytical approaches and findings still remains a challenge for researchers. With some of the leading journals in the tourism field seeking mixed method studies as their focus, there are open avenues for those undertaking rural tourism studies to capitalize on such opportunities. Yet, the challenges of conducting mixed methods research during COVID-19 cannot be ignored. Traditional approaches involving face-to-face data collection for studies employing qualitative and quantitative methods need to be given more thought for both participant and researcher safety. Thus, this chapter should be used as a point of departure for understanding how the post-positivism and mixed methods studies should be designed and operationalized in rural tourism studies.

REFERENCES

Arbogast, D., Butler, P., Faulkes, E., Eades, D., Deng, J., Maumbe, K., & Smaldone, D. (2020). Using social design to visualize outcomes of sustainable tourism planning: A multiphase, transdisciplinary approach. *International Journal of Contemporary Hospitality Management*, *32*(4), 1413–1448.

Ateljevic, I., Harris, C., Wilson, E., & Collins, F.L. (2005). Getting 'entangled': Reflexivity and the 'critical turn' in tourism studies. *Tourism Recreation Research*, *30*(2), 9–21.

Cheng, H., Yang, Z., & Liu, S.J. (2020). Rural stay: A new type of rural tourism in China. *Journal of Travel & Tourism Marketing*, *37*(6), 711–726.

Collins, K.M., Onwuegbuzie, A.J., & Sutton, I.L. (2006). A model incorporating the rationale and purpose for conducting mixed methods research in special education and beyond. *Learning Disabilities: A Contemporary Journal*, *4*(1), 67–100.

Creswell, J.W. (2007). *Qualitative Inquiry & Research Design: Choosing among Five Approaches* (2nd edn). London: Sage Publications Inc.

Creswell, J.W., & Plano-Clark, V.L. (2007). *Designing and Conducting Mixed Methods Research*. Thousand Oaks, CA: Sage Publications Inc.

Dedeoğlu, B.B., Taheri, B., Okumus, F., & Gannon, M. (2020). Understanding the importance that consumers attach to social media sharing (ISMS): Scale development and validation. *Tourism Management*, *76*, 103954.

Denzin, N.K., & Lincoln, Y.S. (eds) (2000). *Handbook of Qualitative Research*. Thousand Oaks, CA: Sage.

Gephart, R.P., Jr. (2004). From the editors: Qualitative research and the Academy of Management Journal. *Academy of Management Journal*, *47*(4), 454–462.

Giddings, L.S., & Grant, B.M. (2007). A Trojan horse for positivism? A critique of mixed methods research. *Advances in Nursing Science*, *30*(1), 52–60.

Gilbert, T. (2006). Mixed methods and mixed methodologies: The practical, the technical, and the political. *Journal of Research in Nursing*, *11*(3), 205–217.

Greene, J.C. (2006). Toward a methodology of mixed methods social inquiry. *Research in the Schools*, *13*(1), 93–98.

Greene, J. (2007). *Mixed Methods in Social Inquiry*. San Francisco, CA: Jossey-Bass.

Greene, J.C. (2008). Is mixed methods social inquiry a distinctive methodology? *Journal of Mixed Methods Research*, *2*(1), 7–22.

Greene, J.C., Caracelli, V.J., & Graham, W.F. (1989). Toward a conceptual framework for mixed-method evaluation designs. *Educational Evaluation and Policy Analysis*, *11*(3), 255–274.

Guba, E.G., & Lincoln, Y.S. (1994). Competing paradigms in qualitative research. In N.K. Denzin & Y.S. Lincoln (eds), *Handbook of Qualitative Research* (pp. 105–117). Thousand Oaks, CA: Sage.

Hesse-Biber, S. (2010). Qualitative approaches to mixed methods practice. *Qualitative Inquiry*, *16*(6), 455–468.

Holtz, P., & Odağ, Ö. (2020). Popper was not a positivist: Why critical rationalism could be an epistemology for qualitative as well as quantitative social scientific research. *Qualitative Research in Psychology*, *17*(4), 541–564.

Jennings, G. (2001). *Tourism Research*. Milton, QLD: John Wiley and Sons Australia, Ltd.

Jick, T.D. (1979). Mixing qualitative and quantitative methods: Triangulation in action. *Administrative Science Quarterly*, *24*, 602–611.

Johnson, R.B., Onwuegbuzie, A.J., & Turner, L.A. (2007). Toward a definition of mixed methods research. *Journal of Mixed Methods Research*, *1*(2), 112–133.

Juschten, M., Page, S., & Fitt, H. (2020). Mindsets set in concrete? Exploring the perspectives of domestic travellers on New Zealand's (auto-) mobility culture. *Sustainability*, *12*(18), 7646.

Khoo-Lattimore, C., Mura, P., & Yung, R. (2019). The time has come: A systematic literature review of mixed methods research in tourism. *Current Issues in Tourism*, *22*(13), 1531–1550.

Kivunja, C., & Kuyini, A.B. (2017). Understanding and applying research paradigms in educational contexts. *International Journal of Higher Education*, *6*(5), 26–41.

Kock, F., Josiassen, A., & Assaf, A.G. (2019). Scale development in tourism research: Advocating for a new paradigm. *Journal of Travel Research*, *58*(7), 1227–1229.

Kuhn, T.S. (1962). *The Structure of Scientific Revolutions*. Chicago, IL: University of Chicago Press.

Li, Y., Zhang, H., Zhang, D., & Abrahams, R. (2019). Mediating urban transition through rural tourism. *Annals of Tourism Research*, *75*, 152–164.

Lincoln, Y.S., & Denzin, N.K. (eds) (2003). *Turning Points in Qualitative Research: Tying Knots in a Handkerchief* (Vol. 2). Lanham, MD: Rowman Altamira.

Lincoln, Y.S., & Guba, E.G. (1985). Establishing trustworthiness. *Naturalistic Inquiry*, *289*(331), 289–327.

Lincoln, Y.S., & Guba, E.G. (2000). Paradigmatic controversies, contradictions, and emerging confluences. In N.K. Denzin & Y.S. Lincoln (eds), *Handbook of Qualitative Research* (2nd edn, pp. 163–188). Thousand Oaks, CA: Sage.

Liu, C., Dou, X., Li, J., & Cai, L.A. (2020). Analyzing government role in rural tourism development: An empirical investigation from China. *Journal of Rural Studies*, *79*, 177–188.

Ma, X.L., Dai, M.L., & Fan, D.X. (2020). Land expropriation in tourism development: Residents' attitudinal change and its influencing mechanism. *Tourism Management*, *76*, 103957.

Molina-Azorín, J.F., & Font, X. (2016). Mixed methods in sustainable tourism research: An analysis of prevalence, designs and application in JOST (2005–2014). *Journal of Sustainable Tourism*, *24*(4), 549–573.

Mordue, T., Moss, O., & Johnston, L. (2020). The impacts of onshore-windfarms on a UK rural tourism landscape: Objective evidence, local opposition, and national politics. *Journal of Sustainable Tourism*, *28*(11), 1882–1904.

Nyaupane, G.P., & Poudel, S. (2012). Application of appreciative inquiry in tourism research in rural communities. *Tourism Management*, *33*(4), 978–987.

Onwuegbuzie, A.J., & Collins, K.M. (2007). A typology of mixed methods sampling designs in social science research. *Qualitative Report*, *12*(2), 281–316.

Onwuegbuzie, A.J., & Combs, J.P. (2011). Data analysis in mixed research: A primer. *International Journal of Education*, *3*(1), 1–25.

Onwuegbuzie, A.J., Witcher, A.E., Collins, K.M., Filer, J.D., Wiedmaier, C.D., & Moore, C.W. (2007). Students' perceptions of characteristics of effective college teachers: A validity study of a teaching evaluation form using a mixed-methods analysis. *American Educational Research Journal*, *44*(1), 113–160.

Onwuegbuzie, A.J., Johnson, R.B., & Collins, K.M. (2009). Call for mixed analysis: A philosophical framework for combining qualitative and quantitative approaches. *International Journal of Multiple Research Approaches*, *3*(2), 114–139.

Palinkas, L.A., Horwitz, S.M., Green, C.A., Wisdom, J.P., Duan, N., & Hoagwood, K. (2015). Purposeful sampling for qualitative data collection and analysis in mixed method implementation research. *Administration and Policy in Mental Health and Mental Health Services Research*, *42*(5), 533–544.

Patton, M.Q. (1990) *Qualitative Evaluation and Research Methods*. Thousand Oaks, CA: Sage Publications, Inc.

Ponterotto, J.G. (2005). Qualitative research in counseling psychology: A primer on research paradigms and philosophy of science. *Journal of Counseling Psychology*, *52*(2), 126–136.

Pratama, A.B., & Wulandari, C. (2019). Villager readiness in rural tourism: A scale development using exploratory factor analysis. *International Journal of Tourism Policy*, *9*(3), 251–261.

Prayag, G. (2019). Mixed methods in tourism: Philosophical assumptions and key research design issues. In R. Nunkoo (ed.), *Handbook of Research Methods for Tourism and Hospitality Management* (pp. 363–372). Cheltenham, UK & Northampton, MA: Edward Elgar Publishing.

Robinson, D. (2021). Rural food and wine tourism in Canada's South Okanagan Valley: Transformations for food sovereignty? *Sustainability*, *13*(4), 1808.

Su, M.M., Wall, G., Wang, Y., & Jin, M. (2019). Livelihood sustainability in a rural tourism destination – Hetu Town, Anhui Province, China. *Tourism Management*, *71*, 272–281.

Tashakkori, A., & Teddlie, C. (eds) (2003). *Handbook of Mixed Methods in Social and Behavioral Research.* Thousand Oaks, CA: Sage.

Wisdom, J.P., Cavaleri, M.A., Onwuegbuzie, A.J., & Green, C.A. (2012). Methodological reporting in qualitative, quantitative, and mixed methods health services research articles. *Health Services Research*, *47*(2), 721–745.

Xu, K., Zhang, J., & Tian, F. (2017). Community leadership in rural tourism development: A tale of two ancient Chinese villages. *Sustainability*, *9*(12), 2344.

Xue, L., & Kerstetter, D. (2018). Discourse and power relations in community tourism. *Journal of Travel Research*, *57*(6), 757–768.

6. Interpretive approaches to research with rural communities: privileging emotion in tourism encounters

Meghan Muldoon

INTRODUCTION

Several years ago, Dr. Donald Reid of the School of Environmental Design and Rural Development at the University of Guelph in Ontario, Canada was approached with a problem. Faced with revenue losses and ageing infrastructure, city council members from a newly amalgamated rural township had agreed that they were unable to support the existing recreational services currently operating in three small towns (population less than 8,500, collectively) (Reid, 2009). Dr. Reid and his colleagues were invited to undertake a community leisure study to assess and rethink their recreational funding at the municipal level, with one potential option being the closure of one of the three ice arenas in the township.

In small-town Ontario, Canada, the hockey arena often becomes a hub of the community's social life, especially during the long winter months. Many act as tourism hubs as well, as families and fans travel to these villages and hamlets to attend hockey tournaments and other events. Many rural arenas were built at a time in the 1960s when rural towns had (relatively) large populations of young baby boomers accessing local recreation options and provincial and municipal governments were flush with resources (Reid, 2009). In the short-distance society (Persson, 1992) transportation options are limited, and local people are accustomed to accessing all of the services they require locally, or not at all (Reid, 2009). As access to personal vehicles, transportation networks, and communication technologies have improved, rural values have blended and overlapped with greater mobilities and the ability to access goods and services elsewhere, leading to what Fuller (1994) described as the Open Society. Ontario's rural landscape has changed significantly since the baby boomer era due to factors including rural to urban migration, factory farming subsuming smaller landholdings, housing development pressures, and government mandates compelling municipal amalgamations in order to share resources and minimize costs.

Dr. Reid and his team employed a number of data collection methods including surveys, usage statistics, document analysis, focus groups, and key stakeholder interviews in order to propose the best solutions for the township. A rational choice model approach was taken to the study, the underlying theory being that humans will rationally make decisions that are of the most benefit to them. Based on all of the data collected, the most logical choice was to close the arena in Community A.

It appears, however, that some of the residents of Community A were forcefully opposed to the loss of their arena and self-mobilized in order to create awareness, have their voices heard, and forestall the recommended closure. At a community meeting following the release of the recommendations, many residents became emotional and protested that "while the facility was under-used it still maintained a strong psychological attachment" (Reid, 2009, p. 523). The community members' pleas were so impassioned that the municipal government opted to shelve the recommendation and developed an ongoing working relationship with the interested citizens in order to come up with alternative solutions.

Clearly, the rational choice model did not adequately encompass the depth of feeling that many in this community have for their arena. While this approach to the study may have resulted in the most logical, cost-effective solution for the municipality, it did not come close to appreciating how the local rural community members *felt* about their arena.

Reid (2009) looked to the sense of place literature in order to better comprehend the response in the community. Smale (2006, p. 370) defines sense of place as "a desire to reconnect with place as an integral part of the human experience … *Place* shifts attention to the subjective lived experience of location, to the profound meanings we ascribe to it, and to the wholly human experience of place." Unlike the positivistic approach of the rational choice model, sense of place is an interpretive, constructivist approach, which "relies heavily on individual and social values" and can help us to understand why the residents of Community A so strongly rejected the rational findings of the original study (Reid, 2009, pp. 526–527). While both (post) positivistic and interpretive approaches to research can bring us important information about a community, interpretivism is needed if we are to understand anything about the emotive, subjective, embodied, and perhaps *ir*rational values and priorities people attach to their lived experiences.

In this chapter I provide an introduction to the interpretive paradigm and consider how it is distinct from but complementary to the positivistic approaches that preceded it. Adopting an interpretive research design means seeking a holistic understanding of a community's and an individual's life-worlds and the meanings and knowledges that are constructed through experiences and shared histories. Finally, I will consider a number of case studies, which have employed interpretive approaches to rural tourism and community research in order to demonstrate the array of rich data and understanding that can result.

INTERPRETIVISM AS ONTO-EPISTEMOLOGY

Interpretivism grows out of the ontological belief that there are fundamental differences between the natural world and the social world, and that therefore different methodological approaches need to be employed in order to understand each (Hughes & Sharrock, 1997). For early proponents of interpretive approaches to research, including Wilhelm Dilthey (1833–1911), Heinrich Rickert (1863–1936),

Max Weber (1864–1920), and those that followed, human interpretations are fundamental to understandings of social phenomena, rather than subjective irrelevancies that must be "bracketed" in order that events and behaviours may be studied objectively (Hughes & Sharrock, 1997). Dilthey emphasized the distinctions between the natural and social sciences and asserted that existing positivistic approaches to data collection were insufficient to capture the substantive richness that encompasses the human experience (Gray, 2004; Hughes & Sharrock, 1997). Rickert claimed that the distinction established by Dilthey between the natural and the social sciences is a fallacy; that because all means of apprehending our natural and socio-cultural worlds are all contingent upon human interpretation, it is not possible to know something as it exists outside human consciousness, but only as it is constructed (Hughes & Sharrock, 1997).

While many philosophical questions have arisen over the years regarding interpretivism's finer points or how it is best employed in various fields of study, some fundamental values hold true. Interpretivism places people and community at the center of their own experience. It emphasizes the lived realities of individuals, including, for the purposes of this volume, rural residents. It accounts for culture and shared histories. The rise in interpretive thinking led to a shifting away from attempting to uncover what is objectively true to apprehending what is *subjectively* true. Take a moment to pause to reflect on how revolutionary this is. For the first time since the beginning of scientific exploration, "the notion that we can remove human meaning from the scientific results of research is no longer the majority opinion" (Bakker, 2009). Interpretivism embraces subjectivities and multi-realities, and is judged by "trustworthiness, credibility, transferability, and confirmability" (Denzin & Lincoln, 2005, p. 24; Rapley, 2018). For interpretivists, positivism falls short because in its rejection of aspects of the human experience that are not observable or replicable, it can only fail to apprehend anything of real importance in the human subjective experience. Social reality is based on the subjective interpretation of actions (Balarabe Kura, 2012).

Two central ontologies often associated with interpretivism, constructivism and social constructionism (often grouped with a third, phenomenology) ascertain that reality is constructed by an individual through their interactions with others in a group, as in the case of the former, or through the mutually constructed meanings created by members of a group, as with the latter. Regardless, the central tenet remains that an individual's understanding of the world around them is a continually evolving construct within the human consciousness based in interactions with their environments. Interpretive research, therefore, attempts to understand phenomena through the meanings that people assign to them (Deetz, 1996). If social reality is based on the subjective interpretation of actions, then researchers seek to understand knowledge based on social reality through detailed understanding and interpretation of the meanings of events and life experiences. They seek to understand social worlds from the perspectives of the residents of the community, as a "subjectively experienced construct" rather than as a mere "collection of facts" (Mottier, 2005).

Whilst interpretive researchers employ a range of research methods, as shall be discussed below, the paradigm is underpinned by observation and interpretation (Aikenhead, 1997). We observe in order to gather information, but then we must also interpret that which we have observed in order to make meaning by drawing on inferences and our own understandings of reality (Aikenhead, 1997). Interpretive research relies upon contextualities – individual, social, national, political, strategic, and economic, which can only be understood through interpretation (Balarabe Kura, 2012). However, the construction of knowledge is not passive – interpretivists recognize that constructed knowledge reflects an individual's goals, experiences, values, history, etc., and that knowledge is thus intentionally constituted (Weber, 2004). In other words, people try to make sense of their life-worlds according to their particular frameworks, their cultures, and their goals (Weber, 2004). And because interpretivism specifically holds that researchers do not stand apart from the people and the phenomena that they study, a central tenet of interpretive research design and evaluation is that data is co-constructed between the researcher and the participant.

As Charmaz (2006, p. 10) noted, researchers are "part of the world we study and the data we collect." This is self-evident, in that we intuitively look to understand the world around us and, in doing so, develop subjective understandings of our experiences (Creswell & Creswell, 2018). Because these meanings are "varied and multiple," researchers must seek out the complexity of perspectives rather than striving to narrow down their frame of understanding to a few categories (Creswell & Creswell, 2018, p. 27). The interpretive researcher, in fact, becomes an *instrument* of measurement, in that their role is to interpret (measure) the phenomena that they observe and wish to understand (Weber, 2004). Researchers understand that their life-worlds affect and are affected by the phenomena that they study – in this sense the researcher and the subject of the research are interdependent (Weber, 2004). It becomes a process of active co-creation of knowledge, wherein the researcher and the research participants become equal co-producers (Creswell, 2014).

The goal of interpretive research is to "rely as much as possible on the participants' views of the situation being studied" (Creswell & Creswell, 2018, p. 27), while at once being attuned to, and cognizant of, the active role that you, as the researcher, are playing in how that research becomes constructed. Both researchers and participants bring a complex set of histories, experiences, expectations, fears, and ambitions which inform how knowledge becomes produced in interpretive research, and which are often obscured to the others participating in the research process. Imbalances of power, gender, race, age, literacy and language, and cultural expectations all may play a role in what knowledge *can* be produced through the research process. While interpretive researchers attempt to understand the ways in which social worlds can have meaning for research participants, those meanings are also being filtered through the researcher's understanding of "social meaning," endlessly complicating our understandings of others' lived experiences.

For this reason, critical self-reflexivity is central to an interpretive research undertaking (Crotty, 1998; Schwandt, 2007). Acknowledging one's social identity and its role in informing how knowledge is co-created through the research does not invalidate the research; rather, it allows the reader to more productively read the data through a better understanding of the social context within which it was produced. The basic production of knowledge is always social, generating from our interactions with other individuals in our communities (Crotty, 1998). The objective of interpretive research is to observe and interpret multiple subjective lived realities from within and through our own situated frameworks as well as the contextualizing environment, culture, history, and social dynamics.

To sum up, in interpretive research:

- Humans engage with their world and make sense of it based on their historical and social perspectives – we are all born into a world of meaning bestowed upon us by our culture (Crotty, 1998).
- There is no "direct, one-to-one relationship between ourselves (subjects) and the world (objects)" (Gray, 2004, p. 23). The world is interpreted through the classification structures in our minds (Williams & May, 1996).
- Reality is fluid and changing and knowledge is constructed jointly and consensually through interactions between the researcher and the participant (Grbich, 2013).
- Multiple realities are presumed, with different people experiencing these differently (Grbich, 2013).
- Understanding motives and actions requires attending to interpretation, social context, and local circumstances (Creswell, 2014; Creswell & Creswell, 2018).
- Failing to acknowledge and understand participants' life-worlds risks failing to describe anything of real social or subjective significance.
- Subjectivities are not neutral and often carry meaning that is socially ascribed and inscribed, leaving places *both* real and imagined (Roberts & Green, 2013).

INTERPRETIVE OBJECTIONS TO LOGICAL POSITIVISM

Much has been made of the rift between (post)positivism and interpretivism and much impassioned criticism has been levelled from both sides. While I have no desire to wade into those waters, I do think it is important to pause here to reflect on some critical differences that exist between the two paradigms. To all of the reasons listed in the preceding section, I would add this concise list, which considers primarily that (post)positivist perspectives rely often on humans behaving *rationally*. As demonstrated in the introduction to this chapter, humans cannot be counted on to behave in a rational manner or in a way that most benefits them (ask any author who has procrastinated in sitting down to write their book chapter). Some of the assumptions that interpretivism brings to its study of human nature and experience are:

- Motives cannot always be inferred from what people do.
- A person's actions can be interpreted differently by different people.

- A person's motivations in undertaking an action may have no bearing on how that action is interpreted by others.
- Meaning and action are inseparable – action is derived from the networks of meaning that allow for a particular action to be understood by others.
- *If* we were able to separate out meaning and action, we would be left reporting on actions that have no particular meaning or social significance.
- Logical positivist research fails to encapsulate the true meaning of human experiences because it treats them as scientific objects, thereby obfuscating the meaning and history and creativity that humans create in our communities.
- A survey can provide researchers with strong probability data but will be unable to capture the deeper meanings that events and phenomena have for individuals (adapted from Hughes & Sharrock, 1997).

INTERPRETIVE APPROACHES TO RURAL AND COMMUNITY-BASED RESEARCH

The powerful impact of the ways in which we interpret our environment through our individual cultural, historical, raced, and gendered lenses was made evocatively evident to me in my own research with peri-urban township residents in Cape Town, South Africa (Muldoon, 2018). As a white, settler Canadian woman, I failed to recognize the deep generational divide that exists between Black South Africans born before and after the end of apartheid. Despite my intellectual understanding of the ever-present significance of race in that country, my own embodied experiences and personal history made it impossible for me to fully apprehend the ways in which race and poverty structure life for people having lived generations in the townships and under apartheid. Further complicating matters, when my interview transcripts were translated from isiXhosa, I saw that one participant remarked to the other women present that they ought to "help this mother's daughter," clearly in reference to my identity as a younger female visitor in their community. In the same meeting, another woman referred to my "promise" to support their women's center once I had returned to Canada – a promise which was never even implied, but an understanding on this woman's part which informed her desire to help me in my research and "tell [me] what [I] wanted to hear." While I am able to speak about these complicating factors and weigh them in my analysis and writing, I have no way of knowing how much or how little of my identity as a white, younger, female tourist/researcher informed what residents told me about their experiences of tourism, nor how the multiple identities and subjectivities of both myself and the research participants informed what I was able to understand about them.

Understanding and acknowledging the complications, conflicting motivations, and the (subconscious) biases that we bring to our research encounters does not imply that interpretive research is somehow less valuable or less rigorous than (post)positivistic approaches to rural and community-based research. Rather, it requires honest and critical self-reflexivity in order to appropriately contextualize the research and the interpretations embedded within the data collection and analysis processes. What follows is a brief discussion of several innovative and critical rural tourism studies which demonstrate the potential scope of interpretive research and the understandings and insights that become possible through the application of an interpretivist framework.

Engaged Acclimatization in Northern Canada

In his work with Indigenous communities in Canada's North, Bryan Grimwood employed engaged acclimatization alongside community-based participatory research (CBPR) in order to foster respectful and meaningful research relationships. Engaged acclimatization requires a commitment on the part of the researcher to spend time in the community and practice open and honest communication, something which has not always been valued in research practices with Indigenous communities worldwide (Grimwood et al., 2012; Smith, 1999). Further, engaged acclimatization "encourages endogenous constructions of knowledge and instills axiology within the research process" (Grimwood et al., pp. 213–214). This approach to relationship building and privileging values and Indigenous knowledges (*Inuit Qaujimajatuqangit* in the Inuit language) is far removed from the research approaches of the past, which "mined" Indigenous communities and imposed European knowledge systems and racist beliefs on Indigenous peoples and lands (Smith, 1999). Engaging in respectful research relationships means nurturing those relationships over the long term and ceding some of the decision-making power and control traditionally conceived as being solely the privilege of the researcher (Grimwood et al., 2012). The use of the term engaged acclimatization implies an action-oriented approach to critical scholarship, which aligns with the political and ethical values and objectives of the host community (Grimwood et al., 2012).

For his research, Grimwood employed participant-elicited photographs in combination with experiential river trips on the Thelon River in Canada's Far North in order to learn about the moral constructs attached to that landscape by tourists and local residents. The images of their time spent on the river were then used to guide interviews and workshops in which they discussed the social meanings attached to that remote and fragile place (Grimwood et al., 2012). Here, ontology and methodology are closely intertwined in acknowledging "the impact that social and material practices of research have on the world" (Grimwood et al., 2012, p. 219). Grimwood's training as a community-based researcher began through a process of engaging with the community and the landscape, attending meetings, making introductions, and taking photographs – not in "collecting data," but rather in immersing himself in the life of the community and beginning to build trusting and respectful relationships (Grimwood et al., 2012). Rather than employing methodologies that demand a rigid adherence to the technical specifics of data collection, engaged acclimatization "prioritizes the notion that research *is* a relationship-forming process" (Grimwood et al., 2012, p. 220). Further, taking on the mantle of the "learner" displaces the traditional researcher pose as the all-knowing expert and allows for humility and adaptability, leading to learning that becomes "indirect, embodied, emotional, and imprecise" (Grimwood et al., 2012, p. 221). An interpretive approach to research is essential in order to begin to understand the complex interplay of ongoing setter colonialism, psychological and physical violence, Indigenous knowledges and traditions, and the relationships with the land which inform present-day realities for people living in Canada's remote and rural North.

Sun Peaks Resort as a Settler Colonial Moral Terrain

In her work, Lisa Cooke takes an autoethnographic approach to exploring the moralities and mobilities embedded within our understandings of a contested settler/Indigenous place – in this instance the Sun Peaks ski resort in British Colombia or, to give it its original name, Skwelkwelk'welt, in the heart of the territories of the Secwépemc nation (Cooke, 2017). Autoethnography is a form of critical interpretive research in which a researcher uses self-reflexivity and writing to examine one's own personal experiences in order to reflect upon wider social, cultural, and political meanings and significances in society. Cooke draws from her experiences working in construction at the resort as a young adult, drawing analogies between her physically paving over the landscape in order to build condominiums and the discursive paving over of Indigenous values and meanings attached to that landscape (Cooke, 2016; 2017). Settler colonialism relies upon the ongoing erasure and historicization of Indigenous identities on the land, and through stories of her work and leisure time spent on the mountain, Lisa illustrates how our physical presences and activities on the land as settlers obscure the deeper violence and discursive erasure of Indigenous peoples as the only legitimate occupiers of the land. Cooke's work is deeply personal and emotive, as when she reflects upon how her best memories of time spent with her son are of skiing on the mountain, and yet she is forced to reckon with her own role as a settler in Canada in the ongoing violence of colonialism. Her work is powerful *because* it is infused with mixed emotions of joy and freedom and shame and guilt, not because she attempted to bracket those emotions in order to understand the "truth" about colonialism and the ski resort.

Tourism in Tadoussac and the Social Production of Climate Risk

Dominic Lapointe and his team work with residents of the small island community of Tadoussac, Québec, to learn more about the effects of climate change and the implications for the tourism industry, the sole industry extant in the community (Lapointe et al., 2019; Lebon & Lapointe, 2016). Participants in this research do not deny the reality of climate change, as they are able to see its effects in the erosion of their shorelines and worry about the impact on whale populations, which are the main tourism draw to the region (Lebon & Lapointe, 2016). Using semi-structured interviews, Lapointe and his team aim to learn about community members' socially produced perceptions of climate risk in order to develop effective adaptation strategies for a changing environment and economy (Lebon & Lapointe, 2016). Perception of risk is not something that can be understood objectively or quantitatively, and yet it is a fundamental aspect of how rural community members understand and are able to plan for future changes that are beyond their control. For rural communities reliant on tourism and facing uncertain environmental futures, understanding perceptions of climate risk is crucial.

The Productive Nature of Shame in the Tourism Encounter

For our final example of interpretivism in action, we turn to the village of Göreme, Turkey and the work of Hazel Tucker. In the paper that I cite here, Tucker shifts our attention away from the traditional authentic/fake binary that is so often the focus in tourism research to consider the emotional and embodied nature of encounters in tourism. As Veijola and Jokinen (1994) were so prescient to remind us, tourism is an embodied experience – it takes place in and through our bodies. In her ethnographic research with female tourism stakeholders in the village, Tucker recounts an encounter that occurred between herself, two European tourists, and a local woman in her home. The feelings of discomfort and shame that this encounter created in her led to a critical reflection on the problematic nature of the tourism encounter as well as on the ways in which she is "always fully and bodily implicated in the creation of tourism knowledge" (Tucker, 2009, p. 446). Further, because much tourism takes place in spaces of the postcolony, tourism encounters are frequently accompanied by power imbalances and Othering due to differences in race, gender, mobilities, and economic circumstances (Caton & Santos, 2009; Hall & Tucker, 2004; Pritchard & Morgan, 2007; Tucker, 2009). Tucker's work in interpreting the nature of her shame in this encounter led to some powerful insights into the nature of power imbalances in places where tourism occurs. To do this work, she draws from Probyn's (2005, p. xviii) observations of ways in which shame can be "productive in how it makes us think again about bodies, societies, and human interaction."

Tucker's critical reflection on her discomfort and in her aligning herself with the tourists in the awkward encounter led to a realization of her own colonizing role – both as a researcher "helping to make or define the world" and as a tourist "consuming and perhaps further appropriating a locality captured by and through tourism" (Tucker, 2009, p. 453). In analyzing her shame, she came to understand that it was rooted in the efforts on the part of herself and the tourists to Other the local woman – as an authentic, traditional cave-dwelling woman unencumbered by economic or political forces (Tucker, 2009). Rather than perceiving her embodied shame as purely negative, however, Tucker uses her writing to uncover the postcolonial potentialities of examining our feelings of shame in the colonizing tourism encounter. As she writes,

> If … travel narratives are discourses of colonialism through which one culture works to represent and to dominate another, then tourism inevitably works to define and fix both the tourists and the toured "other" in a relationship that is always inherently colonial in nature. (Tucker, 2009, p. 455)

She goes on to explain that the quest for authenticity and pre-modern nostalgia in present-day tourism encounters has the potential to invoke feelings of shame in tourists who wish to be ethical consumers of tourism. Only through examining her uncomfortable embodied emotions and making herself vulnerable in her position as not-the-expert was she able to offer these insights into the postcolonial potentiality of shame in the tourism encounter.

CONCLUSION

Tourism and community development – especially as they occur in small and rural places – are invariably imbued with emotion, with history, and with relationships. Tourism is a desired leisure activity for so many *because* of the emotions and embodied experiences that it brings. (Post)positivistic approaches to tourism research can tell us a lot about travel flows, marketing effectiveness, and environmental impacts (among a great many other things), all of which are critically important to our understanding of tourism functions and impacts. However, interpretive approaches are essential to our understanding of how our subjective worldviews, our shared and individual histories, our cultural perspectives, and our embodied identities all play a role in shaping our experiences and understandings of tourism encounters. As tourism researchers, all of these factors shape our abilities to analyze and interpret the humanness of tourism, those things that make tourism joyous and beautiful and complex and heartbreaking. As Ateljevic and Hall (2007, p. 139) remind us, a tourism researcher is "always an interlocutor charged with the final act of interpretation and representation of knowledge." As hopefully the examples above have demonstrated, taking an interpretive approach to tourism research in rural areas does not diminish the validity or impact of the research. Rather, it provides us with beautifully nuanced understandings of our tourism experiences in all their vigor and complexity.

REFERENCES

Aikenhead, G.S. (1997). *A framework for reflecting on assessment and evaluation.* Paper presented at the Korean Education Development Institute international conference, Globalization of Science Education, Seoul, Korea.

Ateljevic, I., & Hall, D. (2007). The embodiment of the macho gaze in south-eastern Europe: Performing femininity and masculinity in Albania and Croatia. In A. Pritchard, N. Morgan, I. Ateljevic, & C. Harris (eds), *Tourism and Gender: Embodiment, Sexuality, and Experience*. Wallingford, UK: CABI.

Bakker, K. (2009). Neoliberal nature, ecological fixes, and the pitfalls of comparative research. *Environment and Planning, 41*, 1781–1787.

Balarabe Kura, S.Y. (2012). Qualitative and quantitative approaches to the study of poverty: Taming the tensions and appreciating the complementarities. *The Qualitative Report, 17*(20), 1–19.

Caton, K., & Santos, C.A. (2009). Images of the Other: Selling study abroad in a postcolonial world. *Journal of Travel Research, 48*(2), 191–204.

Charmaz, K. (2006). *Constructing Grounded Theory: A Practical Guide Through Qualitative Analysis*. Thousand Oaks, CA: Sage Publications.
Cooke, L. (2016). "North" in contemporary Canadian national-cultural imaginaries: A haunted phantasm. *Settler Colonial Studies*, *6*(3), 235–251.
Cooke, L. (2017). Carving "turns" and unsettling the ground under our feet (and skis): A reading of Sun Peaks Resort as a settler colonial moral terrain. *Tourist Studies*, *17*(1), 36–53.
Creswell, J.W. (2014). *Research Design: Qualitative, Quantitative, and Mixed Methods Approaches* (4th edn). Thousand Oaks, CA: Sage Publications.
Creswell, J.W., & Creswell, J.D. (2018). *Research Design: Qualitative, Quantitative, and Mixed Methods Approaches* (5th edn). Thousand Oaks, CA: Sage Publications.
Crotty, M. (1998). *The Foundations of Social Research: Meaning and Perspective in the Research Process*. Thousand Oaks, CA: Sage Publications.
Deetz, S. (1996). Describing differences in approaches to organization science: Rethinking Burrell and Morgan and their legacy. *Organization Science*, *7*(2), 191–207.
Denzin, N., & Lincoln, Y. (2005). *The SAGE Handbook of Qualitative Research*. Thousand Oaks, CA: Sage Publications.
Fuller, T. (1994). Sustainable rural communities in the arena society. In J.M. Bryden (ed.), *Towards Sustainable Rural Communities: The Guelph Seminar Series*. Guelph, ON: University School of Rural Planning and Development.
Gray, D.E. (2004). *Doing Research in the Real World*. Thousand Oaks, CA: Sage Publications.
Grbich, C. (2013). *Qualitative Data Analysis: An Introduction*. Thousand Oaks, CA: Sage Publications.
Grimwood, B.S.R., Doubleday, N.C., Ljubicic, G.J., Donaldson, S.G., & Blangy, S. (2012). Engaged acclimatization: Towards responsible community-based participation in Nunavut. *The Canadian Geographer*, *56*(2), 211–230.
Hall, C.M., & Tucker, H. (2004). *Tourism and Postcolonialism: Contested Discourses, Identities, and Representations*. New York: Routledge.
Hughes, J.A., & Sharrock, W.W. (1997). *The Philosophy of Social Research*. New York: Routledge.
Lapointe, D., Lebon, C., Guillemard, A. (2019). Space in transformation: Public versus private climate change adaptions in peripheral coastal tourism areas – Case studies from Quebec, Canada. *International Journal of Tourism Research*, *22*(2), 238–251.
Lebon, C., & Lapointe, D. (2016). Community well-being between climate risk and tourism development: Contradictions on the shore of the St. Lawrence Estuary. In B.S.R. Grimwood, K. Caton, H. Mair, & M. Muldoon (eds), *Tourism and Wellness: Travel for the Good of All?* (pp. 173–190). Lanham, MA: Lexington Books.
Mottier, V. (2005). The interpretive turn: History, memory, and storage in qualitative research. *Forum Qualitative Sozialforschung/Forum: Qualitative Social Research*, *6*(2).
Muldoon, M. (2018). *Gazing back: A feminist postcolonial lens on tourism in the townships of South Africa*. [Doctoral dissertation, University of Waterloo, Department of Recreation and Leisure Studies, Waterloo, Canada].
Persson, L.O. (1992). Rural labour markets meeting urbanization policy and planning. In T. Marsden, P. Lowe, & S. Whitmore (eds), *Labour and Locality: Uneven Development and the Rural Labour Process* (pp. 68–94). London: David Fulton Publishers.
Pritchard, A., & Morgan, N. (2007). De-centering tourism's intellectual universe, or traversing the dialogue between change and tradition. In I. Ateljevic, A. Pritchard, & N. Morgan (eds), *The Critical Turn in Tourism Studies: Innovative Research Methodologies* (pp. 11–28). Oxford: Elsevier.
Probyn, E. (2004). Shame in the habitus. In L. Adkins & B. Skeggs (eds), *Feminism After Bourdieu* (pp. 224–248). Oxford: Blackwell Publishing.

Rapley, E. (2018). "Seeing the light." Personal epiphanies and moving towards interpretivism: A researcher's tale of exploring teacher pedagogic practice. *Ethnography and Education, 13*(2), 185–203.
Reid, D. (2009). Rational choice, sense of place, and the community development approach to community leisure facility planning. *Leisure/Loisir*, *33*(2), 511–536.
Roberts P., & Green, B. (2013). Researching rural places: On social justice and rural education. *Qualitative Inquiry*, *19*(10), 765–774.
Schwandt, T. (2007). *The Sage Dictionary of Qualitative Inquiry*. Thousand Oaks, CA: Sage Publications.
Smale, B. (2006). Critical perspectives on place in leisure research. *Leisure/Loisir*, *30*(2), 369–382. https://doi.org/10.1080/14927713.2006.9651358.
Smith, L.T. (1999). *Decolonizing Methodologies: Research and Indigenous Peoples*. London: Zed Books.
Tucker, H. (2009). Recognizing emotion and its postcolonial potentialities: Discomfort and shame in a tourism encounter in Turkey. *Tourism Geographies*, *11*(4), 444–461.
Veijola, S., & Jokinen, E. (1994). The body in tourism. *Theory, Culture, and Society*, *11*, 125–151.
Weber, R. (2004). The rhetoric of positivism versus interpretivism: A personal view. *MIS Quarterly*, *28*(1), iii–vii. https://www.jstor.org/stable/25148621.
Williams, M., & May, T. (1996). *Introduction to the Philosophy of Social Research*. New York: Routledge.

7. Decolonial pedagogies as a pathway to transformational learning for rural community and tourism development

Christine N. Buzinde

INTRODUCTION

The intersection between the field of tourism and that of community development is an interesting and yet complex one because the doxa that govern each are in many instances divergent. The tourism industry is often critiqued for its top-down approach and frequent perpetuation of colonialistic ideals, particularly in the global south (see Chambers & Buzinde, 2015), while the field of community development is premised on a grassroots stance aimed at enhancing community empowerment and well-being. It is thus important to interrogate how issues of injustice manifest and are dealt with at the interstices of these two fields. One cannot teach at the nexus of these two fields and circumvent the socio-political realities that comprise our societies as doing so would be an injustice to students whose future depends on gaining tools with which to effectively navigate their current predicaments. Analogously, one cannot engage in community development practice whilst oblivious to the socio-political dynamics that define a given place. From this vantage point, this chapter takes as its mandate the goal of sharing lessons, from extant literature, which can facilitate dialogue around socio-political matters in a manner that abandons colonial frames and rather amplifies decolonial being and thinking. This chapter draws on the concept of *critical consciousness* as a guiding principle for discussions related to inequities occurring at the intersection of tourism and community development (see Freire, 1973/2000). Also drawn from extant literature is the practice of *culture circles*, which as will be discussed, presents practical ways through which the dialogue on critical conscious can take place (see Freire, 1973/2000). It also illustrates the role of the educator/researcher (or community organizer/activist) as facilitator so as to reduce the hierarchical divide between interlocutors and to encourage the type of genuine conversations that emerge when participants feel equal in their ability to contribute to the dialogue.

Although extensively utilized in the field of education, both critical consciousness and culture circles are underutilized in tourism studies, yet they have the potential to enrich class and community discussions and to yield a more civically aware and engaged group of learners and community participants. The field of tourism studies, given its post-positivistic roots (at least in North America) has tended to pursue a relatively apolitical approach to research, teaching and community outreach. However, in an era wherein marginalized and/or rural communities worldwide are awakening to their predicaments and actively demanding change, one cannot teach subject matters related to communities while ignoring these realities. Our students are often products of these realities, whilst some of their peers from more affluent neighborhoods are allies of the cause (for positive change). For instance, in North America, an on-campus assessment of the composition of members pertaining to the Black Lives Matter movement will result in students from a multiplicity of ethnic/racial and socio-economic backgrounds and not just Black students; notably, university campuses are a microcosm of broader society, so similar demographic representations can be found in the broader community. If there is one key thing that critical pedagogy has taught society, it is that effective education needs to make space for the multiple realties of learners to be reflected in the curriculum. From this vantage point, critical consciousness and culture circles can meaningfully engage with students' current realities while allowing for the theoretical material to resonate with their lived experiences. Similarly, comparative engagements can be facilitated by community organizers/activists who are, for instance, involved in capacity-building efforts for any given group of community members.

It is argued in this chapter that awareness creation that is premised on critical consciousness, coupled with the horizontality inherent in the facilitation of culture circles, both adhere to the principles of decolonial thinking. In order to advance this argument, a definition of decoloniality is presented. This is the fabric in which the discussion on critical consciousness and culture circles is woven, to showcase how these two decolonial tools can help tourism educators and community organizers/activists alike to better prepare learners and community members to navigate their current circumstances with awareness and agentic power. Examples of social challenges drawn from the interstices between tourism and rural community development are presented to showcase how decolonial dialogue can be facilitated by deploying the concepts of critical consciousness and the practice of culture circles.

WHAT IS DECOLONIALITY?

To understand the notion of decoloniality one needs to consider the past legacy of colonialism and its present manifestation, neocolonialism (see Loomba, 1998; Majid, 2001; Said, 1994; Spivak, 1988). Colonialism operated under an imperialist framework that relegated non-Western communities to oppression, exploitation, death and depravity, to name a few (Loomba, 1998). After enduring decades of colonialistic rule, several colonized nations obtained independence from their colonial oppressors but certain entanglements to imperialistic power frameworks (e.g., education and economic systems) remained intact in what is broadly referred to as neocolonialism (Mignolo, 2017). Colonialism was also a vehicle through which Western knowledge was packaged and imposed on non-Western communities, resulting in the subjugation of non-Western ways of knowing (see Escobar, 1995; Ocholla, 2007). It is from this vantage point that scholars of critical theory, particularly those whose work focuses on postcolonial communities, have questioned what constitutes knowledge, whose knowledge is legitimized, which ways of knowing are discredited and why (Briggs & Sharp, 2004; Warren & Rajasekara, 1993). The movement to decolonize the academy prioritizes and calls for, amongst other things, the decolonization of the knowledge production process so as to account for other ways of knowing (see Buzinde et al., 2020; Sillitoe, 1998). Decolonizing the academy first and foremost requires the availability and fostering of critical spaces of inquiry through healthy dialogue about the legacies of colonialism as well as the latent and manifest forms of neocolonialism (see Chambers & Buzinde, 2015; Manuel-Navarrete et al., 2021). Furthermore, recognition of the complex ways through which all social agents are (in)directly implicated in neocolonialistic endeavors is a necessary step in interrogating the (un)conscious perpetuation of colonialistic ideals. From this vantage point, *decoloniality* is the process of dismantling social norms affiliated with colonialism. According to Nye (2019), decolonization is about the problematization of the status quo and the enacting of the positive change needed. It is

> about responding to changes that are taking place well beyond the classroom – and also about changes that should be taking place. Decolonization is about changing how people think, talk, and act through a radical engagement with a plurality of voices and perspectives that have been historically marginalized and silenced … The aim for diversity is to accommodate ("making space" for) "alternatives" and differences within an existing scheme which largely remains unchanged. Decolonization is not about "finding space" at the table: it is about changing the room. (p. 44)

Given their varying experiences with the legacies of colonialism, Black, Indigenous, and other communities of color (BIPOC) have engaged in various forms of activism to affirm identity, express agentic power, and advance civil rights. The current global arena is replete with examples of BIPOC communities advocating for civil rights. Contemporary society has witnessed an uprising in postcolonial communities critiquing their circumstances and advocating for change. For instance, in the United States, excessive violence led by the police has spurred communities of color to engage in peaceful protests, under the Black Lives Matter movement. These demonstrations of civic engagement have contributed to the first conviction of a police officer who killed an unarmed Black citizen, and by so doing reversing decades of immunity enjoyed by the police (Anderson, 2021). In Ecuador, Indigenous groups residing in the Amazon protested against oil extraction companies that are destroying massive rain forests on which local Indigenous livelihoods are dependent. Winning the *first ever* court battle in Ecuador against oil companies, the Indigenous groups asserted their sovereignty and regained power over their land and livelihoods (Bilyeau, 2019). In Finland, Sámi Indigenous groups advocated for the end of exploitative cultural tourism practices and they lobbied Parliament to adopt "Principles for Responsible and Ethically Sustainable Sámi Tourism" (Potinkara, 2020, p. 2144). Arguably, these movements required a critical awakening to the predicaments imposed on communities but also an acknowledgment of their agentic power to reject/fight neocolonialist structures. Although occurring in different geopolitical contexts, a commonality in the aforementioned decolonial examples is the ethos, agency, and commitment to reversing injustices and ensuring more equitable outcomes. But what is the relevance of these decolonial triumphs, and other societal trials, in the context of education and community development practice?

THE RELEVANCE OF DECOLONIALITY AND ITS ROLE IN CONTEMPORARY EDUCATIONAL CONTEXTS

Effective educators are adept at identifying current affairs within media accounts and strategically weaving them into pedagogical outputs. In the context of the classroom, such tactics ensure that students master the provided theoretical content while also learning about related real-life applications. For instance, climate change negotiation meetings offer fodder for spirited discussions within an environmental sustainability class. Similarly, civil protests can be pivotal in centering critical discussions on advocacy within a community development class, led by an educator, as well as within the context of community development practice, facilitated by a researcher. After all, one of the main reasons for the existence of higher education institutions is to share timely and constructive feedback on societal life and related policies, where possible (see Arthur & Bohlin, 2005). Thus, equipping students and community participants with the skill sets with which to analyze societal injustices as well as to assess and co-create potential solutions is one of many ways through which institutions of higher education can maintain their commitment to enhancing community livability/well-being.

An important challenge to account for is that institutions of higher education have predominantly existed as spaces in which only one form of knowledge production is viewed as valid (Mignolo, 2017). From this vantage point, what constitutes "real" knowledge within the academy has traditionally been equated with only Western knowledge (Mignolo, 2017). Accordingly, other ways of knowing, such as Indigenous Knowledges, have traditionally not been granted access in the academy and nor within development contexts (see Sillitoe, 1998). Recently, there have been massive movements to recognize other forms of knowing but there is still a lot of change needed in this arena. Advocates of knowledge plurality have presented proposals for more equitable approaches: to doing research in BIPOC communities; to engaging with BIPOC students in the classroom; and, to teaching about different ways of engaging with BIPOC communities (see Manuel-Navarrete et al., 2021). This movement has required the academy to take a critical look at the various ways it continues to perpetuate colonialistic tendencies as well as the multiple avenues it can take to cultivate new decolonial approaches.

One can state with relative certainty that decolonial approaches are of relevance to the field of tourism studies and pivotal to any related applications to community development, particularly in the context of BIPOC and/rural communities. Tourism scholars have discussed the ontological and epistemological dimensions that need to be accounted for in decolonial approaches to research (see Chambers & Buzinde,

2015). Others have contributed knowledge to decolonial approaches to community development in tourism-reliant BIPOC communities (see Buzinde et al., 2021). However, an important but yet under investigated angle related to decoloniality and tourism is pedagogical in nature and is captured in the following question: What decolonial tools can tourism educators who teach community development courses and organizers/activists who mobilize communities deploy to didactically engage students and community participants, respectively? This question comprises the focus of this chapter. Response to the aforementioned question is vital within contemporary society because tourism educators/researchers cannot teach/research community development without cultivating awareness of socio-political positions through critical questioning of societal events. Furthermore, the multiplicity of social upheavals and ubiquity of protests against injustice require educators/researchers to pedagogically/epistemologically engage with, not ignore, the reality around them, particularly in the context of teaching/researching community development-related topics. Furthermore, many of the protesters, for instance in the case of the US-based Black Lives Matter movement, are generation Z and millennials, which is the key population of learners currently pursuing higher education. Accordingly, many in the younger generation are invested in community activism that is focused on changing the current unjust realities, and these ideals should resonate within the classroom environment in order for the learning experience to be meaningful but also transformative.

Example

A few years ago, I had a memorable conversation with a colleague who was teaching a class on community advocacy and leadership, in a sociology department. She was recounting an experience she had had with her students during which she introduced an exercise designed for them to assume the roles of policymakers, affluent community members, and members from under-served communities. To her utter astonishment, the students, most of whom were from affluent backgrounds, refused to embody the role of the under-served rural community members not because they thought it was beneath them, but they allegedly cited lack of socialization experiences with the group in question. Profoundly dismayed and overwhelmed by a sense of culpability, my colleague vowed to augment her class exercises with more hands-on exercises that encouraged students to temporarily assume roles of marginalized populations because she believed that doing so would enhance their sensitivities and awareness of such matters. She was worried that, upon graduation, the students would be charged with working to help people in under-served neighborhoods and yet they were unable to relate to these constituents. She realized in that moment that fostering opportunities for students to venture outside their own cultural bubbles was vital, particularly given that the curriculum passed down to her was devoid of community advocacy tools that would resonate with non-Western and/or rural communities.

WHAT IS CRITICAL CONSCIOUSNESS?

There are certainly a variety of steps that scholars have taken to broaden students' socio-cultural horizons while advancing decolonial thinking within curricula contexts. In this section we will focus on the fostering of *critical consciousness* coupled with hands-on techniques like *culture circles*. According to Heberle et al. (2020, p. 525) "an individual's awareness of oppressive systemic forces in society, a sense of efficacy to work against oppression, and engagement in individual or collective action against oppression" encapsulate the concept of *critical consciousness*. Paolo Freire (1973/2000) was instrumental in spearheading scholarly discussions on critical consciousness or *conscientização* and he believed that fundamental building blocks in learning included learners' ability to critically read unjust social circumstances, particularly because many learners from under-served communities internalize oppression and erroneously believe that their voices and contributions are not worthy. Freire's work documents the positive development and empowerment that emerge amongst learners from under-served communities as they work over time to unpack the socio-political and socio-economic constraints that characterize their existence.

Since the publication of Freire's (1973/2000) prolegomenon, scholars have attributed numerous benefits to the aforementioned concept including its valuable contributions to human development. One particularly noteworthy addition to Freire's research (1973/2000) is the work of Watts et al. (2011) who articulated three distinct dimensions that encompass the concept of critical consciousness, namely, *critical reflection, political efficacy* and *critical action. Critical reflection* denotes didactic practices geared towards educating people about how to interrogate the social systems that relegate certain members of society to the sidelines. According to Hopper (1999), "It means learning to see, in the mundane particulars of ordinary lives, how history works, how received ways of thinking and feeling serve to perpetuate existing structures of inequality" (p. 13). In the context of a tourism and community development class, critical reflection can manifest as exercises that are explicitly designed to encourage students to critically reflect on a case study that is principally focused on issues of inequity. A similar approach can be taken within the context of capacity-building workshops geared towards mobilizing residents residing in vulnerable neighborhoods.

Political efficacy is also referred to as critical motivation and it is described as the opinions one has regarding one's capacity or motivation to remedy the societal ills one perceives to exist in the world. It is the sensation that one is capable of engaging in activities that impact change. Arguably part of the motivation evidenced at the political efficacy stage can be derived from perceptions of identification with a collective such as, for instance, ethnicity, gender, political affiliation or other dimensions. A didactic approach to political efficacy can be the creation of an activity that encourages students, or community members, to focus on their perceived competence to bring about the change they want to see. This can be paired with an asset-based approach to encourage participants (i.e., students or members of a community) to abandon the adoption of a deficiency lens when reflecting on underserved communities and rather adopt strengths-based approaches. Lastly, *critical action* denotes activities pursued on a small or mass scale to effect change against perceived injustices. Scholars describe this stage as the one "that follows the 'problematization' processes in critical social analysis" (Watts & Hipolito-Delgado, 2015, p. 848). An educational approach to critical action is behavior-oriented in that its incorporation is purposefully geared towards encouraging learners to focus on the specific engagements they would participate in to bring about the change needed. According to Watts and Hipolito-Delgado (2015), critical action should include three levels: "personal action, group action and the mass action of social movements" (p. 867). For instance, Malala Yousafzai's work on advocating for female education is exemplary of critical action at the personal level, the residents of Flint, Michigan, protesting against lack of protection (by politicians) of their local water resources is akin to group action, and the #Me Too movement is a quintessential example of mass action social movement. The concept of critical consciousness thus offers a logical outline that educators can follow in facilitating discussions and action around society injustices.

WHAT ARE CULTURE CIRCLES?

Critical consciousness is a concept of raising awareness regarding social inequities, many of which are linked to legacies of colonialism. Combating these legacies requires a grounded and critical understanding of their intricate and neocolonial existence and it also necessitates the adoption of a *decolonial* lens through which agency can be (re)affirmed and strategic actions can be enacted. From this vantage point, fostering critical consciousness is a necessary and important step to pursuing a path towards decoloniality. Research supports the assertion that when critical consciousness is advanced amongst learners, particularly learners from under-served communities, they become more aware of the inequities and better prepared to deploy self agency to remedy social ills rather than adopt a debilitating or fatalistic approach to their reality (see Shedd, 2015; Diemer et al., 2016). In his work, Brazilian author Paulo Freire was interested in use-inspired scholarship that could help peasants in rural communities and other groups on the margins to question the societal structures that relegate them to certain predicaments but also to empower them to bring about the change they needed to see in their immediate surroundings. His encounters with members of marginalized communities were on the streets and impromptu, but they were also structured interventional conversations that he referred to as *culture circles*. These conversations are a manifestation of *praxis* because they intentionally bridge the gap between theory and practices (Souto-Manning, 2010). They prioritize dialogue and amplify the idea that each speaker's voice matters. In the field of education, the concept of transformative pedagogies is central in curricula designed to prepare students for teaching and central to this is amplifying the strengths of all learners. Culture circles set a foundation of knowledge horizontality wherein everyone, including the teacher, who in this context takes on the role of facilitator, has valuable knowledge to impart. This principle can be linked to the asset-based approaches (ABAs) that are central pillars in the study of and engagement in rural community development; that is, the ability to view communities through a lens of strength rather than deficiencies.

The goal of culture circles is for participants to articulate a societal challenge and formulate a solution. According to Souto-Manning (2010), culture circles "change oppressive, unjust relationships … challenge deficit perceptions" and allow participants "to envision and embrace the necessity to promote change" (p. 101). Participants within culture circles do not "necessarily find solutions to each problem but" they engage in conversations that allow for "breaking monological oppressive discourses by considering alternative perspectives" (p. 89). Within the field of education, numerous scholars have adopted culture circles as spaces within which to raise critical consciousness, not just with participants from marginalized groups but with people from a variety of backgrounds. Poignantly, this critical approach to pedagogy is relatively under-utilized in tourism courses focused on community development, yet the very nature of these classes requires acute attention to the socio-politics and circumstances that constitute given communities.

Equipped with knowledge on the concept of critical consciousness, culture circles thus offer educators an interesting example of how to facilitate dialogue through the re-constitution of seating arrangements (i.e., in a circle), the horizontal approach to knowledge (i.e., everyone has something important to contribute), and the focus on societal challenges and related solutions. The circle grants participants opportunities to speak from their lived experiences and allowance for this, particularly for students whose realities have traditionally been omitted from academic contexts, can produce transformative learning outcomes.

WEAVING CRITICAL CONSCIOUSNESS INTO TOURISM COURSES ON COMMUNITY DEVELOPMENT

Tourism educators, and those concerned with community development more broadly, can deploy critical consciousness exercises and culture circles in a variety of ways. For instance, a session on power and tourism development may instigate discussions related to the realities faced by many tourism-reliant marginalized communities whose members feel that they are unable to influence decision making regarding development in their locale. In terms of critical reflection, students can contemplate the structural impediments, both historical and contemporary that have disempowered locals and stripped them of their right to make decisions that impact their community. A situated socio-political understanding of the community in question can facilitate in-depth discussions. Guest speakers from the community in question and site visits can be woven into the curriculum to help provide a strong foundation on which students can then critically discuss the predicament of the community. In some instances, one may have a student from the rural community who can aid in this part of the collective discussion or even assign a research exercise for students to physically visit a community and gather primary and secondary data to inform their socio-political understanding of the circumstances faced by the community in question.

In terms of political efficacy, students can contemplate their capacity to bring about the needed change; they can determine what they believe the change needs to be but also the rural community's ability to (want to) bring about that change. This is the step wherein community development concepts like asset-based approaches (ABAs) can be linked to culture circles so as to encourage students to not view the community solely from a deficiency lens but to contemplate the existence of the solutions as existing within the community and amongst its citizens. In terms of critical action, this is the creative stage during which students can formulate actions that need to be enacted to bring about the change identified in the second step. To encourage transformative learning, educators can build class assignments around the type of actions students choose to take, based on the matter discussed. For instance, student A may decide to focus on *personal action* in the form of a well-argued letter to an elected policymaker in the rural community in question or an opinion piece in a popular news outlet. Student B and other members of the class may decide to adopt *group action* by mobilizing a peaceful protest on campus focused on creating awareness about the predicament of the community in question. Alternatively, there may be a national movement focused on civil liberties that the entire class may choose to participate in as part of *mass action*. Dialogue facilitated by the *culture circles* can happen prior to obtaining detailed information, as well as during and after to assess evolution in students' understanding of complex societal matters but also to evaluate their growth in sensitivity to the issues at hand.

The above-described approach can also be utilized by educators and community organizers/activists after showcasing a documentary wherein the predicament of a given rural community engaged in tourism development is featured. Drawing on the culture circles model, the educator/practitioner can thus facilitate dialogue that uses the three dimensions of critical consciousness as conversational landmarks. It is vital for tourism educators teaching community development as well as community organizers/activists mobilizing residents to encourage the implementation of activities for the critical action portion because this is a quintessential way to teach students, and participants in the broadest sense, the links between theory and practice. In fact, collectively, fostering critical consciousness vis-à-vis culture circles teaches learners (be they students in a class or residents in a rural community) to be civically engaged and critically aware as well as to learn that they can contribute to solution creation. The challenge for the tourism educator and community organizer/activist in the aforementioned processes is to ensure that s/he/they enact the role of facilitator rather than the holder of knowledge who imparts it to the audience. In the case of the educator, this challenge can be compounded when and if the teacher teaches other core classes, such as research methods, which may not necessarily lend themselves to the facilitator role. Accordingly, students may view the role of the educator as discordant and such cases require frank and trust-building conversations with learners to explicate changes in approaches to pedagogy by the same teacher. Similarly, the educator may find that students who typically appear disengaged in classes that follow the traditional hierarchical model are more inquisitive and involved in classes that promote horizontality, as is the case with culture circles. Analogously, a community organizer/activist facilitator who for instance also concurrently assumes the role of a law enforcement officer, which requires a "power over approach" may have a hard time building trust amongst community members, especially in communities that have had high numbers of negative incidents with the police.

CONCLUSION

This chapter concentrates on the nexus between decolonial pedagogies, community development and tourism. A pressing question addressed in this chapter is: What decolonial tools can tourism educators/researchers and/or community organizers/activists deploy to didactically engage students or mobilize community members? In addressing this question, this chapter draws on the premise that decolonial approaches to community development require radical changes in pedagogy (Mignolo & Escobar, 2013) and the creation of classroom/community spaces within which critical consciousness can be engaged to engender new approaches to thinking about self and society (Watts et al., 2011). Such a foundation is needed for participants, be they from marginalized or non-marginalized groups, to start from a relatively similar existential vantage point that allows for mutual understanding and community building (Freire, 1973/2000). The practice of *culture circles* is presented to showcase how it animates dialogues premised on *critical consciousness* raising, within an environment of learning. Orientation to these subject matters has the capacity to decolonize existing colonial structures, to impede/slow the perpetuation of societal inequities, and to support the development of agentic power amongst vulnerable populations.

REFERENCES

Anderson, J. (2021, April 22). White cops have been convicted of killing a black person before, but it's rare. *USA Today*. https://www.usatoday.com/story/news/2021/04/22/white-cops-convicted-of-killing-black-people/7316914002/.

Arthur, J., & Bohlin, K. (eds) (2005). *Citizenship and Higher Education: The Role of Universities in Communities and Society*. Abingdon: Routledge.

Bilyeau, N. (2019, May 24). Indigenous Amazon tribe wins landmark legal battle to halt oil drilling. *The Vintage News*. https://www.thevintagenews.com/2019/05/24/amazon-tribe-oil/.

Briggs, J., & Sharp, J. (2004). Indigenous knowledges and development: A postcolonial caution. *Third World Quarterly*, *25*(4), 661–676.

Buzinde, C.N., Manuel-Navarrete, D., & Swanson, T. (2020). Co-producing sustainable solutions in Indigenous communities through scientific tourism. *Journal of Sustainable Tourism*, *28*(9), 1255–1271.

Chambers, D., & Buzinde, C.N. (2015). Tourism and decolonization. *Annals of Tourism Research*, *51*(1), 1–16.

Diemer, M.A., Rapa, L.J., Voight, A.M., & McWhirter, E.H. (2016). Critical consciousness: A developmental approach to addressing marginalization and oppression. *Child Development Perspectives*, *10*(4), 216–221.
Escobar, A. (1995). *Encountering Development: The Making and Unmaking of the Third World*. Princeton, NJ: Princeton University Press.
Freire, P. (1973/2000). *Education for Critical Consciousness*. New York: Bloomsbury Publishing.
Heberle, A.E., Rapa, L.J., & Farago, F. (2020). Critical consciousness in children and adolescents: A systematic review, critical assessment, and recommendations for future research. *Psychological Bulletin*, *146*(6), 525.
Hopper, K. (1999). John Berger and Eric Holtzman. *Social Policy*, *30*(2), 13.
Loomba, A. (1998). *Colonialism/Postcolonialism*. New York: Routledge.
Majid, A. (2001, April 12–14). Provincial acts: The limits of postcolonial theory. An International Conference Organized by the Postgraduate School of Critical Theory and Cultural Studies, University of Nottingham, and the British Council, Morocco. http://www.postcolonialweb.org/poldiscourse/casablanca/majid1.html.
Manuel-Navarrete, D., Buzinde, C., & Swanson, T. (2021). Fostering horizontal knowledge co-production with Indigenous people by leveraging researchers' transdisciplinary intentions. *Ecology and Society*, *26*(2).
Mignolo, W.D. 2017. Coloniality is far from over, and so must be decoloniality. Afterall: A Journal of Art, Context and Enquiry, *43*(1), 38–45. https://doi.org/10.1086/692552.
Mignolo, W.D., & Escobar, A. (eds) (2013). *Globalization and the Decolonial Option*. New York: Routledge.
Nye, M. (2019). Decolonizing the study of religion. *Open Library of Humanities*, *5*(1), 43. doi:https://doi.org/10.16995/olh.421.
Ocholla, D. (2007). Marginalized knowledge: An agenda for Indigenous knowledge development and integration with other forms of knowledge. *International Review of Information Ethics*, *7*(09), 1–10.
Potinkara, N. (2020). Categories and boundaries in Sámi exhibitions. *Ethnic and Racial Studies*, *43*(12), 2140–2157. https://doi.org/10.1080/01419870.2019.1674347.
Said, E. (1994). *Culture and Imperialism*. New York: Vintage Books.
Shedd, C. (2015). *Unequal City: Race, Schools, and Perceptions of Injustice*. New York: Russell Sage Foundation.
Sillitoe, P. (1998) The development of Indigenous knowledge: A new applied anthropology. *Current Anthropology*, *39*, 223–252.
Souto-Manning, M. (2010). *Freire, Teaching, and Learning: Culture Circles Across Contexts*. New York: Peter Lang.
Spivak, G. (1988). Can the subaltern speak? In C. Nelson & L. Grossberg (eds), *Marxism and the Interpretation of Culture* (pp. 271–313). Urbana, IL: University of Illinois Press.
Warren, D.M., & Rajasekaran, B. (1993). Putting local knowledge to good use. *International Agricultural Development*, *13*(4), 8–10.
Watts, R.J., & Hipolito-Delgado, C.P. (2015). Thinking ourselves to liberation? Advancing sociopolitical action in critical consciousness. *The Urban Review*, *47*(5), 847–867.
Watts, R.J., Diemer, M.A., & Voight, A.M. (2011). Critical consciousness: Current status and future directions. *New Directions for Child and Adolescent Development*, *2011*(134), 43–57.

8. Interpreting the social impacts of tourism-led rural regeneration: the case of the Alps to Ocean (A2O) Cycle Trail[1]

Michael Mackay, Nick Taylor, Harvey C. Perkins and Tracy Nelson

INTRODUCTION

From both a theoretical and an applied perspective, considerable progress has been made in establishing an intellectual basis and set of methods and principles for assessing the social impacts of projects, programmes, policies and plans in rural areas, including new tourism initiatives. Practical application of social impact assessment (SIA) first emerged in the United States in the late 1960s in an effort to anticipate and ameliorate the negative social effects of large-scale natural resource development projects on people, groups and communities, with an eye to maximising the social benefits (Burdge, 2004). SIA has flourished globally since then.

The development of rural tourism initiatives designed to attract visitors to the countryside, and to provide services to them, can change places, communities and their rural way of life. Rural tourism is generally thought of as accounting for all forms of tourism taking place in rural settings such as farm-based agritourism, film tourism, community festivals and events, sport tourism, adventure, hunting and fishing, wine tasting, orchard and vineyard tours, and ecotourism (Mackay et al., 2014). Cycle trail riding is also a key element of rural tourism in a number of jurisdictions and is linked to a range of economic, cultural and environmental regeneration activities, including built heritage and nature conservation, and the entrepreneurial and community-based provision of visitor services that are also available to locals (e.g., cafes). While rural tourism should benefit host communities *and* tourists, unanticipated impacts will always arise and need to be managed.

The first consolidation of tourism impact research can be traced to the 1980s and the writing of Mathieson and Wall (1982) and Pizam and Milman (1986). These pioneering researchers called for tourism impact research to move beyond the dominant economic analyses of the day to systematically examine the "non-economic and intangible effects [of tourism], such as social and cultural ones" (Pizam & Milman, 1986, p. 29). Since then, conceptual, methodological and empirical research on the social impacts of tourism has grown considerably (Deery et al., 2012).

In this chapter, we use the Alps to Ocean Cycle Trail (A2O) as a case study to discuss social impact research as it relates to tourism-led regeneration initiatives set in rural areas. The A2O is situated in the rural Waitaki Valley of New Zealand's

South Island. Our starting point for the research was an interest in the ways that rural communities were attempting to create positive futures for themselves through collective local action in partnership with government agencies and private investors (Perkins et al., 2019). The A2O cycle trail stood out as an important and useful project to study because of its strong community-based origins placed in a wider national and regional policy setting. We note here that a social impact assessment of the A2O had not been conducted in advance of constructing the trail, although there were evaluations of the national trail programme.

We begin with a description of the applied social impact assessment process, and then summarise recent discussions about more conceptual and exploratory social impact research (Sairinen et al., 2021). This is followed by a short commentary on the use of social impact research on tourism-led rural regeneration projects, such as the development of tourist cycle trails in rural regions. We outline how we approached our study of the A2O, including a description of our field methods and analytical process. Our research findings come next. We conclude with recommendations for future studies of the social impacts of cycle trail tourism that will be of interest to community development, tourism and rural regeneration practitioners.

INTERPRETING SOCIAL IMPACTS: TWO APPROACHES

Social Impact Assessment

SIA is an applied process used to predict and assess the intended and unintended social consequences of a proposed intervention, be it a new project, policy, programme or plan (C.N. Taylor et al., 2004). SIA is initially anticipatory, conducted in advance of an activity taking place, and then shifts to a focus on the management of change once underway. SIA ensures that the concerns of people in potentially affected communities are heard and considered by project proponents and effectively addressed as early as possible, and managed appropriately (Chadwick & Glasson, 2018). With the well-being of people and communities front of mind, SIA practitioners ask: what changes will likely arise from the proposed activity or development, what might these changes mean for human populations in both the short and long term, who will likely win and who might lose, what will remain uncertain, and how might we assure the best outcomes for affected people and communities? These questions are all important in the context of rural tourism development.

In some jurisdictions, SIAs are required by statute in advance of formal planning decisions, such as those for windfarms, hydroelectricity projects, waste management plants, prisons, highways, quarries, transmission lines, or tourist roads (C.N. Taylor & Mackay, 2016). As Sairinen et al. (2021, p. 1) observe, this type of work is "largely intended for practical-level planning and management processes" aimed at maximising social well-being while avoiding or mitigating the anticipated negative social effects of major developments.

The SIA process is well documented (e.g., Becker & Vanclay, 2003; Burdge, 2004). It begins with a scoping and baseline study to develop a detailed appreciation of: (1) the project proposal, including its scale and characteristics; (2) the location and spatial reach of the activity (the impact area); (3) the local political and historical context for the project; and (4) the people and communities who live, visit, or do business in the likely affected area, i.e., "all potential interested or affected parties" (Chadwick & Glasson, 2018, p. 520). The resulting baseline acts as a point of departure for subsequent steps in the process.

The next step involves impact prediction and analysis. The identification of probable social effects draws on a range of data sources including comparative case study material, predictive modelling, trend projections, and an array of qualitative data, including expert judgement, local knowledge and community observations (Baines et al., 2003). The aim is to identify, understand, and determine the significance of potential project-induced changes, which commonly "comprise the physical and emotional experiences that people, groups and communities have as consequences of projects and activities that change the way they live, work, relate to one another, relax, and organise themselves" (Parsons et al., 2019, p. 115).

The final steps of the process focus on the design of agreeable mitigation measures to avoid or minimise undesirable social effects, and optimise social outcomes, which may include the development of acceptable compensatory provisions if mitigation is not possible, or consideration and formulation of alternatives around, for example, the timing, foci and location of the proposed activity, programme, or plan (Therivel, 2010). Social impact monitoring supports adaptive management of social change aiming to maximise community well-being.

Social Impact Research

Social impact research is largely an academic pursuit that looks to interpret conceptually the social changes that have eventuated in an area, or are perceived to have transpired over time, as either a direct or indirect consequence of a particular activity or phenomenon (Sairinen et al., 2021). As we discuss below, social impact research has a strong presence in both the rural tourism and destination development literatures.

Social impact research is theoretically driven, retrospective, and exploratory in nature. It is usually not conducted as part of a formal impact assessment process and so does not follow the steps and standardised guidelines laid out by the professional impact assessment community. Without such constraints, social impact research varies considerably with respect to duration, the approach taken and foci – with research procedures fundamentally shaped by the interests, methodological preferences and capability of the individual researchers.

Social impact research complements SIA because it helps to build a strong conceptual and empirical base through the provision of longitudinal, comparative case study material and new and adapted conceptual frameworks for examining social impacts and, by extension, the broader processes underpinning community formation and change. Sairinen et al. (2021, p. 1) argue that there is room for more "research-orientated approaches to studying social impacts" from a scholarly perspective (also see Arce-Gomez et al., 2015). The research into social impacts that underlies this chapter combines an understanding of social impact assessment and social impact research, to evaluate the success of a particular tourism initiative alongside other rural initiates for community regeneration (C.N. Taylor et al., 2021). Similar to Sairinen et al. (2021), we promote a theory-driven understanding of tourism-induced social impacts and rural social change.

INTERPRETING THE SOCIAL IMPACTS OF TOURISM IN RURAL AREAS

Social impact research is the most common approach to studies investigating how tourism directly and indirectly affects host regions and communities for better or for worse (Deery et al., 2012). As Tovar and Lockwood (2008) observe, many tourism impact studies focus on host community perceptions of the effects of tourism development on particular places and/or changing attitudes towards the presence of visitors in a local area (e.g., Haley et al., 2005; Sanagustin-Fons et al., 2018). Specific tourism initiatives and activity types have also been studied from a social impact research perspective, including research on the social impacts of community events and festivals (Fountain & Mackay, 2017; Small, 2007), sporting events (Wise & Harris, 2017), critical tourism infrastructure (Kanwal et al., 2020), cultural heritage sites (Zhuang et al., 2019), and tourist tracks and trails (Reis & Jellum, 2012; Scherrer et al., 2021).

Tourism-led rural regeneration initiatives (Wise, 2018) are a subset of the kinds of activities outlined above, and these have also been studied from a social impact research perspective. These initiatives aim to revive rural places that have declined, or are in stasis, by making better use of extant natural and social capital, often as part of a wider regeneration strategy, and also to improve a locality's image in the eyes of locals, visitors and investors (Wise, 2016; 2018). While the focus of research associated with this form of tourism development tends to focus on economic outcomes, Wise (2018, p. 1) calls for tourism research "to go beyond the focus on economic impacts of tourism-led regeneration … to address the importance of, and need for, critically assessing issues, problems, and solutions surrounding social regeneration resulting from tourism change, developments, or initiatives". This is because these kinds of ventures can also induce a range of social and cultural outcomes for host communities and visitors alike.

Investment in the development of tourist tracks and trails is a prime example of tourism-led regeneration activity in rural areas, and often provides the impetus for a range of other economic, cultural and environment regeneration activities to take shape, and the entrepreneurial and community-based provision of visitor services that are also available to locals. Regional cycle trails are particularly interesting in this regard because they affect multiple small communities and places that are linked together in shared regeneration activity. With respect to social impacts, international researchers have found that cycle trails provide opportunities for generating new regional partnerships, waves of entrepreneurial experimentation, collective actions, stronger regional connections and economies, and a shared sense of achievement (Corning et al., 2012; Scherrer et al., 2021). It is for these reasons that cycle trails have become a common feature, if not the main attraction, of a growing number of rural areas worldwide (Scherer et al., 2021; Reis & Jellum, 2012; P. Taylor, 2015).

INTERPRETING THE SOCIAL IMPACTS OF THE A2O CYCLE TRAIL

Context

The A2O was developed as part of the New Zealand Cycle Trail Project, or *Nga Haerenga*, meaning "the journeys" in Māori. Nga Haerenga comprises an extensive and interconnected national network of 22 publicly accessible on- and off-road bike trails, some newly built, others pre-existing but now formally linked into a growing web of tracks. The trails, or "Great Rides" as they have become known, cut across some of the country's most spectacular rural landscapes while also passing through and connecting many small rural towns (Bell, 2018). Nga Haerenga was first touted at a government-led National Job Summit in February 2009. The government's aspiration was to develop a network of cycle trails that would provide the ultimate outdoor experience for tourists while also delivering positive flow-on benefits to New Zealand's towns and rural communities. A key feature of Nga Haerenga is its dependence on partnerships between the state, territorial authorities and local communities, who are all involved in the planning and development of trails, and additional fund raising for the national cycle network (Tourism Waitaki, 2017).

At just over 300 km, the A2O is the longest continuous trail in the Nga Haerenga network. It is located in the rural Waitaki Valley of New Zealand's South Island (see Figure 8.1). Partial funding was approved for the A2O in September 2010, and by September 2013 the track was mostly created, signposted, rideable, and placed under management by Tourism Waitaki in partnership with the Department of Conservation (Tourism Waitaki, 2017).

The A2O takes visitors on a journey from the country's highest peak, Aoraki Mount Cook, through a setting comprising conservation land, lakes, rivers, small settlements, and farms, to a heritage precinct located on the waterfront of the region's main settlement on the Pacific Coast, Oamaru, population 15,000. The valley is best described as a "global multifunctional" rural space. It is globally connected in multiple ways and the site of primary and other production (e.g., hydro-electricity generation, agriculture, horticulture, viticulture, aquaculture), high amenity consumption spaces (e.g., tourism, commercial recreation and second homes) and nature and heritage conservation, including Māori rock art reserves, and a National Park (Mackay et al., 2014; also see Perkins et al., 2015; Perkins & Rosin, 2018). Together, these underpin the economies of its small towns and settlements. Populations and

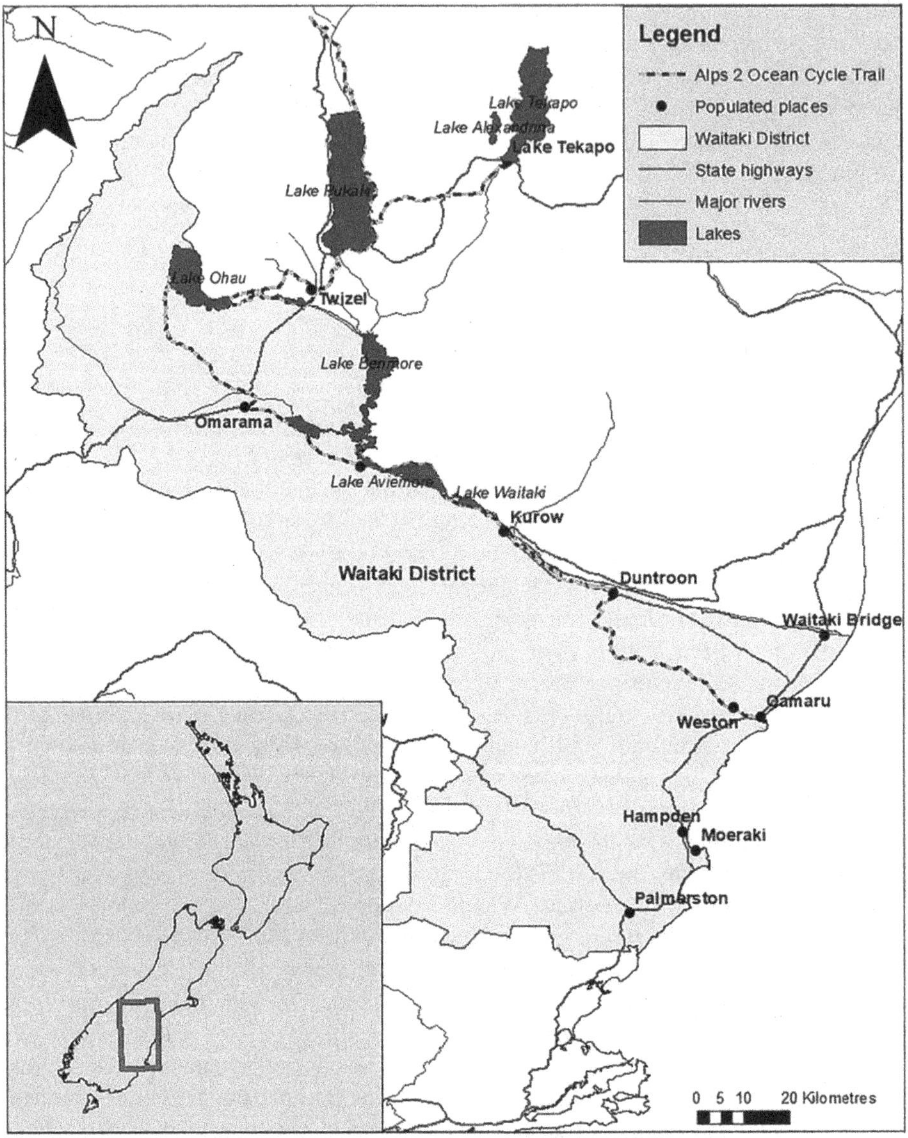

Source: Alps to Ocean trail map sourced from Google Maps. Coastline, populated places, State highways, major rivers and lakes sourced from Land Information NZ and licensed for use under the Creative Commons Attribution 4.0 International license. Waitaki District boundaries sourced from Statistics NZ.

Figure 8.1 Map of the Waitaki Valley

Table 8.1 *Usual resident populations of settlements along the A2O (1986–2018)*

	1986	1996	2006	2018	Change 1986–2018 (%)
Tekapo	276	294	318	561	103
Twizel	1,179	1,179	1,017	1,419	20
Omarama	309	356	231	288	−7
Otematata	414	361	186	177	−57
Kurow	486	412	339	369	−24
Duntroon	140	120	114	105	−25
Oamaru	13,969	13,417	12,687	13,974	0

Notes: Towns listed down the valley. Populations adjusted to 2018 boundaries. 2016 census was postponed to 2018.
Source: Statistics NZ Census data.

changes in populations of Waitaki Valley settlements over a 35-year range are shown in Table 8.1. Most noticeably, the populations of Tekapo and Twizel have increased as they have hosted a range of visitor activities and holiday homes. Down the valley, settlements have struggled to retain their populations.

The A2O stands out as one of the first major tourism-led regeneration initiatives in the Waitaki Valley designed with the intent to deliver positive community outcomes. As one destination manager noted in 2010 when it was announced that the A2O had been shortlisted for government funding: "If we're ultimately successful, we'll be able to create something special for Kiwis and overseas visitors to enjoy as well as providing huge opportunities to build stronger and wealthier communities and a better place to live in" (Stuff News, 2010). Predictions such as these piqued our interest from a social impact research perspective, prompting us to ask: what trail-induced social impacts have transpired in the valley, and in what ways have they affected local people and communities?

Methods

Our approach to interpreting the social impacts of the A2O combined elements of Sairinen et al's (2021) framing of social impact research with retrospective application of SIA to identify key impacts (C.N. Taylor et al., 2021). We followed the same steps that Sairinen et al. (2021) outline for social impact research: (1) research preparation, (2) background studies, (3) fieldwork, and (4) analysis. Our research preparation involved the design of a flexible strategy to guide the research process, including a stakeholder engagement plan, and exploration of appropriate concepts and useful theoretical frameworks, such as rural multifunctionalism and regional regeneration theory.

The second stage of our work involved background research to develop a detailed contextual understanding of the of the A2O, and the likely impacted towns and communities, both nearby and distant (i.e., the impact area). This included developing an appreciation of the historical, environmental and social context of the A2O, and the relevant local, regional and national policy settings such as the District Plan, the Long-Term Community Plan, the District Tourism Strategy and regional development settings.

Our primary fieldwork involved 35 interviews with key informants. We note here the conclusion of Deery et al. (2012) who argued for stronger use of qualitative methods in social impact research, in order to understand the impact of tourism in host communities and at a local level. We talked to people involved in leading the development of the trail, local government politicians and planning and economic development staff, people living in the towns next to or close to the track, heritage advocates, and individual business operators along the A2O, including those owning tourism and commercial recreation ventures.

As per Sairinen et al. (2021), our interviews were exploratory, moving from a rich description of the situation to the development of a grounded theoretically informed interpretation influenced by the literature reviewed above. When themes and impacts areas emerged and stood out, such as heritage development linked to the trail's evolution, we followed up with focused interviews and field trips to see the regeneration activity for ourselves, and to talk to the people making this happen or living nearby. Interview notes and transcriptions were thematically analysed, drawing on the work of Deery et al. (2012) to frame social impacts of tourism on host communities and localities. The main categories of impacts were used as an organisational framework to report the findings that follow.

SOCIAL IMPACTS

Strengthening Local Business, Livelihoods and the Local Economy

Interviewees who had lived in the valley for a long period of time talked a lot about the pre-trail days when busloads of tourists would pass by their small towns and communities on their way to Aoraki Mount Cook National Park and a neighbouring international tourism resort, Queenstown in Central Otago. This is a well-travelled tourist route. The collective view was that for many decades the valley's small towns and communities had missed out on the benefits of tourism so evident in other areas of the country. Most of them believed, however, that the A2O had been truly transformative, noting that the situation had changed significantly with the arrival of cycle tourists. As one interviewee, a long-term resident, put it:

> Before the Alps to Ocean, I didn't see a lot of tourists in town at all, just a backpacker or family out for a day trip or long weekend, but you know what? Now tourists are right on our doorstep, sometimes staying in our houses, and using the town facilities ... they seem to be taking a real interest in our towns and you have to think that the A2O has done that ... definitely.

Another long-term resident, who had also observed the change, put it this way:

> Our little towns suddenly became destinations and we're all tourism experts now! I think that's all because of the [A2O] trail … it's interesting to think that a track can do that, but in the end, it's not rocket science … [the A2O] encourages tourists to take their time and really look around the valley … we didn't realise what we had here, what the value was, until the track was built.

The presence of cycle tourists in the valley had generated a sense of optimism in the local visitor economy. Long-standing tourism and hospitality business owners reported a direct positive effect of trail-related expenditure from A2O users, which had enabled them to offer a wider range of employment opportunities in small towns, particularly in visitor services, such as the cleaning of holiday homes and the maintenance of their grounds. This renewed economic confidence had provided the impetus for local small-scale entrepreneurial experimentation in tourism and, during interviews, our research participants pointed to countless examples of new local businesses serving A2O users. This was linked by our interviewees to a growing sense of pride in the region and a view that the trail had become the incentive for residents to stay in the area and for others to move there, thus altering a perception that local communities were in decline.

> There's all these threads of new economic activity, there's people who get taken out to start the trail and taken out in the bus with the bikes on the back and somebody has made all their lunches … The [cyclists] are picked up and somebody follows them and meets them in case they break down, so there's a whole level of economic activity occurring just in servicing the people who are on the trail. I know new businesses that have started up, like cycle hire. So, people have started up businesses but most of them have been locals who have said "hmm, there's an opening here for me".

Opportunities for Regeneration

The research found that prior to the A2O there were a number of regeneration initiatives underway involving Council and local organisations. These efforts were mainly focused on improvements to the Oamaru Harbour waterfront (at the coastal end of the trail), heritage protection through the re-purposing of old buildings, and place promotion based around geological and landscape features (C.N. Taylor & Mackay, 2016). While regeneration was in train prior to the trail's inception, interviewees suggested that the A2O has significantly boosted motivation for this activity right through the valley.

> We have a lot of great amenities, and the trail has helped people really see them as opportunities for doing new things like converting houses, cottages and old churches into accommodation and it's really taken off. Like there's the old Duntroon Railway Station now accommodation … and then of course you've got the local pub which is now being rebuilt and large amounts of money being spent converting it into cyclist accommodation.

Clearly there were direct links between the presence of visitors on the trail and new businesses starting up in repurposed buildings to service the trail, along its length, and in the town of Oamaru at the end of the trail.

> One investor has converted an old junk store into a café, and bed and breakfast for the A2O. That's just one example of the money being spent rejuvenating old buildings right through the valley … you know beautiful facades and tourists love it, we love it. It's saved a lot of our agricultural heritage actually. And the harbour's been through a tough time. Nobody had any use for it but now it's being rejuvenated for tourists, and of course now we use it a lot too. You have a café down there, penguin colony, the steam train, and it's the end of the A2O trail … like you haven't biked the trail until you've got to the end and that's where it is.

Recreational Opportunities

The trail has also provided new recreational opportunities for members of the host communities and the incentive for groups and farmers along the trail to highlight recreational opportunities. Examples include the protection and restoration of wetlands that cyclists can visit and the opening of an irrigation pond for recreational purposes for local people (Mackay & Perkins, 2019).

There is considerable evidence that involvement in active outdoor recreation enhances physical and mental health (Angus & Associates, 2017). Interviewees made specific reference to these opportunities, noting that local residents were using the trail for physical exercise and social interaction, and gaining access to previously undervalued sites of historical, cultural and natural significance on properly designed, built and signposted surfaces.

> I've seen how it's changed our communities, like it's provided new recreation opportunities for us [the community] and them [tourists] and then from that, you know, locals have started buying bikes. The best bit is that I now have a really good track that gets me and the kids to the lake and some other great areas, so you know we just go to these areas a lot more now.

Opportunities to Increase Social Connection and Collective Action

A further direct social benefit is the connection made by residents of relatively isolated areas to visitors from other parts of the country and around the world. One farmer we interviewed said that the trail had helped her overcome loneliness and a sense of "disconnection to the world". On a similar theme of social connection, some interviewees said that the trail had connected them to local people interested in tourism, and through this interaction new friendships and business support networks had formed. Before the trail, several of the communities were relatively isolated and individual – their own little places socially. We have noted that the strength of these places had declined. There is now a stronger network of social connection and shared interests between people and places along the trail that has provided the basis for collective action along the valley.

> There are lots of things you can do in a community to produce change in an area, and cycle trails are one of them. That's how the Alps to Ocean project started really. It was basically seen as a change maker so a lot of us got behind it for that reason. The widespread support is what made it a success … I can't think of another example where communities, businesses and the council have all come together like that.
>
> What the region achieved collectively was impressive … people working together to develop a cycle network, people working together to give a boost to small communities, a boost to small businesses, a boost to the local job market and a boost to people's morale because, to be honest, some of the smaller towns were struggling.

Emerging Issues

Many respondents were very positive about the trail and its effects on community outcomes. There were, however, also references to emerging issues that can be a focus for understanding the impacts of the trail on the host community and the need to monitor and manage those impacts (Deery et al., 2012). Housing is the first of these issues to consider. Additional research into the effects of accommodation sharing platforms in the area found an increasing influence on a housing market that is under some pressure in respect to the availability and affordability of housing (Campbell et al., 2021). In particular, groups of migrant workers have experienced difficulties in obtaining suitable housing, including people working in the hospitality sector. This pressure on the housing market extends to the availability of accommodation for a variety of purposes.

> One of the offsides of the new accommodation buzz around the Alps to Ocean is that cyclists are now using up accommodation that locals once lived in … like we need tradesmen, but you know, they struggle to find a place to live because people are doing accommodation, so things have changed in the housing space. It's a flow on effect that I don't think anyone really thought of.

We also found that the A2O had caused some issues with visitor carrying capacity and host community perceptions that require active management. Many interviewees recognised that growing numbers of cyclists on the trail had implications for the capacity of local infrastructure, the natural environment and heritage resources. Thus, the dispersal and management of tourists along the whole trail was viewed as one way to alleviate tourist pressure at key sites. Yet there was also an awareness that this could potentially result in uneven development along the trail and distribution of benefits. This includes some townships emerging as nodes for localised use of the trail and other activities versus riding the whole length. Community leaders recognised that rapid change may soon test the adaptive capacity of communities and raise the likelihood of host resistance to further tourism developments, as some residents talked about potential damage to their sense of place, thus undermining the localised relationships that were necessary to initial development of the A2O.

> One of the problems we're seeing is that you actually cannot get any more people into some attractions, but you have to try and cope with the influx or find other ways of dealing with them and still get a buck off them, like passing them onto someone else.
>
> One of the interesting things about the Alps to Ocean is that it's basically 300 kilometres going from the mountains to the sea. That's created some interesting things. Like an emerging challenge is getting people to ride the whole track so that they end up in Oamaru and don't just ride around at the top end. That's hard because there's spectacular natural scenery up there so Oamaru has had to work hard at promoting the town as a place to ride to and visit.

The A2O adds to tensions about the best ways to promote the Waitaki Valley and specific localities and attraction to visitors in a cohesive way (Fitt, 2020). The trail reinforces the description of the Waitaki Valley as a global multifunctional rural space creating an opportunity to use the trail as a means to promote the region as a place for outdoor recreation. There needs to be a balance in the promotion in different aspects of the areas including particular heritage and cultural sites and events and enterprises in a coherent way. The trail provides the opportunity for communities to adopt a common approach to place promotion emphasising sustainable tourism management.

CONCLUSION

Cycle trails result in a range of social impacts and some of these will require ongoing management. Project level SIAs of cycle trails are needed as part of planning and implementing a new trail and they will benefit from social impact research on tourism as discussed in this chapter. Our research also identified that it would be useful for local community organisations and groups to develop capacity to contribute to or do their own social impact assessments. This increase in skills would also be useful for monitoring to generate data sets on the actual effects of trails, for the purpose of future strategic thinking about tourism-led rural regeneration in their areas. In conducting these SIAs it is important to consider direct and indirect social and economic effects, local and regional effects, and positive and negative effects. It is also important to understand the intimate intersection of natural and social environments in these high-amenity natural settings. Caution is necessary as there is a tendency to overlook the potential for negative impacts in the enthusiasm to create a new cycle trail or extend an existing trail. The tyranny of success of a trail is the arrival of more and more visitors to an area requiring ongoing adaptive management, well-informed by analysis of social impacts.

NOTE

1. We gratefully acknowledge funding from the Thriving Regions programme of New Zealand's Building Better Homes Towns and Cities National Science Challenge: Ko ngā wā kāinga hei whakamāhorahora.

REFERENCES

Angus & Associates (2017). *The value of sport and active recreation to New Zealanders.* Report commissioned by Sport NZ. Angus & Associates, New Zealand.

Arce-Gomez, A., Donovan, J.D., & Bedggood, R.E. (2015). Social impact assessments: Developing a consolidated conceptual framework. *Environmental Impact Assessment Review*, *50*, 85–94.

Baines, J., McClintock, W., Taylor, N., & Buckenham, B. (2003). Using local knowledge. In H.A. Becker and F. Vanclay (eds), *The International Handbook of Social Impact Assessment: Conceptual and Methodological Advances* (pp. 26–41). Cheltenham, UK & Northampton, MA, USA: Edward Elgar Publishing.

Becker H .A., & Vanclay, F. (2003). *The International Handbook of Social Impact Assessment: Conceptual and Methodological Advances*. Cheltenham, UK & Northampton, MA, USA: Edward Elgar Publishing.

Bell, C. (2018). 'Great Rides' on New Zealand's new national cycleway: Pursuing mobility capital. *Landscape Research*, *43*(3), 400–409.

Burdge, R. (2004). *A Community Guide to Social Impact Assessment*. Middleton, WI: Social Ecology Press.

Campbell, M., Perkins, H.C., Mackay, M., & Taylor, N. (2021). Accommodation-sharing platforms and small towns in the Covid-19 era. *Lincoln Planning Review*, *11*(1–2), 84–89.

Chadwick, A., & Glasson, J. (2018). Socio-economic impacts 2: Social impacts. In R. Therivel and G. Wood (eds), *Methods of Environmental and Social Impact Assessment* (pp. 515–545). New York: Routledge.

Corning, S.E., Mowatt, R.A., & Chancellor, H.C. (2012). Multiuse trails: Benefits and concerns of residents and property owners. *Journal of Urban Planning and Development*, *138*(4), 277–285.

Deery, M., Jago, L., & Fredline, L. (2012). Rethinking social impacts of tourism research: A new research agenda. *Tourism Management*, *33*(1), 64–73.

Fitt, H. (2020). *Geogastronomy in the Waitaki Whitestone aspiring Geopark: A snapshot of sector perspectives on opportunities and challenges*. LEaP Research Report Number 49, Lincoln University, New Zealand.

Fountain, J., & Mackay, M. (2017). Creating an eventful rural place: Akaroa's French Festival. *International Journal of Event and Festival Management*, *8*(1), 84–98.

Haley, A.J., Snaith, T., & Miller, G. (2005). The social impacts of tourism: A case study of Bath, UK. *Annals of Tourism Research*, *32*(3), 647–668.

Kanwal, S., Rasheed, M.I., Pitafi, A.H., Pitafi, A., & Ren, M. (2020). Road and transport infrastructure development and community support for tourism: The role of perceived benefits, and community satisfaction. *Tourism Management*, *77*, 104014.

Mackay, M., & Perkins, H.C. (2019). Making space for community in super-productivist rural settings. *Journal of Rural Studies*, *68*, 1–12.

Mackay, M., Perkins, H.C., & Taylor, C.N. (2014). Producing and consuming the global multifunctional countryside: Rural tourism in the South Island of New Zealand. In K. Dashper (ed.), *Rural Tourism: An International Perspective* (pp. 41–58). Newcastle upon Tyne, UK: Cambridge Scholars Publishing.

Mathieson, A., & Wall, G. (1982). *Tourism, Economic, Physical and Social Impacts*. London: Longman.

Parsons, R., Everingham, J.A., & Kemp, D. (2019). Developing social impact assessment guidelines in a pre-existing policy context. *Impact Assessment and Project Appraisal*, *37*(2), 114–123.

Perkins, H.C., & Rosin, C. (2018). Tourism, landscapes and biological resources. In E. Pawson (ed.), *The New Biological Economy: How New Zealanders Are Creating Value from the Land* (pp. 137–156). Auckland, New Zealand: Auckland University Press.

Perkins, H.C., Mackay, M., & Espiner, S. (2015). Putting pinot alongside merino in Cromwell District, Central Otago, New Zealand: Rural amenity and the making of the global countryside. *Journal of Rural Studies*, *39*, 85–98.
Perkins, H.C., Mackay, M., Levy, D., Campbell, M., Taylor, N., Hills, R., & Johnston, K. (2019). Revealing regional regeneration projects in three small towns in Aotearoa—New Zealand. *New Zealand Geographer*, *75*(3), 140–151.
Pizam, A., & Milman, A. (1986). The social impacts of tourism. *Tourism Recreation Research*, *11*(1), 29–33.
Reis, A.C., & Jellum, C. (2012). Rail trail development: A conceptual model for sustainable tourism. *Tourism Planning and Development*, *9*(2), 133–147.
Sairinen, R., Sidorenko, O., & Tiainen, H. (2021). A research framework for studying social impacts: Application to the field of mining. *Environmental Impact Assessment Review*, *86*, 106490.
Sanagustin-Fons, V., Lafita-Cortés, T., & Moseñe, J.A. (2018). Social perception of rural tourism impact: A case study. *Sustainability*, *10*(2), 339.
Scherrer, P., Dimmock, K., Lamont, M., & Ripoll González, L. (2021). Rail trails literature: Current status and future research. *Journal of Leisure Research*, *52*(1), 97–119.
Small, K.E. (2007). *Understanding the social impacts of festivals on communities*. [Doctoral dissertation, University of Western Sydney, Australia].
Stuff News (2010, February 12). *National Cycleway Success for Mackenzie Region* [Press release]. https://www.scoop.co.nz/stories/AK1002/S00171.htm.
Taylor, C.N., Bryan, H., & Goodrich, C. (2004). *Social Assessment: Theory, Process and Techniques*. Middleton, WI: Social Ecology Press.
Taylor, C.N., & Mackay, M. (2016). Social impact assessment (SIA) in New Zealand: Legacy and change. *New Zealand Sociology*, *31*(3), 230–246.
Taylor, C.N., Mackay, M., & Perkins, H.C. (2021). Social impact assessment and (realist) evaluation: Meeting of the methods. *Impact Assessment and Project Appraisal*, *39*(1), 1–12.
Taylor, P. (2015). What factors make rail trails successful as tourism attractions? Developing a conceptual framework from relevant literature. *Journal of Outdoor Recreation and Tourism*, *12*, 89–98.
Therivel, R. (2010). *Strategic Environmental Assessment in Action*. London: Earthscan.
Tourism Waitaki (2017). Alps 2 Ocean: One of New Zealand's Great Rides. *NZ Today, April–May Edition*, *73*, 30–39.
Tovar, C., & Lockwood, M. (2008). Social impacts of tourism: An Australian regional case study. *International Journal of Tourism Research*, *10*(4), 365–378.
Wise, N. (2016). Outlining triple bottom line contexts in urban tourism regeneration. *Cities*, *53*, pp. 30–34.
Wise, N. (2018). Editorial: Tourism and social regeneration. *Social Sciences*, *7*(12), 262.
Wise, N., & Harris, J. (2017). *Sport, Events, Tourism and Regeneration*. New York: Routledge.
Zhuang, X., Yao, Y., & Li, J.J. (2019). Sociocultural impacts of tourism on residents of world cultural heritage sites in China. *Sustainability*, *11*(3), 840.

9. Building *rapport à soi*: thinking critically about rural community tourism development[1]

Keith Hollinshead and Heather Mair

INTRODUCTION

This chapter concerns the role and function of critical thinking for those who work in, and study, rural community tourism development (and in cousin areas of open space management/conservation/park development/et cetera). We position critical thinking as a way of evaluating how we act as "judges" in and across contemporary society. We describe this way of thinking as a kind of *rapport à soi* so as to foster critical reflection on power in rural tourism and community development. We seek to encourage those who work in rural areas, in particular "recreation management" and "leisure services" (and related domains) to consider their own work as agents of particular forms of normalisation as to how society "is" or as to how nature "should be seen to be". As such, the chapter is premised on the view that rural areas, as managed and developed, are cultural artefacts, and we aim to guide readers and practitioners in the administration and/or interpretation of recreation/leisure/conservation/whatever rural services in questioning not only what known forms of authority they might hold in regard to the ways in which community tourism (via nature/open space/parks) is projected and promoted, but what other unconscious or under-recognised forms of authority they may also be indulged in on a day-to-day basis in the conduct of their work.

Background: The Role of the Rural Community Tourism-Judge

The term "rural community tourism-judge" parallels the concept of "the tourist-judge" which Keith Hollinshead, the lead author, has been working on in his own work in the UK and Australia on the governmentality of things. For Hollinshead, there is considerable scope in Tourism Studies and in Tourism Management as they pertain to rural areas, in examining the role not so much of well-positioned decision-makers with big budgets working for large authorities or urban areas, but in examining the role, function, and effectivity of each and every "worker"/"practitioner"/"developer" in helping not just provide access to the world of tourism/travel but in actually "making" or "demaking" or "remaking" the world. For Hollinshead, this is all a matter of the agency and authority or worldmaking in its petty (i.e., the seemingly small-time/day-by-day) and incremental (i.e., additively significant) but generally under-examined ways.

For Heather Mair, there has always been a related intention to foster broad critical thinking skills in tourism students, researchers, practitioners, and policymakers and in particular to help these actors challenge the ways the language of rural tourism development becomes normalised and taken for granted. If rural areas are cultural artefacts, then their management and administration are subject to the social, political, and economic forces that shape culture, as well as what we come to know about viable options and opportunities for tourism development.

To these ends, it should be clarified that the term "the rural community tourism-judge" (and hence, here "the recreation-judge"/"the conservation-judge"/"the rural community-judge"/"the whatever judge") is modelled upon Foucault's work on institutional governmentality, that is, upon the opaque power–knowledge decision-making authority of individuals working to tightly prescribed (but under-recognised) dominant views of and about the world. For Foucault, these forms of almost blind institutional agency and surveillance were in the military, in welfare service, and in education. These are also redolent in everyday tourism management, and in the axial domains of public nature and public heritage, et cetera.

RAPPORT À SOI

Foucault (e.g., 1980; 1981; 1988) was a strong believer that no technique or professional skill could ever be fully acquired without personal thought and exercise to accommodate the said "craft": the learning of the *techne tou biou* (the arts of living/working/being) was to him substantively a matter of the training of the self by the self. Frequently in his historico-philosophical career, Foucault strongly advocated the need for individuals to interrogate themselves about their capacities and their contributions to the order of things in life, and this was a notably large issue in his final years, during the early 1980s, when he repeatedly stressed the need for "the return of the self" in contemporary society. To Foucault, man [*sic*!] was all-too-frequently a stranger to himself who did not fully appreciate, for instance, the nature of his [*sic*!] own "social" role in the abstract machinery of the panopticon, nor of the power–knowledge reciprocities that he and every agent of power were necessarily caught up in with his superiors. To Foucault, it was imperative that each individual learnt to develop his/her own self-regardedness (i.e., his/her *rapport à soi*) towards the matters of living, working, and being that he/she was engaged upon (read: captured within). To Foucault, the self-philosophy of kinesthetics was a critical factor of the ethics of existence – where thinking kinaesthetically is that philosophy where one does one's own thinking to work out one's own role/place/function in the way truths became communicated and the way knowledge/power/rights are transmitted through oneself. Influenced by Nietzsche, Foucault endeavoured to conduct this Nietzschean self-rapport regularly in his own life, as he constantly struggled to know, and to become what one is (i.e., what he [Foucault] already was).

Relatedly, the growing tradition of critically oriented tourism scholarship (focused sometimes but not exclusively on rural areas and communities), works to confront and challenge the very taken-for-granted-ness of tourism development. Critical tourism scholars work to challenge the processes by which tourism, frequently positioned as a kind of economic saviour for communities (especially but not exclusively those in rural areas), becomes a shorthand for economic growth and development. Critical tourism scholars have long aimed to not just challenge this shorthand but to explore the implications thereof (see Mair, 2006; 2009; 2011; 2018). Indeed, developing a habit of thinking critically about tourism in all its locations and forms is a valuable, if somewhat daunting task.

As tourism planners, practitioners, and programmers have care for so much of the world's land, resources, and culture expression, that kind of intense and repeatedly engaged *rapport à soi* can be no bad thing if it can engender fewer mindless, axiomatic work routines, and can cultivate a stronger personal commitment to responsible decision-making – perhaps! But who should inspire and communicate these heady kinesthetics in and across the industry, and in communities: whose educational role is that?

CRITICAL AND CREATIVE CONCEPTUAL DEVELOPMENT FOR RURAL COMMUNITY TOURISM

The following are 12 areas of conceptual development in Tourism Studies/Tourism Management. They are important arenas or beachheads of site and space projection in planning and programming activity in contemporary tourism and in rural areas specifically. The 12 selected areas of conceptual development (or approaches) are taken from four authors who are influential in transdisciplinary inspections of tourism. The four authors are:

- Donald Horne – journalist, onetime head of the Australia Council and commentator on matters of public culture;
- Barbara Kirshenblatt-Gimblett – Performance Studies specialist, and analyst of representational activity in tourism and heritage development;
- Larry Helber – landscape architect and designer of tourist resorts and large hotel complexes;
- Nezar AlSayyad – architect and critical observer of the incorporation of viewable tradition in exhibited/constructed "versions" of space and place.

Drawing on these authors, 12 conceptual areas/conceptual approaches are presented and discussed due to their high utility as tools of thinking critically about the world-making effort to shape tourist sites/tourism destinations/travel narratives about rural place or space. The 12 selected areas/approaches of conceptual development are:

- Donald Horne: culture gene banks; site experiencing; *darshana*;
- Barbara Kirshenblatt-Gimblett: performativity; fragmentation; authorisation;

- Larry Helber: critical mix; critical mass; entry statement;
- Nezar AlSayyad: tradition; transition; traditionality.

Table 9.1 serves as an introduction to important areas for the generation today of critical thinking in and through rural community tourism/tourism studies and/or recreation/recreation management. Taking time to review and reflect on these key ideas will engender a habit not only of thinking critically but will help the rural community tourism-judge develop their *rapport à soi.*

Given the 12 conceptual areas or approaches listed above, it is possible to see how these notions can form part of a process for challenging the taken-for-granted-ness of much of rural tourism development "judgement" in rural communities. Further, critically interrogating key notions such as performativity, tradition, and site experiencing (for example) as they are used (and abused!) to form the foundations for tourism development in rural communities (and elsewhere) can open up opportunities for rethinking and challenging not just the "how" and the "what" of rural tourism, but, perhaps even more importantly, the "why" of rural tourism. As Table 9.2 makes clear, Hollinshead's work is vitally important here.

For even the keenest readers and scholars, applying these ideas to develop a *rapport à soi* is no easy endeavour. The next table provides some questions all rural community tourism students/researchers/practitioners/developers/authorities – "judges" – should ask themselves.

Taking time to reflect deeply on these powerful questions is part of the necessary effort to develop one's *rapport à soi* as well as one's ability to become a thinker who is able to critically judge rural community tourism development and knowledge forms.

KEY INSIGHTS ON THE UNDER-SUSPECTED NORMALISATION/NATURALISATION OF THINGS

Clearly, this chapter builds up to the last table (a set of 12 self-rapport questions), which students/researchers/practitioners/developers/authorities of tourism in rural areas can use to gauge the degree to which they may indeed be acting as worldmaking tourism-judges over things. That is, over the projection of peoples, places, and pasts, and thereby over materiality of peoples, places, and pasts over time. We all need to be vigilant about what we make in and through our ordinary daily/weekly/monthly (under-suspected) "practices" of normalisation, and through our matching (under-suspected) naturalising "talk".

Challenging the "Why" of Tourism

Those interested in understanding the trajectory of critical tourism scholarship would benefit immensely from reading the list of Hollinshead's work above. Relatedly, and building on Mair's early investigations of tourism development in two rural

Table 9.1 The 12 conceptual areas or approaches explained

Conceptual Area	Brief Explanation of the Areas/Approaches
Culture gene banks	That set of storehouse narratives/storylines/activities, which a nation draws on (can draw upon) to explain the proud and mighty origins of the people, chiefly in the form of museums, historical sites, libraries, eccentric archives, special places, et cetera. These tales and settings are the fodder of imaginative cultural self-celebration, just as heritage gene banks are for public/national "heritage", and nature gene banks are for "public/national nature" (see Horne, 1992).
Site experiencing	That form of visitation by which an intelligent tourist knows not merely how to crudely "look at" or "score" the site but by which he/she knows how to "read" the site in relation to the prevailing myth-systems being deployed; when a traveller intelligently experiences a site or site, he/she understands how it is being used in and through tourism, and he/she is able to further his/her own cognition of what is seen to be important in that local and global realm (see Horne, 1992).
Darshana	That mix of feelings a tourist has when confronted by a famous landscape, historic site, or celebrated object: those sensations are not just feelings of reverence, they constitute moments of mysteriously ecstatic wonderment as the place is "felt" and absorbed (see Horne, 1992).
Performativity	The fashion in which culture is not just selected/captured and mirrored in a dramatic presentation or display but is actually produced there afresh: the performativity of culture can involve the making of objects/narratives, the remaking of places/storylines or the demaking of events/ meanings, instance (see Kirschenblatt-Gimblett, 1998).
Fragmentation	Fragmentation is the communication of and about a large/grand "cultural", "heritage" or "natural" account via the purposeful (or accidental) selection of particular "small" objects or of singular angles of vision. A boundless litany of things may be used to represent or stand for/explain/reveal a larger cosmology of understanding (see Kirschenblatt-Gimblett, 1998).
Authorisation	Authorisation is the governing way a thing or event is introduced, revealed, or described either by the populations who have historically owned it, or by powerfully positioned/ government/corporate/sanctioning bodies who have been able to assume the right or appropriate "control"/"distribution" of it. The authorisation of a thing may involve the privileging of certain interpretations and of the value/relevance of it, and the suppression of other outlooks on/about it (see Kirschenblatt-Gimblett, 1998).
Critical mix	For Helber, a critical mix is that conglomeration of amenities and services (i.e., the interpretation of sites, the entertainment, the catering, the shopping facilities, plus, plus, plus) that are required to make (perhaps) a day visit to a tourist site work to optimum fashion, or which would make a particular hotel/resort work effectively and efficiently over time, and where the adopted amenities and services are complimentary to each other, and multiplicative for each other (see Helber, 1986a).
Critical mass	For Helber, a critical mass is the level of amenities and services which really should be provided – see critical mix, above – to ensure that the target markets being addressed can use the site or the setting in the optimum fashion. Thus, critical mass is a function of projected "scale" where the arrangement and positioning of the required amenities and services is closely detailed and set up/ designed/fine-tuned in accordance with best operational practice (see Helber, 1986b).
Entry statement	For Helber, a site or resort's entry statement is that mix of viewable features that arriving patrons clearly see or which passers-by clearly notice when it is first encountered. The entry statement consists of an arrangement of arrestive/inductive/orienting phenomena, which closely relate to the sites image-building name, its interpreted themes and its intended "atmosphere". The entry statement is an important element of the design and activity, which helps give the site/resort a unifying signature (see Helber, 1986c).

Conceptual Area	Brief Explanation of the Areas/Approaches
Tradition	The traditions of a population or place are those customs/activities/behaviour that are heavily participated in or associated with it. In the naming of or projection of a tradition, it is generally assumed that practice or pursuit has been popularly supported there for a long time. In fact, the declaration of what decent/proper traditions are is often a highly political and contested act of declaration, and the primary component of "tradition" is constraint (see AlSayyad, 2004).
Transition	A thing-in-transition is the manner in which an assumed custom/activity/behaviour from of old is currently changing in form, style, or use because of some internal reauthorisation of and about it, or (and more commonly) because of some osmotic influence that has gradually or suddenly seeped into that population or territory from elsewhere. In effect, no traditions ever exist – everything is always "in transition"!! (See Rothman, 1998.)
Traditionality	To AlSayyad, the commonly understood differences between "the traditional" and "the modern" is a false dichotomy, and what counts is not so much what the essential tradition "is", but how it is involved in any given age (and thereby also under modernity). The critical task in inspecting the claimed traditionality of a thing/practice/event is how that supposed tradition is identified, manufactured, packaged, and deployed, and who has the authority or power to do that. In this sense, traditionality is a rhetorical and privileging act, which includes ABC or DEF, but which excludes MNO or PQR (see Irazábal, 1994).

Table 9.2 The 12 conceptual areas and relevant readings produced by Hollinshead

Conceptual Area	Journal articles and book chapters
Culture gene bank	On "intelligent tourism" and culture gene banks, see Hollinshead, 1999a.
Site experiencing	On experiencing Walt Disney World, Florida, see Hollinshead, 1998a.
Darshana	On respect and understanding of and about Nova Scotia, Canada, see Hollinshead, 2009a.
Performativity	On tourism as the consciousness industry, see Hollinshead, 2005.
Fragmentation	On micro-understanding of North American "Indians" in/through tourism, see Hollinshead, 1992.
Authorisation	On normalisation/naturalisation – Foucauldian authorisations of people, places, and pasts, see Hollinshead, 1999b.
Critical mix	On the "critical mix" development of theme parks/themed sites, see Hollinshead, 2009b.
Critical mass	On "Disney Authority" [via strategic partnerships] in the leisure/tourism marketplace, see Hollinshead, 1998b.
Entry statement	On postmodern treatments of heritage, see Hollinshead, 2002.
Tradition	On the "new" traditions of diasporic populations, see Hollinshead, 2004a.
Transition	On the power of tourism to represent/declare/project emergent populations, see Hollinshead, 2004b.
Transitionality	On matters of "transitionality" for Aboriginal groups/communities in Australia, see Hollinshead, 2009c.

Canadian communities (2006), the following four critical questions must always be asked, not just when contemplating and making the case for rural community tourism development, but also for understanding the extent to which actions have been driven by a certain way of thinking – of judging – the worldmaking capacity of rural community tourism.

1. Why should tourism development be encouraged or promoted?

Table 9.3 *The 12 conceptual areas: self-test questions applied to "judging" tourism in rural areas*

Conceptual Area	Questions for tourism students/researchers/practitioners/developers/authorities
Culture gene bank	Have you ever inspected the public culture/public nature/public heritage of a rural from the point of view of a counter-narrative (or counter-narratives)?
Site experiencing	Are any of your interpretive activities subversive – that is, sufficiently subversive, from a previously subjugated, arising, or otherwise-important outlook?
Darshana	Can you read rural areas from a different perspective (e.g., from a Māori, a Muslim, or a Buddhist viewpoint)?
Performativity	For non-Indigenous students/researchers/practitioners/developers/authorities: What are the implications of trying to understand what Indigenous people "mean" or "make" when they produce an experience for visitors at place [X/Y/Z]?
Fragmentation	Which population(s)/ethnicity(ies)/nation(s) have you "made" (helped make) recently?
Authorisation	What steps could you take to deal with/make room for "the new" (i.e., the recognition of "the new"/the admittance of "the new"/the interpretation of "the new")?
Critical mix	Should you have your own range of work audited ethnically/politically/cosmologically?
Critical mass	Do you work positively/creatively/axially enough in strategic resource development areas of "access"/"easement"/"borrowing"?
Entry statement	Who do your pre-visit leaflets/your radio advertisements/your site-arrival-entry-statements *erase* (i.e., put off/deny/fail-to-speak to)?
Tradition	Should your organisation help produce a local area memory "gene bank"?
Transition	How are specific ethnic/non-mainstream/emergent populations in your rural area/region collectively transitionalising?
Traditionality	Whose voice (i.e., whose capacity "to speak" or "be heard") have you enabled or empowered in recent years on matters of tradition or transition?

2. What aspects of the world could or should be promoted through tourism? Why?
3. Who is (or should be) responsible for the promotion of tourism development? Why?
4. How should tourism development be promoted? Why?

Further, thinking about, challenging and, yes, understanding how we have come to judge the "why" of tourism (in rural areas and elsewhere) requires understanding the broader political economic context within which knowledge systems producing tourism decisions are situated. Indeed, drawing on early tourism scholarship from critical geography (see Britton, 1991) and more recently political economy (see, for example, Bianchi, 2018; Mosedale, 2011) the concepts introduced above can be grounded in an understanding of the extant political economic context, which undoubtedly shapes the way questions – or judgements – of tradition, performativity, et cetera are taken up and taken for granted. Tourism scholarship is relatively well versed in terms of understanding the way tourism can make and re-make the worlds of rural communities (see Mair, 2009); what is needed, however, is an even deeper appreciation for, and comprehension of, the forces shaping our notions of justice, judgement, what is "normal" or "natural" and, ultimately, what is possible.

UNDERSTANDING JURIDICAL SPACE

The notion of being a judge, invokes, of course, ideas of justice. In terms of justice, Foucault's outlook is pungently Nietzschean: to the later Foucault, the notion of justice is an idea, which has been deployed within different types of societies as a tool of a certain political and economic power or indeed as a sometimes strategic (and sometimes as a non-strategic) weapon against that power: see Patton (2014) here on Foucault's regard for "action upon the action of others". Hence to Foucault, carceral society yields judges of normality everywhere, and they are supported by/support dominant systems of power–knowledge as they carry out their part in the surveillance of their respective domains of administration, justice, and supervision. In the human sciences, for instance, Foucault maintains that knowledge-production and judgement in and of law derive from the same single epistemological–juridical formation. Hence, to Foucault, technologies of justice are readily political technologies of truth, where, as identified by Miller (2000), "the truth does not relate to that which is, but rather to that which happens".

To Foucault, then, juridical space is that site of apparent justice where truth is the outcome of some ordeal of judgement provoked by the rituals of the dominant political economy of truth. In that regard, Foucault maintains that mainstream forms of political power exercise themselves in strong part through the mediation of the given judiciary: like a number of other agencies, juridical bodies may look as if they are independent of the dominant political order when in fact they have much in common with it, and so frequently bolster the same interpretations of truth. Questioning the power of these dominant forms and the interests vested in maintaining that dominance is essential. The enunciative position (i.e., the declaratory points of identity and existence from which individuals – particularly suppressed or subjugated individuals – announce themselves) tends to be fundamentally proscribed in such mediative juridical spaces where teacher-judges, health worker-judges, administrator-judges, whatever-judges, play principal roles in unifying institutional and legal systems of surveillance over populations (and over themselves). One wonders what can be possible in the face of such unification and, in relation to our purpose here, where does tourism fit?

At last, in Tourism Studies, the field is beginning to yield critical inspections of what roughly approximates to the work of *tourism – planner – judges* and *tourism – development – judges*, influenced by the work of Tunbridge and Ashworth (1996) on the use of heritage as a political resource. As noted above, the field is now gaining its first sustained research agenda of the *use* of tourism, where its role as possibly a self-conscious/unselfconscious and an intentional/unintentional vehicle of administrative judicature is examined. As the 12 conceptual areas outlined above illustrate, reaching outside our field for tools can also help us to gain the habits of thought necessary to untangle these dominant forms and interests and to identify and examine other potentialities.

If the influence or force of "tourism-judges" in and through tourism (as for institutional-judges in any field) is to be gauged, it must be assessed in terms of its operational effect in each single institutional setting. The grip of such ongoing organisational or epistemic judgements through tourism cannot in fact be "escaped"; they can only be "redirected". Importantly, tourism scholars such as Crouch (2009) have identified the lay geographic knowledge, which governs cultural geography, and it is right to suspect there is a like form of lay geopolitical knowledge, which governs Tourism Studies. The implications of identification and examination (and ultimately redirection or rethinking) extant rural community tourism judgements, knowledge forms, and interests could be vast.

TOWARDS CREATING A CRITICAL TOURISM-JUDGE WITH A VIBRANT *RAPPORT À SOI*

These essential points and especially the questions listed above, lead us, and we hope you, to ask: what is the juridical geographic or geopolitical knowledge or the juridical rural community tourism knowledge that you have upheld over recent years/recent decades? Where (upon reflection) has it been particularly pregnant or authoritative? Where has it merely reinforced the already existing dominant forms? To what extent have your decisions related to rural community tourism been mere reflections of conscious/unconscious or intentional/unintentional efforts to reinforce the status quo, leave power relations unchallenged and tourism always as it ever was? Lastly, what are the implications of these kinds of tourism judgements? Can you build a *rapport à soi* that opens up opportunities for meaningful and substantive redirection?

Much of Foucault's late work revolved around this ideal of the self-created individual, and he regularly invited "us" to exert control over our own self-constitution. So, to those who work in rural community tourism or more broadly in Tourism Management and/or Tourism Studies, we ask: Are you indeed actively creating your own goals, habits, and practices as a self-shaped and self-realising individual of influence in the field? Can you turn your own work in tourism/travel, or in planning/development/scholarship/whatever into a work of art in and of itself? Can your own agency in Tourism Management/Tourism Studies be or become a distinct matter of aesthetic (and proactive?) style? You must remember that if *you* are engaged in matters of tourism development or community development, whatever, you are *unavoidably (already and always) acting as some form of* ***agent of normalcy***!!

NOTE ON WORLDMAKING AND RURAL COMMUNITY TOURISM

Below we provide some recent views as to what WORLDMAKING is. These statements reveal what WORLDMAKING is ... an immensely powerful force of naturalisation and normalisation ... in and through tourism (in rural communities) and related inscriptive and performative industries:

1 = "what is inculcated through tourism and other acts of symbolism and signification"

2 = "what organisations/authorities/agencies in tourism (and related inscriptive/ projective industries) seek to attain"

3 = "how the real is imagined, and the imaginary is realised"

4 = "how the world and specific locations in it – in place or time – are seen and acted upon"

5 = "how the inherited world is found and cherished ... and then performed and projected"

6 = "how peoples/places/pasts/presents are made ... and/or remade ... and/or demade"

7 = "how reality is collectively imagined ... taking account of the many cultural/ social/psychic/political reverberations of that en groupe exercise in imagination"

8 = "the representation of the world through the believed facts and nurtured fantasies of cultural selection and cultural production"

9 = "the collaborative construction of places and spaces ... and of eras and epochs ... and of heritages and inheritances"

10 = "what institutions/interest groups/individuals aspire towards ... in terms of their own identities/identifications"

SO ... WORLDMAKING IS LIKE BREATHING:
IF YOU ARE ALIVE AND WORKING, YOU ARE INDULGING IN IT!!
YOU DO NOT HAVE THE CHANCE NOT TO!!

CONCLUDING THOUGHTS

This chapter has sought to encourage a kind of two-layered thinking. First, drawing on the work of Foucault and other "non-tourism" scholars, we encouraged you to think about the ways you have intentionally/unintentionally "judged" rural community tourism and to consider how those judgements may be reinforcing dominant ways of thinking about tourism and its potentiality while coterminously suppressing or subjugating others. Next, we offered some questions and ideas to help you develop your *rapport à soi* – a way to encourage self-reflection and critical thinking so as to unmoor you from previous habits of thought about rural community tourism. What comes next is up to you.

NOTE

1. This chapter is based upon a paper given by Keith Hollinshead at the 2010 Natural Values and Natural Areas Symposium, which was hosted by the Centre for Recreation Research at the University of Otago. Reproduced with permission.

REFERENCES

AlSayyad, N. (ed.) (2004). *Consuming Traditions, Manufacturing Heritage: Global Norms and Urban Forms in the Age of Tourism*. London: Routledge.

Bianchi, R. (2018). The political economy of tourism development: A critical review. *Annals of Tourism Research, 70*, 88–102.

Britton, S. (1991) Tourism, capital, and place: Towards a critical geography of tourism. *Environment and Planning D: Society and Space, 9*, 451–478.

Crouch, D. (2009). The diverse dynamics of cultural studies and tourism. In T. Jamal and M. Robinson (eds), *The SAGE Handbook of Tourism Studies* (pp. 82–97). Los Angeles: SAGE.

Foucault, M. (1980). *Power/Knowledge: Selected Interviews and Other Writings, 1972–1977* (C. Gordon et al., eds and trans.). New York: Panthean Books.

Foucault, M. (1981). The order of discourse (I. McLeod, trans.). In R. Young (ed.), *Untying the Text*. London: Routledge and Kegan Paul.

Foucault, M. (1988). *Politics, Philosophy, Culture: Interviews and Other Writings, 1977–1984* (A. Sheridan, trans.). New York: Routledge.

Helber, L. (1986a). Market development strategies for tourist resorts. In K. Hollinshead (ed.), *Tourist Resort Development: Markets, Places and Impacts* (pp. 27–32). Proceedings of National Conference, Artarmon, NSW, Australia, November 1985. Lindfield, NSW, Australia: Centre for Leisure and Tourism Studies, Kuring-gai College of Advanced Education.

Helber, L. (1986b). Integrated approaches to resort development. In K. Hollinshead (ed.), *Tourist Resort Development: Markets, Places and Impacts* (pp. 75–81). Proceedings of National Conference, Artarmon, NSW, Australia, November 1985. Lindfield, NSW, Australia: Centre for Leisure and Tourism Studies, Kuring-gai College of Advanced Education.

Helber, L. (1986c). Site development and project implementation. In K. Hollinshead (ed.), *Tourist Resort Development: Markets, Places and Impacts* (pp. 126–132). Proceedings of National Conference, Artarmon, NSW, Australia, November 1985. Lindfield, NSW, Australia: Centre for Leisure and Tourism Studies, Kuring-gai College of Advanced Education.

Hollinshead, K. (1992). "White" gaze, "red" people – Shadow visions: the disidentification of "Indians" in cultural tourism. *Leisure Studies, 11*(1), 43–64.

Hollinshead, K. (1998a). Disney and commodity aesthetics; A critique of Fjellman's analysis of "distory" and the "historicide" of the past. *Current Issues in Tourism, 1*(1), 58–119.

Hollinshead, K. (1998b). Cross-referential marketing across Walt Disney's "world": Corporate power and the imagineering of nation and culture. *Tourism Analysis, 2*(3), 217–228.

Hollinshead, K. (1999a). Tourism as public culture: Horne's ideological commentary on the legerdemain of tourism. *International Journal of Tourism Research, 1*(4), 267–292.

Hollinshead, K. (1999b). Surveillance of the worlds of tourism: Foucault and the eye-of-power. *Tourism Management, 20*(1), 7–23.

Hollinshead, K. (2002). Playing with the past: Heritage tourism under the tyrannies of postmodern discourse. In Chris Ryan (ed.), *The Tourist Experience: Beyond the Limits* (pp. 172–200). London: Continuum Press.

Hollinshead, K. (2004a). Tourism and third space populations: The restless motion of diaspora peoples. In Tim Coles and Dallen Timothy (eds), *Tourism, Diasporas and Space* (pp. 33–49). London: Routledge.

Hollinshead, K. (2004b). Tourism and new sense. In C. Michael Hall and Hazel Tucker (eds), *Tourism and Postcolonialism: Contested Discourses, Identities and Representations* (pp. 167–204). London: Routledge.

Hollinshead, K. (2005). Tourism as the consciousness industry: Kirshenblatt-Gimblett and the mediative authority of tourism. *Tourism Analysis*, *10*(3), 301–330.

Hollinshead, K. (2009a). "Tourism state" cultural production: The re-making of Nova Scotia. *Tourism Geographies*, *11*(4), 526–545.

Hollinshead, K. (2009b). Theme parks and the representation of culture and nature: The consumer aesthetics of presentation and performance. In Tazim B. Jamal and Mike Robinson (eds), *The SAGE Handbook of Tourism Studies* (pp. 269–289). Los Angeles: SAGE.

Hollinshead, K. (2009c). Tradition and the declarative reach of tourism: Recognizing transitionality—the articulation of dynamic Aboriginal being: A critique of Swain's overview of continuity and adaptation in "Australian" Aboriginal reality, with relevance for the discerning presentation of mutable/in-between peoples elsewhere. Tourism Analysis, 14(4), 537–555.

Horne, D. (1992). *The Intelligent Tourist*. McMahon's Point, NSW, Australia: Margaret Gee Holdings.

Irazábal, C. (1994). Architecture and the production of postcolonial images: Invocations of tradition versus critical transnationalism in Curitiba. In N. AlSayyad (ed.), *The End of Tradition?* (pp. 144–170). London: Routledge.

Kirschenblatt-Gimblett, B. (1998). *Destination Culture: Tourism, Museums, and Heritage*. Berkeley, CA: University of California Press.

Mair, H. (2006). Global restructuring and local responses: Investigating rural tourism policy in two Canadian communities. *Current Issues in Tourism*, *9*(1), 1–45.

Mair, H. (2009). Searching for a new enterprise: Themed tourism and the re-making of one small community. *Tourism Geographies*, *11*(4), 462–483.

Mair, H. (2011). The challenge of critical approaches to rural tourism studies and practice. In I. Ateljevic, N. Morgan, & A. Pritchard (eds), *The Critical Turn in Tourism Studies* (pp. 42–54). London: Routledge.

Mair, H. (2018). Critical inquiry. In R. Nunkoo (ed.), *Handbook of Research Methods in Tourism and Hospitality Management* (pp. 53–62). Cheltenham, UK & Northampton, MA, USA: Edward Elgar Publishing.

Miller, J. (2000). *The Passion of Michel Foucault*. Cambridge, MA: Harvard University Press.

Mosedale, J. (ed.) (2011). *Political Economy of Tourism*. Abingdon: Routledge.

Patton, P. (2014). Foucault and the strategic model of power. *Critical Horizons*, *15*(1), 14–27.

Rothman, H.K. (1998). *Devil's Bargains: Tourism in the Twentieth-Century American West*. Lawrence, KS: University Press of Kansas.

Tunbridge J.E., & Ashworth, G.J. (1996). *Dissonant Heritage: The Management of the Past as a Resource in Conflict*. Chichester: John Wiley.

10. Refusing tourism

tebrakunna country, Emma Lee and Bryan S.R. Grimwood

INTRODUCTION

We write together with kinship as a *trawlwulwuy* woman of *tebrakunna* country, Tasmania, Australia, and as a Settler Canadian learning and living in treaty territories on Turtle Island. We pay our respects to Elders past, present and emerging, and give thanks to the lands and waters of Indigenous Peoples that nourish and sustain us.

Regional and rural development tends to silo Indigenous Peoples as needing development and limits Indigenous agencies to singular modes of being the recipients of Eurocentric economic modes (*tebrakunna* country & Lee & Eversole, 2019). To construct advantage for rural and regional places, Cooke (2007) emphasizes that "community and culture" are missing qualities within economic, governance, and knowledge infrastructures that define development policies. Yet little is known about how to apply the characteristics of community and culture to the complexity of colonized places and people for economic development. This is especially relevant to Indigenous Peoples, where tourism is often regarded as a panacea to the loss of life and lands in the struggle for rights and economic wealth (Altman, 1989).

Recognizing community and culture as important for future rural development, we look to decolonizing theory and practice in tourism as driven by deep desires of the practitioners to compel different worlds, whether they be fairer or more equitable or a tearing down of structural oppressions and discrimination that colonial, Eurocentric systems have created (see e.g., Aikau & Gonzalez, 2019; Chambers & Buzinde, 2015). Practitioners aim to draw attention to the lifehoods of those groups deemed marginal – the Indigenous, the LGBTQ+ and women of colour as examples – and centre the stories and knowledges that shift people and place as frippery and exotic to necessary and privileged. More often than not, tourism is tied to visions of development bound to capital and neoliberal constructs that seek to further extract economic value from the hurt and harm experienced by persons and places that have been colonized. These so-called margins, however, are the centres of resistance where rights are argued on their own terms, contexts, and specificities (Ramdas, 2021).

In this chapter, we look to refusal as part of the resistance toolkit. Refusal, as we relate to it, is an ethic and politic that situates in the discursive and material excesses of colonial power – that is, in spaces where totalizing visions of imperialism and empire are shown and known to be fractured, failing, fleeting. As Tuck and Yang (2014) explain, refusal is a "generative orientation" alert to stories and knowledges other than those of trauma and harm and to the sovereign possibilities invited when limits are placed on colonial frames, including practices of tourism and of social science research. Refusal expands and generates against colonial enclosures and erasures. It is not subtractive, but rather resistance through "a redirection to ideas otherwise unacknowledged or unquestioned" (Tuck & Yang, 2014, p. 239).

For Indigenous Peoples, refusal speaks to the persistent and ongoing struggles for decolonization, sovereignty, and resurgence. According to Simpson (2017), refusal allows Indigenous Peoples "an option for producing and maintaining alternative structures of thought, politics, and traditions away from and in critical relationship to states" (p. 19). For instance, among the Rapa Nui people – the Indigenous stewards and inhabitants of Easter Island – refusal has been a powerful tool to draw attention to the state's oppression through a seemingly benign plan for a Marine Protected Area (MPA). The MPA effectively served to disrupt relationships between Peoples and their "territories of life" (ICCA Consortium, 2018) thereby reinforcing Chilean colonial power with the aim of enclosing the Rapa Nui as nothing more than subjects of "statecraft, international archaeology, global conservation programs and tourism" (Wade Young, 2016, p. 139). Through persistent and diverse political organizing, dialogue, and action, the Rapa Nui unsettled state-led development initiatives and established a self-determined administrative organization to manage the MPA. Here, refusal has been expressed as a worldview that "prevents the Rapa Nui from separating the land, environment, life or nature simply because everything is integrated" (Teave & Cloud, 2014, cited in Wade Young, 2020, p. 8).

In Hawai'i, where the history of bad guests has flowed from US settler state occupation and militarization, Kãnaka Maoli have refused offers of US federal recognition for similar reasons. David Uahikeaikalei'ohu Maile (2019) describes how recent US government attempts to recognize Native Hawaiians as a governing entity are tethered to desires for settling legal claims for past wrongs and curing the harms it perpetrates. By meeting state apologies and visions of reconciliation with a steadfast *'a'ole* ("no"), Kãnaka Maoli reject the state-determined processes of recognition and the subordination this inflicts. 'A'ole turns instead to honour self-determined and sovereign histories of the Kãnaka Maoli and how these are wrapped up in legacies and ongoing stories of struggle for decolonization (Maile, 2019).

What these examples from the Pacific show is that refusal is not just a place of articulating the outrages of genocide, colonization, and lack of redress, but also a place of self-determined love, humour, and respect. bell hooks (1989) speaks of "radical openness" as a generosity of spirit, a nurturing of all that is cherished and loved by those trampled upon by the colonial. In this chapter, we want to story what that love might look like; what the deep bonds of kinship and reciprocity (*tebrakunna* country & Lee, 2019), as expressed in narratives of our experiences being mentored by Indigenous Elders, might illuminate. While refusal is, and should be, taken as a literal word for Indigenous Peoples to decrease colonizing harms, we want to invoke refusal as a positive turn that might allow those of us enrolled in tourism development discourse and practice to move together in respecting Indigenous Peoples, communities, and cultures.

RESTORYING TOURISM

Indigenous cultures are far from homogeneous or universal, yet there are certain cultural precepts that are recognized by Indigenous communities globally. One of these is the role of Elders, who are living archives of knowledge (McKemmish et al., 2019) and mediate ancestral law to communities (Kimmerer, 2013). In different ways and in different geographies, both of us have had our tourism curiosities sparked by Elders who gifted us a sense of belonging and whose knowledge has had lifelong trails, burrows, and tracks into our hearts, minds, and work. We seek to honour these Elders and, in doing so, offer others a place of confidence to write the stories that many of us share, but are too often dissuaded from telling in academia.

In the following narratives – the first written with Emma and the second with Bryan – we want to transform the principles, pedagogy, and practice of tourism development and tourism research by illustrating respectful "listening" of Indigenous Peoples and their stories. This involves not only listening *to* stories, but listening *for* stories (Cruikshank, 2005). Tourism and rural and regional development often subscribe to narratives that perpetuate and circulate harm, and new stories of tourism and development are clearly needed (Grimwood et al., 2019). As Haraway (2016) frames it, "it matters what stories make worlds, what worlds make stories" (p. 12). Storywork is important here, as this is one way to connect with how Indigenous Peoples prioritize "meaningful and useful research for our people" (Archibald et al., 2019, p. 6). Yet storywork has important lessons for many researchers, as the methodology allows us to go beyond binaries and identities (e.g., "Indigenous" and "non-Indigenous") and into a place of shared experience of love, learning, and respectful relations.

So, we want to avoid placing story that laces prevailing norms and discourses within tourism development as the province of Eurocentric and colonizing researchers (aka "the establishment men" referred to by *tebrakunna* country and Lee, 2017). Instead, we find that storywork helps us to travel across temporalities that do not lock us in place or space, but to have mobility, or a "co-motion" of moving together (Muller, 2014), across and within borders and territories. Restorying tourism as travel across time and temporalities orients us to relations *other than.* We find our place in relation to each other through storywork, thus compelling rural tourism development to be more than flat structures of economies and policies but lively and enlivening to and with Indigenous Peoples.

We start with Emma's story with Uncle Laddie and follow this with Bryan's story with Anautalik.

Uncle Laddie Timbery

I became one of the first crop of Indigenous archaeologists in Australia in the mid 1990s and established an archaeology consulting business in Sydney in 1998. I soon came across a cherished Elder, Uncle Laddie Timbery, at La Perouse, Botany Bay, Sydney. Uncle Laddie's family had lived on country for at least 6000 years, since the current sea level, and had probably been there for thousands of years more (Attenbrow, 2010). Even more extraordinary, Uncle Laddie's family had survived the hell of first contact and annexure of country to the British Empire to be replaced by the colonizing enclosure of lands through the ever-expanding and regeneration of cities. Uncle Laddie, who died in 2019, was a world-renowned cultural artist, primarily for traditional boomerang making, but for all the tools of an Indigenous life (Crawford, 2019). I too have a piece of Uncle Laddie's artwork, a string bag woven from bark using a fishing net pattern, tied with a varnished gum nut, displayed with reverence and memory.

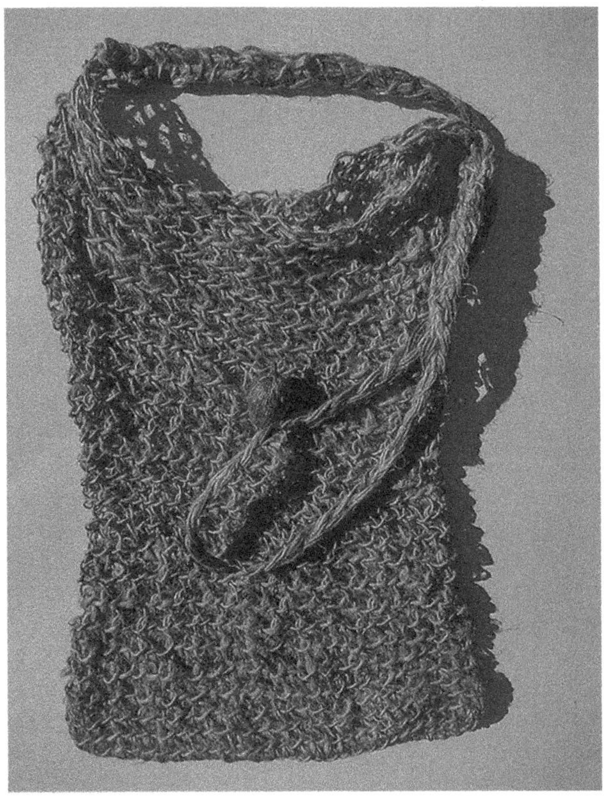

Figure 10.1 A string bag crafted by Uncle Laddie

His extended family sold these artefacts at La Perouse, the homelands and former mission settlement of the Bidjigal peoples, for over 140 years (Nugent, 2012). At first artefacts were sold door-to-door (Nugent, 2012), but by 1902, with the installation of a tramline (Belmessous, 2020), a permanency of place was afforded to sell within their community. Uncle Laddie told me that it was the oldest, continuing, modern Indigenous enterprise in Australia. He also told me this story, one that he had shared with thousands upon thousands of people for decades at his weekend stall at La Perouse that: "A long time ago, a whitefella named Jimmy Cook came here and took all our fish!"

My eyebrows wrinkled in concentration, thinking over the name before realizing that this was the illustrious Captain James Cook, and he was nothing more than a fish thief; Uncle and I bursting into shared laughter. That line has stayed with me and guided thoughts that have taken two decades to begin to articulate.

Uncle keenly understood the power of story-telling as an artist and entertainer. His truths, which flipped and tricked the conventional victory narrative as much as his boomerangs, cut through the idolatry that characterizes Western colonial histories. Those first contact moments with all kinds of strangers – the French, Russians, Dutch and British (Karskens, 2009) – did not just happen "back then", they were happening now in the exchange of cash for culture and the "tourism" value of something exotic, something natural, on the back doorstep of a modern-day city.

Reducing the man's name to a nickname de-places Cook as central to the colonial story while magnifying the intensity of Uncle Laddie's statement of rights: "I stand in my country and this devastation is just part of a tiny moment of our stories and ongoing culture." Uncle Laddie's story-telling is a way to remind the visitors that his family's business is not just selling things, but is making statements about the nature of the exchange. The cash for Uncle Laddie was important to provide an income for his family in this new world on his country, yet was also a vehicle to refuse colonialism. While framed as an art tourism venture, the boomerang, bag or digging stick is a signifier of claim to cultural authority and of not ceding sovereignty of Bidjigal country to the British. To believe the authenticity of the item is to believe in Uncle Laddie's sovereignty to country in the face of Jimmy Cook's triumphant voyage to establish the Australian nation.

Uncle Laddie was known as a generous, caring man who welcomed everyone (Crawford, 2019). A welcome from Indigenous Peoples in Australia is not the equivalent of a "Hello, come on in!" A welcome is both implicit and explicit to assert the power of culture authority and sovereignty (Bodkin-Andrews et al., 2016). However, colonialism "is the ultimate breach of guest protocol" (Aikau & Gonzalez, 2019, p. 8). The peoples who belong to their country maintain their sovereignty and their collegiality with their neighbours through powerful recognition of rights, boundary, and responsibility. A welcome by an Indigenous Australian to their country is an explicit articulation of the status of sovereignty (we have our rights to exercise cultural practices in unceded territory) and an implicit understanding with the person being welcomed that there is reciprocity and kinship that has cultural responsibilities with it (Merlan, 2014). As a visitor, it is incumbent upon us to ensure your safety by, for example, carefully explaining the rules or law and by acting as a big sister or little brother or making sure that behaviour protects our heritage. It is impractical to make it true in today's colonial world and mass populations, but the principal remains the same as the intent. We welcome you to our country and we care for you.

When I met Uncle Laddie in the late 1990s, he used to call whitefellas who came to learn, buy, laugh, and develop extraordinary friendships with him as "Gubba-rigines"

– a neologism of "gubba", meaning whitefella, and "-rigines" from Aboriginal and blackfella. Many Aboriginal and other people may frown upon this language complexity, but Uncle Laddie imbued it with kinship and belonging. Every whitefella who he met became part of his family and this was his welcome to country. He embodied the cultural practices behind his artwork as he defined his sovereignty and belonging to country through the giving of kinship.

As with many Indigenous artists across the world, the term "educator" is often a further characterization. The same is not often said of Western artists. While this distinction I make is minor in comparison to the effects of racism, it does strike at what people like Uncle Laddie do to give us the language of refusing colonization and refusing tourism. At the same time, the economic business of art and tourism sales has become an increasing arena for Indigenous freedoms to turn colonizing and settler narratives around, to tell a truth of what happened (Petsinis, 2021; Smit, 2021). Fish were stolen by these colonizing people (Karskens, 2009), but if they were kin and respected sovereignty, then they would have been considered family to share with. While our cultural practices, traditions, knowledges, and goods have been commodified for a tourism market, this does not equate to a transfer or ceding of our territories, rights, and responsibilities to country and the people in it, whether visitors or not. Every boomerang Uncle Laddie threw and every story that he told of his people's experience and life before and after 230 years ago of colonization was an education in histories that are before our eyes. An Aboriginal man whose family survived colonization, working in the middle of a city, and practising his rights of ownership claimed over millennia is an education – and a privilege to hear and to keep.

Uncle Laddie was the first Aboriginal person I knew who could tell a visceral truth and make a buck and a million friends from it. In the middle of Sydney, the city that began as the launching pad of a colonial nation, a community of peoples at La Perouse refuse both colonialism and tourism by being at the heart of both structures. He did not have to travel far spatially, but he did temporally. Uncle Laddie was part of the Indigenous Australian cosmos that mingles ancestors with plants, animals, stone, and night sky and through which those stories are mediated by our Elders. The cultural underpins of orally transmitted histories means that travel is through connection to country and its people over time, not necessarily a focus on destination, although ritual, seasonal and ceremonial travel over immense distances is a usual feature of Indigenous histories (Curran, 2016).

This type of travel is different from tourism – a distance away from a starting point to experience something new. Uncle Laddie's family, as with many Indigenous Australians, have had to contend with building rights and recognition through a sovereign and cultural lens. There is a continuity of themes across generations in the struggle for rights, while calling for recognition of our deep histories of sovereignty and care for country. Thus, Uncle Laddie has been able to draw upon the ancestral connections to country for a minimum of 6000 years to travel across the past 230 years of colonization and exert his right to refusal through production of cultural artefacts and giving kinship.

Anautalik

The doll was one of many gifts given by *Anautalik*[1] during what was for me the profoundly meaningful, but far too brief, time that I spent with her. Standing roughly 30 centimetres tall and dressed in a white *amautiq*[2] with an intricate beaded motif, mitts lined with wolf fur, and waterproof *kamiks*[3] stitched from softened caribou skin, the doll was one of three that were hand sewn by Anautalik during her participation in a series of doll-making workshops for women in the Kivalliq region of Nunavut. The workshops occurred in 2007 and, if my imperfect memory and online searches are to be trusted, led to the Kivalliq Inuit Doll Festival Exhibition and Story-Telling Traditions in *Kangiqliniq* (Rankin Inlet, Nunavut) and several exhibitions at major Canadian art galleries in 2008. Anautalik presented me with the doll in November 2010, just minutes before I boarded a flight from *Qamani'tuaq* (Baker Lake, Nunavut) to return to my then home in Ottawa. The doll was gifted not to me, but to my first child, Edyn, who had been born just two months prior. With swelling tears and gratitude, I accepted Anautalik's gift as a gesture of welcoming Edyn to the world and embracing her from a distance.

Figure 10.2 *The doll gifted by Anautalik*

I first met Anautalik in July 2008 during the first of several visits I made to Qamani'tuaq in relation to my PhD research. After learning about my interests in studying tourism and Indigenous livelihoods associated with the Thelon River (Grimwood, 2011), Anautalik invited me and a colleague to join her and her family at their cabin, which was located a bumpy, 30-minute ATV ride out of town and on a gentle rise above the Thelon's eastern shoreline near its terminus. Anautalik's knowledge, experience, and skill as an Inuk woman on the land were unmistakable, though hardly flaunted or embellished for the sake of her guests' desires to witness and learn. Anautalik later told me stories of her being raised on the land, just north of the tree line, and that being at her cabin along the Thelon filled her with the sense of being home. From the cabin, Anautalik could access hunting and fishing territory, gather berries, connect with family and ancestors, meet and share stories with other Inuit, drink clean water – practices often attributed to fostering and sustaining cultural livelihoods in the Canadian north. It was also the site where Anautalik raised the traditional Inuit tent she had constructed out of caribou skin. Anautalik designed and built the tent to pass on the techniques to other Inuit women in Qamani'tuaq. It was a gift, I learned, to her community for their use in healing, story-telling, schooling, and food preparation (CCA, 2006). But Anautalik and her husband also welcomed *Qablunaat*[4] visitors, me included, into the warmth of the tent and the stories of ancestors, culture, land, and resilience it conveyed. Like the doll, the tent was a creation to express and practise sovereign relations to kin and to place, and to welcome guests into that web.

Figure 10.3 Anautalik's caribou skin tent

My initial visit to Anautalik's cabin set the stage for several other land- and water-based journeys that I shared with her within the Thelon riverscape during my PhD research. Reflecting on these journeys now, it's clear how foundational they were to our friendship and the collaborative spirit of my project. Anautalik was guide and mentor – on the tundra and in Qamani'tuaq – helping me navigate and build the necessary relationships with people and place for responsible, respectful, and reciprocal research. She showed me how to see the movement of caribou far off in the distance, how to lash down a *qamutik*,[5] how to build a small fire for brewing tea in a landscape barren of trees. A fluent speaker and skilled writer in both Inukitut and English, Anautalik's gift of language enabled linguistic borders to be crossed. She transformed research objectives and interview questions written in English into invitations for Inuktitut-speaking Elders to share knowledge and interpret stories. Sipping tea or preparing food in her kitchen, she listened with me to the audio recordings of these interviews and translated the words and worlds of Elders into something I could comprehend. Anautalik also knew how to read and write the landscape. For instance, she would create and identify *inuksuit* as part of a network that communicates among Inuit where to find game, locate caches, navigate, or signal hazardous conditions. Part of what it means to be Inuk, as I learned from Anautalik, is to know how to give and receive the gifts conveyed by inuksuit.

Anautalik passed away unexpectedly in 2011, during the time I was writing up my dissertation. Like all life on this wonderful world that many of our species seemingly want to waste, Anautalik's life was a gift. What distinguishes her from most others I've encountered is that she used this gift to bestow her own gifts. This is how I remember Anautalik. Her devotion to family expressed in sewing clothing from caribou skins to ensure their warmth during winter hunts. Her care for community practised in her regular work supporting other Elders or in the invitations she extended to young neighbours to join her on the land. Her commitment to conveying and creating Inuit culture entwined in her art. Her ambition to build bridges and learn across difference propelling her willingness to collaborate not only with me but also several other Qablunaat researchers. This is how I remember Anautalik, giving her gifts to those so fortunate to receive them.

WEAVINGS, OPENINGS

Decolonization is "an unfinished, excessive historical process" (Clifford, 2013, p. 6) and requires creative projects where cultural practices and land relations beyond colonial command are fostered and nourished (Mackey, 2015; Tuck & Yang, 2012). It is a process that must be imagined to be practised (Veracini, 2011). The refusal of colonial conventions that infiltrate tourism development and research practices is imperative to activating these decolonial politics and responsibilities. Refusal orients to what is, or could be, other than colonial. Refusal helps to weave openings to other possible worlds.

In this chapter, we use storywork to invite and generate tourism worlds that foster Indigenous life. Refusing conventions in tourism development, pedagogy, and research discourse, we anchor this chapter to our stories with honoured Elders. What we locate are several insights and expressions that exceed colonial violence. For instance, the selling and sharing of art is not about commodification but about asserting sovereignty and speaking truth to power. The expression of Indigenous "welcome" – to place, to new life, to learning *with* – is similarly an expression of that power. But welcome also translates into care for guests, which involves teaching guests about Indigenous law and appropriate behaviour and conduct. This is why land-based learning is so crucial in our respective experiences with Uncle Laddie and Anautalik. To know law, to know culturally sanctioned and appropriate conduct, requires learning with country.

And art, welcome, and learning travel across space and time. They land in the hearts and hands of guests who carry them home. Sovereignty is, therefore, stretched across space as it is enacted in and through story. Moreover, the storywork with Uncle Laddie and Anautalik show how these practices of art, welcome, and learning convey the intergenerational building of Indigenous rights and recognition. They are practices storied in generations of resistance and sovereignty and resilience. They are artefacts and infrastructures that travel across temporalities to bridge relations. Uncle Laddie and Anautalik are respected for navigating and negotiating these worlds. They are both deeply rooted in culture and place, a sense of who they are and where they come from, and yet they have the skills (humour, story-telling, language) to make messes of colonial frames and conventions. They refuse to be locked into how cultures or how worlds are defined and determined to be. Like coyote, they navigate worlds to expose cracks or foster kinship.

Our stories with Elders who have influenced our theory and practice are also a mode for us to express gratitude. In her book *Braiding Sweetgrass*, botanist and Potawatomi scholar, Robin Kimmerer (2013), reflects on the cultures of gratitude that tether together many different Indigenous Peoples from across the world. For Kimmerer (2013), gratitude is a uniquely human capacity to acknowledge the multitude of gifts the natural world provides and that sustain human life. In contrast to prevailing consumer society, "gratitude cultivates an ethic of fullness" (2013, p. 111) whereby we recognize abundance (as opposed to unmet desires) and our place within reciprocal relationships.

Reciprocal relationships are at the core of Indigenous principles-based behaviours that influence tourism, education, or economic development (Graci et al., 2021). To conceive of Indigenous principles, pedagogy, and practice in tourism is to firstly understand that gratitude convenes reciprocity. Reciprocity enacts agency in tourism through three principles: hosting (receiving and welcoming people), connecting (meaningful relationships for mutual recognition), and sharing (promoting truth) (Curtin & Bird, 2021). However, reciprocity extends beyond relationships with people and encompasses the connections to country, territories, and the lands and waters that bind Indigenous Peoples and the responsibility from the ancestors to care for country. For Māori in Aotearoa New Zealand, this life force is *kaitiakitanga*

(Ransfield & Reichenberger, 2021), for the *Matbat* and *Matlou* People of West Papua, it is *sasi* (land) and *sasi laut* (sea) (Prasetyo et al., 2020), for Anangu in central Australia it is *tjukurpa* (Judd, 2019) – a multitude of affiliations for Indigenous Peoples centred upon life.

Reciprocity, welcome, gratitude, and Elder stories are openings to recognize a different tourism pedagogy – how we teach and learn not *for* Indigenous Peoples, but rather *from* and *with* Indigenous Peoples. Matunga (2017) writes that if theories cannot or will not comprehend Indigenous experiences, then "the only option left is to ditch them" (p. 641). Filling the pedagogy void requires us to work with Indigenous Peoples in "respectfully recouping the fragments of memory scattered from coloniality; designing effective modes of gathering and communicating back to community those fragments of memory; and enabling decolonising options of redirective designed/ing action" (Schultz, 2018, p. 89). Good tourism pedagogy requires Indigenous-led research. Indigenous-driven research that "draws on rich, personal and often intimate perspectives based on deliberately reflective and participatory approaches" (Nielsen & Wilson, 2012, p. 6) anchors the work with legitimacy, authenticity, and meaningful use, yet is often missing from theory. From Whitford and Ruhanen's (2016) analysis of 400+ papers on Indigenous tourism, caring for country is a constant thread, but Indigenous-driven research is not. We cannot have a decent tourism pedagogy without the practice of Indigenous-driven research and the principles of Indigenous axiologies.

Autonomy for Indigenous Peoples in tourism development is hard work. It is not just the struggle for self-determination and sovereign rights away from colonizing structures to retain lands and power (Morphy, 2009), but also the refusal against the clichés, stereotypes and Othering that characterizes tourism and tourism research of Indigenous Peoples (Viken et al., 2021). When Indigenous Peoples act against the rural development grain and dependency tropes, especially when paying attention to the fractured histories of colonization that have separated families, families from territories, and territories from the ancestral and creator beings (Smith, 2012), then accusations of inauthenticity from non-Indigenous people are inevitable. To state, as an Indigenous person, that knowledge has been lost or stolen, that gaps can be filled with modern interpretations, or that to be colonized is part of the Indigenous life (Maaka & Andersen, 2006; Smith et al., 2016), is dangerous. In Boukhris' (2017) appraisal of the Black Paris project, fear is a constant in a tour guide's work, where the

> discourse constantly oscillates between the conscious claim of the project and the fear, the doubt, the desire for justification. This "fear of speaking to the French people of the blind spots of history" becomes more palpable when the group is predominantly composed of "a white senior audience". (p. 696)

Accusation, reprisal, fear, and de-centring are common elements within the trauma of the colonized when attempting truth-telling through tourism. Indeed, there are few spaces and places for cultural safety in conducting Indigenous tourism (*tebrakunna* country and Lee, 2018). This is why storywork is important to rural tourism development: stories are personal and are a means to armour the individual against the racial hectoring when the speaker is both embodied and present. Storywork gives refusal life, it makes whole the disparate parts of the struggle for recognition and rights. It can generate relations that allow us to gaze back upon power or orient towards more loving horizons.

NOTES

1. "Anautalik" is a self-selected pseudonym. Anautalik described the Inuktitut term to mean the force of a judge's mallet.
2. A parka, typically worn by Inuit women, with a large hood for carrying a baby.
3. Winter boots.
4. People who are not Inuit.
5. A sled designed for travelling on snow and ice behind a skidoo or dog team.

REFERENCES

Aikau, H.K., & Gonzalez, V.V. (eds) (2019). *Detours: A Decolonial Guide to Hawai'i*. Durham, NC: Duke University Press.

Altman, J. (1989). Tourism dilemmas for aboriginal Australians. *Annals of Tourism Research*, *16*(4), 456–476.

Archibald Qum Xiiem, J., Bol Jun Lee-Morgan, J., & De Santolo, J. (2019) *Decolonizing Research Indigenous Storywork as Methodology*. London: Zed Books.

Attenbrow, V. (2010). *Sydney's Aboriginal Past: Investigating the Archaeological and Historical Past* (2nd edn). Sydney: UNSW Press.

bell hooks (1989). Choosing the margin as a space of radical openness. *Framework: The Journal of Cinema and Media*, *36*, 15–23.

Belmessous, S. (2020). Commemorating French colonialism in Australia: On the Lapérouse monument in Sydney. *History Australia*, *17*(3), 471–488.

Bodkin-Andrews, G., Bodkin, F., Andrews, G., & Whittaker, A. (2016). "Mudjil'dya'djurali Dabuwa'wurrata" (how the white waratah became red): D'harawal storytelling and Welcome to Country "controversies". *AlterNative*, *12*(5), 480–497.

Boukhris, L. (2017). The Black Paris Project: The production and reception of a counter-hegemonic tourism narrative in postcolonial Paris. *Journal of Sustainable Tourism*, *25*(5), 684–702.

CCA (Canada Council for the Arts) (2006). *Provincial and Territorial Profiles, 2005–2006: Grants to Nunavut*. The Canada Council for the Arts. https://publications.gc.ca/collections/collection_2016/canadacouncil/K21-69-2006.pdf.

Chambers, D., & Buzinde, C. (2015). Tourism and decolonisation: Locating research and self. *Annals of Tourism Research*, *51*(1), 1–16.

Clifford, J. (2013). *Returns: Becoming Indigenous in the Twenty-First Century*. Cambridge, MA: Harvard University Press.

Cooke, P. (2007). To construct regional advantage from innovation systems first build policy platforms. *European Planning Studies*, *15*, 124–146.

Crawford, R. (2019, July 29). Highly respected Aboriginal elder Laddie Timbery passes away aged 78. *South Coast Register*. https://www.southcoastregister.com.au/story/6297398/aboriginal-elder-laddie-timbery-to-be-farewelled-this-wednesday/.

Cruikshank, J. (2005). *Do Glaciers Listen? Local Knowledges, Colonial Encounters, And Social Imagination*. Vancouver: UBC Press.

Curran, G. (2016). Travelling ancestral women: Connecting Warlpiri people and places through songs. In P.K. Austin, H., Koch, & J. Simpson (eds), *Language, Land & Song: Studies in Honour of Luise Hercus* (pp. 403–418). London: EL Publishing.

Curtin, N., & Bird, S. (2021). "We are reconciliators": When Indigenous tourism begins with agency. *Journal of Sustainable Tourism*. doi:10.1080/09669582.2021.1903908.

Graci, S., Maher, P.T., Peterson, B., Hardy, A., & Vaugeois, N. (2021). Thoughts from the think tank: Lessons learned from the sustainable Indigenous tourism symposium. *Journal of Ecotourism, 20*(2), 189–197.

Grimwood, B.S.R. (2011). *Picturing the Thelon: Natures, ethics, and travel within an Arctic riverscape*. [Doctoral Dissertation, Carleton University, Ottawa, ON, Canada].

Grimwood, B.S.R., Stinson, M.J., & King, L. (2019). A decolonizing settler story. *Annals of Tourism Research, 79*, Article 102764.

Haraway, D.J. (2016). *Staying with the Trouble: Making Kin in the Chthulucene*. Durham, NC: Duke University Press.

ICCA Consortium (2018). *Territories and areas conserved by Indigenous Peoples and local communities*. Accessed October 27, 2021 from https://www.iccaconsortium.org/index.php/discover/.

Judd, B. (2019). Kapi Wiya: Water insecurity and aqua-nullius in remote inland Aboriginal Australia. *Thesis Eleven, 150*(1), 102–118.

Karskens, G. (2009). *The Colony: A History of Early Sydney*. Sydney: Allen & Unwin.

Kimmerer, R.W. (2013). *Braiding Sweetgrass: Indigenous Wisdom, Scientific Knowledge, and the Teachings of Plants*. Minneapolis, MN: Milkweed Editions.

Maaka, R., & Andersen, C. (2006). *The Indigenous Experience: Global Perspectives*. Toronto: Canadian Scholars Press.

Mackey, E. (2015). *Unsettled Expectations: Uncertainty, Land and Settler Decolonization*. Halifax: Fernwood Publishing.

Maile, D.U. (2019). 'A'ole is our refusal. In H.K. Aikau, & V.V. Gonzalez (eds), *Detours: A Decolonial Guide to Hawai'i* (pp. 193–199). Durham, NC: Duke University Press.

Matunga, H. (2017). A revolutionary pedagogy of/for Indigenous planning. *Planning Theory & Practice, 18*(4), 640–644.

McKemmish, S., Chandler, T., & Faulkhead, S. (2019). Imagine: A living archive of people and place "somewhere beyond custody". *Archival Science, 19*(3), 281–301.

Merlan, F. (2014). Recent rituals of Indigenous recognition in Australia: Welcome to Country. *American Anthropologist, 116*(2), 296–309.

Morphy, F. (2009). Enacting sovereignty in a colonized space: The Yolngu of Blue Mud Bay meet the native title process. In D. Fay & D. James (eds), *The Rights and Wrongs of Land Restitution: "Restoring What Was Ours"* (pp. 99–122). Toronto: Routledge.

Muller, S. (2014). Co-motion: Making space to care for country. *Geoforum, 54*, 132–141.

Nielsen, N., & Wilson, E. (2012). From invisible to Indigenous-driven: A critical typology of research in Indigenous tourism. *Journal of Hospitality and Tourism Management, 19*(1), 67–75.

Nugent, M. (2012). An economy of shells: A brief history of La Perouse Aboriginal women's shell-work and its markets, 1880–2010. In N. Fijn, I. Keen, C. Lloyd, & M. Pickering (eds), *Indigenous Participation in Australian Economies II: Historical Engagements and Current Enterprises* (pp. 211–228). Canberra: ANU Press.

Petsinis, A. (2021, July 28). 3 Australian First Nations artists on truth-telling through their work. *Vogue*. https://www.vogue.com.au/vogue-living/arts/3-australian-first-nations-artists-on-truthtelling-through-their-work/image-gallery/94407e165d3fe94efa199b7c985b89af.

Prasetyo, N., Carr, A., & Filep, S. (2020). Indigenous knowledge in marine ecotourism development: The case of *Sasi Laut*, Misool, Indonesia. *Tourism Planning & Development, 17*(1), 46–61.
Ramdas, K. (2021). Negotiating LGBTQ rights in Singapore: The margin as a place of refusal. *Urban Studies, 58*(7), 1448–1462.
Ransfield, A.K., & Reichenberger I. (2021). Māori Indigenous values and tourism business sustainability. *AlterNative: An International Journal of Indigenous Peoples, 17*(1), 49–60.
Schultz, T. (2018). Mapping Indigenous futures: Decolonising techno-colonising designs. *Strategic Design Research Journal, 11*(2), 79–91.
Simpson, A. (2017). The ruse of consent and the anatomy of "refusal": Cases from Indigenous North America and Australia. *Postcolonial Studies, 20*(1), 18–33.
Smit, S. (2021, June 25). Tasmanian statue project a lesson in truth-telling. *National Indigenous Times*. https://nit.com.au/tasmanian-statue-project-a-lesson-in-truth-telling/.
Smith, L.T. (2012). *Decolonizing Methodologies: Research and Indigenous Peoples* (2nd edn). London: Zed Books.
Smith, L.T., Maxwell, T.K., Puke, H., & Temara, P. (2016). Indigenous knowledge, methodology and mayhem: What is the role of methodology in producing Indigenous insights? A discussion from mātauranga Māori. *Knowledge Cultures, 4*(3), 131–156.
tebrakunna country & Lee, E. (2017). Performing colonisation: The manufacture of Black female bodies in tourism research. *Annals of Tourism Research, 66*, 95–104.
tebrakunna country & Lee, E. (2018). Black female cultural safety: What is wellness for us? In B.S.R. Grimwood, H.M. Mair, K. Caton, & M. Muldoon (eds), *Tourism and Wellness: Travel for the Good of All?* (pp. 1–20). Lanham, MA: Lexington Books.
tebrakunna country & Lee, E. (2019). Reset the relationship: Decolonising government to increase Indigenous benefit. *Cultural Geographies, 26*(4), 415–434.
tebrakunna country & Lee, E., & Eversole, R. (2019). Rethinking the regions: Indigenous peoples and regional development. *Regional Studies, 53*(11), 1509–1519.
Tuck, E., & Yang, K.W. (2012). Decolonization is not a metaphor. *Decolonization: Indigeneity, Education & Society, 1*(1), 1–40.
Tuck, E., & Yang, K.W. (2014). R-words: Refusing research. In D. Paris & M.T. Winn (eds), *Humanizing Research: Decolonizing Qualitative Inquiry with Youth and Communities* (pp. 223–248). Thousand Oaks: SAGE.
Veracini, L. (2011). Introducing Settler Colonial Studies. *Settler Colonial Studies, 1*(1), 1–12.
Viken, A., Höckert, E., & Grimwood, B.S.R. (2021). Cultural sensitivity: Engaging difference in tourism. *Annals of Tourism Research, 89*, Article 103223.
Wade Young, F. (2016). Unsettling the moral economy of tourism on Chile's Easter Island. In M. Mostafanezhad, R. Norum, E.J. Shelton, & A. Thompson-Carr (eds), *Political Ecology of Tourism: Communities, Power, and the Environment* (pp. 134–149). New York: Routledge.
Wade Young, F. (2020). Unsettling the boundaries of Latin America: Rapa Nui and the refusal of Chilean settler colonialism. *Settler Colonial Studies*, doi:10.1080/2201473X.2020.1823751.
Whitford, M., & Ruhanen, L. (2016). Indigenous tourism research, past and present: Where to from here? *Journal of Sustainable Tourism, 24*(8–9), 1080–1099.

PART III

KEY CONCEPTUALIZATIONS

11. Gender and rural tourism

Katherine Dashper

INTRODUCTION

The United Nations (2018) notes that the world is urbanising rapidly. In 1950, only 30 per cent of the global population lived in urban areas, growing to 55 per cent by 2018. This has profound consequences for rural regions and communities, which suffer from numerous problems including population decline, the waning of agriculture and associated loss of incomes, the effects of climate change and ageing populations, to name but a few. Tourism offers a possible counterpoint to some of these challenges, creating new jobs and entrepreneurship opportunities, and attracting investment and inward migration to help sustain rural communities (see Dashper, 2014a). However, the benefits of rural tourism development are not felt equally by all rural communities or everyone within a given community. Many groups and individuals are unable to enjoy the promised benefits of tourism development due to a variety of factors such as limited involvement in planning, low education, lack of capital (economic, social, cultural, and political) and complex social, cultural, and religious factors. Many of these issues are gendered and rural women in particular are less likely to benefit directly from rural tourism development than are men in the same communities. Rural communities are deeply gendered in multiple and complex ways, and this is reflected, sometimes exacerbated, in the context of rural tourism. As Chambers and Rakic (2018) comment, tourism is beset by gender inequality and change has been patchy, uneven, and too slow. This is particularly true of rural tourism.

However, as Argent (2011, p. 184) points out, there are 'many different rurals'. Indeed, the UN notes that global urbanisation rates differ markedly across geographic regions; for example, northern America is the most urbanised region, with 82 per cent of the population living in urban areas, compared with approximately 50 per cent in Asia and just 43 per cent in Africa (United Nations, 2018). Gender is an important aspect of rural tourism in all regions across both the Global North and South, but the nuances of gender relations and their impacts on rural tourism development and practices vary considerably, influenced by a variety of social, cultural, political, economic, religious, geographic, and ecological factors.

The chapter begins by outlining some of the ways in which rural communities are gendered before considering some of the complexities in rural tourism when examined through a gender-aware lens. Pritchard (2018) suggests that gender is still considered a marginal issue in tourism research and practice, and so this chapter contributes to efforts to argue for the centrality of gender for understanding all aspects of (rural) tourism and for working towards greater gender equality. As Alarcón and Cole (2019, p. 903) note, 'without gender equality, there can be no sustainability', and so gender needs to be a foundational consideration in any (rural) tourism research and development.

GENDER AND RURAL COMMUNITIES

Rural societies are deeply gendered. In a wide variety of national and social contexts, rural communities are tied to farming and agriculture, economically and socially (Little, 2002). Even when the economic importance of agriculture has declined, farming and relationships with the land remain symbolically important for shaping rural relations and identities. The field of rural sociology has demonstrated the persistence of a gendered ideology that positions men at the head of a 'natural hierarchy', performing outward-facing roles in the community and working mainly in farming and production, with women largely confined to domestic roles and (re)production of family and community life (Saugeres, 2002; Little & Panelli, 2003). 'Real men' in rural societies can cope with physical extremes and hardship, and exercise control over their own bodies, the elements, the land and, by extension, women (Bye, 2009). Bryant and Pini (2009) argue that in rural Australia the dominance of farming histories and families shapes rural communities. It is male lineage that dominates and defines farming families, and women's identities get subsumed under those of men.

The importance of the land, and of nature and agriculture, is a fundamental feature of rural societies and shapes gender relations and practices in and beyond tourism. Rural masculinities are often (re)produced in relation, often opposition, to nature and privilege bravery, fitness, strength, fearlessness, and stoicism. Practical knowledge is highly valued, and the body is often favoured over the mind. In such ways rural masculinities are the antithesis of femininity and tend to denigrate normative feminine traits like emotion and sensitivity (Bye, 2009). In contrast, rural femininities are constructed in relation to familial bonds, and women's positions have often been restricted to supportive and nurturing roles in the family and wider community. Women have always worked on the land alongside men, but this labour has often been rendered invisible and unvalued (Shortall, 2006; Bryant & Pini, 2009). Where women do work closely with the land, such as by producing food and protecting water supplies for their families and communities, this knowledge and expertise is usually not valued as such, dismissed as part of an inherent synergy between women and nature and contrasted to the mastery that characterises men's engagement with the land and environment. In such ways, rural communities are often based on the privileging of family-based heteronormativity, with conservative gender norms that constrain acceptable roles, positions and behaviours for men and women (Little, 2007). Rural societies are presented as different to urban ones, with distinctive values and social relationships.

However, rural communities are experiencing a prolonged period of significant economic, social, and cultural change that is profoundly affecting rural societies and, by extension, gender relations (Dashper, 2014b). Mechanisation and technological and other innovations have led to a decline in the economic significance of farming in many areas, shifting the nature of rural work and thus the very bases of rural life. One consequence of this is the migration of younger people to urban areas in search of new employment, as well as better standards of living and more exciting social and cultural opportunities (Bye, 2009). In some rural areas, this has left behind a rapidly ageing population with limited opportunities to sustain the rural community. In others, there has been an influx of new affluent residents from urban areas, often through the purchase of second homes, attracted by the appeal of rural life (Pitkänen, 2008; Sutherland, 2012). This has shifted the fabric of rural communities, leading to gentrification, which further marginalises young people who can no longer afford to live in the community, increasing the problem of out-migration further.

Bock (2015, p. 731) notes that a major problem for the sustainability of rural areas is 'the rural exodus of women'. This is particularly problematic for more remote areas, which differ substantially from urban regions in terms of facilities, services and opportunities and so are less attractive places for young people to remain. However, some regions that are rural in terms of physical characteristics, such as landscape and the historical dominance of agriculture, are manifesting increasingly urban characteristics in terms of facilities and lifestyles (Kauppila, 2011; Möller et al., 2014). The distinctions between 'rural' and 'urban' are increasingly blurred, but this further marginalises remote regions with poor access, services, facilities, and income opportunities. Rural regions are increasingly being forced to diversify in response to these changes, and tourism offers one opportunity to try and create new employment and entrepreneurship opportunities that might support the sustainability of rural societies.

TOURISM AND RURAL COMMUNITIES

Tourism has often been presented as a 'magic panacea' (Miller et al., 2010, p. 10) for the problems that many rural regions face; a way to stimulate local economies, diversify employment and business opportunities, and support and maintain vibrant communities. The multifunctional countryside where production and consumption run side by side (McAreavey & McDonagh, 2011) is 'a rural world where the certainties of agricultural production as the traditional "rural hub" are giving way to a much more polyvalent rural scene and regulatory structure' (Marsden, 1998, p. 107). This provides opportunities for tourism development.

Sharpley and Sharpley (1997, p. 7) define rural tourism as 'a state of mind whereby tourists are actively seeking a leisure experience in a rural environment'. Rural regions often have rich and varied resources, which Garrod et al. (2006) call 'countryside capital', and which can be capitalised on as assets in rural tourism, including natural (e.g., wildlife), built (e.g., settlements) and social (e.g., local traditions) resources. Not all rural places will have valuable and attractive assets to draw upon for tourism development, and so the quality of rural tourism experiences depend upon on the quality of countryside capital that supports them.

A variety of rural regions around the world have benefited from tourism development. Henriksen and Halkier (2014) show how food has been developed as a tourist attraction in rural Denmark. Dimitrovski et al. (2012) found that tourism can help preserve regional identity, heritage, and the environment, as well as encouraging young people to stay in the community in rural Serbia. Rural tourism has been a catalyst for industrialisation and economic growth in the Mediterranean basin (Gülcan et al., 2009), and for revitalising villages in rural China (Gao & Wu, 2017) and Japan (Vafadari et al., 2014). Tourism has stimulated small business development in many regions (Kallmuenzer & Peters, 2018; Mura & Ključnikov, 2018).

Rural tourism is often touted as a powerful way of overcoming poverty in the Global South. Azmi (2014) illustrates the willingness of local communities to engage in and support rural tourism development in Sri Lanka, while Akyeampong (2011) found that local residents in Ghana were also positive about the potential of rural tourism to help improve their lives. Tourism can also be an important factor in sustaining traditional practices of indigenous communities, as Aminudin (2014) demonstrated in her research with indigenous communities in the tropical rain forests of Malaysia.

However, rural tourism is not in fact a panacea and the reality is not universally positive, often resulting in conflict and resentment as the benefits are not shared amongst community members (Miller et al., 2010). Frequently the bulk of benefits, particularly in economic terms, go to transnational corporations and non-local business elites. As Torres and Mornsen (2005) note about rural tourism development in Mexico, tourism may sometimes exacerbate poverty. Although tourism can bring valuable work opportunities to rural regions with few other options, the jobs are often low paid and seasonal (Wilson et al., 2001). Without community involvement in planning and development, the benefits of rural tourism are unlikely to be felt by most local community members, which can lead to feelings of disempowerment and resentment, sometimes resulting in conflict (Sulle et al., 2014).

There is thus wide evidence to illustrate the ambiguities of rural tourism development for local communities, facilitating new opportunities for some whilst at the same time marginalising others. When examined through a gender-aware lens, these ambiguities take on extra significance as it becomes apparent that rural tourism is a gendered phenomenon, both reflecting and shaping wider gender norms and practices.

GENDER, TOURISM AND RURAL COMMUNITIES

Tourism is a highly gendered sector, although it is rarely acknowledged as such in research, practice, and development. Ferguson and Alarcón (2015) suggest that gender issues are often sidelined, and dominant paradigms, such as those about sustainability and development, are rarely challenged or considered through a gender-aware lens. They argue that gender perspectives, equality and women's empowerment need to be embedded in all aspects of tourism. Organisations like the UNWTO often fail to recognise the centrality of gender equality to achieving wider objectives, like the Sustainable Development Goals (SDG) (Alarcón & Cole, 2019). This hampers efforts to achieve broader goals, such as poverty alleviation and climate protection, as gender equality is a necessary step towards a more just and equitable world. The following sections detail some important issues in relation to rural tourism development when viewed through a gender-aware lens, focusing on three key areas: work; active involvement of women; and intersectional analyses.

Rural Tourism, Work and Gender

As noted above, one of the purported benefits of rural tourism development is that it can bring new jobs to regions through diversifying away from agriculture and production. While both men and women can and do work in tourism, many such jobs are dominated by women (Costa et al., 2017a). Petrzelka et al. (2005) suggest that research on rural tourism tends to homogenise groups, ignoring or trivialising women's experiences. In many areas, tourism is seen as 'women's work', while men continue to do farming and physical work, and the marginalisation of rural women contributes to a lack of focus on their experiences as workers in the planning and implementation of rural tourism development. Costa et al. (2017b, p. 66) argue that 'the concept of "work" is implicitly gendered and reinforces the separation between the public and private sphere'. This is linked to the division of labour in many societies, where 'women's work' is seen to be that allied to care, sustenance, and social reproduction. Many jobs in tourism replicate these features, such as cooking, cleaning, and hosting. However, as these are seen as roles more suitable for women than men, and more akin to the domestic sphere than the public world of work, many tourism jobs are poorly paid and of low status. Tourism work can thus provide employment opportunities for many women but may exacerbate existing divisions of labour and reinforce gendered hierarchies that value men's work over that of women (see also Je et al., 2020). These gendered divisions, while manifesting different nuances in different contexts, are evident in a wide range of regions throughout both the Global North and the Global South.

Gentry's (2007) research illustrates this, as she found that tourism development in Belize did offer local women new employment opportunities. However, at the same time it entrenched gender inequalities related to pay, lack of education and status. Women may have been provided new opportunities to work outside the home, but this did not detract from expectations that women should be solely responsible for domestic labour as well, leading to a 'double workday'. Tourism work thus provided limited empowerment for these women, as their minimal economic gains were accompanied by punishingly long days and few gains in terms of status and autonomy.

Tucker (2022) noted similar ambiguities in her longitudinal research in rural Turkey. Over an extended period of engagement with a specific community, Tucker noticed some improvements in women's involvement in tourism work and associated access to higher levels of education. However, in many ways this helped entrench women in low-paid positions, with the double burden of home working as well as paid employment.

Cultural context is important to understanding the paradoxical nature of women's involvement in rural tourism work. Tucker and Boonabaana's (2012) research in Turkey and Uganda found that, in both communities, gender relations shifted over time, but slowly. Women in both areas experienced restrictions on their involvement in work and public spaces, as well as low pay and a lengthened workday, as they remained responsible for domestic work even when they were working in tourism businesses outside the home.

In addition to providing opportunities to work in tourism businesses, rural tourism development can also provide entrepreneurship opportunities for local residents. In some regions, women have capitalised on this through the production of crafts and souvenirs to sell to tourists, something that can be fitted in around their ongoing care responsibilities in the family and wider community (Phommavong & Sörensson, 2014). However, this work remains poorly paid and low in status, and so can contribute to further entrenching gender inequalities in such communities.

Figueroa-Domecq et al. (2020) explored the ways in which tourism entrepreneurship is gendered, through research conducted in Spain. They found that female entrepreneurs were more likely to consider sustainability issues and try to embed social and environmental sustainability into their entrepreneurship. In contrast, male entrepreneurs placed less emphasis on sustainability, focusing on economic rather than social and cultural issues. This suggests that the gendering of tourism entrepreneurship may have far-reaching consequences for many rural communities as social and environmental factors – which are important to the sustainability of rural regions and societies – are more likely to be considered by women, whose work is positioned as lower status and attracts lower pay. There may therefore be less attention and fewer resources, including economic investment, directed towards initiatives that try to ensure rural tourism development is oriented to supporting local communities and natural environments.

Recognising rural tourism work as gendered does not mean focusing solely on its impacts on women. Brandth and Haugen (2005) explored how shifts from traditional logging to tourism-related work in rural Norway shaped local gender relations. Rural men must now combine traditional rural masculinity – physical toil, hunting, etc. – with more urban and feminised elements related to tourism – service, hospitality, etc. Brandth and Haugen suggested that as rural men became part of the tourism product it seemed to become more important to them to show they can master the rural better than urban men. In their non-work time away from tourists, many of the men in their study sought leisure activities and environments, which enabled them to show their difference from both urban men and from women, emphasising their ongoing involvement in traditionally masculine practices such as hunting. In many ways, far from challenging traditional gender relations by incorporating so-called feminised traits into local masculinities, the development of tourism work actually reinforced differences between masculine and feminine identities, as well as between rural and urban masculinities. Rural tourism development thus impacts on gender relations in local communities in various complex ways.

New and enhanced work opportunities are one of the most celebrated benefits of tourism for rural communities, hailed as a way to overcome many of the problems associated with the decline of agriculture and loss of work. However, work is gendered and in rural communities, tourism-related employment and entrepreneurial opportunities should be examined through a gender lens in order to consider who may benefit and who may lose out, and if and how rural tourism development may either entrench or reshape local gender relations.

Involving Women in Rural Tourism

In order for rural tourism to benefit local communities it is important to ensure that those communities are involved in development activities from the beginning. However, too often the idea of 'local community' representation is homogenised, failing to account for differences within communities and often resulting in the further marginalisation of women (Petrzelka et al., 2005; Phommavong & Sörensson, 2014). The gendering of rural communities means that the interests of agriculture and production – masculine domains – are often seen as the most important community issues and the interests and voices of women community members are not considered.

A failure to involve women, and female-dominated interests, can mean that rural tourism development policies are skewed towards male interests. An example of this can be seen in the development of equestrian tourism. Equestrian tourism is a female-dominated domain in many European countries and a failure to consider the needs and requirements of the equestrian sector in rural development plans can be seen in part as a lack of inclusion of women in local decision making. For example, Andersson and Lehtola (2011) found that the existing regulatory frameworks in rural Finland worked to sideline the interests of the female-dominated equine sector in discussions and planning for rural development. This marginalisation of women's interests may not have been deliberate but was an outcome of the gendered structures in place that prioritised male interests as representative of community interests. Unless gender is integrated into all aspects of the design and implementation of rural tourism development, such outcomes are likely to continue, and women's interests and priorities may not be recognised (Ferguson & Alarcón, 2015).

The exclusion of women's voices and interests may occur due to community norms and gender power structures. Lenao's (2014) discussion of rural tourism development at Lekhubu Island, Botswana, highlighted the importance of the *kgotla* as a space and forum through which to consult and involve local community members in tourism development. However, the *kgotla* reflects the patriarchal nature of Botswanan society and so privileges elderly male voices and marginalises women, young people, disabled people, and other minority groups. In such ways, women can easily be excluded from involvement in rural tourism development through existing community gender power structures, meaning that decisions and development policies and practices may not take women's interests and situated knowledge into account, strengthening existing gender power structures and relations further.

The exclusion of women from tourism development, and the silencing of women's concerns and priorities, is problematic in terms of social justice and ensuring that rural tourism development benefits everyone in local communities. However, it also risks overlooking valuable knowledge and expertise about the local area and natural resources, knowledge that could be crucial for sustainable tourism development. Scheyvens (2000) argues that in many societies, particularly in rural Asia, Africa, and the Pacific, women and girls have greater interactions with the natural environment than do men, as they are responsible for the collection of water, fodder, and fuel for the community. This not only gives women greater knowledge of the natural environment, but also a strong interest in protecting it. This knowledge could be mobilised more effectively in rural development efforts to the greater benefit of local community members, tourists, and the natural environment. Local gender power relations often limit women's involvement in decision making, and a lack of gender focus in development policies and strategies means that this valuable knowledge is not being used to full advantage.

Kato (2019) notes the value of the intimate knowledge of the environment that some women develop, through her examination of the role of *ama* – women divers in Japan. She argues that 'women's power lies in their knowledge and their ways of knowing' (Kato, 2019, p. 940) and this could be mobilised to support sustainable tourism development. The *ama* have a close and embodied relationship with the ocean, which enables them to notice environmental changes, to develop sensitive practices, and to respect the natural environment for long-term sustainability. However, due to gendered structures, and hierarchical knowledge and value systems, this embodied expertise is overlooked and, with it, important insights on how best to balance tourism development with environmental protection.

Mkono et al. (2021) provide an example of a rural tourism project that does explicitly focus on involving women through their examination of social media responses to an all-female anti-poaching unit in Zimbabwe. The goal of the project is to empower women and draw on their situated knowledge of the local environment, wildlife, and culture. Although the project is not without problems in relation to both race and gender, as it is led by a white man who imposes some strict requirements on the female rangers (such as veganism, which is alien to local cultures), it does illustrate the subversive potential of actively involving women in rural tourism in order to tackle long-standing problems, such as poaching.

It is widely accepted that for rural tourism to benefit local communities, to minimise conflict and ensure local support, local people need to be involved all aspects of development from planning to implementation. However, the 'community' is not a homogeneous group, and applying a gender lens to tourism development and research reveals that women are often excluded from planning and decision making, and their knowledge and expertise are overlooked. This works to further entrench gender inequalities in and through rural tourism. Gender power relations often intersect with other forms of inequality to further marginalise women in rural tourism development.

Intersectional Approaches to Gender and Rural Tourism

The focus of this chapter is on *gender* and rural tourism, and gender equality is an essential aspect of efforts towards more sustainable, just, and equitable tourism development (Alarcón & Cole, 2019). However, while direct and explicit focus on gender is needed in both research and practice, gender does not operate in isolation from other axes of power and so intersectional analyses are also important.

Numerous studies show how, in different contexts and in slightly different ways, gender intersects with class, ethnicity, and a variety of other factors to shape the experiences of men and women within rural tourism. Phommavong and Sörensson (2014) examine the intersections of ethnicity and gender in rural tourism in Laos. They focus on Akha women, an ethnic group that has an elder, clans and indigenous cultural beliefs, such as in ancestral sacrifice and spirits, which have strong gendered implications. The beliefs and practices of the Akha community influence gendered divisions of labour in tourism and limit women to certain culturally acceptable roles, such as massage, which are low paid and low status. Cultural beliefs influence education within the community, and girls and women are generally less well educated than men and boys, and so less able to work in better paid positions such as guiding. Phommavong and Sörensson (2014) suggest there is a contradiction between cultural preservation in community-based tourism projects and gender equality in that the ethnic-specific practices seen as authentic and attractive by tourists often contribute to the preservation of unequal gender relations and practices that prevent women from gaining many of the economic and social benefits of rural tourism development (see also Trupp & Suntana, 2017).

Religion, and its cultural implications, also often intersects with gender to shape rural tourism development and experiences. Alrwajfah et al. (2020) illustrate how women in the Middle East and Muslim countries may experience particular barriers to involvement in rural tourism development. Drawing on research conducted in Jordan, they suggest that social stigma can be a powerful barrier as tourism work is seen as inconsistent with local Muslim traditions. Muslim women avoid mixing with foreign men (tourists) and so tourism work can actually damage a woman's reputation, meaning that women are unlikely to be involved or benefit directly from rural tourism development. Tucker and Boonabaana (2012) found that both Islam and Christianity intersect with gender to shape women's involvement (or lack of) in rural tourism, prescribing expected roles for women and invoking stigma and shame to limit women's involvement in tourism work.

Intersectional analyses are important for exploring the complexities of gender and rural tourism in a variety of geographic, social and cultural contexts, and draw attention to the complexity of gender as it shapes rural tourism.

CONCLUSION

Rural tourism is gendered, and both reflects and helps (re)produce the gendering of rural societies more broadly. Research and practice have tended to overlook the importance of gender as a fundamental aspect of rural tourism, and a lack of gender focus has often led to the marginalisation of women and girls. A wide variety of factors contribute to this, including the gendered nature of tourism work, the lack of involvement of women in all stages of rural tourism development and the intersections of gender with other axes of power in marginalising women in and through rural tourism. The issues identified in this chapter are by no means an exhaustive overview and further research is needed to understand better 'the complex and interlocking factors that result in the continued disempowerment of women in tourism and that have defied any sustainable transformation' (Chambers & Rakic, 2018, p. 81). What is apparent is that gender must be a core consideration in all rural tourism development activities, and at all stages of implementation, if greater gender equality is to be achieved and the benefits of rural tourism are to be maximised for all members of the community.

REFERENCES

Akyeampong, O.A. (2011). Pro-poor tourism: Residents' expectations, experiences and perceptions in the Kakum National Park Area of Ghana. *Journal of Sustainable Tourism, 19*(2), 197–213.

Alarcón, D.M., & Cole, S. (2019). No sustainability for tourism without gender equality. *Journal of Sustainable Tourism, 27*(7), 903–919.

Alrwajfah, M.M., Almeida-García, F., & Cortés-Macías, R. (2020). Females' perspectives on tourism's impact and their employment in the sector: The case of Petra, Jordan. *Tourism Management, 78*, 104069.

Aminudin, N. (2014). Sustainable tourism development and indigenous communities in Malaysia. In K. Dashper (ed.), *Rural Tourism: An International Perspective* (pp. 267–282). Newcastle upon Tyne: Cambridge Scholars Publishing.

Andersson, K., & Lehtola, M. (2011). Regulating the new equine industry in Finland. Wicked problems, governance models and gendered power structures. *Sociologia Ruralis, 51*(4), 387–403.

Argent, N. (2011). Trouble in paradise? Governing Australia's multifunctional rural landscapes. *Australian Geographer, 42*(2), 183–205.

Azmi, F. (2014). Will the 'dancing waves' bring harmony to the lives of the poor? Making a case for pro poor tourism development planning in Arugam Bay, Sri Lanka. In K. Dashper (ed.), *Rural Tourism: An International Perspective* (pp. 189–209). Newcastle upon Tyne: Cambridge Scholars, 189–209.

Bock, B.B. (2015). Gender mainstreaming and rural development policy; the trivialisation of rural gender issues. *Gender, Place & Culture, 22*(5), 731–745.

Brandth, B., & Haugen, M.S. (2005). Doing rural masculinity – from logging to outfield tourism. *Journal of Gender Studies, 14*(1), 13–22.

Bryant, L., & Pini, B. (2009). Gender, class and rurality: Australian case studies. *Journal of Rural Studies, 25*(1), 48–57.

Bye, L. M. (2009). 'How to be a rural man': Young men's performances and negotiations of rural masculinities. *Journal of Rural Studies, 25*(3), 278–288.

Chambers, D., & Rakic, T. (2018). Critical considerations on gender and tourism: A postscript. *Tourism, Culture and Communication, 18*(1), 81–84.

Costa, C., Bakas, F.E., Breda, Z., & Durão, M. (2017a). 'Emotional' female managers: How gendered roles influence tourism management discourse. *Journal of Hospitality and Tourism Management, 33*, 149–156.

Costa, C., Bakas, F.E., Breda, Z., Durão, M., Carvalho, I., & Caçador, S. (2017b). Gender, flexibility and the 'ideal tourism worker'. *Annals of Tourism Research, 64*, 64–75.

Dashper, K. (ed.) (2014a). *Rural Tourism: An International Perspective*. Newcastle upon Tyne: Cambridge Scholars Publishing.

Dashper, K. (2014b). Rural tourism: Opportunities and challenges. In K. Dashper (ed.), *Rural Tourism: An International Perspective* (pp. 1–20). Newcastle upon Tyne: Cambridge Scholars.

Dimitrovski, D.D., Todorović, A.T., & Valjarević, A.D. (2012). Rural tourism and regional development: Case study of development of rural tourism in the region of Gruţa, Serbia. *Procedia Environmental Sciences, 14*, 288–297.

Ferguson, L., & Alarcón, D.M. (2015). Gender and sustainable tourism: Reflections on theory and practice. *Journal of Sustainable Tourism, 23*(3), 401–416.

Figueroa-Domecq, C., Kimbu, A., de Jong, A., & Williams, A.M. (2020). Sustainability through the tourism entrepreneurship journey: A gender perspective. *Journal of Sustainable Tourism, 30*(7), 1562–1585.

Gao, J., & Wu, B. (2017). Revitalizing traditional villages through rural tourism: A case study of Yuanjia Village, Shaanxi Province, China. *Tourism Management, 63*, 223–233.

Garrod, B., Wornell, R., & Youell, R. (2006). Re-conceptualising rural resources as countryside capital: The case of rural tourism. *Journal of Rural Studies, 22*(1), 117–128.

Gentry, K.M. (2007). Belizean women and tourism work: Opportunity or impediment? *Annals of Tourism Research, 34*(2), 477–496.

Gülcan, Y., Kuştepeli, Y., & Akgüngör, S. (2009). Public policies and development of the tourism industry in the Aegean region. *European Planning Studies, 17*(10), 1509–1523.

Henriksen, P.F. & Halkier, H. (2014). Feeding countryside tourists: Exploring collaborative actor strategies in rural Denmark. In K. Dashper (ed.), *Rural Tourism: An International Perspective* (pp. 250–256). Newcastle upon Tyne: Cambridge Scholars Publishing.

Je, J.S., Khoo, C., & Yang, E.C.L. (2022). Gender issues in tourism organisations: Insights from a two-phased pragmatic systematic literature review. *Journal of Sustainable Tourism, 30*(7), 1658–1681.

Kallmuenzer, A., & Peters, M. (2018). Entrepreneurial behaviour, firm size and financial performance: The case of rural tourism family firms. *Tourism Recreation Research, 43*(1), 2–14.

Kato, K. (2019). Gender and sustainability – exploring ways of knowing – an ecohumanities perspective. *Journal of Sustainable Tourism, 27*(7), 939–956.

Kauppila, P. (2011). Cores and peripheries in a northern periphery: A case study in Finland. *Fennia – International Journal of Geography, 189*(1), 20–31.

Lenao, M. (2014). Packaging culture and heritage for tourism to improve rural lives at Lekhubu Island, Botswana. In K. Dashper (ed.), *Rural Tourism: An International Perspective* (pp. 172–188). Newcastle upon Tyne: Cambridge Scholars Publishing.

Little, J. (2002). Rural geography: Rural gender identity and the performance of masculinity and femininity in the countryside. *Progress in Human Geography, 26*(5), 665–670.

Little, J. (2007). Constructing nature in the performance of rural heterosexualities. *Environment and Planning D: Society and Space, 25*(5), 851–866.

Little, J., & Panelli, R. (2003). Gender research in rural geography. *Gender, Place and Culture, 10*(3): 281–289.

Marsden, T. (1998). New rural territories: Regulating the differentiated rural spaces. *Journal of Rural Studies*, *14*(1), 107–117.
McAreavey, R., & McDonagh, J. (2011). Sustainable rural tourism: Lessons for rural development. *Sociologia Ruralis*, *51*(2), 175–194.
Miller, E., Van Megen, K., & Buys, L. (2010). Impacts and opportunities: Resident's views on sustainable development of tourism in regional Queensland, Australia. *Journal of Tourism Challenges and Trends*, *31*, 175–194.
Mkono, M., Rastegar R., & Ruhanen, L. (2021). Empowering women to protect wildlife in former hunting tourism zones: A political ecology of Akashinga, Zimbabwe. *Journal of Sustainable Tourism*. doi:10.1080/09669582.2021.1900205.
Möller, P., Thulemark, M., & Engström, C. (2014). Urbanity and rurality in a tourism context: Exploring the myth of vivid cities and sleepy villages. In K. Dashper (ed.), *Rural Tourism: An International Perspective* (pp. 22–40). Newcastle upon Tyne: Cambridge Scholars Publishing.
Mura, L., & Ključnikov, A. (2018). Small businesses in rural tourism and agro tourism: Study from Slovakia. *Economics & Sociology*, *11*(3), 286–300.
Petrzelka, P., Krannich, R.S., Brehm, J., & Trentelman, C.K. (2005). Rural tourism and gendered nuances. *Annals of Tourism Research*, *32*(4), 1121–1137.
Phommavong, S., & Sörensson, E. (2014). Ethnic tourism in Lao PDR: Gendered divisions of labour in community-based tourism for poverty reduction. *Current Issues in Tourism*, *17*(4), 350–362.
Pitkänen, K. (2008). Second-home landscape: The meaning(s) of landscape for second-home tourism in Finnish Lakeland. *Tourism Geographies*, *10*(2), 169–192.
Pritchard, A. (2018). Predicting the next decade of tourism gender research. *Tourism Management Perspectives*, *25*, 144–146.
Saugeres, L. (2002). The cultural representation of the farming landscape: Masculinity, power and nature. *Journal of Rural Studies*, *18*(4), 373–384.
Scheyvens, R. (2000). Promoting women's empowerment through involvement in ecotourism: Experiences from the Third World. *Journal of Sustainable Tourism*, *8*(3), 232–249.
Sharpley, R., & Sharpley, J. (1997). *Rural Tourism: An Introduction.* London: International Thomson Business Press.
Shortall, S. (2006). Gender and farming: An overview. In B.B. Bock & S. Shortall (eds), *Rural Gender Relations: Issues and Case Studies* (pp. 19–26). Wallingford: CABI Publishing.
Sulle, E., Banka, H., & Ntalwila, J. (2014). Wildlife tourism experiences: Case studies from rural Tanzania. In K. Dashper (ed.), *Rural Tourism: An International Perspective* (pp. 210–228). Newcastle upon Tyne: Cambridge Scholars Publishing.
Sutherland, L.A. (2012). Return of the gentleman farmer? Conceptualising gentrification in UK agriculture. *Journal of Rural Studies*, *28*(4), 568–576.
Torres, R., & Momsen, J. (2005). Planned tourism development in Quintana Roo, Mexico: Engine for regional development or prescription for inequitable growth? *Current Issues in Tourism*, *8*(4), 259–285.
Trupp, A., & Sunanta, S. (2017). Gendered practices in urban ethnic tourism in Thailand. *Annals of Tourism Research*, *64*, 76–86.
Tucker, H. (2022). Gendering sustainability's contradictions: Between change and continuity. *Journal of Sustainable Tourism*, *30*(7), 1500–1517. doi:10.1080/09669582.2020.1839902.
Tucker, H., & Boonabaana, B. (2012). A critical analysis of tourism, gender and poverty reduction. *Journal of Sustainable Tourism*, *20*(3), 437–455.
United Nations (2018). *The speed of urbanization around the world.* Population Facts. United Nations Department of Economic and Social Affairs. Retrieved March 31, 2021, from https://population.un.org/wup/Publications/Files/WUP2018-PopFacts_2018-1.pdf.

Vafadari, K., Cooper, M., & Nakamura, K. (2014). Rural tourism and regional revitalisation in Japan: A spotlight on Satoyama. In K. Dashper (ed.), *Rural Tourism: An International Perspective* (pp. 337–360). Newcastle upon Tyne: Cambridge Scholars Publishing.
Wilson, S., Fesenmaier, D.R., Fesenmaier, J., & Van Es, J.C. (2001). Factors for success in rural tourism development. *Journal of Travel Research*, *40*(2), 132–138.

12. Food, tourism, and rural community development: close encounters of the nourishing kind?

Jennifer Sumner

INTRODUCTION

Food is a vital human need and rural communities have always been at the centre of food production. As public interest in food has grown over the last few decades, tourism has stepped up to capture the zeitgeist. Tourists have been drawn to rural communities for such food-related experiences as farm stays, wine tours, food festivals and fall fairs. The impacts of this interface have been mixed, leading to questions of what contributes to rural community development and what detracts from it.

This chapter will examine tourism and rural community development through the lens of food. It will begin with a brief discussion of food, then move to food-related tourism, with a focus on rural communities. The chapter will then address issues of rural community development and the effects of food-related tourism on this development, concluding with some recommendations for improving this food–community–tourism interface.

FOOD

Food is central to human existence – without it, we cannot survive. But food is much more than merely sustenance. It is also

> a symbol, a product, a ritual object, an identity badge, an object of guilt, a political tool, even a kind of money. Food determines how tall we are, how healthy, the extent of our civic peace, the sorts of jobs we hold, the amount of leisure we enjoy, the crowding of our cities and suburbs, what we look for in life, how long we look to live – all of that and much more. (Reardon, 2000, p. 1)

As a result of its multifaceted aspects, food attracts the attention of people for a wide variety of reasons and many seek an "up close and personal" experience with food that they cannot replicate at home or in the supermarket. Some travel to Paris to partake of French cuisine; others fly to Rome to enrol in cooking classes. As Telfer and Hashimoto (2004, p. 159) have observed, the foodways of every country directly reflect its geography, history, culture and people, and in this way "provide a special sensory window into the culture of a country for tourists who are willing to taste". In short, food represents one of the essential elements of any tourist experience and the recognition of the role of food in tourism is growing.

FOOD TOURISM

Food tourism brings together two dynamic terms to create an evocative concept that combines planned travel with intentional eating. When discussing food tourism, Hall and Sharples (2004) have argued that it is important to emphasize the difference between tourists who simply consume food during their travels and those whose interest in food influences their activities, behaviours and destination. With this in mind, they have defined food tourism as "visitation to primary and secondary food producers, food festivals, restaurants and specific locations for which food tasting and/or experiencing the attributes of [a] specialist food production region are the primary motivating factor for travel" (p. 10). Over time, such descriptions have undergone a "cultural turn" to include discussions of food and culture, with an emphasis on physical experiences motivated by a desired engagement with local foods (Ellis et al., 2018). As such, food tourism explores new cultures and lifestyles, which can help to preserve and safeguard local traditions through a heavy reliance on experience in the place where food is produced (Berno & Fusté-Forné, 2020).

Place is central to food tourism – tourists visit a place to experience the food. For Hall and Sharples (2004, p. 10), a critical factor of food tourism is its spatial fixity – it is literally "the consumption of the local and the consumption and production of place". Ellis et al. (2018) go one step further, seeing food tourism as a presentation of history and place – like haggis and Scotland or kimchi and South Korea. While food tourism can and does occur in urban settings, this chapter will emphasize food tourism as a presentation of rural histories and rural places.

FOOD TOURISM IN RURAL COMMUNITIES

Rural communities are ground zero for food production. Whether farming communities, fishing communities or isolated communities, food is the focus around which many were built. As these communities experience the ongoing depredations of the neoliberal economy, food stands out as a new direction for rural development and, as Welsh and MacRae (1998) have explained, a nexus for rural/urban relations. One of the forms these relations can take is food tourism.

Like rural tourism in general, food tourism is a diverse activity underpinned by a sense of rural place (Jepson & Sharpley, 2015) and understood as a rural regeneration and conservation tool (Lane & Kastenholz, 2015). Rural communities offer a cornucopia of food tourism opportunities, from farm tours and artisanal kitchens to rural restaurants and wineries. These opportunities are based on the concept of terroir – "the notion that the taste of food is affected by a region's soil, its climate, its landscape, its hours of sunlight – that the place where food is made is inextricably linked to what it is" (Elton, 2010, p. 167). While terroir has always been seamlessly built into the food associated with place-based rural communities, it has been facing erasure with the advent of the placeless industrial food of the neoliberal economy. This threat, in turn, is being countered by the local food movement – a coalition of people and organizations interested in the production and consumption of local food. Among its relocalization strategies, it promotes local food production and encourages farmers' markets (Broadway, 2017).

Keeping in mind that food tourism is both the experience of consumption and the consumption of experience (Hall & Sharples, 2004), five examples will help to illustrate the sheer range of rural food tourism: farmers' markets, 100 mile stores, food trails, food festivals and farm-to-table restaurants.

Farmers' Markets

Although their origins are lost in time, farmers' markets are a permanent fixture of contemporary rural and urban foodscapes. In North America, their continued existence was seriously compromised by the rise of supermarkets in the second half of the 20th century, but markets have experienced a resurgence through government initiatives and what Cummings et al. (1998, p. 7) described as the desire for a shopping experience "closer to the food producer and the community – a more personal approach". Whether seasonal or year-round, farmers' markets are on the front lines of the (re)localization of food and understood as part of the local food supply chain. In this way, food tourism actively supports the local food movement and its associations with freshness, proximity and sustainability.

At their most basic, farmers' markets involve the selling food and other items directly to consumers by those who grew, reared or produced these goods (LaTrobe, 2001) (although some markets have expanded to include "resellers" – those who buy from farmers or food terminals and then sell at markets). Toronto's St. Lawrence Market, Ottawa's ByWard Market and the Halifax Seaport Farmers' Market are just some of the Canadian examples of venues where farmers bring their produce to consumers in the city. Consumers also enjoy travelling to rural markets as a day trip or as part of a longer excursion. In addition to providing opportunities to buy fresh produce and interact with farmers, the atmosphere of farmers' markets is particularly appealing to visitors looking for authentic food experiences, reflecting local lives and livelihoods and often offering a festive environment in which to experience the local culture (Jolliffe, 2006).

The St. Jacobs Market in St. Jacobs, Ontario, provides an illustration of food tourism in rural communities, although not without problems, as will be discussed later. Established in 1975 by a group of farmers, the market has become a destination for tourists from Canada, the United States and Europe (Davis & Desmond, 2013). Open all year round, it offers fresh local fruit and vegetables, quality meats and cheeses, locally produced maple syrup, honey, preserves, baking, authentic cuisine from many cultures, imported produce, various gluten-free and organic options, local crafts and housewares (St. Jacobs Market, 2020). The market's website notes:

> Still home to the largest population of Old Order Mennonites in Canada, you'll often see farmers on local roads traveling to the market by horse & buggy. In season, say hello to small farm animals, enjoy some kiddie rides, or ride a heritage railway to nearby St. Jacobs. (St. Jacobs Market, 2020)

Similarly, the York Farmers' Market in the village of Thornhill, Ontario, hosts full-time vendors whose wares attract visitors from nearby Toronto. Built in 1953, it offered a central location where local farmers could sell their produce, as well as baked goods, preserves and crafts. While the demographics of the village have changed in the intervening years, the market has remained much the same, although it is now open four days per week (York Farmers' Market, 2020), attesting to its staying power as a site of food tourism.

In a more remote location, the Fireweed Community Market in Whitehorse, which celebrates Yukon culture and local food, represents part of the emerging food tourism trends in the Circumpolar North (de la Barre & Brouder, 2013). The market is run by the Fireweed Community Market Society, a non-profit organization with a volunteer board, part-time and seasonal staff, and over 170 members whose fees support the market. According to its website, the Fireweed Community Market Society promotes

> the production and consumption of Yukon products. We offer authentic experiences to residents and tourists by supporting local entrepreneurs and small businesses and directly connecting consumers with producers of quality goods that are locally grown, raised, wild-harvested, or hand-crafted. (Fireweed Community Market Society, 2020)

All three examples highlight the dual function of an increasing number of rural farmers' markets: to serve the local population – both producers and consumers – and to attract food tourism.

100 Mile Stores

Another outcome of the growing local food movement is 100 mile stores, which took their inspiration from Alisa Smith and J.B. MacKinnon's (2007) 100-mile diet, a food regimen composed entirely of products found within 100 miles of the consumer. These innovative retail outlets only offer food products from within a 100-mile radius, drawing on the output of farmers and processors from the region. Mair and Sumner (2019) argue that 100 mile stores herald a change in the ways people relate to food. "While farmers' markets can include products brought in from around the world, and tourist shops may offer some local products as part of their retail mix, 100 mile stores are exclusively dedicated to local food" (p. 148). In other words, they deliberately buy food grown, raised, and/or processed by local producers, making them prime destinations for food tourism. The 100 Mile Store in the rural community of Creemore, Ontario, provides an example of such a destination.

Opened in 2008, the 100 Mile Store stocks local vegetables, fruit, nuts, meat, dairy and eggs, as well as value-added products such as jams, breads, soaps, desserts and food-related crafts – all sourced from suppliers who are situated within approximately 100 miles of the store. The store complements the rural tourism opportunities already offered by Creemore. For example, Mair and Sumner (2019) describe how the village hosts festivals, maintains its heritage buildings, features upscale boutiques, boasts a number of pubs and restaurants, and offers accommodation, as well as being a fully functioning rural community with a post office, hardware store, grocery store, weekly newspaper and the renowned Creemore Springs Brewery. In addition, it is the gateway to skiing, snowshoeing, cycling and hiking opportunities in the region. These amenities attract tourists and the 100 Mile Store contributes to the burgeoning food tourism in the area.

Food Trails

Also referred to as food routes or taste trails, food trails are designated circuits that feature different places of alimentary interest, such as wineries, bakeries, dairies, restaurants or farm shops. They can feature one product (e.g., butter tarts), a family of products (e.g., cheeses) or a range of products (e.g., local cuisine/specialties) that define a region. Broadway (2017) explains how food trails can be used to market a particular destination, increase the number of visitors or promote local products and/or producers. They can also take different forms, including "bus tours, walking tours around markets and food establishments, and self-guided tours" (p. 472).

Food trails can offer a wide range of opportunities for rural communities. For example, Meyer-Czech (2004, p. 156) suggests that they can help to position and brand a region and create a feeling of togetherness among its inhabitants: "strategic partnerships are formed, farmers can sell more products, and gastronomy and accommodation benefit by increasing numbers of tourists". Broadway (2017) agrees, noting that food trails support local economies, sustain traditional artisanal food production and processing, and enhance the visitor experience.

The Bala Cranberry Festival (2021) in the province of Ontario includes a cranberry route, described as a culinary trail "with 19 crantastic stops across Muskoka". This food trail contributes to the overall success of the festival and to the rural community of Bala. According to its website, the festival not only extends the tourist season but also improves the quality of life in the community by providing financial assistance to organizations and individuals in need. This has resulted in support for hundreds of local community groups and non-profit associations, as well as the disbursement of over $100,000 in scholarships (ibid.).

Food is also part of the Great Spirit Circle Trail on Manitoulin Island, which provides an opportunity to bring people together in a communal fashion to learn about and experience Indigenous foodways (Indigenous Tourism Ontario, 2020). An example of Indigenous rural tourism, it includes picking, boiling and tasting teas made of local plants, while hearing stories of the land, or making bannocks over an open fire and sampling local berries (Great Spirit Circle Trail, 2019).

Food Festivals

Food festivals involve communities publicly celebrating and promoting local food and identity (Lee & Arcodia, 2011, p. 356). As such, they offer food and entertainment in a festive atmosphere – an ideal environment for rural food tourism. Food festivals may be commercial endeavours, but they also "contribute to the conservation and development of regional identities by promoting local food assets, bringing people together, and linking local resources with global markets" (Kwiatkowski 2018, p. 217).

While food festivals in urban settings (such as A Taste of Little Italy or Taste of the Danforth in Toronto) can feature foods from many cultures, food festivals in rural settings showcase local food and local food producers, who can talk to people about local products, sustainable production methods, and terroir. Specific foods can also be highlighted at these food festivals, as in the garlic festival in Stratford or the Bala cranberry festival (both located in Ontario), where tourists can learn everything they ever wanted to know about the history, importance and preparation of the star attraction.

Some food festivals go beyond food and entertainment to send a critical message. One example was the grassroots food festival called Foodstock, which was organized in 2011 to raise money and awareness to resist the "mega-quarry" planned for over 2300 acres of prime farmland in south-central Ontario. Over 28,000 people gathered in the rain to sample local foods prepared by chefs from across Canada and to register their resistance to farmland destruction. Dubbed by the Canadian Chefs' Congress (Good Life Vancouver, 2014) as "the largest culinary protest in the history of Canada", this food festival helped to raise awareness about the future of food system sustainability. In the words of celebrity chef Michael Stadtlander, who helped to organize the event, "We have to protect land that can grow food. ... We're just at the dawn of localism, and after seeing the drought, hurricanes and floods that have happened in the United States, I think we had better secure where our food comes from" (quoted in Bain, 2011).

Farm-to-Table Restaurants

Farm-to-table restaurants are a form of destination dining, which involves travel to a particular location – near or far – in order to eat at an establishment that has a reputation for a unique dining experience. Farm-to-table restaurants offer locally grown food in either an urban or a rural setting. For example, Alice Waters' pioneering farm-to-table restaurant, Chez Panisse, is located in the city of Berkeley, California. Like the rest of the farm-to-table movement, these restaurants are responding to the malaise engendered by the global corporate food system and replacing the status quo with something more healthy, fair, just and delicious (Benjamin & Virkler, 2016). While the formats may vary from fine dining to popular price and fast-casual, these authors propose that all farm-to-table restaurants share a few core principles: a focus on fresh, wholesome, flavourful food; a commitment to supporting local farmers and producers; a commitment to supporting sustainable farming and animal husbandry practices; and an understanding of the importance of supporting the local economy (p. 97).

Rural communities are a logical location for farm-to-table restaurants. An establishment near the rural community of Singhampton, Ontario, provides a glimpse of how a farm-to-table restaurant operates. Chef Michael Stadtlander's farm-to-table restaurant offers a rural fine dining experience that attracts food tourists from around the world. As the owner of Eigensinn Farm, Stadtlander loves both cooking and gardening. He grows vegetables and herbs, raises pigs, goats, sheep, rabbits, chickens, ducks and geese, forages for wild plants, makes maple syrup and extracts honey from two beehives. Described as being perhaps the chef with the deepest commitment to a uniquely Canadian cuisine (Jacobs, 2010), Stadtlander operates from a basic starting point:

> Our greatest philosophy is growing the vegetables the way we like to see them, fully ripe, fully grown, and you will never find any GMO on our farm. Our pigs take 14 months and not 6 to grow, our fowl are free range. They take a half a year to maturity, and yes, we lose about 20% to predators. All our grain is certified organic. (Eigensinn Farm, 2020)

According to Heinzman (2010, p. 204), Stadtlander is famous for creating extraordinary meals made up only of local fare and served in his living room, with city people paying "a pretty penny to experience this unconventional form of fine dining".

Less expensive options also exist, but operate from basically the same philosophy. For example, The Globe (2020) farm-to-table restaurant in the rural community of Rosemont, Ontario, declares that "The Globe's cuisine features the best of seasonal, fresh, local food and beverages. We are proud to produce the majority of our items on-site, including our rotisserie chicken, in-house cured meats, pickled veggies and preserves, and baked bread directly from our bakery."

These examples of food tourism do not occur in a vacuum; each one affects, in positive and/or negative ways, the development of the rural communities in which they occur.

FOOD TOURISM AND RURAL COMMUNITY DEVELOPMENT

Food and tourism are intimately connected with rural community development. The type and amount of food tourism can have cascading effects – planned and unplanned – not all of which are positive. To provide context for this discussion of food tourism and rural community development, we begin with the work of Herman Daly (1990), an ecological economist and former senior economist with the World Bank. He has pointed out that although development is often understood as growth, they are two separate concepts. For Daly,

> *to grow* means "to increase naturally in size by the addition of material through assimilation and accretion." *To develop* means "to expand or realize the potentialities of – that is, to bring gradually to a fuller, greater, or better state." In short, growth is quantitative increase in physical scale, while development is qualitative improvement or unfolding of potentialities. (p. 1)

Following Daly, rural community development involves the qualitative improvement or the unfolding of potentialities of rural communities. This does not necessarily entail growth, unless this growth contributes to some kind of qualitative improvement or enables the unfolding of potentialities. For example, adding a number of fast-food chain restaurants to a rural community to meet an anticipated increase in tourists would not qualify as development, but adding a variety of restaurants that serve local food and are owned and operated by different local people would.

With this distinction in mind, we turn to Canadian economist John Loxley (2007), whose work focused on community economic development (CED), with an emphasis on helping economic actors learn to conceptualize and develop their economic activity beyond the narrow confines of being simply individuals who operate within the market. CED initiatives work to foster "inclusive economic revitalization, access to capital for business development, local ownership of resources, job creation, poverty reduction, and environmental stewardship" (Marks & Toye, 2020, p. 17).

As a major theorist of CED, Loxley (2007) was particularly interested in examining different types of linkages, such as backward linkages and forward linkages. Backward linkages evaluate how one sector's *demands* can benefit and strengthen another sector (for example, the demand of the local restaurant sector for the products of local maple syrup producers). Forward linkages evaluate how one sector's *outputs* can connect to other sectors (for example, how the local restaurant sector sells premade lunches to a local tour operator to include in a tour package). In this way, we see how one component in the rural community matrix can link its activities forward and backward to create an economically stronger community. Or, in Daly's terms, how one component in the rural community matrix can link its activities forward and backward to provide qualitative improvements or unfold potentialities in the community. To paraphrase Sumner et al.'s (2014) work on co-operative food systems, the richer the linkages, the greater the impact on rural communities – linkages are a way in which the effects of rural food tourism can be identified, built and measured.

Loxley (2007) was also interested in the concept of leakages. While linkages measure how sectors are connected, leakages measure how the opportunity for value creation in various sectors is being lost (Sumner et al., 2014). In the case of rural communities, it means, for example, how value leaks out to non-local interests. For example, rural restaurants could order from urban food distributors, not local suppliers. Or wineries could sell souvenirs made in China in their boutiques, not locally made crafts.

Loxley (2007) added a final caveat to his analysis of linkages and leakages. He emphasized that we cannot think of these concepts simply in terms of products – we must include what he termed "supply-side factors of production" (p. 61), such as labour, capital and technology, which can have either local or corporate associations. In other words, "we must also be thinking in terms of local labour, local capital, and local technical capacity being nurtured and developed" (Sumner et al., 2014, p. 54).

Sumner et al. (2014) inject a third concept into the discussion of community economic development: leveraging. This term emerges from Loxley's idea that

> communities need both to articulate to government bodies at every level the value of "small scale production to meet local need" in economic terms in order to secure support through subsidies, and also, crucially, to articulate these demands as a united movement with the capacity for "collective action". (Sumner et al., 2014, p. 54)

In other words, rural community groups have to leverage their potential power collectively to influence public policy, as well as strengthen linkages and plug leakages.

Linkages, leakages, and leveraging are important concepts to consider when assessing how rural food tourism contributes to rural community development, particularly because, as Friedmann (1993) has observed, food has long been associated with wealth and power. In order to resist the incursions of wealth and power and aim for the qualitative improvement or the unfolding of potentialities in rural communities, it is important for those involved in rural community development to look beyond individual economic actors and envision a network of increasing linkages and decreasing leakages, while working collectively within and among rural communities to leverage their power and lobby all levels of government to influence public policy and secure support for rural food tourism.

The case of St. Jacobs, Ontario, provides a salutary warning of how rural food tourism can contribute to destroying the quality of life in a rural community – an example of growth over development, and a cautionary tale of diminished linkages, increased leakages and lopsided leveraging.

Begun as a rural service centre almost 200 years ago, St. Jacobs became a thriving community catering to the needs of Old-Order Mennonites, subsequently experienced long-term decline for many decades, and then reversed its fortunes with the establishment of the Stone Crock Restaurant in 1975 (Dahms, 1991), specializing in Pennsylvania Dutch cuisine. This kickstarted a phase of growth that transformed St. Jacobs into a thriving tourist centre, which in turn promoted more growth. In terms of food tourism, the Stone Crock Restaurant hosted 1000 customers on a good day within two years of opening (Mitchell, 1998) and the farmers' market grew to host one million visitors per year (Deschamps, 2015) – in a town with a population of 1500 people. By the early 1990s, Dahms (1991, p. 1) noted that the community "is excessively busy, some residents resent the tourist influx, and many Mennonites now go elsewhere to shop". As one of the local residents lamented, "tourism has spoiled what was once a tranquil village. It took our small town atmosphere away and left us with a circus", creating significant tensions between visitors and local residents that manifested around issues of parking, traffic congestion, exploitation of the Mennonites, loss of atmosphere, disrespect for property, safety concerns and litter (Mitchell, 1998, p. 281).

In his study of this rural community, Dahms (1991) speculated that over-development might ruin the basis of St. Jacobs' tourist appeal. Following Daly (1990), we can understand that St. Jacobs was experiencing growth, not development. There was certainly quantitative increase in opportunities for food tourism and in numbers of food tourists, but little qualitative improvement in the lives of most of the people who lived in St. Jacobs. Put another way, this growth did not significantly increase the number of linkages, or decrease the number of leakages, or leverage the power of the community. Most of the transformation in this rural community has been led by one individual, outside investors backed a railway link from the nearby city of Kitchener to St. Jacobs, and public-sector involvement in the changes was lacking (Mitchell, 1998; Mitchell & Coghill, 2000).

As Mitchell and Coghill (2000) observed, St. Jacobs evolved to meet the demands of the visiting population to the detriment of local consumers. Driven to a large extent by food tourism, this evolution threatened to destroy the basis upon which the tourism project itself was built, referred to by Mitchell (1998) as the rural idyll – an image of rural life that is happy, healthy and problem free, with a closely knit community nestled in the natural environment.

The example of St. Jacobs helps us understand how rural food tourism can easily reflect the deficiencies of tourism in general. Higgins-Desbiolles (2018) points out that tourism today has a problem because it is addicted to growth, which is incompatible with sustainability. This growth fetish, she observes, results in tourism killing tourism. She argues that "sustainable tourism necessitates a clear-eyed engagement with notions of limits that the current culture of consumerism and pro-growth ideology precludes" (p. 159). To achieve this, she advocates for reclaiming tourism for its higher purposes, by regulating and managing it for a better-balanced accounting with respect to fairness, ecological limits, human benefit and sustainable futures.

Food tourism can follow this advice by eschewing the unsustainable growth imperative and committing to development focused on fairness, ecological limits, human benefit and sustainable futures. Such development could be achieved through increasing linkages, decreasing leakages and encouraging leveraging, which can lead to qualitative improvement and the unfolding of potentialities for everyone in rural communities – locals and tourists alike.

CONCLUSION

American essayist Wendell Berry (1990) has maintained that eating is an agricultural act, and food tourism in rural communities embodies this statement. Whether tasting cheeses, sampling meats, tucking into farm cooking, hunting for butter tarts or sipping wine, food tourists crave a close encounter with food in its natural habitat. These encounters, however, are not without consequences, as the evidence from St. Jacobs shows. To ensure these consequences are positive, those involved in rural food tourism must keep in mind Higgins-Desbiolles' (2018) advice to reclaim tourism, just as rural food tourists are reclaiming food. After all, "what, why and how we eat says something about ourselves, why we travel and the society we live in" (Hall & Sharples, 2004, p. 2) and we want that something to be more than petty personal preferences, for the sake of a more meaningful present and a more resilient future.

REFERENCES

Bain, J. (2011, September 17). Chefs, farmers unite for Foodstock. *Toronto Star*. https://www.thestar.com/life/food_wine/2011/09/17/chefs_farmers_unite_for_foodstock.html?rf.

Bala Cranberry Festival (2021*). 2020 Cranberry Route*. https://www.balacranberryfestival.on.ca/read-more.

Benjamin, D., & Virkler, L. (2016). *Farm to Table: The Essential Guide to Sustainable Food Systems for Students, Professionals, and Consumers*. White River Junction, VT: Chelsea Green Publishing.

Berno, T., & Fusté-Forné, F. (2020). Imaginaries of cheese: Revisiting narratives of local produce in the contemporary world. *Annals of Leisure Research*, *23*(5), 608–626.

Berry, W. (1990). *What Are People For?* San Francisco, CA: North Point Press.

Broadway, M.J. (2017). 'Putting place on a plate' along the West Cork Food Trail. *Tourism Geographies*, *19*(3), 467–482.

Cummings, H., Galin, K., & Murray, D. (1998). *Farmers' markets in Ontario and their economic impact*. School of Rural Planning and Development, University of Guelph. http://hcaconsulting.ca/wp-content/uploads/2017/03/1999-Farmers-Markets-Ontario-Economic-Impact.pdf.

Dahms, F. (1991). St. Jacobs, Ontario: From declining village to thriving tourist community. *Ontario Geography*, *36*, 1–13.

Daly, H.E. (1990). Toward some operational principles of sustainable development. *Ecological Economics*, *2*, 1–6.

Davis, B., & Desmond, P. (2013, September 2). Community rallies around destroyed St. Jacobs market building. Waterloo Region Record. https://www.therecord.com/news/waterloo-region/2013/09/02/community-rallies-around-destroyed-st-jacobs-market-building.html.

de la Barre, S., & Brouder, P. (2013). Consuming stories: Placing food in the Arctic tourism experience. *Journal of Heritage Tourism*, *8*(2–3), 213–223.

Deschamps, T. (2015, June 12). St. Jacobs market rises from the ashes after 2013 Blaze. *Toronto Star*. https://www.thestar.com/news/gta/2015/06/12/st-jacobs-market-rises-from-the-ashes-after-2013-blaze.html.

Eigensinn Farm (2020). *Eigensinn farm*. https://www.stadtlanderseigensinnfarm.com/.

Ellis, A., Park, E., Kim, S., & Yeoman, I. (2018). What is food tourism? *Tourism Management*, *68*, 250–263.

Elton, S. (2010). *Locavore: From Farmers' Fields to Rooftop Gardens – How Canadians Are Changing the Way We Eat*. Toronto: HarperCollins.

Fireweed Community Market Society (2020). *Fireweed Community Market*. https://fireweedmarket.ca/.

Friedmann, H. (1993). After Midas's feast: Alternative food regimes for the future. In P. Allen (ed.), *Food for the Future: Conditions and Contradictions of Sustainability* (pp. 213–233). New York: John Wiley and Sons, Inc.

Good Life Vancouver (2014). *Spring Thaw with Canadian Chefs' Congress*. http://goodlifevancouver.com/spring-thaw-with-canadian-chefs-congress/.

Great Spirit Circle Trail (2019). *Food and tea*. https://www.circletrail.com/food-and-tea.

Hall, C.M., & Sharples, L. (2004). The consumption of experiences or the experience of consumption? An introduction to the tourism of taste. In C.M. Hall, L. Sharples, R. Mitchell,

N. Macionis, & B. Cambourne (eds), *Food Tourism Around the World* (pp. 1–24). New York: Routledge.
Heinzman, A. (2010). *The New Entrepreneurs: Building a Green Economy for the Future*. Toronto: House of Anansi Press.
Higgins-Desbiolles, F. (2018). Sustainable tourism: Sustaining tourism or something more? *Tourism Management Perspectives*, *25*, 57–160.
Indigenous Tourism Ontario (2020). *Position paper on growing Indigenous food tourism in Northern Ontario*. https://indigenoustourismontario.ca/wp-content/uploads/2020/11/ITO-Food-Tourism-Position-Paper-.pdf.
Jacobs, H. (2010). Structural elements in Canadian cuisine. *Cuizine*, *2*(1). https://doi.org/10.7202/039510ar.
Jepson, D., & Sharpley, R. (2015). More than a sense of place? Exploring the emotional dimension of rural tourism experiences. *Journal of Sustainable Tourism*, *23*(8–9), 1157–1178.
Jolliffe, L. (2006). Linking farmers' markets and tourism in New Brunswick, Canada. *Geotour*, *5*, 150–156.
Kwiatkowski, G. (2018). The multifarious capacity of food festivals in rural areas. *Journal of Gastronomy and Tourism*, *3*(3), 217–227.
Lane, B., & Kastenholz, E. (2015). Rural tourism: The evolution of practice and research approaches – towards a new generation concept. *Journal of Sustainable Tourism*, *23*(8–9), 1133–1156.
LaTrobe, H. (2001) Farmers' markets and local rural produce. *International Journal of Consumer Studies*, *25*(3), 181–192.
Lee, I., & Arcodia, C. (2011). The role of regional food festivals for destination branding. *International Journal of Tourism Research*, *13*, 355–367.
Loxley, J. (2007). *Transforming or Reforming Capitalism: Towards a Theory of Community Economic Development*. Halifax: Fernwood Publishing.
Mair, H., & Sumner, J. (2019). Critical thinking for sustainability at the Creemore 100 Mile Store. *Journal of Rural and Community Development*, *14*(3), 146–163.
Marks, R., & Toye, M. (2020). Community economic development: A viable solution for COVID-19 recovery. *Canadian Journal of Nonprofit and Social Economy Research*, *11*(2), 16–19.
Meyer-Czech, K. (2004). Food trails in Austria. In C.M. Hall, L. Sharples, R. Mitchell, N. Macionis, & B. Cambourne (eds), *Food Tourism Around the World* (pp. 149–157). London: Routledge.
Mitchell, C.J.A. (1998). Entrepreneurialism, commodification and creative destruction: A model of post-modern community development. *Journal of Rural Studies*, *14*(3), 273–286.
Mitchell, C.J.A., & Coghill, C. (2000). The creation of cultural heritage landscape: Elora, Ontario, Canada. *The Great Lakes Geographer*, *7*(2), 88–105.
Reardon, P.T. (2000, June 11). "We are what we ate." *Chicago Tribune*. http://articles.chicagotribune.com/2000-05-11/entertainment/0006170192_1_hunger-food-french-revolution.
Smith, A., & MacKinnon, J.B. (2007). *The 100-Mile Diet: A Year of Local Eating*. Toronto: Random House.
St. Jacobs Market (2020). Farmers' market. https://stjacobsmarket.com.
Sumner, J., McMurtry, J.J., & Renglich, H. (2014). Leveraging the local: Co-operative food systems and the Local Organic Foods Co-op Network in Ontario, Canada. *Journal of Agriculture, Food Systems and Community Development*, *4*(3), 47–60.
Telfer, D.J., & Hashimoto, A. (2004). Food tourism in the Niagara region: The development of a nouvelle cuisine. In C.M. Hall, L. Sharples, R. Mitchell, N. Macionis, & B. Cambourne (eds), *Food Tourism Around the World* (pp. 158–177). London: Routledge.
The Globe (2020). *The Globe Restaurant*. Available at https://www.thegloberestaurant.ca.

Welsh, J., & MacRae, R. (1998). Food citizenship and community food security: Lessons from Toronto, Canada. *Canadian Journal of Development Studies*, XIX, 237–255.
York Farmers' Market (2020). *York Farmers' Market*. https://www.yorkfarmersmarket.com/.

13. Heritage, heritage interpretation and rural tourism

Bernard Lane

INTRODUCTION

Rural tourism and heritage interpretation are closely related. Research shows that natural heritage, and human-made landscape heritage are both major attractions. Sustainable tourism has especially close links to effective and principled interpretation. Interpretation can be a vital part of the tourism management process, guiding visitor flows, encouraging time spent at destinations, and giving visitors a better understanding of their heritage experience (Cooper, 1991; Bramwell & Lane, 1993).

Back in the early 1950s, the US National Parks Service asked a former journalist, novelist and playwright who had written about National Parks and their value to advise them on how best to interpret their National Parks to their fast-growing numbers of visitors. That person was Freeman Tilden. He was an inspired choice. The fruits of his work appeared in a book called *Interpreting Our Heritage* in 1957. The fourth edition appeared in 2007. The fourth edition remains in print; it is still the bedrock of all interpretation principles worldwide. Its ideas apply to all rural areas, not just to national parks (Tilden, 2007).

Freeman Tilden was born in 1883 and lived until 1980. Amongst many pages, he wrote two remarkable sentences, both of which are valuable for sustainable rural tourism planners.

First, as Tilden (2007, p. 65) wrote, "Through interpretation, understanding; through understanding, appreciation; through appreciation, protection". This statement both influenced and reflects on what has become a fundamental policy belief of many tourism planners, that if tourists develop a close attachment to an area, they are more likely to protect it (Ramkissoon et al., 2012; Cheng & Wu, 2015).

Second, Tilden is credited with writing:

> The early Greek philosophers looked at the world about them and decided that there were four elements: fire, air, water, and earth. But as they grew a little wiser, they perceived that there must be something else. These tangible elements did not comprise a principle; they merely revealed that somewhere else, if they could not find it, there was a soul of things – a Fifth Essence, pure, eternal, and inclusive. (http://npshistory.com/publications/sontag/tilden.htm)

At first glance, the meaning of that rather romantic statement is a little unclear, but it refers to the intangible but complex attractions that many destinations have for visitors, which make those destinations "must see" places, bringing people back time and time again. It links to the earlier discussion of place attraction. Knowing more about a place can perhaps give access to a destination's inner soul, to understand their sometimes misunderstood "sense of place".

Tilden discusses six basic principles for rural interpreters (Tilden, 2007, pp. 34–35). They are given below:

1. Any interpretation that does not somehow relate to what is being displayed or described to something within the personality or experience of the visitor will be sterile. The importance of linking interpretation to personalities, and to real life experiences is an extension of this principle.
2. Information, as such, is not interpretation. Interpretation is revelation based on information. But they are entirely different things. However, all interpretation includes information. The difference between information centres and interpretation centres is key.
3. Interpretation is an art, combining many arts (see Ham, 1992).
4. The chief aim of interpretation is not instruction but provocation (*the "hands on" interpretation concept is closely related*, see Ham, 1992).
5. Interpretation should aim to present a whole rather than a part and must address itself to the whole man [*sic*] rather than any phase. This was a remarkable foresight into the holistic planning concepts of the 1990s (see Veverka, 2011).
6. Interpretation addressed to children (say up to the age of 12) should not be a dilution of the presentations to adults but should follow a fundamentally different approach. To be at its best, it will require a separate programme.

HERITAGE INTERPRETATION AND RURAL TOURISM

This chapter examines how heritage interpretation and heritage tourism can be defined and how, together, they can create sustainable product development opportunities in rural tourism situations. It reviews the tasks involved in that process, including assessing the market for heritage interpretation, using heritage audits and strategies, and how heritage interpretation can be delivered to rural tourists. It discusses the question of how conventional museums and the ecomuseum concept can fit into the delivery of heritage interpretation. The contemporary and fast-growing idea of the multi-purpose visitor centre is considered. Finally, the chapter discusses the many questions surrounding the governance of both heritage and heritage interpretation in rural tourism management.

What is Heritage?

That is a difficult question. It is related to history, but it is not history. Heritage includes everything that people want to save from the past. It is strongly influenced by conservation, which in turn is influenced by the fear of loss of features from the past. It includes human-made objects and buildings. It includes cultural practices of many different kinds. And it includes natural heritage – animals, plants, landscapes, and geological features. There can be great disagreement about what should be included in any list of an area's heritage: see the sections below about heritage audits and strategies. Heritage can be very political and divide people and whole communities. Language can also become heritage, from regional dialects to recognised written languages such as Scottish Gaelic, which are now treasured and being taught again after years of decline. Interpretation Canada, a non-profit group of Canadian interpreters, founded in the 1970s, created the following definition of heritage interpretation in 1976:

> Heritage interpretation is any communication process designed to reveal the meanings and relationships of cultural and natural heritage to the public, through first-hand involvement with an object, artefact, landscape or site. (https://interpretationcanada.wildapricot.org/about)

What is Heritage Tourism?

This inevitable follow-up question is answered by McKercher and Du Cros (2002) in their encyclopaedic and well-argued book on cultural tourism. Heritage tourism is part of the growing activity called cultural tourism, which occurs as a self-standing activity, and also as *part* of other multi-aspect forms of tourism, including heritage-linked days out during sun, sea and sand holidays. Heritage tourism is driven by rising levels of education, by the growing power of television's insatiable hunger for travel programmes, and the heritage industry's need for cash flow: heritage retention and conservation is expensive!

But Why Do Tourists Visit *Rural* Areas?

Lane (1994), defining rural tourism, listed 20 types of rural holiday then available. Walking holidays came top of the list. It was followed by cycling/cycle touring, landscape appreciation, rural heritage studies and small towns. No reliable international statistics were then available for how many visitors took which type of rural holiday, and that remains the case today. Reliable information that covers all rural tourism areas is extremely difficult to collect because there are very many different types of rural area, and many tourists are influenced by visiting multiple types of attraction.

This was revealed in detail by Carneiro et al. (2015) who interviewed 49 rural tourists in depth – some were domestic, some international – and all were visiting central Portugal. Vegetation type and colour, and built heritage, emerged as key experience elements, but they were followed by a complex group of largely non-visual stimuli, including sounds, smells, tastes, people contact, "nature" in its wider sense, and the past. Other researchers have had similar findings (see for example Lindemann-Matthies et al., 2010; Surová and Pinto-Correia, 2008). There is no doubt that heritage is a major part of the complex mix of attractions that bring visitors to rural areas, joining health, exercise, adventure tourism, peace and quiet, etc., as the series of niche markets that collectively support rural tourism's business activities.

The Tasks for the Heritage Interpreter

The complex nature of rural heritage can be summarised as being strongly linked to landscape, small-scale buildings and settlements, nature (in its many forms), living history (also in many forms), including people, farming, and a series of relict activities and features. Those relict activities are especially important. They provide potential hands-on experiences, a coveted experience in these days of virtual onscreen experiences. They range from visiting archaeological remains, learning to make arrow heads from flint, to tasting regional foods, and traversing heritage railways behind historic steam locomotives. The task for the heritage interpreter is to explain heritage, put it into the visitor's personal context, and to make those processes interesting, clear, and enjoyable. Thus, heritage buildings can be, through interpretation, made to feel as though their builders have just left, and that their historic owners and users are still alive and might suddenly reappear. Demonstration is, therefore, essential, including, for example, the act of bringing in the horse, or firing the boiler of a steam locomotive yourself. Participation and animation are key components.

Since Freeman Tilden's days, the concept of interpretation has been widened. It is now used as a tourism management device to influence visitor behaviour. The siting of visitor centres helps influence visitor flow patterns. The interpreter's message is recognised to be politically charged in that and other senses. And the interpretation system itself can become a destination attraction in its own right. Marketing (and de-marketing) can be linked to the interpretive system (see below). The biggest names in heritage interpretation have now become Sam Ham, from the University of Idaho, USA and John Veverka, originally an academic, but for the last 35 years an international consultant in heritage interpretation (see Ham, 1992; Veverka, 2011).

We also now know much more about the many types of interpretation, about the use of themes and story lines, about linking to real-life human stories, and to technicalities such as the three and five rule (not less than three pieces of information in any one story and not more than five; most people have limits to their comprehension!). All sorts of technology have been introduced, from interactive displays to QR codes and mobile phone apps. Well-trained and even quite old humans are, however, still remarkably useful, and have lower capital costs. Both technology and humans need skilled and informed management.

Heritage audits are a useful tool for anyone trying to create a successful heritage tourism product and strategy. The audit, especially if done publicly, using public meetings and/or local press coverage, can be both revealing, and educational for the community: many consultants carrying out such audits discover heritage that the community – and tourism stakeholders – often did not know about. Heritage consultants know what to look for, where to look and who to ask about the past. This is partly because archaeologists, naturalists, geologists, and historians are very often unaware of the concept of heritage tourism, its potential economic importance, and the public's fascination for discovering the past and its hidden, or lost, stories. Audits can also be politically important, drawing the attention of planners and both public and private sector decision makers to issues, areas, items and practices that need conservation, that may need funding, and that could reap financial benefits if well presented (see: https://www.communities-ni.gov.uk/publications/heritage-asset-audit; and https://www.abartaheritage.ie/what-we-do/sustainable-tourism-through-heritage-tourism-audit-services/).

Heritage strategies can become important once a thorough audit has been carried out. Strategies have a number of roles. They help work out what should be prioritised for conservation, what can be a viable heritage tourism story, and, importantly, how specific heritage stories should best be told. They can also be politically important because they can suggest heritage that needs to be publicly recognised locally as well as by potential visitors. And they can help bring about community support for local heritage (see Dragouni et al., 2018).

The Highland Interpretive Strategy Project (1994) devised and published a strategy for the heritage of the Highlands of Scotland in 1997. It was revised and expanded in 2001. It is a very valuable, well-written and carefully researched document (Carter, 2001) and is available online at: https://jamescarter.cc/wp-content/uploads/2014/09/A_Sense_of_Place_James_Carter.pdf. Unfortunately, little has been published about its implementation. It has been much discussed by members of Interpret Europe (https://interpret-europe.net/), at their regular conferences. Interpret Europe is a membership-based organisation, registered as a charity in Germany. It has more than 800 members from more than 48 countries. Interpret Europe's mission is: "Empowering all who inspire meaningful connections with Europe's natural and cultural heritage to shape our common future" (www.interpret-europe.net). Interpret Europe is a member of the Global Alliance for Heritage Interpretation (www.gahi.online) – an umbrella organisation for other heritage organisations (for other strategy examples and issues, see Bord Failte, 1992; Phillips & Tubridy, 1994; Stewart et al., 2001; Ducros, 2017).

DELIVERING HERITAGE TOURISM

The traditional method of delivering heritage tourism was by printed, usually outdoor, panels at key points, and sometimes also by panels within buildings. Tubb (2003) reviewed and evaluated this method in the UK's Dartmoor National Park and found that it did not succeed very effectively in transmitting its messages. There was too much information, which was not given in a way that stimulated interest. But this method is widely used because it is low cost, it is traditional, and it requires little administration.

A second traditional method is not low cost and requires much more administration. It relies on human guides (see Bryon, 2012; Cohen, 1985; Weiler & Black, 2015; Xu et al., 2013). Research shows that, if given good training, human guides can be very effective indeed. There is a second win for this system. Older humans can avoid loneliness by becoming a heritage guide. And costs can be reduced by seeking volunteers (Orr, 2006).

It seems likely that the largest and one of the best-organised users of volunteer guides is the Museum of Applied Arts and Sciences (MAAS) in Sydney, Australia. The MAAS group includes the well-known Power House Science Museum that has over 200 volunteer guides (the complexity of the administrative task can be gauged by searching for Volunteer with the Sydney Observatory Visitor Services Team on Google and go to https://maas.museum › app › sites). Light (1995) reviews how visitors use interpretive media for heritage tourism, although this research is now rather dated.

A recent innovation in heritage tourism delivery is the use of websites on smart phones, which allow the delivery to include a range of voices, music, and other sounds. These systems can also locate where visitors are and guide them along a route (try an urban example: https://soundcloud.com/culturenight/stream-of-consciousness-by-tom-lane-with-olwen-fouere-self-guided-walking-tour). Guidance can be especially important in rural areas. Finally, in some rural areas, there can be mobile heritage enterprises (see www.thetravellinghistorycompany.com for an example).

Museums

In rural areas there are few museums, except in small towns. They can be valuable, but there are often governance problems, especially because there can be old-fashioned beliefs to work with. In some cases, the museum itself can be a museum of outdated or contested beliefs. There can be instances where museum directors see their role as primarily one of curating the exhibits for all time, with little interest in visitor numbers or interpretation. The author recalls visiting a small-town museum in western Poland, near Gdansk, while drawing up a rural tourism strategy for the region. At the time, free enterprise was relatively new in Poland. I discussed ways for the Director to make his museum more attractive to visitors. He became enraged at the idea and said in a sudden burst of English: "I will not make my museum into a circus!" In free enterprise England there are also problems. The author was recently hired to suggest ways of boosting tourist numbers visiting a small English county town. It had four museums. Three were publicly owned, one was privately owned. It was suggested that they formed a museums partnership for marketing purposes, exploiting the existence of four museums within one mile, borrowing the slogan of Frankfurt's Museum Mile from Germany. This suggestion was dismissed. The best-endowed and most popular museum did not feel the need to assist the others. And the idea of the private sector museum having a relationship with the public sector's museums was disliked by all four. But some museums are ahead in their heritage tourism thinking (see Liu & Lan, 2021).

The Ecomuseum Concept

Ecomuseums are not museums of the type discussed above. They are areas or regions with distinct economic and conservation aims, rather than buildings. The concept of ecomuseums is well described by Bowers (2016). In the 1980s, it was a fashionable rival term to that of sustainable tourism. But it was a term mainly used in France and was limited to rural areas. Those two factors have limited its worldwide popularity. It should not, however, be forgotten; as Bowers argues, "Ecomuseological principles can be used to manage heritage resources and economic development by focusing on, for example, holistic interpretation and information sharing, placing equal attention on heritage resources, and monitoring changes to the region over time" (2016, p. 1). Some of its ideas lend themselves well to rural heritage tourism, especially its belief in community participation, democratic decision taking, and the conservation of local culture.

VISITOR CENTRES: CONCEPTS AND PRACTICE

One of the key ideas within the concept of sustainable tourism is that of visitor education. The theory is that the knowledgeable visitor is both a better and a happier visitor; respecting physical and cultural environments, adopting sustainable travel practices, making repeat visits, buying local goods, recommending friends to visit. And since the 1980s, authorities across the world have worked on how best to educate visitors. A range of ideas has been tried. These have included tourism-friendly ranger services, printed and web-based materials, media work, financial incentives and disincentives, school-based educational work, modified marketing, and the development of visitor centres. It is important to note at the outset that none of these ideas is perfect for all situations, and that most work best as part of a broad range of complementary initiatives.

The visitor centre concept seems to have originated in North America in the 1970s, typically as part of the US National Parks Service programmes. Variants of the visitor centre concept include:

- Tourist information centres offer accommodation booking and other services, but no interpretation. The information role of visitor centres is discussed in Ballantyne et al. (2009).
- Heritage centres offer information about the physical and socio-cultural development of regions, with a specialist interest in the past. Heritage centres are closely related to museums, but typically offer a more hands-on approach to the past and to historic artefacts, while offering only a limited curatorial and conservation role.
- Activity centres offer both on-site activities and booking facilities for the attractions and activities in the region.
- Retail and restaurant centres are devoted to the products of a region while also offering some visitor centre advice and promotion facilities.

- Hubs for parking and public transport, which may have some visitor and interpretation centre roles.
- Administrative centres for a region offering a range of visitor and host community services on the same accessible site.
- Mobile facilities offer a range of facilities, which can be transported around a region or national park, to seasonal sites and/or to events.

By the end of the twentieth century "pure" visitor centres were rare: most were multifunctional to some extent. The advantages of the visitor centre concept can be summed up as follows:

- They present tangible evidence of help, concern, and activity.
- They can be used (in some cases) to channel uninformed visitors towards them, and they can help guide and channel visitors onwards from them.
- They are tangible, capital investments, capable of attracting public sector funding, grants and support: they can be opened by politicians and can be used as memorials to politicians. Both of the latter can be valuable for getting public sector involvement.
- They can contain administrative offices, keeping administrators aware of visitor needs and visitors aware of the activity of the administrators.
- They can be used as a way of boosting tourism locally.
- They can also be used as a way of boosting tourism's economic impact by employing local people and selling local goods.
- They can be used in a multi-purpose role (see earlier), one aspect of which can be that of a visitor accommodation base situated at the visitor centre.

But visitor centres can also have disadvantages:

- They are expensive to build.
- They can be expensive to run in terms of both financial costs and staff time.
- They can be situated at inappropriate places (often due to pressures to use redundant buildings or cheap or publicly owned land).
- They may be inflexibly designed, making multi-purpose use difficult. Many architects understand little about interpretation or visitor centres, leading to inappropriate designs.
- They may have displays that are not attractive to the market (see Tubb, 2003).
- Fixed displays can be expensive to update.
- The local community may be alienated by the capital costs and top-down approach of the centre.
- The concept is not universally applicable and may be outdated by the idea of decentralised interpretation, using many local multi-use centres, the environment itself, mobile facilities and hand-held visitor-led interpretation devices.

Current Issues in Visitor Centre Development

The development of visitor centres needs to be undertaken carefully. Some key issues and needs include:

- The need to link visitor centres to the concept of heritage regions, rather than their role as stand-alone entities (Brown, 1996).
- The use of centres for multiple tourism-related purposes. Carballo and León, (2018) describe a centre that also acts as an art gallery and as a centre for cultural performances.
- The need to create multi-purpose centres capable of being used for retail, hospitality, community, accommodation, and other purposes. This idea can help to attract capital and reduce running costs.
- The need to involve local people as volunteers, and/or as franchisees of businesses within the centres.
- The need to use the centres as showcases of architectural conservation, or equally as examples of quality new build, perhaps even avant-garde buildings using local materials. Or alternatively, to showcase traditional rural building techniques, reviving or conserving those techniques. This latter process can help to maintain existing traditional buildings – by raising their perceived value in the eyes of the public and by giving work to those who know how to build and repair in traditional ways.
- The need to have "action exhibits", demonstrations, events, "living history", festivals, etc. linked to the centres.
- The concept of the visitor centre as base for walking and other trails, cycle hire, audio guide hire, etc.
- The need to link to smaller centres of information spread through local communities.
- The need to develop a heritage strategy with "story lines" linked to the heritage regions concept.
- The role of the centre as a tourism management device, guiding visitor flows.
- Ideas about using the centre as a magnet, or as a distribution centre for tourists.
- The possibility to run the centre as a public–private–community partnership.
- There is an ongoing discussion about the role of human as opposed to automatic "guidance" systems. Human beings are returning to fashion after a period when computer-driven automatic displays were in the ascendancy. And human volunteers – often older people – can gain a sense of self-worth and esteem from working a few hours in a centre helping people.

TRAILS, SENSORY TRAILS AND SLOW TOURISM

Heritage trails have been created for many years, either using existing routes, or by setting up totally new routes, or new routes linking together existing rights of way by opening new links. An example is given below from south Somerset in south-west England.

The Liberty Trail

The 40-kilometre pedestrian Liberty Trail (https://www.visitsouthsomerset.com/SiteAssets/Files/liberty_trail_pdf.pdf) was created by an informal three-way partnership between a historian specialising in the seventeenth century and working for a county museum, a district tourism officer working for a largely agricultural district within that county and trying to develop rural tourism to diversify the district's economic base into tourism, and volunteers from the area who would be able to maintain signposts and heritage interpretation material along the route. The trail used a number of existing footpaths, but required new signing, marketing and local support, together with both launch and long-term funding to maintain the signs, marketing, support and governance.

The seventeenth century was a very important period in the political history of England. Republicans opposed the monarchy and sought liberty from the monarchs, the King was arrested and beheaded, an alternative head of state was proposed and took over. France and the Netherlands considered invading England for economic gain, and so-called religious non-conformists who wished to change the national religion (the Church of England) threatened to rebel to secure their religious freedom. The Liberty Trail was designed to illustrate and explain how the people on the route chosen reacted to this difficult but exciting period.

The Trail would normally entail at least two overnight stays. It would bring participants into direct contact with the experiences of politically active seventeenth-century people and the buildings, artefacts and landscapes that those people encountered. Launched as a heritage trail in the 1980s and supported by local regeneration funding, it was successful in attracting tourists, but little is known about how successful it was. Like many heritage trails, there was no money available to assess its impacts. The details were available as a free printed brochure; they are now available on the web.

Long-term funding has become politically and economically impossible because of England's political and economic problems. But the trail could be re-launched with new web-based interpretation and coverage on national television: walking programmes have become very popular and are relatively low cost to produce.

Sensory Trails

Sensory trails are a specialist development from conventional trails. They are designed to cater for those with visual and walking problems who need help to use interpretation. Trails tend to be short and have relatively simple interpretation systems.

A national list of all sensory trails in England and details of how to create more such trails can be found at https://www.sense.org.uk/get-support/arts-sport-and-wellbeing/sense-sport/our-programme/sensory-walks/. In addition, an interesting specific trail, north of Bristol, England, is described at https://www.forestryengland.uk/westonbirt-the-national-arboretum/self-led-sensory-walk-westonbirt.

Slow Tourism

As a concept, slow tourism grew in the late 1990s. It was linked to the emergence of the Slow Food concept, first in Italy and eventually worldwide (see Dickinson & Lumsdon, 2010; Lumsdon & McGrath, 2011). It has been a concept welcomed by academic commentators, but few tangible real-world working examples have emerged. An offshoot of this idea, slow travel corridors, encourages the use of heritage trail networks to help create linear, sustainable tourism destinations, and they have been proposed for a number of protected areas. There have been several proposals to incorporate heritage trails with heritage railways, using both pre-recorded and live commentaries.

HERITAGE GOVERNANCE

The governance of heritage is complex and that makes heritage tourism complex too. Heritage is owned by many people and organisations, both in the private sector and in the public sector. Creating partnerships to both conserve and to use heritage, and to agree on a strategy, is not easy (see Bramwell & Lane, 2000). Many political problems can arise about aspects of the interpretation of events in the past (see for example, Wong, 2013; Ducros, 2017; or consider an Australian example: Staiff et al., 2002).

Heritage tourism marketing can also raise political issues (Chhabra, 2009). There are many reasons for that, but fundamentally it is about money. Heritage conservation costs a lot of money. Whitelaw et al. (2014) explain this in detail and list various ownership and governance arrangements that deal with those questions. Tourism management costs money too, as does the provision of heritage interpretation. Governance, which is closely linked to ownership, determines how those costs are paid and by whom in each case. This is described by Leask et al. (2002), who found that the role of the non-profit public sector currently contrasts sharply with that of the cash-conscious private sector. Public sector heritage tends to be accessible without charge, has interpretation, but has little marketing. Private sector heritage usually charges for access, charges for interpretation, and uses extensive and often sophisticated marketing.

Heritage interpretation is an additional tourism product. It follows that a marketing plan must almost always be implemented for it (Chhabra, 2009; Donohoe, 2012). This has important implications for heritage tourism governance, partnerships and the relationship between public sector and private sector organisations. It also means that much more information needs to be collected and analysed about the market's reactions to heritage, to its interpretation, and to how its costs are paid.

Finally, it must be mentioned that the governance of heritage interpretation determines what is interpreted, and the political approach taken in that interpretation. That usually depends on the political beliefs of the heritage owners. Changing attitudes to high levels of wealth in the past, and to, for example, colonialism, can be increasingly important.

THE FUTURE

It is always difficult to predict the future. Whatever the future will be, however, heritage interpretation is likely to require more imaginative and flexible planning than ever before. It can avoid the mistakes of the past and go for well-sited, well-integrated, human-led, flexible, market-driven concepts, which can stress the role of heritage conservation and local development. And it must seek to avoid the potential problems that both rural tourism and heritage tourism can bring, including the purchase of local properties by wealthy city people, driving out local people with low incomes, and driving out local culture at the same time.

Murray et al.'s edited 2015 book, *The Interpreter's Big Book of Disasters*, should therefore become required reading. We tend to seek out best practice and try to recommend "good" examples, although there are many circumstances where good examples cannot work well. The study of bad practice can be very valuable. It can even encourage the difficult process of thought.

REFERENCES

Ballantyne, R., Hughes, K., & Ritchie, B.W. (2009). Meeting the needs of tourists: The role and function of Australian visitor information centers. *Journal of Travel & Tourism Marketing, 26*(8), 778–794.

Bord Failte (1992). *Proceedings of the Second Conference on the Development of Heritage Attractions in Ireland.* Dublin: Bord Failte.

Bowers, J. (2016). Developing sustainable tourism through ecomuseology: A case study in the Rupununi region of Guyana. *Journal of Sustainable Tourism, 24*(5), 758–782.

Bramwell, B., & Lane, B. (1993). Interpretation and sustainable tourism: The potential and the pitfalls. *Journal of Sustainable Tourism, 1*(2), 71–80.

Bramwell, B., & Lane B. (eds) (2000). *Tourism Partnerships and Collaboration: Politics, Practice and Sustainability.* Clevedon: Channel View Publications.

Brown, V. (1996). Heritage, tourism and rural regeneration: The Heritage Regions Programme in Canada. *Journal of Sustainable Tourism, 4*(3), 174–182.

Bryon, J. (2012). Tour guides as storytellers – from selling to sharing. *Scandinavian Journal of Hospitality and Tourism, 12*(1), 27–43.

Carballo, R.R., & León, C.J. (2018). The influence of artistically recreated nature on the image of tourist destinations: Lanzarote's art, cultural and tourism visitor centres and their links to sustainable tourism marketing. *Journal of Sustainable Tourism, 26*(2), 192–204.

Carneiro, M.J., Lima, J., & Lavrador Silva, A. (2015). Landscape and the rural tourism experience: Identifying key elements, addressing potential, and implications for the future. *Journal of Sustainable Tourism, 23*(8–9), 1217–1235.

Carter, J. (ed.) (2001). *A Sense of Place: An Interpretive Planning Handbook.* Inverness: The Tourism and Environment Initiative. https://jamescarter.cc/wp-content/uploads/2014/09/A_Sense_of_Place_James_Carter.pdf.
Cheng, T.-M., & Wu, H.C. (2015). How do environmental knowledge, environmental sensitivity, and place attachment affect environmentally responsible behavior? An integrated approach for sustainable island tourism. *Journal of Sustainable Tourism, 23*(4), 557–576.
Chhabra, D. (2009). Proposing a sustainable marketing framework for heritage tourism. *Journal of Sustainable Tourism, 17*(3), 303–320.
Cohen, E. (1985). The tourist guide: The origins, structure and dynamics of a role. *Annals of Tourism Research, 12*(1), 5–29.
Cooper, C. (1991). Interpretation: A destination management and marketing tool. *Insights A*, July, 1–8. London: English Tourist Board.
Dickinson, J.E., & Lumsdon, L. (2010). *Slow Travel and Tourism.* London: Earthscan.
Donohoe, H.M. (2012). Sustainable heritage tourism marketing and Canada's Rideau Canal world heritage site, *Journal of Sustainable Tourism, 20*(1), 121–142.
Dragouni, M., Fouseki, K., & Georgantzis, N. (2018). Community participation in heritage tourism planning: Is it too much to ask? *Journal of Sustainable Tourism, 26*(5), 759–781.
Ducros, H. (2017). Confronting sustainable development in two rural heritage valorization models. *Journal of Sustainable Tourism, 25*(3), 327–343.
Ham, S.H. (1992). *Environmental Interpretation – A Practical Guide for People with Big Ideas and Small Budgets.* Golden, CO: North American Press.
Lane, B. (1994). What is rural tourism? *Journal of Sustainable Tourism, 2*(1–2), 7–21.
Leask, A., Fyall, A., & Garrod, B. (2002). Heritage visitor attractions: Managing revenue in the new millennium. *International Journal of Heritage Studies, 8*(3), 247–265.
Light, D. (1995). Visitors' use of interpretive media at heritage sites. *Leisure Studies, 14*(2), 132–149.
Lindemann-Matthies, P., Briegel, R., Schüpbach, B., & Junge, X. (2010). Aesthetic preference for a Swiss alpine landscape: The impact of different agricultural landuse with different biodiversity. *Landscape and Urban Planning, 98*, 99–109.
Liu, P., & Lan, L. (2021). Museums as multisensorial site: Story co-making and the affective interrelationship between museum visitors, heritage space, and digital storytelling, *Museum Management and Curatorship, 36*(4), 403–426.
Lumsdon, L.M., & McGrath, P. (2011). Developing a conceptual framework for slow travel: A grounded theory approach. *Journal of Sustainable Tourism, 19*(3), 265–279.
McKercher, B., & du Cros, H. (2002). *Cultural Tourism: The Partnership Between Tourism and Cultural Heritage Management.* New York: Haworth Hospitality Press.
Murray, P., Martin, C., & Laurin, A.Y. (eds) (2015). *The Interpreter's Big Book of Disasters.* Red Deer, AB, Canada: Interpretation Canada.
Orr, N. (2006). Museum volunteering: Heritage as "serious leisure". *International Journal of Heritage Studies, 12*(2), 194–210.
Phillips, A., & Tubridy, M. (1994). New supports for heritage tourism in rural Ireland. *Journal of Sustainable Tourism, 2*(1–2), 112–129.
Ramkissoon, H., Weiler, B., & Smith, L. (2012). Place attachment and pro-environmental behaviour in national parks: The development of a conceptual framework. *Journal of Sustainable Tourism, 20*(2), 257–276.
Staiff, R., Bushell, R., & Kennedy, P. (2002). Interpretation in National Parks: Some critical questions. *Journal of Sustainable Tourism, 10*(2), 97–113.
Stewart, E.J., Glen, M.H., Daly, K., & O'Sullivan, D. (2001). To centralise or disperse – a question for interpretation: A case study of interpretive planning in the Brecks. *Journal of Sustainable Tourism, 9*(4), 342–355.

Surová, D., & Pinto-Correia, T. (2008). Landscape preferences in the cork oak Montado region of Alentejo, southern Portugal: Searching for valuable landscape characteristics for different user groups. *Landscape Research, 33*(3), 311–330.

The Highland Interpretive Strategy Project (1994). *Highland Interpretive Strategy*. Inverness: The Highland Interpretive Strategy Project.

Tilden, F. (2007). *Interpreting Our Heritage*. Chapel Hill, NC: University of North Carolina Press.

Tubb, K.N. (2003). An evaluation of the effectiveness of interpretation within Dartmoor National Park in reaching the goals of sustainable tourism development. *Journal of Sustainable Tourism, 11*(6), 476–498.

Veverka, J. (2011). *Interpretive Master Planning* (2 volumes). Edinburgh: Museums Etc.

Weiler, B., & Black, R. (2015). *Tour Guiding Research: Insights, Issues and Implications*. Bristol: Channel View Publications.

Whitelaw, P.A., King, B.E.M., & Tolkach, D. (2014). Protected areas, conservation and tourism – financing the sustainable dream. *Journal of Sustainable Tourism, 22*(4), 584–603.

Wong, Cora Un In (2013). The sanitization of colonial history: Authenticity, heritage interpretation and the case of Macau's tour guides. *Journal of Sustainable Tourism, 21*(6), 915–931.

Xu, H., Cui, Q., Ballantyne, R., & Packer, J. (2013). Effective environmental interpretation at Chinese natural attractions: The need for an aesthetic approach. *Journal of Sustainable Tourism, 21*(1), 117–133.

14. Climate change

Mark C.J. Stoddart and Yixi Yang

INTRODUCTION

Tourism is often positioned as a sustainable pathway for rural community development. Rural tourism promises connections with nature and wildlife, as well as encounters with the historicity and authenticity of rural communities (Baldacchino, 2010; Stoddart & Graham, 2016). For community leaders and planners, tourism promises diversification beyond natural resource extraction (George et al., 2009). It is a form of community economic development based on exporting experiences and memories of unique environments and communities, which has been described as an "experience economy" (Thrift, 2001). However, tourism also has environmental costs. Tourism travel – whether by airplane, car or bus, or ferry or cruise ship – has a significant carbon footprint and contributes to global climate change. The carbon intensity of tourism travel is an issue that is often under the surface and unacknowledged in the positive discourse of tourism sustainability (Cohen et al., 2011; Peeters et al., 2019; Scott et al., 2016). This presents a puzzle: Is tourism under climate change a sustainable development pathway, or is it just another carbon-intensive industry? As we show in this chapter, when we bring climate change into the equation, the image of tourism as a sustainable pathway for rural community development gets more complicated.

In addition to reviewing previous research literature in these areas, we draw on our own research related to tourism development. The project *The Oil-Tourism Interface and Social-Ecological Change in the North Atlantic* examines intersections between tourism development and oil development in sites across the North Atlantic region, including in Denmark, Iceland, Newfoundland and Labrador (Canada), Norway, and Scotland. We examine issues of climate impacts on tourism landscapes, the carbon intensity of tourism travel, and potential responses by the tourism sector to climate change. This project uses a mixed-methods approach to data collection and analysis, including fieldwork, interviews, and analyses of media content, policy documents, and websites produced by the tourism sector, energy sector, government agencies, and environmental organizations.

This chapter proceeds through three sections. First, we provide an overview of the notion of tourism as a sustainable pathway for rural community development. Second, we examine the interaction between tourism and climate change, in terms of climate change impacts on rural tourism environments, as well as the carbon footprint of tourism mobility. Third, we examine the potential for decarbonizing tourism. This involves technological experiments and socio-cultural tipping points that could reshape how tourism works as a form of rural community development.

These positive tipping points may help ease some of the tensions between tourism and the imperative to respond to climate change.

TOURISM AS A SUSTAINABLE PATHWAY FOR RURAL COMMUNITY DEVELOPMENT

The notion of tourism as a mechanism for sustainable community development appears repeatedly in the UN Sustainable Development Goals (SDGs). As a global framework for sustainable development, the Agenda 2030 was adopted by all 193 UN member states at the 2015 Sustainable Development Summit. Several of the 17 SDGs reference tourism's role as a catalyst for sustainable development. Tourism appears in targets for Goal 8 on inclusive and sustainable economic growth, Goal 12 on sustainable consumption and production, and Goal 14 on the sustainable use of oceans and marine resources (United Nations, 2015). Furthermore, the UN General Assembly declared 2017 as the International Year of Sustainable Tourism for Development and recognized the potential for tourism to accelerate the SDGs in five key areas, including sustainable economic growth, social inclusiveness, environmental protection and climate change, cultural diversity, and peace and security (World Tourism Organization & United Nations Development Programme, 2017). Academic studies similarly discuss whether and how tourism can be leveraged to help communities make progress towards the SDGs (Bramwell et al., 2017; Fletcher et al., 2019).

Beyond direct contributions to individual SDGs, tourism also has the cross-cutting potential to contribute to all 17 goals and connect economic, environmental, and social sustainability. For example, Turpie et al. (2015) report how the Government of Zambia's agroforestry actions under the UNFCCC's REDD+ program (Reducing Emissions from Deforestation and Forest Degradation) have aligned with the local ecotourism business (SDG 8, decent work and economic growth), erosion control and sediment retention (SDG 15, life on land), carbon storage (SDG 13, climate action), and food security with pollination services (SDG 2, zero hunger). The case of agroforestry in Zambia represents the potential of tourism as a venue where the three components of sustainable development – environment, economy, and society – can positively interact with each other. On the environmental side, nature-oriented tourism provides incentives for host communities to protect local environments and raises tourists' environmental awareness about wildlife, biodiversity, and ecosystems. On the economic side, tourism is a force for poverty alleviation in rural communities as it increases direct employment and supports local small- and medium-sized enterprises, which has spillover effects that benefit a range of businesses in the local economy.

The tourism studies literature points to a range of potential social–environmental benefits of tourism. At the most basic level, tourism can be a driver of economic diversification for rural communities that traditionally relied on resource extraction, such as forestry, mining, oil extraction, or fisheries. Tourism development promises another economic lifeline to rural communities, so they are less reliant on natural resource extraction (Baldacchino, 2010; George et al., 2009; Reed & Gill, 1997). This provides communities with opportunities to transform and reinvent ways to make use of local culture and the environment based on an "experience" economy (Thrift, 2001).

Our research on tourism development across the North Atlantic region provides further examples of these dynamics. For many rural host communities across the North Atlantic, tourism development was pursued as an economic diversification or recovery strategy alongside traditional (and often declining) fisheries economies. Figure 14.1 uses social network analysis techniques to visualize relationships among themes in our project data. Grey circular nodes are thematic coding categories. White triangular nodes represent our case study regions. To simplify the diagram, we have set thresholds for numbers of co-occurrences, which brings out the most frequent themes for each case. Key nature-based tourism attractors include a range of unique natural environments and wildlife such as: glaciers, volcanic landscapes, whales, icebergs, puffins and seabirds, fjord landscapes, mountains, and seals. Across multiple cases, valued modes of interaction with the environment include hiking, boat tours, and cycling.

Tourism development can have environmental benefits by providing a rationale for protecting natural areas or wildlife that serve as attractors for visitors. For example, many Canadian National Parks were initially established as tourism-oriented economic anchors and revenue generators for nearby communities, where ecological protections were often secondary to meeting the interests of tourists until relatively recently (MacEachern, 2001; Reichwein, 2014). Elsewhere, Mellon and Bramwell (2016) examine the Lake District National Park in the UK and find that sustainable tourism policies are being increasingly integrated into park planning and management in order to better ensure co-benefits between the ecological protections offered by parks and contributions to community economic wellbeing through tourism. As host communities see the economic benefits from tourism to the park, there is increasing social buy-in for environmental protection. Several researchers have also focused on whale watching as a wildlife-oriented tourism activity that has transformed host community relationships with whales. In places like Iceland and the Azores (Portugal), whales are no longer taken as targets for hunting but have become valued tourism attractors. If managed correctly with standards to ensure the wellbeing of whales, whale watching can contribute to their protection and can serve as bridge between the tourism sector and environmental organization interests (Cunningham et al., 2012; Neves, 2010; Silva, 2015).

These themes from the tourism studies literature are also seen in our research on the North Atlantic region (recall Figure 14.1). National Parks, nature reserves, and other protected areas are frequently invoked as important tourism attractors. Tourism helps maintain social and economic supports for nature protection in these regions and leads to a re-evaluation of relationships with local wildlife, including iconic tourism attractors such as whales, puffins, and seals. For example, an Icelandic environmental NGO representative notes:

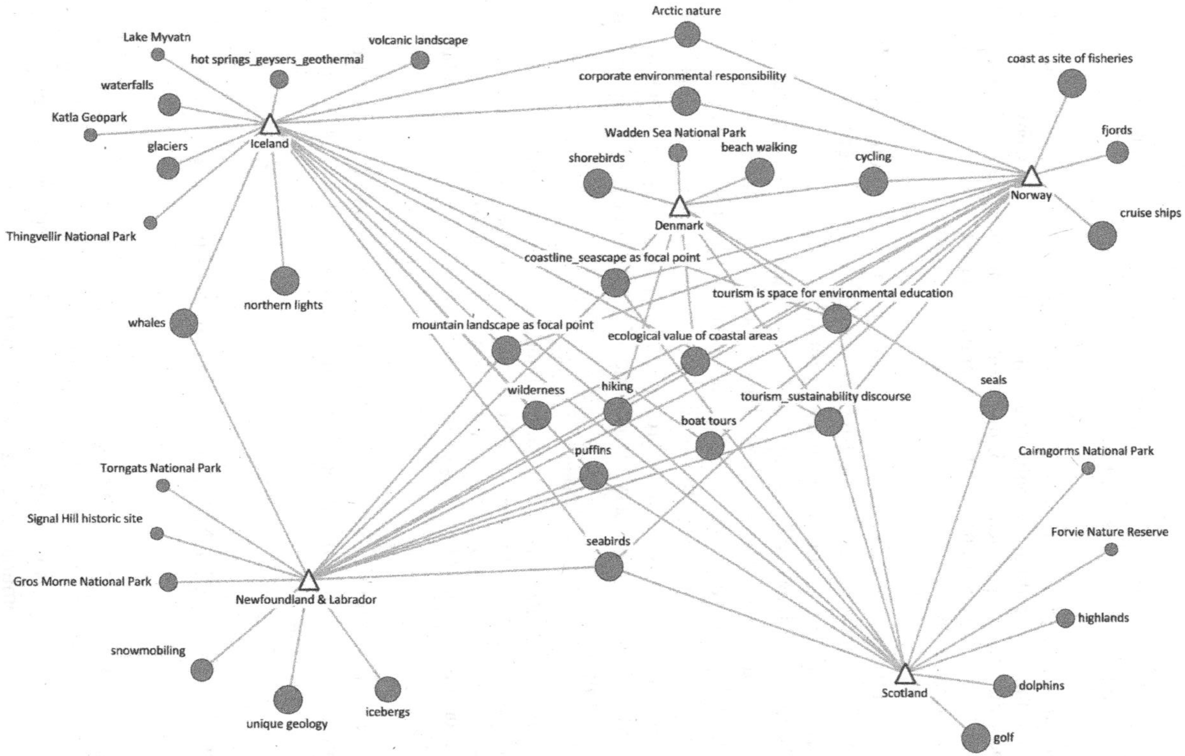

Note: Grey circular nodes = thematic categories; white triangular nodes = case study regions. Node size adjusted for degree centrality. Filtered for threshold of 20 or more coding references in data.

Figure 14.1 *Case study regions and nature-based tourism*

> But I think in general we are starting as Icelanders to see tourism in a different way. We see that it's [Icelandic nature] so valuable and people want to come here to look at this. And so, all of a sudden, we have these political parties being interested in nature conservation or at least they're willing to look at it in a different way. (Iceland NGO interviewee, 2016)

An Icelandic tourism operator expresses similar views on the synergy between nature-based tourism and environmental protection as follows:

> We'll support reforestation in certain areas and things like that. I guess protecting the wilderness, for us, is the biggest thing ... controlling trail damage. We have an environmental fund and I know that the last payment out of that went into trail preservation ... But what we campaign against is, like, building power lines over parts of the highlands, which reduce the wilderness and the wilderness experience. (Iceland tourism sector interviewee, 2016)

Similarly, the notion that tourism is a space for environmental education, particularly in building ecological literacy through first-hand experience of local environments, also comes up frequently across our case study regions.

As rural landscapes and local wildlife are revalued through tourism development, the importance of tourism has been invoked in opposition to controversial new proposals for natural resource extraction, such as developing offshore oil near the Lofoten Islands (Norway) and around the Gulf of St. Lawrence (Newfoundland and Labrador). For example, a Norwegian environmental NGO participant talks about the role tourism plays in the controversy over oil development in the area around the Lofoten Islands as follows:

> I think that if it wasn't for the tourism it would be much harder to protect this area against the oil industry because that is kind of ... when you mention Lofoten for people, they already know the area, it's very famous. And they can't believe Norway wants to open this area for oil drilling because of the waste, nature, and everything. (Norway NGO interviewee, 2016)

In these cases, tourism economies are positioned as more sustainable forms of rural development, in opposition to oil exploration and extraction, which is defined as unsustainable by comparison.

INTERACTIONS BETWEEN CLIMATE CHANGE AND TOURISM

Climate change has far-reaching impacts on the environments and the livelihoods of rural tourism host communities. These include direct damage to local tourism resources and hospitality operations, as well as indirect impacts on the geography and seasonality of tourist demand patterns. A substantial part of the global tourism industry operates in coastal regions and islands, which are directly exposed to climate risks, such as sea level rise, ocean acidification, coastal erosion, and damage to coastal infrastructure from storm surges and increasingly intense storms. As projected in the IPCC *Fifth Assessment Report*, coastal and low-lying areas, particularly those in subtropical and tropical regions, will increasingly experience submergence, flooding, and erosion (Pachauri et al., 2014). The degradation of coastal landscape features will have long-term impacts on rural communities that depend on coastal tourism for economic wellbeing. Accelerating coral reef bleaching has already impacted iconic tourism attractors in places like Australia and Hawaii (Scott et al., 2016). In the years between 2016 to 2018 alone, large coral reef systems such as the Great Barrier Reef (Australia) have lost as much as 50% of their shallow-water corals (Hughes et al., 2017). Cultural heritage important to tourism is also at high risk from climate change. A global analysis projects the loss of 47 out of 720 UNESCO Cultural World Heritage sites to rising sea levels under 1 °C of warming, with this increasing to 110 sites with 2 °C and 136 with 3 °C of global average temperature increase (Marzeion & Levermann, 2014).

Host communities that depend on specific climatic conditions will also be challenged by changing seasonal weather patterns. A typical example is the rural mountain communities that orient around skiing and snowboarding. Warmer and wetter winters will negatively affect natural snow-dependent operating conditions, thus decreasing the value of winter tourism businesses and impacting local employment (Elsasser & Messerli, 2001; Steiger et al., 2017). Conversely, changing summer weather patterns, characterized by more frequent and intense heatwaves, may make warm-water tourism destinations less appealing for visitors and shift travel patterns towards higher-latitude destinations (Gössling et al., 2012). Under 1.5 °C of warming, summer tourism in coastal Mediterranean regions would likely be reduced, with overnight stays in places like Spain and Cyprus predicted to decrease by 8% and 2%, respectively (Jacob et al., 2018).

The impacts of climate change on rural tourism host communities will vary by location. While constraining tourism development in some destinations, the changing climate may increase travel to other destinations. While still a niche destination for many, there is growing interest in Arctic tourism, as increasing numbers want to visit this rapidly changing region before it is radically transformed by climate change. As Sisneros-Kidd et al. (2019) note, this is a form of "last chance tourism" that draws visitors because the "landscapes, seascapes, or the species that inhabit these places are at risk of disappearance due to global forces like climate change" (p. 1259). This is seen in more remote rural locations like Greenland, the Westfjords region of Iceland, or Svalbard (Norway), which are emerging tourism destinations (James et al., 2020; Ren et al., 2016; Sisneros-Kidd et al., 2019). This spills over into an increasing general interest in tourism to the high north to gain first-hand experience of melting glaciers, declining polar bear populations, and other signs of ecological change. Relatedly, melting sea ice and longer ice-free seasons in the Arctic Ocean and northern reaches of the North Atlantic have led to increasing cruise ship travel through northern routes around Greenland, Iceland, or Norway (Hall et al., 2008).

The example of increasing Arctic tourism highlights the irony of the relationship between tourism and climate change. While climate change makes northern destinations more accessible and desirable to visitors, access to these destinations relies on networks of carbon-intensive airplane and/or cruise ship travel. The carbon footprint of tourism travel subsequently amplifies the negative impacts of climate change on these environments, accelerates the transformations of northern ecosystems, and poses adaptation challenges for the rural host communities in the Arctic region, thus creating a "paradox" for Arctic tourism host communities (Dawson et al., 2010; Sisneros-Kidd et al., 2019).

Tourism's carbon footprint is a challenge that lurks beneath the surface of the pro-environmental self-image of tourism as a pathway for rural community development. Pro-environmental tourism discourse often takes a very localized view, emphasizing the immediate environmental benefits for host communities. As James et al. (2020) observe in their study of cruise ship tourism in Greenland and Iceland, a short-term economic growth orientation to tourism development often overshadows environmental concerns, or holistic approaches to sustainability that consider impacts beyond local host communities. This localized view avoids serious consideration of the carbon footprint of carbon-intensive transportation networks that provide tourism connectivity (Sharpley, 2020). However, these connections are increasingly highlighted in the tourism studies literature. Cohen et al. (2011) were among the first to describe the "flyer's dilemma" that the short-term satisfaction of tourism travel comes with long-term climate consequences. However, they find that many travellers downplay the contribution of tourism air travel to climate change. Similarly, Klein (2010) argues that the carbon intensity of cruise ship tourism is not being adequately addressed within the cruise sector. The failure to grapple with cruise ship contributions to climate change contradicts the self-image of the cruise sector as a "responsible steward of the ocean environment" (p. 114).

Elsewhere, Scott et al. (2016) provide an assessment of the 2015 Paris climate change agreement through a tourism lens. As an outcome of the 2015 Conference of the Parties (COP) meetings, this non-binding international agreement seeks to cap global warming at 2 °C above pre-industrial levels. The carbon footprint of air travel is a vexing challenge for the global tourism sector for meeting the goals of the Paris Agreement. Peeters et al. (2019) examine a range of "transport futures" and find that projected growth in tourism-related air travel will make the goals of the 2015 Paris climate change agreement unattainable. An ecologically desirable transport future, by contrast, would require a reduction in both the number of trips and distances travelled by air. Furthermore, climate mitigation strategies that increase the cost of carbon-intensive travel will potentially reshape patterns of tourism travel, such as encouraging more people to travel closer to home and avoid long-range travel (Gössling et al., 2012).

Our research on tourism development across the North Atlantic also illuminates the tensions inside tourism's image as a sustainable development pathway for rural communities. Across our data, tourism is generally viewed as a form of sustainable development. Where environmental issues come up, these are more often localized issues related to waste management, overdevelopment, or trail erosion and damage to vegetation due to overcrowding or insufficient infrastructure. However, some of our interview participants explicitly demarcate tourism as a locally sustainable form of development from broader questions about its extra-local carbon footprint. For example, when asked if they see tourism as an environmentally sustainable development pathway, a Scottish government interviewee responds:

> Well, I suppose more and more people are coming in by plane, which environmentally – there's issues there around emissions. I suppose that's true of every city which has a tourism sector. Beyond that, it's difficult to see [the negative impacts], I think. Environmental regulations are being tightened all the time, I think more and more people are conscious about these things. (Scotland government interviewee, 2017)

Where the carbon footprint of tourism is acknowledged and discussed, air travel is flagged as the main challenge that tourism needs to address to respond to climate change. By focusing on air travel as the biggest culprit, the locus of action shifts away from individual operators or tourism associations to a relatively small group of large corporations and national air carriers. Overall, however, few tourism stakeholders engage in substantial reflection about tourism's contributions to climate change, or its potential role in taking action to address climate change.

DECARBONIZING TOURISM?

The imperative to respond to climate change touches all spheres of society. This includes tourism, despite its alignment with sustainability discourse and the UN SDGs. Writing about the climate science–policy interface, Gross (2010) notes that ignorance and surprise are inherent to climate change. Gross argues that the best path forward is to engage in modularized "public experiments" that can be institutionalized in an evolutionary way to address climate change while recognizing the likelihood of surprise and unintended effects. As Gross writes,

> As people try out things, they better understand what they are doing as well as what they plan for the future, which is part of the recursive practice of experimenting. Since ignorance can never be delimited to zero, an experimental approach appears to be a useful long-term strategy for ecological design. (pp. 173–174)

While the carbon intensity of airplane travel or cruise ship travel is likely to remain a vexing challenge for the foreseeable future, there are a range of technological experiments already unfolding within tourism that, if scaled up, could make a substantial difference.

There are also socio-cultural trends that have the potential to transform the relationship between tourism and climate change. The notion of positive tipping points is important here, which is defined as "emergent properties derived from complex systems dynamics that allow rapid transformations in individual and collective practices so as to reach evolutionary-like solutions to the present socio-climate quandary" (Tàbara et al., 2018, p. 12). These are positive technological, political, socio-cultural, or economic changes that have systemic ripple effects and transform the social conditions within which people make choices and act. For example, Tàbara et al. (2018) highlight the "consolidation of the IPCC" as a political tipping point for communicating the scientific consensus on climate change, which had ripple effects on global dialogue about climate change, as well as national climate policymaking. In the case of tourism, if technological experiments and positive socio-cultural tipping points begin to converge and cascade, then there is the possibility for decarbonizing tourism and better aligning the sector with its self-image as a pathway for sustainable rural community development.

Our research on tourism development in the North Atlantic provides examples of decarbonization experiments. The most frequent examples of tourism responses to climate change focus on the electrification of vehicles. Some tourism operators in Iceland are adapting hybrid or electric buses or cars for road tours. There are also experiments in the electrification of boat tour operations. One example is the company North Sailing, which operates "carbon neutral" whale watching tours off the northern coast of Iceland using hybrid wind and electric boats. In their website, North Sailing notes that they hope this hybrid boat tour technology can be scaled up in order to reduce tourism's climate change impacts (North Sailing, n.d.). An Icelandic government interviewee highlights North Sailing in talking about the potential for tourism to respond to climate change as follows:

> I keep mentioning them, North Sailing, it's a truly amazing company. And they're true to their vision and they're doing a fantastic job in the electric arena around those boats. Because it might have a huge impact globally on both small and large vessels … And just imagine the impact that would have, to have electric vessels go between the fjords in Norway, for example, or on all the European rivers instead of boats that are run on carbon fuel. (Iceland government interviewee, 2016)

Another example comes from the Norwegian coastal boat operator Hurtigruten, which transports tourists through Norwegian fjords, coastal communities, and northern islands. They collaborate with the environmental NGO Bellona to reduce GHG emissions through the electrification of ships and the use of algae-based biofuels as alternatives to fossil fuels. As they claim on their website, "The future is emission-free. It's just a matter of time" (Hurtigruten, n.d.).

Going beyond the actions of individual operators, Norway's official travel guide website, Visit Norway, promotes the use of electric cars and Norway's well-developed charging infrastructure as a form of pro-environmental tourism (Visit Norway, 2017). As an example that tries to grapple with the challenge of air travel, Icelandair has set environmental policy goals that commit the company to adopting green technology to improve fuel efficiency and reduce fuel use. The company also works with the Kolviður Iceland Carbon Fund to support Icelandic reforestation programs as a localized form of carbon offsets that counterbalance the carbon cost of flights to this relatively remote destination (Icelandair, n.d.).

If they are successfully scaled up, technological innovations could reduce the carbon footprint of the tourism sector by improving resource efficiencies or promoting lower-carbon alternative technologies. These innovations promise technological solutions that do not disturb the social practices and patterns of tourism travel.

By contrast, there are potential socio-cultural tipping points that may decarbonize tourism by reshaping tourism practices and changing how visitors travel. Several social movements have mobilized around the climate emergency, including the youth-led Fridays for Future, Extinction Rebellion, and fossil fuel divestment movements, all of which seek to undermine the moral legitimacy of the fossil fuel sector and articulate visions for decarbonized futures (Hestres & Hopke, 2020; Mangat et al., 2018; Stuart et al., 2020). The discursive tactic of flight shaming, which asks people to limit their air travel, has had some impact, especially in Europe where population densities are higher and travel distances are often shorter. One impact of flight shaming is to encourage tourists to travel closer to home, or to engage in "slow tourism" with fewer but longer trips. There have also been environmental movement campaigns targeting fossil fuel company sponsorship of museums and other cultural sites that serve as important tourism attractors. For example, social movement campaigns like Liberate Tate problematized British Petroleum's sponsorship of British museums and galleries, which are prominent cultural tourism sites (Motion, 2019; Uldam, 2017). Activists created oil spill-related art that was performed in the Tate and amplified to the public through social media channels. This makes the connections between oil and cultural influence visible to bystander publics. These social movement campaigns ask art and culture institutions to sever their ties with fossil fuel companies. The "cultural justice" (Motion, 2019) work of these movements may also influence tourists' perspectives and practices in relation to these places.

The tourism studies literature has also turned its attention to conversations about tourism and the degrowth movement, although most of this conversation focuses on over-touristed urban destinations. The degrowth movement questions relentless economic expansion as the guiding logic of consumer capitalism. One of the major implications of degrowth is to question models of tourism development that rely on assumptions of growing tourism numbers year after year – assumptions that remain embedded in the discourse of the UN Sustainable Development Goals (Fletcher et al., 2019; Higgins-Desbiolles et al., 2019). Rather than being anti-tourism, degrowth asks host communities and governments to engage in the work of envisioning what kinds of tourism they want to cultivate for long-term social, environmental, and economic wellbeing. As Milano et al. (2019) put it, this represents a shift in thinking from a "growth for development" orientation to a "degrowth for liveability" orientation to tourism development (p. 1870). For city destinations like Barcelona that suffer from over-tourism, key tactics include de-marketing, or reducing marketing budgets in order to reduce the public visibility of a destination; shifting resources from marketing to tourism planning and management; increasing citizen representation in tourism governance; and improving the conditions of tourism sector workers (Milano et al., 2019). Elsewhere, Chassagne and Everingham (2019) examine the implementation of degrowth and *Buen Vivir* principles in the rural context of Cotacachi County, Ecuador. This is a nature-oriented tourism region that includes the Cotacachi-Cayapas reserve, where bottom-up tourism development initiatives are oriented around principles of community wellbeing, reciprocity, and decolonization. In this literature, the focus shifts from a quantitative growth model of tourism promotion and development to a reorientation towards models like community-based tourism development, or models of slow tourism, where the driving question is how tourism can best be leveraged to help host communities meet their own development goals (Chassagne & Everingham, 2019; Fletcher et al., 2019; Higgins-Desbiolles et al., 2019). These alternative models have implications for decarbonizing tourism insofar as they highlight the potential for reduced numbers of overall trips and visitors, which would be offset by slower forms of travel that have more positive economic and social impacts on a per-visitor basis.

Finally, without minimizing the harm and suffering caused by the COVID-19 pandemic, this global public health event also has the potential to serve as an unintended socio-cultural tipping point. The global tourism economy has been growing almost non-stop for decades, with a seismic shift in the 1990s following the end of the Cold War and the acceleration of multiple forms of globalization, including the perception (at least among affluent social groups) that hyper-mobile tourism travel is a natural part of everyday social life. The COVID-19 pandemic pulled the rug out from under these assumptions and created a jarring pause in global travel. For rural tourism host communities, the lockdowns and travel restrictions of 2020 quickly revealed the "fragility" of tourism-dependent economies. In discussing COVID-19 and its impacts on islands, Sindico and Ellsmoor (2020) note that the disruption created by the pandemic may also provoke reflection by policymakers and tourism stakeholders about how community relationships with tourism could be reconfigured through the post-COVID recovery to ensure more sustainable or resilient forms of tourism development going forward.

CONCLUSION

Tourism is often held up by policymakers and stakeholders as a sustainable development pathway for rural communities. Tourism development has the potential to offer social–environmental benefits, including providing an economic rationale for protecting natural areas and wildlife. Tourism development also comes with fewer local environmental impacts compared with extractive resource economies, such as forestry, mining, or oil extraction. However, pro-sustainable tourism discourse asks us to ignore the carbon footprints of travel networks of airplanes, cruise ships, ferries, boat tours, buses, and cars that connect tourism sites and host communities to each other and to visitor source markets that are spread around the globe. Our intent here is not to play *gotcha* and pull back the curtain to reveal that tourism fails to live up to its pro-environmental self-image. Rather, we encourage tourism policymakers, practitioners, and researchers to move beyond binary ways of thinking about tourism as either sustainable or unsustainable. There is a spectrum of tourism sustainabilities. Whether tourism can be leveraged to advance sustainable rural community development and meet the UN Sustainable Development Goals is not pre-given but is a question of how it is structured. We provided an overview of the complex and often contradictory relationship between tourism and climate change. However, we also highlighted emergent technological experiments and potential socio-cultural positive tipping points that may help resolve some of the tensions inherent to the relationship between tourism and climate change. If these experiments and positive tipping points can converge and scale up across tourism host communities, then there is potential to create transformative solutions that will allow tourism to better fulfil its promise as a form of sustainable rural community development.

REFERENCES

Baldacchino, G. (2010). *Island Enclaves: Offshoring Strategies, Creative Governance, and Subnational Island Jurisdictions*. Montreal & Kingston: McGill-Queen's University Press.

Bramwell, B., Higham, J., Lane, B., & Miller, G. (2017). Twenty-five years of sustainable tourism and the *Journal of Sustainable Tourism*: Looking back and moving forward. *Journal of Sustainable Tourism, 25*(1), 1–9. doi:10.1080/09669582.2017.1251689.

Chassagne, N., & Everingham, P. (2019). Buen vivir: Degrowing extractivism and growing wellbeing through tourism. *Journal of Sustainable Tourism, 27*(12), 1909–1925. doi:10.1080/09669582.2019.1660668.

Cohen, S.A., Higham, J.E.S., & Cavaliere, C.T. (2011). Binge flying: Behavioural addiction and climate change. *Annals of Tourism Research, 38*(3), 1070–1089. doi:10.1016/j.annals.2011.01.013.

Cunningham, P.A., Huijbens, E.H., & Wearing, S.L. (2012). From whaling to whale watching: Examining sustainability and cultural rhetoric. *Journal of Sustainable Tourism, 20*(1), 143–161. doi:10.1080/09669582.2011.632091.

Dawson, J., Stewart, E., & Scott, D. (2010). Climate change and polar bear viewing: A case study of visitor demand, carbon emissions and mitigation in Churchill, Canada. In C.M. Hall & J. Saarinen (eds), *Tourism and Change in Polar Regions: Climate, Environments and Experiences* (pp. 89–103). London: Routledge.

Elsasser, H., & Messerli, P. (2001). The vulnerability of the snow industry in the Swiss Alps. *Mountain Research and Development, 21*(4), 335–339, 335. doi:10.1659/0276-4741(2001)021[0335:TVOTSI]2.0.CO;2.

Fletcher, R., Murray Mas, I., Blanco-Romero, A., & Blázquez-Salom, M. (2019). Tourism and degrowth: An emerging agenda for research and praxis. *Journal of Sustainable Tourism, 27*(12), 1745–1763. doi:10.1080/09669582.2019.1679822.

George, E.W., Mair, H., & Reid, D.G. (2009). *Rural Tourism Development: Localism and Cultural Change*. Bristol: Channel View Publications.

Gössling, S., Scott, D., Hall, C.M., Ceron, J.-P., & Dubois, G. (2012). Consumer behaviour and demand response of tourists to climate change. *Annals of Tourism Research, 39*(1), 36–58. doi:10.1016/j.annals.2011.11.002.

Gross, M. (2010). *Ignorance and Surprise: Science, Society, and Ecological Design*. Cambridge, MA: MIT Press.

Hall, C.M., Müller, D.K., & Saarinen, J. (2008). *Nordic Tourism: Issues and Cases*. Bristol: Channel View Publications.

Hestres, L.E., & Hopke, J.E. (2020). Fossil fuel divestment: Theories of change, goals, and strategies of a growing climate movement. *Environmental Politics, 29*(3), 371–389. doi:10.1080/09644016.2019.1632672.

Higgins-Desbiolles, F., Carnicelli, S., Krolikowski, C., Wijesinghe, G., & Boluk, K. (2019). Degrowing tourism: Rethinking tourism. *Journal of Sustainable Tourism*, *27*(12), 1926–1944. doi:10.1080/09669582.2019.1601732.

Hughes, T.P., Kerry, J.T., Álvarez-Noriega, M., Álvarez-Romero, J.G., Anderson, K.D., Baird, A.H., et al. (2017). Global warming and recurrent mass bleaching of corals. *Nature*, *543*(7645), 373–377. doi:10.1038/nature21707.

Hurtigruten (n.d.). *Hurtigruten inngår samarbeid med Bellona* [Hurtigruten enters into a partnership with Bellona]. https://www.hurtigruten.no/om-oss/nyheter/hurtigruten-inngar-samarbeid-med-bellona/.

Icelandair (n.d.). *Terms and Conditions for Carbon Calculator*. https://www.icelandair.com/en-ca/support/terms-and-conditions/carbon-calculator-terms/.

Jacob, D., Kotova, L., Teichmann, C., Sobolowski, S.P., Vautard, R., Donnelly, C., et al. (2018). Climate impacts in Europe under +1.5°C global warming. *Earth's Future*, *6*(2), 264–285. doi:10.1002/2017EF000710.

James, L., Olsen, L.S., & Karlsdóttir, A. (2020). Sustainability and cruise tourism in the Arctic: Stakeholder perspectives from Ísafjörður, Iceland and Qaqortoq, Greenland. *Journal of Sustainable Tourism*, *28*(9), 1425–1441. doi:10.1080/09669582.2020.1745213.

Klein, R.A. (2010). The cruise sector and its environmental impact. In S. Christian (ed.), *Tourism and the Implications of Climate Change: Issues and Actions* (Vol. 3, pp. 113–130). Bingley: Emerald Group Publishing Limited.

MacEachern, A. (2001). *Natural Selections: National Parks in Atlantic Canada, 1935–1970*. Montreal & Kingston: McGill-Queen's University Press.

Mangat, R., Dalby, S., & Paterson, M. (2018). Divestment discourse: War, justice, morality and money. *Environmental Politics*, *27*(2), 187–208. doi:10.1080/09644016.2017.1413725.

Marzeion, B., & Levermann, A. (2014). Loss of cultural world heritage and currently inhabited places to sea-level rise. *Environmental Research Letters*, *9*(3), 034001. doi:10.1088/1748-9326/9/3/034001.

Mellon, V., & Bramwell, B. (2016). Protected area policies and sustainable tourism: Influences, relationships and co-evolution. *Journal of Sustainable Tourism*, *24*(10), 1369–1386. doi:10.1080/09669582.2015.1125909.

Milano, C., Novelli, M., & Cheer, J.M. (2019). Overtourism and degrowth: A social movements perspective. *Journal of Sustainable Tourism*, *27*(12), 1857–1875. doi:10.1080/09669582.2019.1650054.

Motion, J. (2019). Undoing art and oil: An environmental tale of sponsorship, cultural justice and climate change controversy. *Environmental Politics*, *28*(4), 727–746. doi:10.1080/09644016.2017.1416904.

Neves, K. (2010). Cashing in on cetourism: A critical ecological engagement with dominant e-NGO discourses on whaling, cetacean conservation, and whale watching. *Antipode*, *42*(3), 719–741. doi:10.1111/j.1467-8330.2010.00770.x.

North Sailing (n.d.). *North Sailing takes action – goes from ecofriendly to carbon neutral*. https://www.northsailing.is/carbon-free-2015/.

Pachauri, R.K., Allen, M.R., Barros, V.R., Broome, J., Cramer, W., Christ, R., et al. (2014). *Climate Change 2014: Synthesis Report. Contribution of Working Groups I, II and III to the Fifth Assessment Report of the Intergovernmental Panel on Climate Change*. Geneva: IPCC.

Peeters, P., Higham, J., Cohen, S., Eijgelaar, E., & Gössling, S. (2019). Desirable tourism transport futures. *Journal of Sustainable Tourism*, *27*(2), 173–188. doi:10.1080/09669582.2018.1477785.

Reed, M.G., & Gill, A.M. (1997). Tourism, recreational, and amenity values in land allocation: An analysis of institutional arrangements in the postproductivist era. *Environment and Planning A: Economy and Space*, *29*(11), 2019–2040. doi:10.1068/a292019.

Reichwein, P. (2014). *Climber's Paradise: Making Canada's Mountain Parks, 1906–1974.* Edmonton, AB: University of Alberta Press.

Ren, C., Bjørst, L.R., & Dredge, D. (2016). Composing Greenlandic tourism futures: An integrated political ecology and actor-network theory approach. In M. Mostafanezhad, R. Norum, E.J. Shelton, & A. Thompson-Carr (eds), *Political Ecology of Tourism: Community, Power and the Environment* (pp. 284–301). London: Routledge.

Scott, D., Hall, C.M., & Gössling, S. (2016). A report on the Paris Climate Change Agreement and its implications for tourism: Why we will always have Paris. *Journal of Sustainable Tourism, 24*(7), 933–948. doi:10.1080/09669582.2016.1187623.

Sharpley, R. (2020). Tourism, sustainable development and the theoretical divide: 20 years on. *Journal of Sustainable Tourism, 28*(11), 1932–1946. doi:10.1080/09669582.2020.1779732.

Silva, L. (2015). How ecotourism works at the community-level: The case of whale-watching in the Azores. *Current Issues in Tourism, 18*(3), 196–211. doi:10.1080/13683500.2013.786027.

Sindico, F., & Ellsmoor, J. (2020). *COVID-19: A Global Island Response.* Policy Brief No. 14, University of Strathclyde & Island Innovation.

Sisneros-Kidd, A.M., Monz, C., Hausner, V., Schmidt, J., & Clark, D. (2019). Nature-based tourism, resource dependence, and resilience of Arctic communities: Framing complex issues in a changing environment. *Journal of Sustainable Tourism, 27*(8), 1259–1276. doi: 10.1080/09669582.2019.1612905.

Steiger, R., Scott, D., Abegg, B., Pons, M., & Aall, C. (2017). A critical review of climate change risk for ski tourism. *Current Issues in Tourism, 22*(11), 1343–1379. doi:10.1080/13683500.2017.1410110.

Stoddart, M.C.J., & Graham, P. (2016). Nature, history, and culture as tourism attractors: The double translation of insider and outsider media. *Nature and Culture, 11*(1), 22–43. doi:10.2307/26430610.

Stuart, D., Gunderson, R., & Petersen, B. (2020). The climate crisis as a catalyst for emancipatory transformation: An examination of the possible. *International Sociology, 35*(4), 433–456. doi:10.1177/0268580920915067.

Tàbara, J.D., Frantzeskaki, N., Hölscher, K., Pedde, S., Kok, K., Lamperti, F., et al. (2018). Positive tipping points in a rapidly warming world. *Current Opinion in Environmental Sustainability, 31*, 120–129. doi:https://doi.org/10.1016/j.cosust.2018.01.012.

Thrift, N. (2001). Still life in nearly present time: The object of nature. In P. Macnaghten & J. Urry (eds), *Bodies of Nature* (pp. 34–57). London: SAGE Publications Ltd.

Turpie, J., Warr, B., & Ingram, J.C. (2015). *Benefits of Forest Ecosystems in Zambia and the Role of REDD+ in a Green Economy Transformation.* Nairobi, Kenya: United Nations Environment Programme (UNEP). https://www.uncclearn.org/sites/default/files/inventory/flyer-benefits_of_forest_ecosystem-zambia.pdf.

Uldam, J. (2017). Social media visibility: Challenges to activism. *Media, Culture & Society, 40*(1), 41–58. doi:10.1177/0163443717704997.

United Nations (2015). *General Assembly Resolution A/RES/70/1. Transforming Our World: The 2030 Agenda for Sustainable Development.* https://undocs.org/en/A/RES/70/1.

Visit Norway (2017, March 16). Close to Norway by electric car. Retrieved March 16, 2017, from https://www.visitnorway.no/media/nyheter-fra-norge/tett-pa-norge-med-elektrisk-bil/.

World Tourism Organization & United Nations Development Programme (2017). *Tourism and the Sustainable Development Goals – Journey to 2030.* https://www.e-unwto.org/doi/epdf/10.18111/9789284419340.

15. Tourism: vulnerability and resiliency in rural regions

David W. Marcouiller

INTRODUCTION

Disasters and other regional perturbations create situations that exacerbate social and economic hardship. The extent of the hardship and the time it takes to return to normalcy are reflective of underlying vulnerabilities. These vulnerabilities are often most severe among the lowest skilled and least advantaged in society. Poverty, family strife, and historic racism are key drivers of vulnerability (Chakraborty et al., 2014; Kim et al., 2018). Vulnerability also varies by location along the urban–rural continuum (Cutter et al., 2016). Rural and remote regions face limited opportunities that allow for resiliency to shocks and thus experience unique vulnerabilities.

Recent global influenza pandemics highlight the vulnerability of rural regions dependent on visitor spending. While not new, this vulnerability has history that precedes the COVID-19 pandemic (Hall et al., 2020; Gössling et al., 2020). Wars, volcanic eruptions, Ebola outbreaks, famines, flooding events, and other hazards have long plagued the abilities that motivate leisure travel demand and its offshoot tourism (Jiang & Ritchie, 2017; Orengo-Serra & Sánchez-Jauregui, 2021).

The tourism and hospitality sectors are not monolithic concepts; impacted sectors flexibly accept traveler spending. Often carried out strategically, travel-sensitive sectors react to changing regional demands and income streams (Bronner & de Hoog, 2017; Sheppard & Williams, 2016). Dependent on disposable incomes, tourism and the hospitality sector are often first in line for spending reductions by leisure travelers (households). Tyrrell and Johnston (2008) discuss economic and fiscal resiliency as a set of flexible trade-offs that reflect travel decisions between domestic and international destinations. As clearly pointed out by Watson and Deller (2022), substitutes exist for travel decisions; short distance versus long distance (local versus inter-state or international). This provides differences in regional economic resilience metrics. Further, this local/non-local distinction affects the scope of what we typically define as tourism.[1]

The resiliency of tourism reflects an ability of traveler decisions and spending to rebound to pre-event levels following a perturbation. In essence, economic resiliency of tourism examines how growth paths in hospitality sectors dependent on traveler spending deviate in response to some external shock (Watson & Deller, 2022). The literature specific to resiliency with respect to hospitality is growing (N.A. Brown et al., 2017).

Vulnerability, on the other hand, reflects social, economic, and environmental elements that support the underlying phenomena of tourism. While necessary, resiliency alone is insufficient to derive usable conclusions about the sustainability of tourism within regional structures. Vulnerability goes hand in hand with resiliency and requires incorporation of economic, social, cultural, and environmental attributes (Kim & Marcouiller, 2015b; R.M. Brown, 2017). For example, one important element of vulnerability with respect to tourism is the tendency for hospitality sectors to employ large numbers of low-wage, seasonal, and part-time employees relative to other sectors of the economy. This raises integrative tourism planning issues that relate to regional job quality and the provision of livable, stable, consistent employment opportunities that include advancement and professional development. This is particularly true as dependence on these forms of employment dominates labor markets found in less economically diverse rural regions.

Outcomes are important. In this light, it is good to remember that outcomes of regional planning involve a variety of beneficial attributes that are elusive to track empirically. An appropriate outcome of both integrative planning for tourism and climate change involves making disenfranchised communities who are vulnerable to rapid change less vulnerable. This has little to do with resiliency as currently defined and empirically tracked in the existing literature. As a matter of fact, assessment of vulnerability is a precursor to examining resiliency. Indeed, outcomes of integrative tourism planning address making the vulnerable less vulnerable, not necessarily more resilient to their historic reliance on distinctly tourism-related income streams. Distribution of benefits and costs are central to this discussion.

In this chapter, I outline issues of academic discourse with respect to the vulnerability of tourism development in rural, and often, remote regions. This is contrasted and amplified with the literature on tourism resilience. Relying on the published literature, I will focus on key factors behind vulnerability and develop a conceptual framework leading to more robust and integrative planning approaches. Rural resilience with respect to tourism in the post-industrialized and increasingly amenity-driven future will require evaluation, consideration of alternatives, and a more progressive approach to decision-making.

Following this introduction, the chapter is organized into three subsequent sections. First, I outline the current issues of rural development associated with rural tourism from the perspective of their role in economic growth and change. I then contrast vulnerability of rural communities within the context of how rural tourism has transformed amenity-rich rural regions. I conclude this chapter with a section that both summarizes key points and ties these points to relevant public policy implications.

REGIONAL ECONOMIC SUSTAINABILITY AND GROWTH

Rural regions often suffer from a lack of economic diversity and have been subject to severe booms and busts. Simply stated, a limited number of options exist for employment of factor resources and subsequent income-generation opportunities. Dependency on a limited number of economic sectors reduces the variety of options for employment. Market vicissitudes in sectoral activity create periods where supply and demand reflect dramatic wealth creation. At the same time, when these market conditions decline, these same regions suffer from dramatic economic downturns accompanied by societal and economic hardship. When sectoral demands decline, regional economic activity wanes and can stop completely.

This is somewhat unique to rural regions relative to more urban and economically diverse regions (Hamilton et al., 2008; Bowe & Marcouiller, 2007). While there are a variety of theories that speak to this, one popular explanation derives from the lack of economic agglomeration in rural regions and the subsequent lack of pecuniary externalities that are found as urbanization increases (A.J. Scott & Storper, 2003; Hoover & Giarratani, 2020). Demand and supply linkages in close proximity provide cost savings for production and sales through both urbanization and localization economies. These pecuniary externalities are simply less prevalent in rural regions. Thus, resiliency in economic activities is likewise limited to the comparative advantages of rural regions, namely land and amenity resources.

More diversified regional economies tend to be more economically resilient and able to bounce back from intermittent perturbations. This is clear from the work of Watson and Deller (2022) who found that more diverse regional economies were significantly more resilient. In other words, those economies more dependent on tourism were found to have lower levels of economic resiliency. Further, the same study suggested that regions with higher levels of poverty tended to be less resilient. Given the lack of economic diversity, rural incomes tend to be lower than urban counterparts and poverty rates tend to be higher.

The comparative advantage of rural regions lies in land-based factors of production. Increasingly, underlying natural amenity attributes are intrinsically land based and serve as important motivators for travel and its offshoot, tourism. This provides the entrepreneurial stimulus required to supply the indirect goods and services that tourism provides to meet this demand, namely, the personal service and retail sectors found within the hospitality grouping of economic activities. Indeed, tourism is all about the entrepreneurial stimulus that takes advantage of natural land-based amenity resources. These resources are a complex mixture of jointly produced outputs that utilize both private and publicly available resources.

To be sure, leisure-based traveler spending (tourism) has played an important role in diversifying the economies of traditional resource-based regions. Forestry, mining, and agricultural regions possess important externalities upon which tourism demands rest. As traditional resource bases decline, this leisure-based traveler spending (tourism) is often left as a dominating influence in these regions. Some might even go so far as to use the term "tourism dependent". Thus, tourism can be viewed as a new alternative form of resource-dependency in and of itself.

While tourism comes in many forms and is based on the underlying amenity base, these resources tend, like land, to be fixed in place and are regionally specific. The substitutability of travel type discussed by Bronner and de Hoog (2016; 2017),

Sheppard and Williams (2016) and Watson and Deller (2022) can be manipulated to a certain extent, but there are historic precedents that serve as baselines. These can be severely limiting in rural regions; remoteness is the basis of this statement. Market centers of tourism demand are often simply too far distant to allow for robust travel demands.

Households that rely on the retail and service sectors related to tourism for employment can find themselves without options when the source of receipts for tourism businesses declines. Further, the income streams from tourism have long been significantly below prevailing regional wage rates (Lee & Kang, 1998; Bernhardt et al., 2003; Marcouiller & Xia, 2008). This is most likely due to low pay rates, sub-minimum wage laws, and the reliance on tip income. Combine this with the general lack of benefits and highly seasonal nature of tourism demands and it is an obvious conclusion that tourism is limiting with respect to sustainable income sources for households. These basic shortcomings of tourism income dependence make rural households more vulnerable.

Again, with respect to sustainable household incomes, the notion of returning to the status quo following a disturbance is of little comfort to those who struggle to maintain a household. The fact that travel demands and tourism-sector activity resume following disasters, as measured through resiliency analysis, says little about the vulnerability of rural populations. Recent experience from the COVID-19 pandemic in the United States suggests that labor shortages are leading to increased wages and opportunities in other (non-tourism) sectors. This stimulates a decline in those seeking employment in tourism and exacerbates the labor shortage in tourism. This can be a good thing for labor markets. Making vulnerable populations less vulnerable requires an improvement in the conditions, income streams, and benefits derived from resumed travel demands. Simply stated, hospitality firms need to make up for this with dramatically improved wage and benefit structures. This is likely to increase the cost of leisure travel, reducing the demand for tourism.

Social capital that reflects the extent to which local decision-making assets affect regional economic change is important with respect to both vulnerability and economic resiliency. There are a growing number of studies that suggest that the ability and timing of economic response to natural disasters is significantly related to social capital. These studies are primarily found in the community vulnerability and disaster resiliency literature (cf. Adger, 2003; Tierney, 2014; Miles, 2015; Kim et al., 2021). More work is required to identify the role of social capital in tourism resiliency. It may be the case that the role of social capital in resiliency differs for tourism by disaster type.

Kumar (2021) and Jones (2009) suggest that tourism sector investments tend to have higher risk profiles due to significant potential downturns. From a regional hospitality sector perspective, Watson and Deller (2022) likewise find that higher levels of tourism dependency result in lower levels of regional economic resiliency. Given the depth and length of downturns, potential tourists reduce discretionary spending, directly impacting hospitality sectors.

Region-specific vulnerability of tourism-sensitive sectors is most acute where geologic and climatic catastrophes combine with high levels of dependence (Scott et al., 2019). This is particularly true for island economies (e.g., Caribbean Small Island Developing States), given global climate change (Mycoo, 2018; Scott et al., 2019; Becken et al., 2017), and coastal regions, given the myriad of potential disaster scenarios (Nguyen et al., 2018). This notion of vulnerability is further exacerbated by human-caused issues brought on by terrorism and war. This increases the extent and severity of vulnerability (Liu & Pratt, 2017).

Economic vulnerability to tourism is likewise a function of tourism type. Canh & Thanh (2020) suggest that domestic tourism results in lower vulnerability indices as compared to international tourism. Cellini and Cuccia (2015) find that structural issues and regional economic dependency on tourism result in a reduction in that region's resilience to economic downturns such as the Great Recession in Italy. This is confirmed by the results of Watson and Deller (2022) in the United States. In areas where tourism is mostly made up of local and regional residents, the prevalence of a tourism sector is likely to act as a stabilizing force in the local economy. However, in places that depend on tourists coming from further away, the prevalence of a tourism sector is likely to have the opposite effect and make the regional economy less resilient to economic shocks.

VULNERABILITY OF REGIONS

The future belongs to those who prepare for it. This statement, attributed to Ralph Waldo Emerson, speaks volumes to the topic of this chapter. Exposure to sudden change affects households as they exist in time and place. Preparedness, while important, requires forward thinking, a certain level of resources, and a tacit recognition of risk. Disasters, broadly defined, bring to the fore the simple fact that preparedness for sudden shocks at the household level is not widely practiced. This is particularly true among lower-income households in rural regions.

An overview of the concepts that link exposure effects within the context of vulnerability and resiliency are outlined in Figure 15.1. Given prescience, ample resources, and an ability to assess and act to match expectations with risk, exposures can produce an ability to rebound and succeed. The investment of individuals to be self-reliant upon exposure is not distributed evenly among households, which exist within a variety of vulnerability realms that include economic, social, cultural, and environmental attributes.

While individual wealth (or the lack thereof) plays a central role in the financial vulnerability of households, many of the social, cultural, and environmental attributes are supra-individual and exist in networks of others in a collective fashion. Take, for example, the knowledge required to form both prescience of the future and an assessment of risk. These are fundamental core intellectual elements linked directly to education and research, and their outreach. The cultural mores that support a value in education and research are often at a level above the individual household.

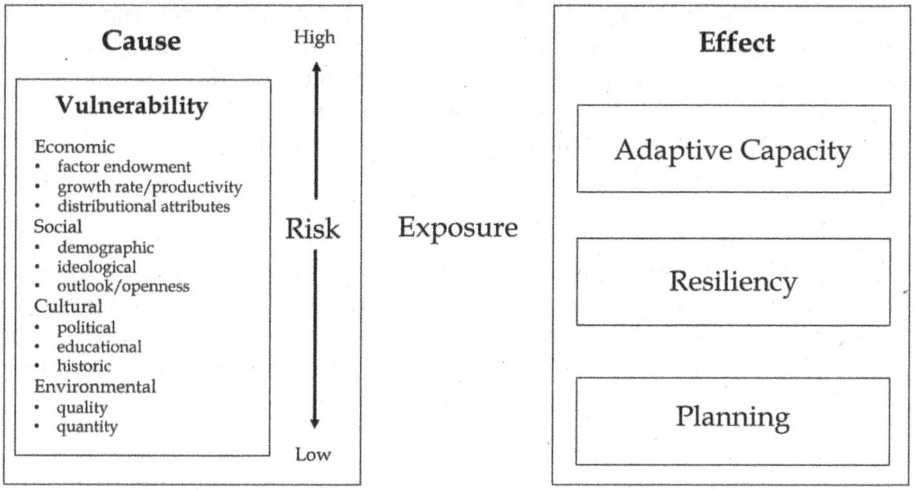

Source: Adapted from Kim and Marcouiller (2015a).

Figure 15.1 Vulnerability and its causal relationship with resiliency

This is where collective action is important. Proactive public sector planning can be assistive but is not uniformly distributed throughout space. The linkages between vulnerability, adaptive capacity, and resilience with respect to disasters are central to sustainability (Gallopín, 2006).

The intransigence of regional vulnerability is exacerbated by the self-selection process. Those who have the means to move out of risk, do so. This concentrates vulnerability in regions exhibiting risk. Many of those who are left behind make up the workforce for restaurants, hotels, resorts, and travel. These workforces often exist on the lowest rung of the skill–career ladder and thus have limited options. Rather, the vast number of jobs created are much less skills based. Youth lacking skills, single-headed households, transitory employees, low-income elderly, and others lacking opportunity comprise the tourism workforce. Increasingly, the general lack of preparation for retirement also serves as the basis for employment in retail and service sectors. These are the most vulnerable with respect to downsizing of firm activity due to disruption (pandemics and other disasters).

Further, tourism is driven forward by entrepreneurs. To be sure, there are those who choose careers and entrepreneurial opportunities in the hospitality sector. These are the primary beneficiaries of tourism. This can be both rewarding and quite risky. Restaurateurs, in particular, face unique challenges in start-ups. Also, many in the so-called "gig" economy are self-employed subcontractors. This substantially reduces the responsibility of corporate interests who increasingly require this type of labor to be contractual and "at arm's-length" of the corporation itself, thus limiting the need for benefits such as time off, healthcare, and retirement provisions. While there are profits to be made from traveler spending, success can be elusive.

Remoteness hampers the sustainability of rural households. Remote rural regions suffer from cycles of booms and busts indicative of resource dependency and a general lack of economic diversity. Tourism dependency is no different. Booms can be generally positive, but as dependency takes hold, busts can be devastating to rural households. Again, these are central elements of vulnerability.

Confounding empirical results of resiliency (cf. Watson & Deller, 2022; Sheppard & Williams 2016) suggest underlying issues of complex relationships that relate to vulnerability. Those regions most reliant on tourism receipts are the most vulnerable. At the very least, increased diversity fosters improved resilience. Again, looking at resilience without vulnerability provides only partial explanations. Isolating resilience without vulnerability characteristics leads to confounding and often spurious interpretations when measured through an integrative tourism planning perspective. Further, it leads to murky conclusions, opaque to scientific process.

CONCLUSIONS, POLICY IMPLICATIONS, AND FURTHER RESEARCH NEEDS

In this chapter, I have addressed the uniqueness of regional vulnerability brought about by tourism development in rural and remote regions. Relying on the published literature, I focused on key factors behind vulnerability and developed a conceptual framework leading to more robust planning approaches. Rural resilience with respect to tourism in the post-industrialized and increasingly amenity-driven future requires evaluation, consideration of alternatives, and a more progressive approach to decision-making.

Theoretical development is needed to better understand how tourism-sensitive business activity relates to regional development goals and aspirations. As an outcome, viewing vulnerability rather than resilience would provide a longer-term, more sustainable approach. Jobs created by tourism tend to be of lower quality, paying wages well below prevailing regional wages of other sectors. They tend to be seasonal, tip dependent, and lacking benefits. This leads to the simple conclusion that rural households, which are dependent on tourism-sensitive jobs, tend to be in jeopardy of increased vulnerability; unable to prepare for disturbances brought on by disasters.

Further, another area of theory building needs to embrace the stark distributional benefit differentials of tourism. The primary benefits of tourism accrue to owners of tourism-sensitive firms. Regional tourism policies brought on by motivating factors associated with entrepreneurial activity focus on the cheerleading of regional marketers. This is a hallmark of tourism boosterism. To this point, if some travel is deemed positive for tourism businesses, more is obviously better. Sustainability, as a concept, is often not within the purview of marketers; something especially important in tourism development. A shift to reducing the vulnerability of rural households through more integrative tourism planning would view both benefits and costs of tourism in an objective fashion.

A continual boosterism approach neglects this but rather uses employment growth as the key benefit. Jobs in tourism-sensitive firms are consistently low-wage, lacking benefits and subject to severe seasonality. Travel and disposable income spending generate disproportionately large numbers of low-quality jobs. Further, dependency on these income streams, given declines in demand for travel, create enhanced vulnerability of these households.

Regional economic policy is an important avenue for improved sustainability of rural household incomes. Abolishing the sub-minimum wage for servers who are dependent on tip income and establishment of a livable minimum wage would be a step in the right direction. This increased income from tourism jobs would help alleviate vulnerability by providing a resource base with which to improve preparedness for future downturns or disasters. Arguments against the livable wage are focused on entrepreneurial success. Not discounting this, an obvious statement needs to be made in support of a national (or federal) minimum wage that would increase the cost of travel equally across regions.

Regional (city or county) efforts would cause extra-regional spillovers and competitive inequalities across regions. Even state-level efforts at establishing livable minimum wages would create inequalities in competitiveness in regions near state borders. This effort of livable wages needs to be national to level the playing field. Indeed, this would increase the costs of travel-dependent spending and perhaps hamper resiliency. But again, reducing the vulnerability of rural households, if focused on as a precursor to resiliency, appears to be a more appropriate outcome.

Joint production of the amenity base is another arena of public policy that requires consideration. Much of the demand for travel is based on activities on public lands that surround tourism centers. The management of public lands for recreation is certainly not costless. Likewise, the costs of developing publicly funded infrastructure in tourism destinations could be addressed through increased use of user fees or taxes borne by travelers. As clearly pointed out by Brasington (2022), public investments in parks and recreation result in widespread benefits that extend beyond hospitality sectors.

As climate change brings about increased severity and more frequent weather-related disasters and other regional disruptions, the time to act with more integrative tourism planning is upon us. This policy action needs to be evidence-based. Outcomes that focus on making rural households less vulnerable serve as an improved empirical metric. There continues to be a strong need for further empirical research that develops sound proxies for tourism impacts that account for both vulnerability and its causal relationship to resiliency.

NOTE

1. To be sure, as travel decisions focus on more local pursuits, the term "tourism" becomes a moot subject. Local recreational pursuits (as decisions become local versus non-local) now are a part of local quality-of-life and outdoor recreation assets of communities, not

travel. Travel is a basic definitional attribute of the term "tourism". Local quality-of-life has been a vibrant topic in the literature for the past half-century (as noted by Dissart & Deller, 2000 and Bronner & de Hoog, 2016), but it remains outside the scope of "tourism" itself. Tourism entails some form of non-local travel. Purely local recreational pursuits, while associated with leisure, relate to the local overall economic condition that extends beyond the scope of *tourism* resilience and/or vulnerability.

REFERENCES

Adger, W.N. (2003). Social capital, collective action, and adaptation to climate change. *Economic Geography*, *79*(4), 387–404. https://doi.org/10.1111/j.1944-8287.2003.tb00220.x.

Becken, S., Mahon, R., Rennie, H.G., & Shakeela, A. (2014). The tourism disaster vulnerability framework: An application to tourism in small island destinations. *Natural Hazards*, *71*(1), 955–972.

Bernhardt, A., Dresser, L., & Hatton, E. (2003). The coffee pot wars: Unions and firm restructuring in the hotel industry. In E. Appelbaum, A. Bernhardt, & R.J. Murnane (eds), *Low-Wage America: How Employers Are Reshaping Opportunity in the Workplace* (pp. 33–76). New York: Russell Sage Foundation.

Bowe, S., & Marcouiller, D.W. (2007). Natural resources and the tourism–timber tradeoff: Issue of regional dependency and economic diversity. *Forest Policy and Economics*, *9*, 653–670.

Brasington, D. (2022). Local economic growth and local government investment in parks and recreation, or five cheese pizzas for $2.6 million. *Journal of Regional Science*, *62*(1), 81–95. doi:10.1111/jors.12554.

Bronner, F., & de Hoog, R. (2016). Crisis resistance of tourist demand: The importance of quality of life. *Journal of Travel Research*, *55*(2), 190–204.

Bronner F., & de Hoog, R. (2017). Tourist demand reactions: Symmetric or asymmetric across the business cycle? *Journal of Travel Research*, *56*(7), 839–853.

Brown, N.A., Rovins, J.E., Feldmann-Jensen, S., Orchiston, C., & Johnston, D. (2017). Exploring disaster resilience within the hotel sector: A systematic review of literature. *International Journal of Disaster Risk Reduction*, *22*, 362–370.

Brown, R.M. (2017). "The tourists still come, but they don't buy as much as before": Vulnerability and resilience in two bay island communities in the wake of the global financial crisis. *The Qualitative Report*, *22*(4), 1112–1136.

Canh, N.P., & Thanh, S.D. (2020). Domestic tourism spending and economic vulnerability. *Annals of Tourism Research*, *85*, 102876.

Cellini, R., & Cuccia, T. (2015). The economic resilience of tourism industry in Italy: What the "great recession" data show. *Tourism Management Perspectives*, *16*, 346–356.

Chakraborty, J., Collins, T.W., Montgomery, M.C., & Grineski, S.E. (2014). Social and spatial inequities in exposure to flood risk in Miami, Florida. *Natural Hazards Review*, *15*(3), 04014006.

Cutter, S.L., Ash, K.D., & Emrich, C.T. (2016). Urban–rural differences in disaster resilience. *Annals of the American Association of Geographers*, *106*(6), 1236–1252.

Dissart, J.C., & Deller, S.C. (2000). Quality-of-life in the planning literature. *Journal of Planning Literature*, *15*(1), 135–161.

Gallopín, G.C. (2006). Linkages between vulnerability, resilience, and adaptive capacity. *Global Environmental Change*, *16*, 293–303.

Gössling, S., Scott, D., & Hall, C.M. (2020). Pandemics, tourism and global change: A rapid assessment of COVID-19. *Journal of Sustainable Tourism*, *29*(1), 1–20.

Hall, C.M., Scott, D., & Gössling, S. (2020). Pandemics, transformations and tourism: Be careful what you wish for. *Tourism Geographies*, *22*(3), 577–598.

Hamilton, L.C., Hamilton, L.R., Duncan, M., & Colocousis, C. (2008). *Place Matters: Challenges and Opportunities in Four Rural Americas*. Durham, NH: Carsey Institute.

Hoover, E.M., & Giarratani, F. (2020). An introduction to regional economics. In S. Loveridge (ed.), *The Web Book of Regional Science*. Morgantown, WV: Regional Research Institute. rri.wvu.edu/web-book.

Jiang, Y., & Ritchie, B.W. (2017). Disaster collaboration in tourism: Motives, impediments, and success factors. *Journal of Hospitality and Tourism Management*, *31*, 70–82.

Jones, P. (2009). Impact of the global recession on the hospitality and tourism industry. *Tourism and Hospitality Research*, *9*(4): 363–367.

Kim, H., & Marcouiller, D.W. (2015a). Urban vulnerability and resiliency to natural disasters: An integrative tourism planning perspective. In P. Fillion, G. Sands, & M. Skidmore (eds), *Cities at Risk: Planning for and Recovering from Natural and Human Urban Disasters* (pp. 159–188). Aldershot, UK: Ashgate.

Kim, H., & Marcouiller, D.W. (2015b). Considering disaster vulnerability and resiliency: The case of hurricane effects on tourism-based economies. *Annals of Regional Science*, *54*(3), 945–971.

Kim, H., Marcouiller, D.W., & Woosnam, K.M. (2018). Rescaling social dynamics in climate change: The implications of cumulative exposure, climate justice, and community resilience. *Geoforum*, *96*, 129–140.

Kim, H., Marcouiller, D.W., & Woosnam, K.M. (2021). Multilevel climate governance, anticipatory adaptation, and the vulnerability–readiness nexus. *Review of Policy Research*, *38*(2), 222–242.

Kumar, D. (2021). European travel and leisure sector and uncertainties: A risk spillover analysis. *Tourism Economics*, August. https://doi.org/10.1177/13548166211035954.

Lee, C.K., & Kang, S. (1998). Measuring earnings inequality and median earnings in the tourism industry. *Tourism Management*, *19*(4), 341–348.

Liu, A., & Pratt, S. (2017). Tourism's vulnerability and resilience to terrorism. *Tourism Management*, *60*, 404–417.

Marcouiller, D.W., & Xia, X. (2008). Distribution of income from tourism sensitive employment. *Tourism Economics*, *14*(3), 545–565.

Miles, S.B. (2015). Foundations of community disaster resilience: Well-being, identity, services, and capitals. *Environmental Hazards*, *14*(2), 103–121. https://doi.org/10.1080/17477891.2014.999018.

Mycoo, M.A. (2018). Beyond 1.5°C: Vulnerabilities and adaptation strategies for Caribbean Small Island Developing States. *Regional Environmental Change*, 18, 2341–2353.

Nguyen, D.N., Imamura, F., & Iuchi, K. (2018). Barriers towards hotel disaster preparedness: Case studies of post 2011 Tsunami, Japan. *International Journal of Disaster Risk Reduction*, *28*, 585–594.

Orengo-Serra, K.L., & Sánchez-Jauregui, M. (2021). Coping with earthquakes and COVID-19: A perspective of customer relationship management. *Estudios Gerenciales*, *37*(159), 318–331.

Scott, A.J., & Storper, M. (2003). Regions, globalization, development. *Regional Studies*, *37*(6–7), 579–593.

Scott, D., Hall, C. M., & Gössling, S. (2019). Global tourism vulnerability to climate change. *Annals of Tourism Research*, *77*, 49–61.

Sheppard, V.A., & Williams, P.W. 2016. Factors that strengthen tourism resort resilience. *Journal of Hospitality and Tourism Management*, *28*, 20–30.

Tierney, K. (2014). *The Social Roots of Risk: Producing Disasters, Promoting Resilience*. Stanford, CA: Stanford University Press.

Tyrrell, T.J., & Johnston, R.J. (2008). Tourism sustainability, resiliency and dynamics: Towards a more comprehensive perspective. *Tourism and Hospitality Research*, *8*(1), 14–24.

Watson, P., & Deller, S. (2022). Tourism and economic resilience. *Tourism Economics*, *28*(5), 1193–1215. doi.org/10.1177/1354816621990943.

16. Degrowing rural tourism development: thinking globally to save the local

Alexander Safonov and C. Michael Hall

INTRODUCTION: TOURISM SUSTAINABILITY AND ENVIRONMENTAL CONCERNS

Tourism is becoming less sustainable than ever (Hall, 2013; Gössling & Peeters, 2015). Tourism systems, consisting of a generating region, a transit region through which tourists travel to and from a destination and a destination itself, imply that impacts occur not only at a destination level but throughout all the stages of travel with the effects being uneven over space and time. While tourism development has long been argued by policymakers as making a positive contribution to locations and their sustainable development (Gössling & Peeters, 2007; Gössling et al., 2009), it remains a considerable contributor to carbon emissions (Lenzen et al., 2018), with the impacts of climate change already being experienced by many locations, including rural areas in which agricultural production is affected by shifts in temperatures and the frequency and intensity of weather events (Howden et al., 2007; Wiréhn, 2018; Raza et al., 2019). Moreover, other impacts of tourism are often overlooked. Tourism activities induce changes in land, water and energy use, biodiversity loss, and pollution (Gössling & Peeters, 2015; Lew, 2014), with many of these things slowly accreting and accumulating over time. Changes in land use often receive limited attention, for example, and when considered, the focus historically has tended to be on commercial accommodation or resort development. Thus, attention to the effects of second home occupancy and development or the more recent short-term rental of properties on sharing platforms, is often much more limited in assessments of the impacts of tourism in rural areas, which is surprising given that they often provide more overnight beds than the formal accommodation sector (Hall & Müller, 2018).

Tourism can hardly be considered environmentally friendly. Even the ecotourism that promotes 'eco' and environmentally favourable holidays produces substantial emissions (Hall, 2009; 2014). The majority of tourism-related CO_2 emissions come from air travel between and within the developed countries, with the main concern being long-haul flights. However, along with air travel, car-based tourism is also a significant contributor to climate change (Lenzen et al., 2018). Moreover, various government subsidies play a substantial role in encouraging air travel, allowing the occasional offers of tickets lower than the price of air fuel (Gössling, 2020). Ironically, the growth of budget airlines and cheap air travel has been significant for the growth of rural tourism given the emergence of secondary airports as budget airline destinations and their role as gateways to second homes and other tourism opportunities (Hall & Müller, 2018).

The contemporary context of sustainable tourism, as noted above, raises substantial questions with respect to how rural tourism development can actually be sustainable and the role of community development approaches within such a process. Arguably, from the perspective of sustainability, a core issue with respect to a community focus is that it is, almost by definition, local in orientation. Yet, given that the demand for resources required for tourism continues to grow over time (Gössling & Peeters, 2015), can a community focus on tourism development provide a pathway to sustainable tourism? This is the question to which this chapter seeks to respond.

This chapter is divided into three main sections. First, it seeks to place rural tourism within the context of the notion of a tourism system and how this might be understood with respect to sustainability. Second, it discusses the impact of COVID-19 and whether this might lead to a reduction in tourism demand. Third, it focuses on issues of growth and the prospects of degrowth.

SUSTAINABILITY AND COMMUNITY DEVELOPMENT IN THE TOURISM SYSTEM

The resource needs of tourism systems enable the understanding of the impacts of tourism as well as sustainable development concerns. As it is usually defined, sustainable development includes the three dimensions of economic, social and environmental sustainability, with organisations such as the United Nations World Tourism Organization (UNWTO) suggesting that a balanced approach to sustainability means that these three dimensions are used to weigh up the sustainability of tourism development (UNWTO & UNEP, 2005; UNWTO & UNDP, 2017). For example, the UNWTO–UNDP report on tourism and the Sustainable Development Goals (SDGs) defined sustainable tourism in relation to impacts as 'tourism that takes full account of its current and future economic, social and environmental impacts' (UNWTO & UNDP, 2017, p. 17). Economic sustainability concerns the long-term viability of firms and activities, generating prosperity for communities in a cost-effective manner. Social sustainability tends to be focused on equal opportunities and human rights within society. It focuses on strengthening local communities, maintaining well-being support systems, recognising different cultures, providing benefit distribution and poverty reduction. Environmental sustainability is primarily concerned with resource conservation and management, including non-renewable and finite resources, which implies the minimisation of pollution, appropriate land and water use, and conservation of biological diversity. All three dimensions are deeply entwined with one another (UNWTO & UNEP, 2005). As the UNWTO and UNDP (2017) suggest, without any irony (Hall, 2019), following their review of national reporting on the SDGs and tourism, 'Evidently, if tourism is not well managed, it can have a negative impact on people, the planet, prosperity and peace' (p. 31).

The problem of the issue of balance in relation to tourism and the SDGs, and sustainable tourism overall, is that it raises basic questions about the significance of biophysical processes for tourism production systems. According to Hall (2019), this therefore requires addressing dualistic notions about the natural world and its relationship to socio-economic processes, including sustainable tourism and community development. As Mansfield (2003) questioned, 'How can we treat nature, biophysical processes, and the "organic" as analytically significant without treating them as a mechanistic function or a materiality that is outside of social existence?' (p. 10). Yet, arguably, this is what the standard notion of a tourism system does by segregating the environment from the stages of the tourism journey and casting it as a passive backdrop. Figure 16.1 provides the 'classic' representation of a tourism system (Hall & Page, 2010) in which the trip, as well as the tourist-generating region and the destination, including the destination community, are surrounded by the environment. However, the multi-scale nature of the tourism system (from the local to the global) is generally only weakly acknowledged in the tourism studies literature, or otherwise ignored (Hall, 2005).

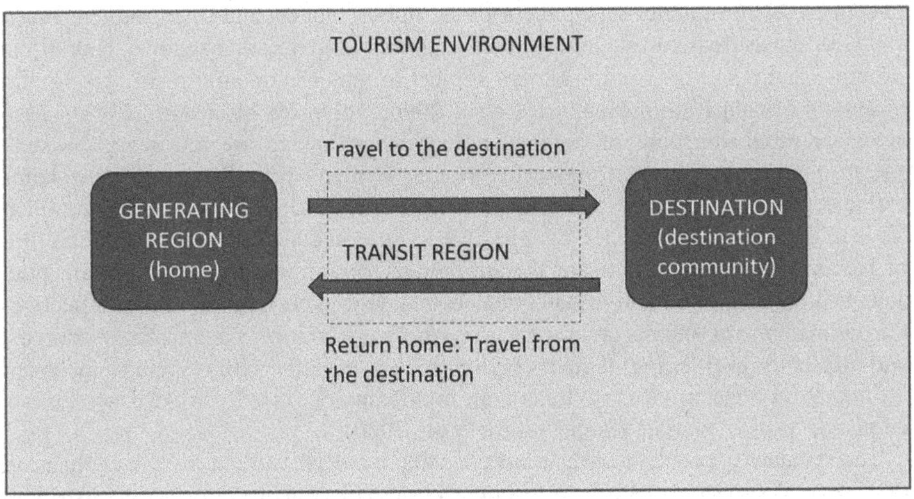

Figure 16.1 *The elements of a tourism system*

Yet the environment is critical here in at least two ways. First, it reinforces the way in which tourism-related pollution and emissions can affect the entire environment, not just at the destination. Second, and perhaps even more importantly, the environment is finite. It has a boundary: Planet Earth. This means that the environmental effects of tourism do not just disappear but instead, 'mount up', in exactly the same way that CO_2 accrues in the atmosphere or plastics in the oceans. In such a situation, notions of 'balance' surely make no sense given that there *is* only one Earth. Nevertheless, despite having only one planet and supposedly recognising the dangers of biodiversity loss and climate change, the tourism industry, along with many others, remains committed to growth trajectories with respect to visitors, numbers of travellers, and their expenditure. Although their 'rational' and 'efficient' allocation in space and time remains a major issue (WTTC and McKinsey & Company, 2017; Hall, 2019).

Rural tourism and community-based tourism have long been a focus in the development of rural areas (Zielinski et al., 2020). Rural tourism usually incorporates different types of rural-related tourism: ecotourism, nature tourism, activity tourism, agritourism, wine tourism, adventure tourism and other types of tourism activities outside main urban areas, with a low density of population. Sustainability and sustainable consumption are a fundamental focus of rural tourism development, with community-based approaches usually being regarded as a mechanism to ensure greater sustainability. It is often argued that, if appropriately undertaken, rural tourism meets the sustainability requirements by providing the opportunity to create jobs, increase incomes, grow local economies, and simultaneously strengthen local communities, protect the environment, conserve natural resources and cultural heritage (Blackstock, 2005; Johnson, 2010; Dodds et al., 2018; An & Alarcón, 2020).

However, rural tourism has a Janus-faced nature when it comes to development. The nature of rural areas makes them prone to both under- and over-tourism. From one side, these areas can be less accessible to tourist-generating regions, lack infrastructure and resources, and are often subject to major economic restructuring and change, particularly in agriculture (Mair, 2006; Torres & Momsen, 2011). These areas are often the focus of regional policies that seek to use tourism to encourage economic development and generate employment. Significantly for the latter element, rural tourism is often perceived as having low skill requirements to establish basic tourist activities, which do not require long-term education and could therefore be based on the hospitality of local communities, their traditions, craft and lifestyle, and absorb the employment loss in other sectors. However, from another perspective, the high volume of tourists creates excessive pressure on local systems and resources, and affects the well-being of such communities, especially with respect to localised inflation and housing costs in areas with high amenity values – what is sometimes described as over-tourism (Sæþórsdóttir et al., 2020).

The standard approach to over-tourism is that it can be 'solved' by the application of a rational decision-making and policy framework that seeks to better distribute tourism demand and supply in time and space so that negative externalities and impacts on affected communities are at an acceptable level. For the World Travel and Tourism Council (WTTC), this is regarded as a problem of 'coping with success' (WTTC & McKinsey & Company, 2017). Lead international agencies, such as the UNWTO and WTTC, together with larger corporate actors, industry associations and destination marketing organisations, are key actors in the promotion of the value of the market, growth, competition and the 'management' of the over-tourism problem (Hall, 2019). At the official closing ceremony of the International Year of Sustainable Tourism for Development 2017, which discussed the 'roadmap' for advancing tourism's contribution to the 2030 Agenda for Sustainable Development, Gloria Guevara, President and CEO, WTTC, commented, 'Sustainability remains the bedrock of our activity. We will continue to drive the conversation on planning for and managing tourism growth' (UNWTO, 2017). For such agencies the problem is not defined in terms of there being too much emphasis on tourism growth, rather it is seen as being insufficiently managed. For example, the UNWTO–UNDP (2017) notes that 'Both countries and companies lack frameworks to capture, aggregate and report on the full economic, social and environmental impacts of tourism' with respect to 'Improving performance by measuring impact and sharing knowledge' (p. 14). Furthermore, Taleb Rifai, Secretary-General of the UNWTO, argued for a greater role for business in sustainability, suggesting that

> The 2030 Agenda for Sustainable Development with its 17 [SDGs] sets the path that we all must embrace. … the private sector, which is the key player in tourism, … is beginning to recognise that the SDGs offer true business opportunities as sustainable business operations can spur competitiveness and increase profit. (UNWTO–UNDP, 2017, pp. 6–7)

As Dauvergne and Lister (2013, p. 25) observe:

> Turning sustainability into eco-business … is altering the nature of environmentalism, increasing its power to accelerate some forms of change, but limiting what is on the table to question, challenge, and alter. Sustainability as an idea can be radical: not just calling for changes in the rules of the game (i.e. market dynamics), but also to the game itself (i.e. the global economy).

Indeed, the overall sense of crisis engendered by the SDGs, sustainable tourism, over-tourism, and the more recent supposed under-tourism of the COVID-19 pandemic, only appear to encourage the UNWTO and the WTTC to advocate 'more of the same' strategies, even though they are not empirically working with respect to creating a more sustainable tourism (Hall, 2019).

Rural tourism suits the economic sustainability agenda as rural areas tend to convey positive images of sustainability, in the sense that they are non-urban spaces, but they are often seeking the economic benefits of tourism with an assumption that by their very nature, the environmental aspects of rural areas will not be unduly harmed. From a tourism development perspective this could be construed as 'Brundtland-as-usual' (Hall, 2019; Hall & Seyfi, 2021) in the sense that Brundtland's ambiguity allows business and policymakers to promote sustainable development by using Brundtland's support for rapid growth to justify 'sustainable growth' or, in the case of the UNWTO and others, 'sustainable tourism' in which growth is very much embedded (Hall, 2019).

One response to this situation is that of degrowth. The concept of degrowth or décroissance has been slowly entering the lexicon of tourism research since the late 2000s. The literal translation of 'décroissance' is decrease or reduction. From an economic development perspective this is not stagnation or a recession, even though such a situation may reduce some use of natural resources and accompanying emissions and pollution. Instead, any reduction is a necessary step to reach 'right-size' economies and establish steady-state economies that balance consumption and production in the context of not running down natural capital. Therefore, degrowth is not an opposing strategy to growth, it is about the growth within the system capacity, i.e. the environment of the tourism system. Steady-state tourism systems deal with qualitative growth through such indicators as the quality of life and well-being rather than quantitative expansion of tourism to the detriment of natural capital (Hall, 2009; 2010; Hall & Seyfi, 2021). These are concepts that have substantial resonance with community development ideas (Blackstock, 2005), but which are regarded as needing to be applied at a much wider scale.

The degrowth literature is rich in critique and debate regarding the negative effects of a growth-oriented economy as well as particular components of what a degrowth economy might look like (Hall et al., 2021). However, degrowth is not a specific theory but is instead an umbrella term or, more provocatively, a 'missile term' (Hickel, 2021) with which to attack growthism (Hall & Wood, 2021). The notion of degrowing tourism is therefore focused squarely on ensuring that tourist numbers and tourism infrastructure and travel do not negatively impact social and environmental well-being. Nevertheless, it is important to recognise that while being extremely relevant at a community scale, it is also applied at other economic scales right up to the national, given that that is a set environmental limit (Hall et al., 2021). Although there is a lack of analysis with respect to how a transition from growth to degrowth might be achieved and the nature of the changes required (Hall & Wood, 2021), in part because opportunities to do so have been extremely limited, the COVID-19 pandemic has been regarded by some as a transformative opportunity to 'right-size' tourism.

WAS THE PANDEMIC AN OPPORTUNITY FOR DEGROWTH?

The COVID-19 pandemic provided the context for the investigation of tourism systems under conditions of unprecedented disruption to transport networks and restrictions on international tourist mobility. The impact of the pandemic was a response to the contemporary, highly connected world marked by international networks of transport and capital (Gössling et al., 2020; Hall et al., 2020; Hall & Seyfi, 2021). Moreover, the high connectivity of countries contributed to the spread of the virus and reinforced the significance of the inclusion of high mobility and connectivity in conversations surrounding tourism's environmental impacts due to the movement of biota and disease. The role of international tourism as a channel for the spread of disease was well recognised in some quarters well before COVID-19, with the COVID-19 pandemic only reinforcing the lack of anticipatory action, strategy and preparedness by the tourism sector to deal with such events (Hall, 2020; Kim et al., 2021).

While international tourism was paused, the pandemic was widely believed by many commentators to be a catalyst for long-term changes to the behaviour of tourists and the harmful impacts of tourism development (e.g., Everingham &

Chassagne, 2020; Higgins-Desbiolles, 2020; Lew et al., 2020; Romagosa, 2020). When nature started taking back by reclaiming urban spaces during lockdowns, it was argued that these circumstances would lead the way to the ecological grief of lost natural resources (Crossley, 2020). Generally, the opinions on the future of travel and tourism were optimistic (Lew et al., 2020), emphasising the opportunities to reset tourism (Higgins-Desbiolles, 2020) and slow consumption (Everingham & Chassagne, 2020; Romagosa, 2020). It was also predicted that the pandemic would produce more environmentally mindful tourists (Stankov et al., 2020), and that the learnt lessons would assist in mitigating the climate change impacts (Botzen et al., 2021).

In the beginning, there was a strong belief that the COVID-19 pandemic would change the perceptions of the values towards travelling local and less with slowing down consumption (Higgins-Desbiolles, 2020). Although such speculations were full of hope, they lacked robust evidence. What could be stated with confidence was that the pandemic provided an opportunity to reflect on the growth in volume, reliance on state aid, environmental impacts and how much aviation is really needed (Gössling, 2020). It was perhaps frivolous to suggest that people would stick to restricted travelling. For example, business travellers, taking advantage of software for online meetings, considered it as a temporary solution and were back to travelling as soon as possible (Manca et al., 2021). Sudden restrictions influenced short-term behaviour but did not demonstrate fundamental changes in behaviour patterns. Vaccination and certificates added a layer of documentation to international travel, but did not fundamentally change the way people travel or reduce consumption drastically. Generally, COVID-19 had little effect on the future projections of global emissions in 2030 and beyond (Reilly et al., 2021) and on the willingness of people to travel.

The restrictions allowed the promotion of domestic tourism in response to the absence of international tourists, especially in rural and remote areas. Domestic tourism has therefore come to the forefront of many government policies and interventions with respect to maintaining the tourism economy. This focus may help find positive solutions to reduce emissions and promote place-based tourism and economic development (Hall et al., 2020). However, the domestic tourism orientation is potentially only temporary and will depend on how long international travel is relatively restricted and how long the economy shows limited growth, given that many governments will be concerned to generate employment, increase tax revenues, and reduce their support for COVID-19 affected industries, such as tourism and hospitality.

The changes to the tourism system resulting from the pandemic will be uneven in space and time, affecting international tourism-dependent rural and remote areas greatly. The necessary reorientation to more sustainable local tourism will therefore be a priority in such destinations (Hall et al., 2020). However, such transitions are difficult because domestic tourists may need to be persuaded to visit their own country and rural regions first and, from a system perspective, shifts from outbound to domestic tourism in many countries are highly likely to transfer tourism-related negative externalities from one location to another (Hall, 2014). The current tourism system is therefore not designed to slow down consumption. Instead, after a disruption, the system returns to the 'normal' state of functioning, often encouraged by government interventions (Ioannides & Gyimóthy, 2020).

THE GROWTH OBSESSION

The justification for tourism development and subsequent growth comes from the economic benefits that tourism brings. However, despite the mainstreaming of sustainability, environmental and social issues may only receive limited attention within the context of economic development or may be excluded from economic measurement of the value of tourism development, with subsequent devaluing of community and environmental concerns in policymaking. For example, many governments appear relatively content that large aggregators encourage tourists to travel and book accommodation in many areas, increasing demand through marketing campaigns, and receive massive discounts to grow and satisfy stakeholders, while being located in other countries with corresponding economic leakage from local economies (Gössling & Higham, 2021).

Despite the acknowledged importance of sustainability, the UNWTO continued to rely on growth in tourist numbers and spending to tackle the pandemic's consequences (Butcher, 2021). The value of the transfer of consumption to other parts of economy is often not considered in assessments of the effects of the pandemic on tourism. Destinations have their goal in growing tourist numbers, with very few seeking to introduce measures to reduce them even though the rhetoric of reduction may be employed (Gössling et al., 2013; 2016). At the aggregate level, tourism is measured through tourism satellite accounts that consider outputs of products and services and which tend to maximise attributions of the economic benefits of tourism, while inadequately measuring well-being, environmental impacts, or sustainability.

Paradoxically, while economic growth is seen as a part of the problem, producing emissions and other harmful impacts, it is also believed that economic growth is the solution to these issues (Butcher, 2021). Thus, it is argued that social and economic sustainability can only be achieved through increased visitor numbers. The logic is simple; high tourist numbers and greater loyalty lead to higher income and assistance in solving socio-economic problems (e.g., López-Sanz et al., 2021). The argument is nearly always that higher tourist numbers mean higher spending, which converts into personal, business, and regional incomes that allow socio-economic and environmental problems to be solved. But such measures do not fully account for the real costs of hosting such numbers of tourists.

The ultimate problem is that what is measured dictates what is done, which can lead to false policy choices (Stiglitz, 2009). Stiglitz (2009) notes that the government's outputs are measured by the inputs as there is no way to do it otherwise. That implies that when a government spends more, the output rises irrespective of the efficiency of such spending. If a government decides to tackle environmental pollution, that could lower GDP growth (Stiglitz, 2009). Therefore, the degradation of natural resources and the environment cannot be measured through Tourism Satellite Accounts. Moreover, GDP is not a good proxy for quality of life or well-being.

However, the development of rural tourism is still about the growth and potential benefits it brings even though it may also be damaging in the longer term (Milano et al., 2019). Rather than helping rural communities to develop and improve their lives, consumption is simply being shifted to somewhere else in space and time alongside all the issues that come with it. For instance, the COVID-19 outbreak revealed that places that were struggling with over-tourism then began complaining about under-tourism (Hall & Seyfi, 2021). At a global scale fundamentally nothing changed; consumption and its externalities were simply being moved around. Therefore, the question of balance lies at the core of addressing rural tourism sustainability. The aspirational goals of reducing emissions and simultaneously increasing tourist numbers present a paradox in the sustainable growth agenda (Gössling et al., 2013), with the main issue being that a growth-oriented economy does not know how to degrow (Kallis et al., 2012).

Degrowth challenges the conventional and well-established growth-based market relationships. To slow growth means to think about current market mechanisms (Trainer, 2012). Degrowing is not inherent in the current tourism system. Without changes in thinking, all offerings appear to be in favour of the current growth-based economic approaches. Degrowing and right-sizing rural tourism development should start with the understanding of tourism impacts and that tourism is a part of a larger socio-economic system (Hall, 2015). Tourism development needs to move away from a linear economy to a circular one to demonstrate more sustainable consumption and production behaviour. However, this process requires fundamental changes in thinking to shift to more efficient and sufficient approaches to the production and consumption sides of tourism. In some cases, a staycation is an option within this approach, i.e. to stay and consume locally around the place of residence. Considering longer stays at a destination and reducing travel would also be of benefit. But the impact of such measures on some communities may still be substantial, given the sheer number of people who travel.

CONCLUSIONS

Tourists are not ready to give up international travel or longer-distance domestic travel. Mass awareness of climate change or over-tourism in rural areas does not translate to mass changes in personal travel motivations to change behaviour to become more environmentally friendly. Very few tourists purposefully damage a destination environment. Individual tourists do not recognise their impact as affecting the environment because they do not understand how tourism affects the environment on a global scale. However, tourists *do* believe that they assist local economies by creating jobs and spending money. The implementation of market-oriented policies in the absence of regulation is not enough (Gössling et al., 2016). Policies and approaches in managing tourism are primarily based on the assumption that individuals are willing to change behaviour rather than stopping tourists or tourism per se (Hall, 2014). Knowing that environmental problems exist or that resources are finite or that species go extinct does not change behaviour. In fact, the last chance to see an environment or species before it disappears may even be an attraction.

Community development is an important issue. Governance and policy intervention modes reinforce each other and need to be considered all together to be adequately understood. Communities, whether defined geographically or via other commonalities, are important locations for change. But the very nature of communities means that the change and the adoption of more sustainable ways of doing are only local. A sustainable community can be surrounded by a sea of unsustainability, meaning that, at the wider scale, tourism remains as unsustainable as ever.

This chapter is ostensibly about rural tourism, communities and development within the context of sustainability. It very much is, except it is not. To understand the long-term prospects for the sustainability of a rural community and its development means to understand its positionality in the tourism system and the wider global system beyond. This is not to suggest that community-based tourism development is meaningless, but from a global perspective, it is miniscule. And whatever local advances in sustainability and biodiversity conservation that may be gained, for example, can easily be driven away by the inexorable effects of global environmental change. If not immediately then, given tourism BAU (business as usual, or as Hall (2019) describes it, 'Brundtland as usual'), in the warming climate and the increased frequency and intensity of climatic events of the future. Rural communities and their development are bounded by the wider system and the limits of the environment.

The need for change has to be acknowledged and acted upon by businesses, governments and tourists and, just as significantly, by tourism educators and researchers. Steady-state tourism is about encouraging qualitative development, quality of life, social and ecological well-being (Hall & Seyfi, 2021). If degrowth is pursued, the focus should be on the quality of life, basic human needs, societal change, reduced dependence on economic activity, non-materialism, equity, the cultural difference. Tourism desperately needs to be reimagined in a manner that focuses more on social and environmental dimensions than on economic in a balanced approach to tourism development because, in the long term, that actually makes good economic sense. Community development may assist this process from the bottom up but, just as importantly, if not more so, it needs to come from the top down. Unfortunately, such a vision does not exist. While the tourism industry fiddles, the planet burns.

REFERENCES

An, W., & Alarcón, S. (2020). How can rural tourism be sustainable? A systematic review. *Sustainability*, *12*(18), 7758. https://doi.org/10.3390/su12187758.

Blackstock, K. (2005). A critical look at community based tourism. *Community Development Journal*, *40*(1), 39–49. https://doi.org/10.1093/cdj/bsi005.

Botzen, W., Duijndam, S., & van Beukering, P. (2021). Lessons for climate policy from behavioral biases towards COVID-19 and climate change risks. *World Development*, *137*, 105214. https://doi.org/https://doi.org/10.1016/j.worlddev.2020.105214.

Butcher, J. (2021). Covid-19, tourism and the advocacy of degrowth. *Tourism Recreation Research*, 1–10. https://doi.org/10.1080/02508281.2021.1953306.

Crossley, É. (2020). Ecological grief generates desire for environmental healing in tourism after COVID-19. *Tourism Geographies*, *22*(3), 536–546. https://doi.org/10.1080/14616688.2020.1759133.

Dauvergne, P., & Lister, J. (2013). *Eco-Business: A Big-Brand Takeover of Sustainability*. Cambridge, MA: MIT Press.

Dodds, R., Ali, A., & Galaski, K. (2018). Mobilizing knowledge: Determining key elements for success and pitfalls in developing community-based tourism. *Current Issues in Tourism*, *21*(13), 1547–1568. https://doi.org/10.1080/13683500.2016.1150257.

Everingham, P., & Chassagne, N. (2020). Post COVID-19 ecological and social reset: moving away from capitalist growth models towards tourism as Buen Vivir. *Tourism Geographies*, *22*(3), 555–566. https://doi.org/10.1080/14616688.2020.1762119.

Gössling, S. (2020). Risks, resilience, and pathways to sustainable aviation: A COVID-19 perspective. *Journal of Air Transport Management*, *89*, 101933. https://doi.org/https://doi.org/10.1016/j.jairtraman.2020.101933.

Gössling, S., & Higham, J. (2021). The low-carbon imperative: Destination management under urgent climate change. *Journal of Travel Research*, *60*(6), 1167–1179. https://doi.org/10.1177/0047287520933679.

Gössling, S., & Peeters, P. (2007). 'It does not harm the environment!' An analysis of industry discourses on tourism, air travel and the environment. *Journal of Sustainable Tourism*, *15*(4), 402–417. https://doi.org/10.2167/jost672.0.

Gössling, S., & Peeters, P. (2015). Assessing tourism's global environmental impact 1900–2050. *Journal of Sustainable Tourism*, *23*(5), 639–659. https://doi.org/10.1080/09669582.2015.1008500.

Gössling, S., Hall, C.M., & Scott, D. (2009). The challenges of tourism as a development strategy in an era of global climate change. In E. Palosuo (ed.), *Rethinking Development in a Carbon-Constrained World* (pp. 100–119). Helsinki: Ministry for Foreign Affairs.

Gössling, S., Scott, D., & Hall, C.M. (2013). Challenges of tourism in a low-carbon economy. *Wiley Interdisciplinary Reviews. Climate Change*, *4*(6), 525–538. https://doi.org/10.1002/wcc.243.

Gössling, S., Ring, A., Dwyer, L., Andersson, A.C., & Hall, C.M. (2016). Optimizing or maximizing growth? A challenge for sustainable tourism. *Journal of Sustainable Tourism*, *24*(4), 527–548. https://doi.org/10.1080/09669582.2015.1085869.

Gössling, S., Scott, D., & Hall, C.M. (2020). Pandemics, tourism and global change: A rapid assessment of COVID-19. *Journal of Sustainable Tourism*, *29*(1), 1–20. https://doi.org/10.1080/09669582.2020.1758708.

Hall, C.M. (2005). *Tourism: Rethinking the Social Science of Mobility*. Harlow: Pearson.

Hall, C.M. (2009). Degrowing tourism: Décroissance, sustainable consumption and steady-state tourism. *Anatolia*, *20*(1), 46–61. https://doi.org/10.1080/13032917.2009.10518894.

Hall, C.M. (2010). Changing paradigms and global change: From sustainable to steady-state tourism. *Tourism Recreation Research*, *35*(2), 131–143. https://doi.org/10.1080/02508281.2010.11081629.

Hall, C.M. (2013). Framing behavioural approaches to understanding and governing sustainable tourism consumption: Beyond neoliberalism, 'nudging' and 'green growth'? *Journal of Sustainable Tourism*, *21*(7), 1091–1109. https://doi.org/10.1080/09669582.2013.815764.

Hall, C.M. (2014). *Tourism and Social Marketing*. London: Routledge. https://doi.org/10.4324/9780203854259-18.

Hall, C.M. (2015). Economic greenwash: On the absurdity of tourism and green growth. In M.V. Reddy & K. Wilkes (eds), *Tourism in the Green Economy* (pp. 361–380). Abingdon: Routledge.

Hall, C.M. (2019). Constructing sustainable tourism development: The 2030 agenda and the managerial ecology of sustainable tourism. *Journal of Sustainable Tourism, 27*(7), 1044–1060. https://doi.org/10.1080/09669582.2018.1560456.

Hall, C.M. (2020). Biological invasion, biosecurity, tourism, and globalisation. In D. Timothy (ed.), *Handbook of Globalisation and Tourism* (pp. 114–125). Cheltenham, UK & Northampton, MA, USA: Edward Elgar Publishing.

Hall, C.M., & Müller, D. (eds) (2018). *The Routledge Handbook of Second Home Tourism and Mobilities*. London: Routledge.

Hall, C.M., & Page, S. (2010). The contribution of Neil Leiper to tourism studies. *Current Issues in Tourism, 13*(4), 299–309. https://doi.org/10.1080/13683500.2010.482652.

Hall, C.M., & Seyfi, S. (2021). COVID-19 pandemic, tourism and degrowth. In C.M. Hall, L. Lundmark, & J.J. Zhang (eds), *Degrowth and Tourism* (pp. 220–238). London: Routledge.

Hall, C.M., & Wood, K.J. (2021). Demarketing tourism for sustainability: Degrowing tourism or moving the deckchairs on the *Titanic*? *Sustainability, 13*, 1585. https://doi.org/10.3390/su13031585.

Hall, C.M., Scott, D., & Gössling, S. (2020). Pandemics, transformations and tourism: Be careful what you wish for. *Tourism Geographies, 22*(3), 577–598. https://doi.org/10.1080/14616688.2020.1759131.

Hall, C.M., Lundmark, L., & Zhang, J.J. (eds) (2021). *Degrowth and Tourism: New Perspectives on Tourism Entrepreneurship, Destinations and Policy*. Abingdon: Routledge.

Hickel, J. (2021). What does degrowth mean? A few points of clarification. *Globalizations, 18*(7), 1105–1111. https://doi.org/10.1080/14747731.2020.1812222.

Higgins-Desbiolles, F. (2020). The 'war over tourism': Challenges to sustainable tourism in the tourism academy after COVID-19. *Journal of Sustainable Tourism, 29*(4), 551–569. https://doi.org/10.1080/09669582.2020.1803334.

Howden, S.M., Soussana, J.F., Tubiello, F.N., Chhetri, N., Dunlop, M., & Meinke, H. (2007). Adapting agriculture to climate change. *Proceedings of the National Academy of Sciences, 104*(50), 19691–19696. https://doi.org/10.1073/pnas.0701890104.

Ioannides, D., & Gyimóthy, S. (2020). The COVID-19 crisis as an opportunity for escaping the unsustainable global tourism path. *Tourism Geographies, 22*(3), 624–632. https://doi.org/10.1080/14616688.2020.1763445.

Johnson, P.A. (2010). Realizing rural community-based tourism development: Prospects for social economy enterprises. *Journal of Rural and Community Development, 5*(1–2), 150–162.

Kallis, G., Kerschner, C., & Martinez-Alier, J. (2012). The economics of degrowth. *Ecological Economics, 84*, 172–180. https://doi.org/https://doi.org/10.1016/j.ecolecon.2012.08.017.

Kim, M.J., Bonn, M., & Hall, C.M. (2021). What influences COVID-19 biosecurity behaviour for tourism? *Current Issues in Tourism, 25*, 21–27. https://doi.org/10.1080/13683500.2021.1883558.

Lenzen, M., Sun, Y.-Y., Faturay, F., Ting, Y.-P., Geschke, A., & Malik, A. (2018). The carbon footprint of global tourism. *Nature Climate Change, 8*(6), 522–528. https://doi.org/10.1038/s41558-018-0141-x.

Lew, A.A. (2014). Scale, change and resilience in community tourism planning. *Tourism Geographies, 16*(1), 14–22. https://doi.org/10.1080/14616688.2013.864325.

Lew, A.A., Cheer, J.M., Haywood, M., Brouder, P., & Salazar, N.B. (2020). Visions of travel and tourism after the global COVID-19 transformation of 2020 [Editorial]. *Tourism Geographies, 22*(3), 455–466. https://doi.org/10.1080/14616688.2020.1770326.

López-Sanz, J.M., Penelas-Leguía, A., Gutiérrez-Rodríguez, P., & Cuesta-Valiño, P. (2021). Sustainable development and consumer behavior in rural tourism: The importance of image and loyalty for host communities. *Sustainability, 13*(9), 4763. https://doi.org/10.3390/su13094763.

Mair, H. (2006). Global restructuring and local responses: Investigating rural tourism policy in two Canadian communities. *Current Issues in Tourism, 9*(1), 1–45. https://doi.org/10.1080/13683500608668237.

Manca, F., Sivakumar, A., Pawlak, J., & Brodzinski, N.J. (2021). Will we fly again? Modeling air travel demand in light of COVID-19 through a London case study. *Transportation Research Record.* https://doi.org/10.1177/03611981211025287.

Mansfield, B. (2003). Fish, factory trawlers, and imitation crab: The nature of quality in the seafood industry. *Journal of Rural Studies, 19*(1), 9–21. https://doi.org/10.1016/S0743-0167(02)00036-0.

Milano, C., Novelli, M., & Cheer, J.M. (2019). Overtourism and degrowth: A social movements perspective. *Journal of Sustainable Tourism, 27*(12), 1857–1875. https://doi.org/10.1080/09669582.2019.1650054.

Raza, A., Razzaq, A., Mehmood, S.S., Zou, X., Zhang, X., Lv, Y., & Xu, J. (2019). Impact of climate change on crops adaptation and strategies to tackle its outcome: A review. *Plants, 8*(2), 34. https://doi.org/10.3390/plants8020034.

Reilly, J.M., Chen, Y.H.H., & Jacoby, H.D. (2021). The COVID-19 effect on the Paris agreement. *Humanities and Social Sciences Communications, 8*(1), 16. https://doi.org/10.1057/s41599-020-00698-2.

Romagosa, F. (2020). The COVID-19 crisis: Opportunities for sustainable and proximity tourism. *Tourism Geographies, 22*(3), 690–694. https://doi.org/10.1080/14616688.2020.1763447.

Sæþórsdóttir, A.D., Hall, C.M., & Wendt, M. (2020). From boiling to frozen? The rise and fall of international tourism to Iceland in the era of overtourism. *Environments, 7*(8), 59. https://doi.org/10.3390/environments7080059.

Stankov, U., Filimonau, V., & Vujičić, M.D. (2020). A mindful shift: An opportunity for mindfulness-driven tourism in a post-pandemic world. *Tourism Geographies, 22*(3), 703–712. https://doi.org/10.1080/14616688.2020.1768432.

Stiglitz, J.E. (2009). GDP fetishism. *The Economists' Voice, 6*(8), 1–3.

Torres, R., & Momsen, J. (eds) (2011). *Tourism and Agriculture: New Geographies of Consumption, Production and Rural Restructuring.* Abingdon: Routledge.

Trainer, T. (2012). De-growth: Do you realise what it means? *Futures, 44*(6), 590–599. https://doi.org/https://doi.org/10.1016/j.futures.2012.03.020.

UNWTO (2017, December 19). *A roadmap towards 2030: The legacy of the International Year of Sustainable Tourism or Development 2017* [Press release]. PR 17137.

UNWTO & UNDP (2017). *Tourism and the Sustainable Development Goals – Journey to 2030.* Madrid: UNWTO.

UNWTO & UNEP (2005). *Making Tourism More Sustainable: A Guide for Policy Makers.* Paris: UNWTO/UNEP.

Wiréhn, L. (2018). Nordic agriculture under climate change: A systematic review of challenges, opportunities and adaptation strategies for crop production. *Land Use Policy, 77*, 63–74. https://doi.org/10.1016/j.landusepol.2018.04.059.

WTTC and McKinsey & Company (2017). *Coping with success: Managing overcrowding in tourism destinations.* WTTC.

Zielinski, S., Jeong, Y., Kim, S., & Milanés, C.B. (2020). Why community-based tourism and rural tourism in developing and developed nations are treated differently? A review. *Sustainability, 12*(15), 5938. https://doi.org/10.3390/su12155938.

17. Understanding a multifunctional rural tourism landscape: a case study of the Upper Hunter Valley, NSW, Australia

Sinead Francis-Coan and Annæ Buchmann

INTRODUCTION

Tourism has a complex, interdependent relationship with the landscapes and communities in which it is situated (Richards & Hall, 2003). And for rural tourism, natural resources are not only crucial tourism assets but also have to further facilitate other economic and even non-economic activities. In other words, rural landscapes are subject to multiple influences and social, environmental, political and economic pressures (Anderson et al., 2016; Argent et al., 2013). Consequently, there are multiple perspectives on how tourism relates to landscapes and communities, and indeed, what constitutes appropriate use(s) of rural landscapes. Failure to address these differences of perspective, especially regarding forms of land use that work against tourism and balanced land use in the area, have led to public debate and protest, and even legal actions (taken by various stakeholders). In fact, in a recent Upper Hunter election, it was revealed that 57.4% of locals supported a moratorium on new coal mines (The Australia Institute, 2021) despite most of the high-profile candidates campaigning strongly in support of coal mining (Jupp, 2021). These developments in turn have created an image and branding issue for the region that has started impacting its tourism marketing. It has been argued that tourism entwines people, places and things through complex translation and negotiation (Van der Duim, 2007). This situation becomes increasingly complex when considering the additional processes that are occurring simultaneously in relation to other rural land uses in a shared location.

In this exploratory study, the authors sought to understand the complexity of a chosen rural landscape through examining different interpretations of 'proper behaviour' in regard to land use planning and management in the Upper Hunter Valley, located in New South Wales, Australia, that in turn feed land use conflict.

Commonly, the idea of holistic destination management asks that diverse stakeholder opinions are considered. This study consequently investigates different perceptions and understandings and how these are consequently expressed in different interpretations of landscape and its development (or not). In doing so, this explorative study refers to the social construction model (Greider & Garkovich, 1994) to examine the similarities and differences between perspectives of the Upper Hunter Valley region as a contested rural tourism destination. The assumption is that a better understanding of how conflicts arise may facilitate a more productive debate about future land uses in the region by revealing shared interests that could provide a basis for collaborative land use planning and development.

Rural landscapes are often defined in contrast to urban spaces, or through geographical boundaries drawn for a range of government or administrative purposes (Halfacree, 1993; McManus, 2008). A common understanding of 'rural' is typically associated with agricultural activity (Friedland, 2002). This, for example, has been shown by social representation theory, which detangles how systems of values, ideas and practices are socially formed and communicated and in turn influence perception (Moscovici, 2001). Over time, representations evolve, and today's discussions have expanded the idea of rurality to incorporate further qualities such as open spaces, clean air, close-knit communities, and relaxation (Halfacree, 2007; Sherval & Hardiman, 2014). Such qualities relate to how an area is socially constructed by particular groups or individuals (Greider & Garkovich, 1994); and what, and how, meaning is projected onto a landscape. In short, society's idea of what a 'rural' landscape is and could be has expanded considerably.

However, the idea of the rural 'idyll' (commonly associated with terms such as green, clean, and pristine) endures and continues to appeal to visitors, residents and businesses including tourism operators (Argent et al., 2014). This appeal has also supported significant migration of urban residents to rural areas in industrialised countries. This phenomenon has been labelled 'amenity migration' (Argent et al., 2014) and illustrates a shift in societal expectation for landscapes to perform an ever-growing variety of functions (Holmes, 2006), and to modernise. The latter author describes such change as a 'multifunctional' rural transition, where rural landscapes accommodate new production values, consumption values and protection values (Holmes, 2006), which are reflected in the types of land uses a landscape supports. The established combination of the values depends on the specific landscape, and a negotiation of what is desired or deemed possible (Holmes, 2006).

In the Upper Hunter Valley, wine tourism and other forms of rural tourism have been incorporated into the area's economic fabric (O'Neill, 2000). Today, tourism ventures include wine-related tourism, scenic hot-air ballooning and a range of active tourism enterprises, such as horseback riding, skydiving, birdwatching, camping, rafting and bushwalking (see Destination NSW, 2016). A successful tourism cluster has been formed and is commonly attributed to proactive governmental support and the successful networking of local business owners (March & Wilkinson, 2009). Such networking involves collaborations between the viticulture and tourism sectors, and more recently, interest from the equine sector in tourism (Hunter Valley Thoroughbred Association & Hunter Valley Wine and Tourism Association, 2015; March & Wilkinson, 2009). Such collaboration hints at similarities between the way these groups understand their shared landscape. In particular, it suggests a basis of shared knowledge, history, and values from which collaborations are possible.

Still, the Upper Hunter Valley receives comparatively less government support than other Australian wine-tourism destinations such as the Margaret River (Western Australia) or the Barossa Valley and McLaren Vale (South Australia), which are also protected by the Character Preservation (Barossa Valley) Act 2012 and Character

Preservation (McLaren Vale) Act 2012 respectively (Hunter Valley Thoroughbred Association & Hunter Valley Wine and Tourism Association, 2015). Consequently, networks are under personal and financial pressure and have broken down in the past.

In this chapter, we propose that with a better understanding of different positions, one can anticipate power relationships and management decisions, including their perception in a particular region, and the likelihood of support for selected management tactics and strategies. This could create a more open discussion regarding the negotiation of land use in the area and especially its transition towards a more touristic landscape. This makes the Upper Hunter Valley region a great case study for examining how understanding differing attitudes and decisions could improve efficiency through a wider acceptance of short- and long-term intentions for its tourism destination planning, development, and governance.

UNDERSTANDING HOW LANDSCAPES ARE CONSTRUCTED

The meaning attributed to landscapes has been theorised as not inherently objective, but also as subjective as it is seen and projected by those to whom a particular landscape carries significance or relevance (Greider & Garkovich, 1994) in a process of social construction (Gailing & Leibenath, 2015). Social construction acknowledges the potential for multiple constructions of meaning for a common object or phenomenon based on different knowledge, different history, different values and beliefs (Jovchelovitch, 2001). Social construction can be alternatively defined as 'symbolic interactionism' (Deaux & Philogene, 2001). This alternative definition speaks to the visual and verbal 'code' through which constructions are formed, disseminated, and reformed. In this context, symbols both result from and inform the knowledge, values and beliefs of particular individuals that become linked through this process of construction and communication (Jovchelovitch, 2001). In this way, the constructions are socially shared and communicated. This process is framed by a shared interpretation of historical events that provides a context for how new phenomena, such as changes to a rural landscape, are to be interpreted (Wagner et al., 1999).

Of course, in Australia, the reading of landscapes can also acknowledge or indeed ignore association with Indigenous Australian culture and the later colonising cultures indicating different historical engagements with the landscape (Jordan et al., 2009). Such dynamics harbour potential for the development of incompatible or contrasting constructions, as is increasingly explored in research.

Indicatively, one study investigated how differing perceptions of landscape change with the general aspiration to convey them to planners, thus improving planning outcomes (Anderson et al., 2016). In that case study, the variation in perspective seemed to relate to how the symbolism of a forest informed different understandings of a landscape. There are other case studies seeking to understand why and how differing perspectives on land use may create ongoing conflict. For example, opposing social groups in the Italian wine region of Chianti disputed the positioning of regional boundaries (Brunori & Rossi, 2007), leading to a standstill in infrastructure planning. For these stakeholders, the regional boundary lines appeared to symbolise or stand for what was to be considered as being authentic of the region, which in this case was further complicated by the marketability of the Chianti brand (Brunori & Rossi, 2007). Another study of land use in Taiwan's Yushan National Park revealed differing social, environmental and economic outcomes for the various stakeholders that significantly disadvantaged local Indigenous people, thus creating ongoing conflict (Lai et al., 2013).

The above examples reveal historical differences that relate to how landscapes have been symbolised in the process of attributing them with meaning. Still, in a predominantly European context, the concept of Integrated Landscape Initiatives (ILIs) provides an example of collaborative land management (Garcia-Martin et al., 2016). Such ILIs are 'projects, programs, platforms, initiatives, or sets of activities that foster a broad range of landscape values [such as the production, consumption and protection values in the participants' constructions] and contribute to the personal and social fulfilment and well-being of people' (Garcia-Martin et al., 2016, p. 44). Such initiatives have been particularly successful in heterogeneous rural areas with major agriculture and winemaking, as can be found in the Upper Hunter Valley.

THE UPPER HUNTER VALLEY OF AUSTRALIA AS A CASE STUDY

The Upper Hunter Valley is an area of approximately 1,320 km^2, about 250 km north of Sydney, which has supported agriculture (early 18th century), mining (1870s), winemaking (1825) and equine breeding (1830s) (McManus, 2008). Nowadays, the mining, equine and viticulture/tourism sector continue to represent three of the biggest economic players in the region (NSW Department of Trade and Investment, 2015). As a result, at different times these industries may work alongside or against each other's practices and priorities (McManus et al., 2000). Collaboration includes marketing the area, though economic competition persists as coal, wine and horse breeding are commodified for domestic and export markets (NSW Department of Trade and Investment, 2015).

Research shows that collaboration has specifically been a driving force in boosting tourism in the area (March & Wilkinson, 2009). But after 200 years of generally complementary economic activity and an absence of ongoing and high-profile public conflict, increased competition for natural resources has increased demand for modification of 'traditional' land use(s). Similarly, the area has seen an increase in public debate about differing values and experiences (e.g. McCarthy, 2014), exacerbated by changes in demand and loss of undeveloped land. At times, these differing opinions inhibit or contradict each other, and conflicting development and practices occur. Of course, such change is occurring in rural landscapes all across Australia, initiated by changing population patterns, as well as economic activities and practices (Argent et al., 2014; McGuirk & Argent, 2011). Regardless, there is a need to accommodate

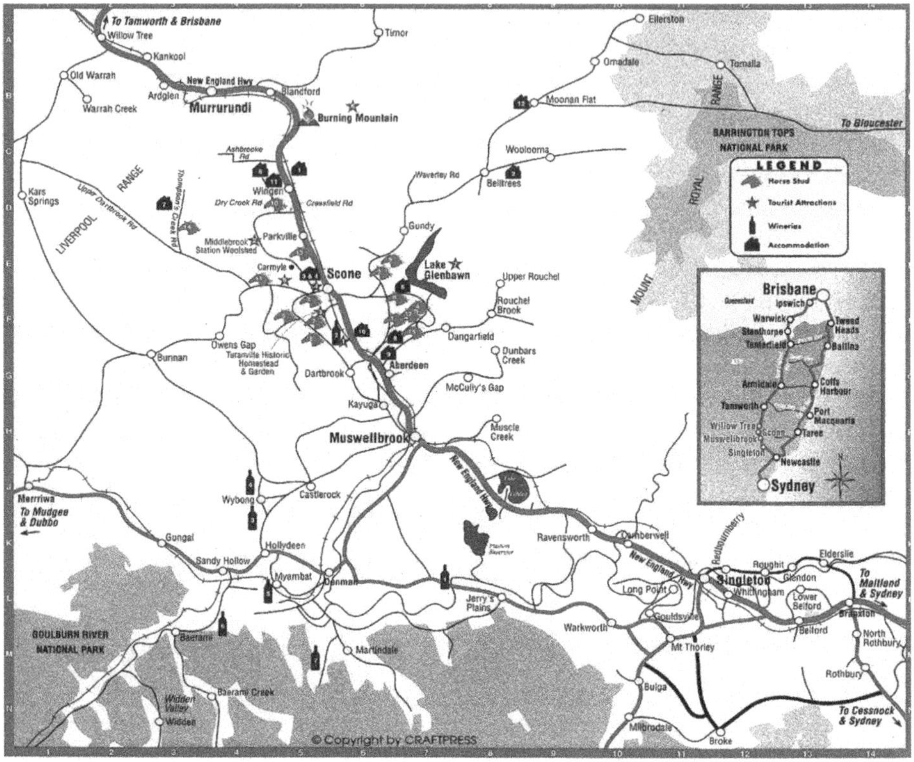

Source: Copyright by Upper Hunter Shire Council. Reproduced with permission.

Figure 17.1 Tourist map of the Upper Hunter Valley (n.d.)

increasing and diversifying land uses (Argent, 2011). Notably, in the Upper Hunter Valley this challenge and renegotiation have largely coincided with a mining 'boom' that began in the late 1980s to early 1990s, which also saw the mass introduction of open-cut coal mining (Cottle, 2013; McManus, 2008).

Overall, the area is best described by its peri-metropolitan character. The peri-metropolitan regions are characterised by being located 30 to 60 minutes from a large metropolitan centre like Sydney, with a hotly contested combination of production, consumption and protection values, and overall substantial economic activity (Holmes, 2006). These areas are often used for resource extraction, waste disposal and water supply in connection with metropolitan demand for agricultural production, service, consumption and rural resources (Argent, 2011; Holmes, 2006; McGuirk & Argent, 2011). Currently, the Upper Hunter Valley experiences significant public debate about land use in media campaigns, direct protest action and ongoing legal disputes (Francis-Coan et al., 2017), all aimed at interested parties achieving their preferred combination of land use.

And yet, research conducted in the Upper Hunter Valley region has so far been restricted to issues of health (Albrecht et al., 2007; Higginbotham et al., 2010), social justice (McManus & Connor, 2013), and environmental and economic sustainability (Connor et al., 2009; McManus, 2008); with these concerns typically linked to the mining boom and open-cut coal mining, which expedite changes in the topography of the landscape (Cottle, 2013). Indeed, a common theme across this research has been the proposed incompatibility of open-cut coal mining with other land uses such as farming and grazing, but also with less obtrusive and smaller-scale economic development in the area and with the landscape aesthetics in general. Today there is ongoing community opposition, particularly to open-cut coal mining in the area (Connor et al., 2009; Connor et al., 2008), and media reports often relay details of legal disputes between community groups and mining companies (Higginson, 2014), winemakers and mining/energy companies (Thompson, 2012) or thoroughbred breeders and mining companies (McCarthy, 2014). Given these disputes involve local stakeholder groups, it is challenging to envisage a holistic and collectively managed rural tourism landscape.

CHOOSING A SOCIAL-CONSTRUCTIVIST APPROACH

The methodology of this explorative study follows a social-constructivist approach (Bryman & Bell, 2011) recognising multiple perspectives about a social issue. Firstly, key literature related to contested land use in the Upper Hunter Valley (including Higginbotham et al., 2010; McManus & Connor, 2013; McManus et al., 2000; Wearing et al., 2014) was reviewed to clarify key issues in the public debate and the major readings of the local landscape attributed to the mining sector, viticulture/tourism sector, equine sector and community groups. The state government was also considered to incorporate a perspective from an influential decision-making position in terms of land use regulation, and was thus a significant player in key decision-making. In contrast, other stakeholders, such as the Indigenous-led Wannaruah Land Council, did not frequently feature in literature or print media, further indicating their marginalised status, resulting in their perspectives being mostly excluded from public discussion. In other words, the voices of Indigenous Australians and custodians continue to be ignored and marginalised in mainstream debates.

In order to understand the local stakeholders and their viewpoints and key arguments, the authors systematically reviewed local print media, specifically to identify commentators who had expressed their opinions of the Upper Hunter Valley in the public sphere (mostly in newspaper and radio contributions). This resulted in four female and nine male people being contacted and consequently consenting to in-depth interviews. These selected participants were interviewed through face-to-face and telephone-conducted meetings that lasted for an average of one hour. The researchers used a semi-structured interview guide to increase consistency across different interviews and perspectives (Creswell, 2014; Ryan, 2005). In summary, the questions covered such topics as what the Upper Hunter Valley meant to the participants, what changes they had noticed over time and how they judged those changes. All interviews were recorded and the transcripts analysed using the NVivo11 qualitative analysis software (Bryman & Bell, 2011). The resulting analysis was guided by a coding system derived from previously identified comments made in print media, and considered aspects such as environmental concern, references to production, consumption and protection values, and references to economic interests. NVivo11 was used to identify and sort where each code occurred in the interview transcripts, which identified themes that will now be discussed.

PLURALITY AND DIVISION IN THE SHARED LANDSCAPE

This explorative study produced over 13 hours of interviews resulting in about 100 pages of transcripts. The following section describes and analyses the findings, sorted by major themes, revealing examples of the differing perceptions of the local environment as conveyed by direct participant quotations. The range of understandings presented incorporated non-human entities and human–non-human interactions. Further to this, the significance of non-human entities differentiated the perspectives across a spectrum. And while the study sample was small, the findings were nevertheless informative and indicative of an area worthy of further research. The major themes as expressed in the 13 interviews are introduced below:

'The Hunter Region as a Working Landscape'

Two of the interviewees revealed an understanding of the Upper Hunter Valley landscape as a 'working landscape' that was mainly seen as a resource provider. These participants were typically affiliated (employed) with the mining sector and had mostly a family history of generational employment in mining work. These participants described their involvement in negotiations with other stakeholders in relation to mining operations and mining impacts, indicating an awareness of different perspectives. This perspective is exemplified in comments such as the following: 'I know what's involved in all the technical work and I see – people in sort of really highly skilled jobs undertaking that work' (M1, personal interview, October 11, 2016).

M1's comment above illuminates skill and technical expertise, which was a theme observed across the idea of the 'working landscape'. Other themes that were identified by analysis included: a landscape providing financial benefit, a landscape of resources, working families, and continuity of the working landscape that requires established legitimacy. The latter was, for example, expressed thus: 'Some of our economic contributions, the fact that we pay from a coal point of view about $200m a year in royalties' (M2, personal interview, October 11, 2016).

The working landscape was further expressed in public discussions by referring to towns such as Singleton and Muswellbrook as 'mining towns' (Evans, 2008; Munro,

2012), which describes a specific kind of working landscape relating to mining. Such naming, along with reference to figures such as the volume of coal exported, or the amount of royalties paid by mining to the NSW government, provided examples of the symbolic vehicles through which the working landscape was formed and carried, and ultimately manifested. In short, the landscape of the Hunter Valley was read from an economic point of view as a working landscape and resource base.

'The Hunter Region as a Landscape of Provision'

Similarly, the Upper Hunter Valley was also more generally understood as a landscape of wider provision distinct from 'just mining', while still related to the resource extracting sector. The three participants that described the Hunter as a landscape of provision were aligned with governmental bodies, with two being elected representatives at state government level based locally and in Sydney, and one being a local departmental staff member. Provision referred in particular to the acknowledgement of the significance of the provision of energy and coal as well as a place for local community, as expressed by participant G2:

> For us it [the Upper Hunter Valley] means a region, or a subregion depending on how you define it, in transition. It's one that as we all know has, you know, relied heavily on resources and energy production for the region both in terms of coal production and in power generation but the long term, and that is absolutely going to remain a part of … the region's economic future. (G2, personal interview, September 16, 2016)

G2's comments illuminate themes such as: a landscape responsive to future needs; a landscape that provides economic benefits; a source of energy; and a managed landscape, which carried across the landscape of provision, and was a connection between rural and urban landscapes. The notion of landscape of provision was carried through imagery such as the large power stations in the Upper Hunter Valley and 'coal chain' logistics and infrastructure network that treats and transports coal from the point of extraction to the port of Newcastle. The coal chain in particular became a symbol of economic success and engineering prestige when used in the contexts referred to by these participants. However, other points of views described the Hunter Valley in sharp contrast to these perspectives, in distinctively non-economic terms.

'The Hunter Region as a Rural Idyll'

Another understanding of the Upper Hunter Valley landscape identified was that of a 'rural idyll'. This description was expressed directly by four participants, who were linked to either viticulture/tourism or the equine sector. These participants described a long history between the area and their respective sectors, in contrast to 'just mining'. Interestingly, those associated with the equine sector spoke of few connections with Australian networks and instead emphasised connections with global, high-profile equine regions such as Newmarket in the United Kingdom and

Kentucky in the United States. However, participants from the viticulture sector spoke more frequently of cooperating with the local or general Australian viticulture sector. Significantly, all participants described their involvement in associations that promote and advocate their sectors, and typically in opposition to mining, for example the Hunter Thoroughbred Breeders Association and the Hunter Valley Wine and Tourism Association. Further themes within this discussion of the rural idyll included a clean, green place that produces quality wine and tourism experiences, a strong sense of community, a landscape that presents challenges and opportunities, and continuity of the rural idyll. And certain, hard to grasp qualities were also expressed, for example, by participant VT2: 'The macro-climate combined with the geology of the Hunter Valley creates a unique style of wine that cannot be replicated anywhere else in the world' (VT2, personal interview, September 8, 2016).

Understanding the Hunter Valley as productive and yet a rural idyll is a complex construction and also incorporates themes including prestige derived from the rural idyll; a landscape that supports interconnectivity between people; a vulnerable landscape; a landscape shaped by human influence and contribution; and a changing landscape that requires human adaptation. It also touches upon ideas of sense of place, as exemplified in the following statement: 'The thoroughbred breeders have been there for over 200 years. It's an iconic part of Australian history, Australian culture. There are significant landmarks, and the beauty of it is unparalleled, which is why people go there' (E1, phone interview, September 6, 2016).

The common use of terms such as 'Wine Country' and 'Horse Capital of Australia' and iconic branding propositions such as 'Hunter Valley Shiraz Wine' were also seen as examples of the symbolism used to maintain and promote the idea of a rural idyll in the wider public sphere.

'The Hunter Region as a Living Landscape'

For others, the Upper Hunter Valley was more generally seen as 'simply' a living landscape. Two participants were local residents and members of various progress associations (a historical form of localised community advocacy groups) and further community groups local to the Upper Hunter Valley. These participants had lived in the area for decades and were well-established, by contrast to more recently arrived 'treechangers', including those that continue to flee the cities since the Covid-19 pandemic (ABS, 2020). For the already more established residents, the living landscape encompassed a diversity of life forms that acknowledged non-anthropogenic aspects, as expressed by participant C1:

> It's got major protected natural areas with the Wollemi National Park, and Barrington Tops, and Manobalai Nature Reserve, and the Goulburn River National Park. It's also very naturally diverse, because it's a crossover of a number of major bioregions. The biodiversity in the Upper Hunter is extremely important and quite unique. (C1, personal interview, September 13, 2016)

Themes within the living landscape included a place of personal significance, a natural landscape, a threatened landscape. and a heterogeneous community that includes Indigenous people. The idea of a living landscape was further described in contrast to changes the participants considered to be inconsistent with this perception: 'Green space and green trees and, as it used to be, clean air, which we certainly don't have now; the pollution's quite frightening up here, it really is' (C3, personal interview, September 27, 2016).

In other words, the idea of a living landscape was seen as contradicting the idea of simply being a resource provider for certain industry sectors. The living landscape was further described through imagery such as 'home' or 'unique' and through geographic features that spoke to the specific qualities of the Upper Hunter Valley, which connects to the discussion of 'sense of place'. This term seeks to encompass the person-to-place bond and is also referred to as 'place attachment, place meaning, place identity, place dependence, sense of belonging, and sense of community' (Nelson et al., 2020, p. 237). As in the case of the other constructions, this imagery helped to inform and communicate the idea of a living landscape.

THE MANY READINGS OF THE HUNTER REGION LANDSCAPE(S)

The Upper Hunter Valley region has long been dominated by agriculture, mining and forestry, but also conservation (Argent et al., 2014) and has changed considerably over the last decades. In particular, tourism has emerged, while there has been a marked decline of traditional production functions including agriculture. Concurrently parts of the community have begun championing alternative aesthetic and cultural qualities to support opportunities for rural tourism (Holmes, 2006). In other words, the role and function(s) of this rural landscape have become somewhat 'fluid' as old forms of land use are being renegotiated and new expressions of land use are tentatively introduced. This dynamic indicates a development of knowledge; for example, knowledge of business potential in tourism and perhaps increased knowledge surrounding the impacts and causes of anthropogenic climate change.

The resulting, and at times volatile, conflict in the public and even legal domains indicates a diversity of (and controversy in) views on what constitutes appropriate land use. For example, previous research has shown that some residents have expressed concern about health impacts in the area (Connor et al., 2009; Cottle, 2013), focusing on water use (Connor et al., 2008), and controversial decisions such as the approval of mine expansions (McCarthy, 2014; Munro, 2012). For those with a family history in mining, for example, a mine expansion may be welcomed. Alternatively, those who are attracted to the area's various rural amenities or rely on these for economic gain, as in the case of tourism, may not view mining expansion as a positive development.

The findings of this explorative study confirm the existence of diverse perceptions of a single shared and yet contested landscape. The varied understandings were exemplified through the opinions and quotations of 13 participants based on their different knowledge, beliefs and a different history of engagement with the area, and while not necessarily representative due to the small sample size, nevertheless provide insight.

In particular, these examples provided insight into how the symbols that represented these constructions at times appeared to be in direct conflict with each other, as in the case of 'home' and figures describing the volume of coal exported. In other words, the perspectives explored in this primary research reflected some of the sentiments expressed in other qualitative research conducted in the same geographical area and other comparable geographical areas. Furthermore, while each construction displayed elements of production, the associated consumption and protection values are indicative of a multifunctional rural transition (Holmes, 2006), and they manifested quite differently. For example, notions of the working landscape encompassed positive sentiments towards the mechanisms of resource extraction and technical expertise required for successful operation, whereas the living landscape concept included a voice for concerns regarding environmental degradation, potentially linked to the working landscape. The conceptualisation of the landscape as one of provision shares elements with the working landscape, such as economic and financial benefits, linking those that shared this construction. On the other hand, the rural idyll seems unique in its focus on continued land use, showing commonalities with the working landscape and landscape of provision, but also emphasising particular aesthetics such as open spaces and clean air, and thus displaying similarities with the living landscape concept.

Overall, participants describing the living landscape and rural idyll expressed a degree of uncertainty in their capacity to continue engaging in the landscape in consistency with the constructions they attributed to it, which is a tendency not shared by participants who described the working landscape or a landscape of provision. Moreover, the findings revealed significant commonalities between the symbolism that represented the working landscape and the landscape of provision, allowing the associated values a position of prominence and power. Significantly, some perspectives, such as any Indigenous readings, were largely absent from the public debate of changes to the Upper Hunter Valley landscape. This situation powerfully demonstrates the ongoing struggle of disempowerment and dispossession suffered by Indigenous Australians and once again indicates power imbalances that need special consideration when discussing land use conflict. Future studies should expand the pool of interviewees to add further depth to this line of enquiry.

CONCLUSION

Landscapes have values and meaning. Our probing confirms that the Hunter Valley is 'seen and understood' in a variety of readings. And while this is an exploratory study with limited interviews, outcomes confirm the importance of understanding local opinions towards landscape and its management. Foremost, the study also provides insight into the ongoing changes to how people live in and from the landscape and the influence this has on their perception. For example, it has been argued that people respond more negatively to changes they perceive to threaten their capacity to engage with their environment and resources in consistency with the construction(s) they ascribed to it (Anderson et al., 2016). The significant increase in open-cut coal mining in the Upper Hunter Valley exemplifies one such visually significant change (Cottle, 2013). And indeed, the change from underground to open cut mines exacerbates and polarises the public debate (Connor et al., 2009). Furthermore, this development emphasises the observable manifestation of production values in relation to land use. Production values manifest in mining and energy production complement the working landscape and landscape of provision but are less observable in the rural idyll and the living landscape. This point of difference provides one explanation for public opposition to mostly open-cut coal mining from community and members of the equine and viticulture/tourism sector, but not from the government.

Overall, in the context of multifunctional rural transition, conflict suggests a manifestation of production, consumption, and protection values that doesn't support all forms of suggested land use. Although a contested combination of values is typical of peri-metropolitan zones, it is possible that commonalities between the working landscape and landscape of provision, as predominantly held by decision-makers, increase the perceived threat to the rural idyll and the living landscape in the Hunter region.

This research revealed interrelated and yet distinct understandings of the Upper Hunter Valley that reveal notable similarities and differences. The key suggestion is that similarities between the constructions present a possible basis for collaborative and less controversial land use strategies. Consequently, the concept of Integrated Landscape Initiatives (ILIs) may provide guidelines (Garcia-Martin et al., 2016). Either way, it is arguable that the success of any initiatives depends upon collaboration between different actors in the area, which appears to be hindered by the presence of conflict over land use. Here, the introduction of a variety of ILIs could facilitate opportunities for varied perspectives to be in some way reflected in future land use debates and decisions.

RECOMMENDATIONS AND OUTLOOK

In this chapter, we argue that conflicting ways to read and understand the Upper Hunter Valley landscape have considerable influence on local land use planning and management, and thus the future of this rural landscape. Currently, production values overshadow the consumption and protection values that are significant to the rural idyll and living landscape. This is partially explained in the unequal economic significance of these sectors (NSW Department of Trade and Investment, 2015). Still, tourism research shows that effective tourism planning and development should be collaborative in nature and consult widely with local stakeholders (Moscardo, 2011).

Based on these learnings and the findings of this explorative study, the following four recommendations are made:

1. Monitor Combination of Production, Consumption and Protection Values

Our approach does link land use conflict and the absence of a balance of production, consumption and protection values that are compatible with different constructions. As a consequence, it is suggested that by understanding key land use in the area and its associated values, we can also identify opportunities for less prominent values to also be reflected in land use. For example, more resources for conservation programmes could manifest protection values or tourism initiatives that facilitate conservation. By monitoring the combination of values, governments and tourism planners can address the dominant manifestation of production values currently reflected in Upper Hunter Valley land use; for example, through the development of a metric that facilitates a balance between these land uses and grapples with non-economic uses and meanings.

2. Incorporate Multifunctionalism into Land Use Planning and Policy

Land use planning and policy is increasingly complex and diverse. Directly arising from Holmes's work and this study's findings is the suggestion that a shifting balance of multiple uses of rural land should be acknowledged in and incorporated effectively into policy to facilitate successful multifunctionalism (Holmes, 2006). Examples in Queensland, such as the result of discussions with the Indigenous Wik people and pastoralists, demonstrate how policy can ensure that one form of rural land use does not diminish or is not mutually exclusive from other forms of rural land use (Godden, 1999; Queensland Department of Environment and Resource Management, 2019). Such adjustments in policy settings could allow production, consumption and protection values all to manifest in land use in the area, thus theoretically becoming compatible with a wide range of constructions ascribed to the landscape.

3. Respond to Concerns Regarding How Changes in Land Use May Affect Ideal Construction of Rural Landscapes

As noted, successful tourism governance adopts a consultative and collaborative approach by involving stakeholders (Moscardo, 2011). It is the aim of groups and individuals to circulate their constructions through their group and in the public sphere so that they are supported and sustained (Howarth, 2006). Land use conflict suggests a perceived lack of public endorsement of the desired land use that is compatible with particular constructions and increasing difficulty for perseverance with these constructions. A shift in the public messaging from government and tourism planners that would explicitly acknowledge the impact of changes in the landscape may provide some reassurance to those who feel their constructions are threatened and open the way for dialogue. This process may be assisted by the explicit inclusion of and reference to the representative symbols associated with different constructions as this research demonstrates that these symbols represent a group or individual's history, shared knowledge, values and beliefs.

4. Further Research

It is noteworthy that despite Australia's location in the Asia-Pacific region, Australian understandings of 'rural' strongly resemble British perceptions of rural traditions and agricultural customs (Dixson, 1999; Jordan et al., 2009). Future studies should examine possible cross-cultural differences in perceptions of rural to better inform tourism planners of what may appeal to a range of visitors to rural tourism destinations. And the voices of Indigenous Australians must be empowered to influence the understanding and management of the landscape.

As it is, the current land use conflict in the Upper Hunter Valley has negative social, economic and environmental impacts for most interested parties and at times negatively impacts upon the destination's image. This chapter hopes to provide an argument for the need to acknowledge similarities and differences of the various perspectives to facilitate an integrated debate on what an inclusive rural tourism landscape could look like.

REFERENCES

ABS (Australian Bureau of Statistics) (2020). *Regional Population.* https://www.abs.gov.au/statistics/people/population/regional-population/2018-19.

Albrecht, G., Sartore, G.-M., Connor, L., Higginbotham, N., Freeman, S., Kelly, B., Stain, H., Tonna, A., & Pollard, G. (2007). Solastalgia: The distress caused by environmental damage. *Australasian Psychiatry, 15*(Supplement 1), 95–98. doi:10.1080/10398560701701288.

Anderson, N.M., Ford, R.M., & Williams, K.J.H. (2016). Contested beliefs about land-use are associated with divergent representations of a rural landscape as place. *Landscape and Urban Planning, 157*, 75–89. doi:http://dx.doi.org/10.1016/j.landurbplan.2016.05.020.

Argent, N. (2011). Trouble in Paradise? Governing Australia's multifunctional rural landscapes. *Australian Geographer, 42*(2), 183–205.

Argent, N., Tonts, M., Jones, R., & Holmes, J. (2013). A creativity-led rural renaissance? Amenity-led migration, the creative turn and the uneven development of rural Australia. *Applied Geography, 44*, 88–98. doi:10.1016/j.apgeog.2013.07.018.

Argent, N., Tonts, M., Jones, R., & Holmes, J. (2014). The amenity principle, internal migration, and rural development in Australia. *Annals of the Association of American Geographers, 104*(2), 305–318.

Brunori, G., & Rossi, A. (2007). Differentiating countryside: Social representations and governance patterns in rural areas with high social density: The case of Chianti, Italy. *Journal of Rural Studies, 23*, 183–205. doi:10:1016/j.jrurstud.2006.10.001.

Bryman, A., & Bell, E. (eds) (2011). *Business Research Methods.* Oxford: Oxford University Press.

Connor, L., Higginbotham, N., Freeman, S., & Albrecht, G. (2008). Watercourses and discourses: Coalmining in the Upper Hunter Valley, New South Wales. *Oceana, 78*(1), 76–90.

Connor, L., Freeman, S., & Higginbotham, N. (2009). Not just a coalmine: Shifting grounds of community opposition to coal mining in Southeastern Australia. *Ethnos: Journal of Anthropology*, *74*(4), 490–513.

Cottle, D. (2013). Land, life and labour in the sacrifice zone: The socio-economic dynamics of open-cut coal mining in the Upper Hunter Valley, New South Wales. *Rural Society*, *22*(3), 208–216. doi:10.5172/rsj.2013.22.3.208.

Creswell, J. (2014). *Research Design: Qualitative, Quantitative and Mixed-Method Approaches*. Thousand Oaks, CA: Sage Publications.

Deaux, K., & Philogene, G. (2001). *Representations of the Social*. Oxford: Blackwell Publishers.

Destination NSW (2016). *Upper Hunter*. https://www.visitnsw.com/destinations/hunter/upper-hunter.

Dixson, M. (1999). *The Imaginary Australian: Anglo-Celts and Identity, 1788 to the Present*. Sydney: UNSW Press.

Evans, G. (2008). Transformation from 'Carbon Valley' to a 'post-carbon society' in a climate change hot spot: The coalfields of the Hunter Valley, New South Wales, Australia. *Ecology and Society*, *13*(1), 39.

Francis-Coan, S., Buchmann, A., & Lai, P.-H. (2017). Exploring land use conflict from a social representations point of view: A case study of the Upper Hunter Valley, NSW, Australia. In C. Lee, S. Filep, J.N. Albrecht, & W.J.L. Coetzee (eds), *CAUTHE 2017: Time for Big Ideas? Re-thinking the Field for Tomorrow* (pp. 805–808). Dunedin, New Zealand: Department of Tourism, University of Otago.

Friedland, W.H. (2002). Agriculture and rurality: Beginning the 'final separation'? *Rural Sociology*, *67*(3), 350–371. doi:10.1111/j.1549-0831.2002.tb00108.x.

Gailing, L., & Leibenath, M. (2015). The social construction of landscapes: Two theoretical lenses and their empirical applications. *Landscape Research*, *40*(2), 123–138.

Garcia-Martin, M., Bieling, C., Hart, A., & Plieninger, T. (2016). Integrated landscape initiatives in Europe: Multi-sector collaboration. *Land Use Policy*, *58*, 43–53.

Godden, D. (1999). Attenuating Indigenous property rights: Land policy after the Wik decision. *Australian Journal of Agricultural and Resource Economics*, *43*(1), 1–33.

Greider, T., & Garkovich, L. (1994). Landscapes: The social construction of nature and the environment. *Rural Sociology*, *59*(1), 1–24.

Halfacree, K.H. (1993). Locality and social representation: Space, discourse and alternative definitions of the rural. *Journal of Rural Studies*, *9*(1), 23–37. doi:http://dx.doi.org/10.1016/0743-0167(93)90003-3.

Halfacree, K.H. (2007). Trial by space for a 'radical rural': Introducing alternative localities, representations and lives. *Journal of Rural Studies*, *23*(2), 125–141. doi:http://dx.doi.org/10.1016/j.jrurstud.2006.10.002.

Higginbotham, N., Freeman, S., Connor, L., & Albrecht, G. (2010). Environmental injustice and air pollution in coal affected communities, Hunter Valley, Australia. *Health & Place*, *16*(2), 259–266. doi:http://dx.doi.org/10.1016/j.healthplace.2009.10.007.

Higginson, S. (2014). Environment and planning: Legal challenges like Bulga key to safeguarding against corruption. *LSJ: Law Society of NSW Journal*, *1*(1), 74.

Holmes, J. (2006). Impulses towards a multifunctional transition in rural Australia: Gaps in the research agenda. *Journal of Rural Studies*, *22*(2), 142–160.

Howarth, C. (2006). A social representation is not a quiet thing: Exploring the critical potential of social representations theory. *British Journal of Social Psychology*, *45*, 65–86.

Hunter Valley Thoroughbred Association, & Hunter Valley Wine and Tourism Association (2015). *Time to protect the Hunter Valley's state significant agricultural lands and industries*. http://www.htba.com.au/_p/nws3/B49A6781-30A1-44DB-A101-9C7FB0E19B3D/hunter-thoroughbred-breeders%2C-wine-and-tourism-associations-call-for-bipartisan-commitment-to-protect-the-hunter-valley.

Jordan, K., Krivokapic-Skoko, B., & Collins, J. (2009). The ethnic landscape of rural Australia: Non-Anglo-Celtic immigrant communities and the built environment. *Journal of Rural Studies*, *25*(4), 376–385. doi:http://dx.doi.org/10.1016/j.jrurstud.2009.05.004.

Jovchelovitch, S. (2001). Social representations, public life, and social construction. In K. Deaux & G. Philogene (eds), *Representations of the Social* (pp. 165–182). Oxford: Blackwell Publishers.

Jupp, C. (2021) Meet the Upper Hunter by-election candidates. *The Land*. https://www.theland.com.au/story/7236026/meet-the-candidates-as-the-upper-hunter-race-heats-up/.

Lai, P.-H., Hsu, Y.-C., & Nepal, S. (2013). Representing the landscape of Yushan National Park. *Annals of Tourism Research*, *43*, 37–57.

March, R., & Wilkinson, I. (2009). Conceptual tools for evaluating tourism partnerships. *Tourism Management*, *30*, 455–462.

McCarthy, J. (2014, November 11). New mine plan revives battle for Upper Hunter: Coal v Foals. *The Newcastle Herald*. http://www.theherald.com.au/story/2689895/new-mine-plan-revives-battle-for-upper-hunter-coal-v-foals/.

McGuirk, P., & Argent, N. (2011). Population growth and change: Implications for Australia's cities and regions. *Geographical Research*, *49*(3), 317–335. doi:10.1111/j.1745-5871.2011.00695.x.

McManus, P. (2008). Mines, wines and thoroughbreds: Towards regional sustainability in the Upper Hunter, Australia. *Regional Studies*, *42*(9), 1275–1290. Doi:10.1080/00343400701827394.

McManus, P., & Connor, L.H. (2013). What's mine is mine(d): Contests over marginalisation of rural life in the Upper Hunter, NSW. *Rural Society*, *22*(2), 166–183. doi:10.5172/rsj.2013.22.2.166.

McManus, P., O'Neill, P., & Loughran, R. (eds) (2000). *Journeys: The Making of the Hunter Region*. St Leonards, NSW: Allen & Unwin.

Moscardo, G. (2011). Exploring social representations of tourism planning: Issues for governance. *Journal of Sustainable Tourism*, *10*(4–5), 423–436.

Moscovici, S. (2001). Why a theory of social representation? In K. Deaux & G. Philogène (eds), *Representations of the Social: Bridging Theoretical Traditions* (pp. 8–35). Oxford: Blackwell Publishing.

Munro, S. (2012). *Rich Land, Wasteland – How Coal Is Killing Australia*. Sydney: Pan MacMillan.

Nelson, J., Ahn, J.J. and Corley, E. (2020) Sense of place: Trends from the literature. *Journal of Urbanism: International Research on Placemaking and Urban Sustainability*, *13*(2), 236–261.

NSW Department of Trade and Investment (2015). *Economic Profile Hunter*. http://www.trade.nsw.gov.au/__data/assets/pdf_file/0006/67704/4.-Hunter_Regional-Economic-Profile_Feb2015.pdf.

O'Neill, P. (2000). The gastronomic landscape. In P. McManus, P. O'Neill, & R. Loughran (eds), *Journeys: The Making of the Hunter Region* (pp. 158–185). St Leonards, NSW: Allen & Unwin.

Queensland Department of Environment and Resource Management (2019). *Guide to the Pastoral ILUA template*. https://www.resources.qld.gov.au/__data/assets/pdf_file/0003/1450308/ilua-guideline.pdf.

Richards, G., & Hall, D. (2003). *Tourism and Sustainable Community Development*. London: Routledge.

Ryan, C. (2005). Ethics in tourism research: Objectivities and personal perspectives. In B.W. Ritchie, P. Burns, & C. Palmer (eds), *Tourism Research Methods: Integrating Theory with Practice* (pp. 9–20). Wallingford: CABI Publishing.

Sherval, M., & Hardiman, K. (2014). Competing perceptions of the Rural Idyll: Responses to threats from coal seam gas development in Gloucester, NSW, Australia. *Australian Geographer*, *45*(2), 185–203.

The Australia Institute (2021). *Upper Hunter Polling: Majority of Voters Agree with Turnbull Call for Moratorium on New Coal Mines*. https://australiainstitute.org.au/post/upper-hunter-polling-majority-of-voters-agree-with-turnbull-call-for-moratorium-on-new-coal-mines/.

Thompson, F. (2012, January 15). Winemakers put heat on O'Farrell. *The Newcastle Herald*. http://www.theherald.com.au/story/442099/winemakers-put-heat-on-ofarrell/.

Van der Duim, R. (2007). Tourismscapes – an actor-network perspective. *Annals of Tourism Research*, *34*(4), 961–976.

Wagner, W., Duveen, G., Farr, R., Jovchelovitch, S., Lorenzi-Cioldi, F., Markova, I., & Rose, D. (1999). Theory and method of social representations. *Asian Journal of Social Psychology*, *2*, 95–125.

Wearing, S.L., Schweinsberg, S., Lai, P.-H., & Lyons, K. (2014). A discussion of coal seam gas in Australia's Hunter Valley wine tourism area. *Australasian Parks and Leisure*, *17*(2), 49–52.

18. Creatively regenerating St. Jacobs Village

Tawsif Dowla and Karla A. Boluk

INTRODUCTION

The focus on growth in tourism has led to concerns regarding the sector being a victim of its own success. The two most-cited precursors of such tourism impasses are over-tourism (i.e., Phi, 2019; Butler, 2020) and unsustainable growth (i.e., Wheeller, 1993; Boluk et al., 2019). Rural communities are particularly vulnerable to over-tourism and unsustainable growth because of their limited capacity. They are susceptible to overcrowdedness and the disruption of over-tourism on local residents' activities in environmental, social, and even economic terms (Butler, 2020).

The tourism sector is driven by an economic orientation pursuing growth at all costs and, as such, has obstructed the progress of sustainability (Becken, 2017; Boluk et al., 2017; 2019). Tourism destinations are typically led by businesses and governments who are focused on profit generation and keen to respond to the consumption desires of travelers (Weeden & Boluk, 2014). An emphasis on production and consumption has hindered attending to local destination stakeholders' rights, needs, interests, and environments where tourism occurs (Higgins-Desbiolles et al., 2019; Boluk et al., 2021). Tourism growth may lead to congestion, resentment on behalf of host communities, environmental vulnerability, and a decline in visitor satisfaction, repeat visits and/or referrals (i.e., Pollock, 2019). Recognizing the harm generated by tourism allows us to see manifestations of carelessness and its adverse impacts on places, communities, and environments (Carnicelli & Boluk, 2022). Acknowledging such negligence requires a radical reform in the way tourism is planned and practiced.

Concerns regarding carrying capacity (O'Reilly, 1986), recently termed over-tourism (i.e., Phi, 2019), that is, too many people visiting the same place simultaneously, have led to calls for new paradigms to respond to the wicked problem of mass tourism. Specifically, scholars have engaged in critical discussions on degrowth practices in tourism (Hall et al., 2020), responding to the current climate crisis, and the critical role of tourism's social entrepreneurs in driving change in destination communities (i.e., Boluk & Mottiar, 2014; Mottiar & Boluk, 2017; Mottiar et al., 2018; Sheldon et al., 2017).

A new definition was recently offered by a team of scholars recognizing the imperative to drive tourism decision-making with a social justice orientation and, specifically, recenter the rights of local communities above the rights of tourists (Higgins-Desbiolles et al., 2019). Offering this novel rethinking of tourism led to a new vocabulary to support social and ecological justice, recognizing the need to socialize tourism, that is, the process of learning to behave in ways that are appropriate to society (Higgins-Desbiolles et al., 2021). Such thinking is clearly required given the inimical impacts generated by the tourism sector.

A new emergent paradigm, regenerative tourism, offers a radical reform, proposed to recenter host communities with intentionality, emphasizing restoring harm and creating conditions for flourishing and thriving communities. As a guiding philosophy or mindset, regenerative tourism responds to scholars who have called for new ways of engaging and practicing tourism embracing a social and care ethic (Boluk & Panse, 2022; Carnicelli & Boluk, 2022; Pollock, 2019). This is important in guiding the process of planning, developing, offering, and thinking about tourism design in, and with, communities (i.e., Dredge, 2017).

Recent scholarship has signaled that the pause in travel due to the global pandemic may be viewed as a silver lining offering the space and time to rethink, retool, and redesign tourism (see Ateljevic, 2020; Sheldon, 2022) in support of just, caring, and radical pathways. A review of the literature highlights a few critical examples in theory and praxis signaling processes for engaging in tourism that will foster regenerative tourism benefits for the communities and places where tourism occurs (i.e., Everingham & Chassagne, 2020). This chapter aims to contribute to our ongoing exploration of the potential of St. Jacobs Village, in Ontario, Canada in becoming a destination true to its shifting multicultural and multi-ethnic identity. We aspire to explore the prospects for utilizing the emerging creative turn in St. Jacobs' tourism and community revitalization discourse following regenerative principles.

To set the scene, we will commence with a review of the literature organized into two sections. First, we will reflect on the pertinent literature on creative and regenerative tourism. Second, we will provide a sketch of St. Jacobs and its evolution as a heritage destination through different stages outlined in prominent tourism literature. The literature review is followed by a statement of methodology and analysis of the data emitted from the voices of St. Jacobs. Upon a brief reflection on such voices in our analysis, we will conclude with some key considerations so that St. Jacobs, as well as other rural communities, may realize their tourism and revitalization agenda. We hope this work will open up avenues for future researchers to develop coherent approaches for tourism development (or redevelopment), mainly in rural communities, small towns, or islands in peripheral regions, responsive to local communities' concerns and interests, and equitable participation.

LITERATURE REVIEW

Creative Tourism

The emergence of the creative tourism concept can be traced back to the mid 1990s (Richards, 2005). One of the early examples of interest in creative tourism came to the limelight during the EUROTEX project (1997 to 1999), aiming to bolster textile-related cultural tourism in disadvantaged peripheral regions in Europe (Richards, 2005). The key objective of the EUROTEX project was to foster the demand and supply side of crafts production by (a) actively marketing local products to tourists and (b) developing new outlets for craft sales and securing local jobs (Richards, 2005; 2009). The EUROTEX team realized that when tourists appreciate the works and skills involved in producing craft products, they are usually willing to pay a comparatively higher price for the high-value handmade products against cheaper mass-produced items in the vicinity (Richards, 2009). Such a trend aligns with the emergence of the experience economy where contemporary consumers seek experience over products (Pine & Gilmore, 2013) or doing over being and gazing.

Tourists' appreciation for high-value authentic products is further supplemented, the EUROTEX team revealed, if producers tap into tourists' disposition for active participation in making products themselves instead of passively observing production techniques (Richards, 2005). Tourists' participation in co-creating authentic experiences is what underlines the emergence of "creative tourism" (Richards, 2009), which Richards and Raymond (2000) defined as "tourism which offers visitors the opportunity to develop their creative potential through active participation" (p. 18).

The pioneering scholars of creative tourism recognized creative tourism as an extension of cultural tourism (Binkhorst & Den Dekker, 2009; Richards, 2005; Richards & Raymond, 2000; Richards & Wilson, 2007), which saturated the tourism market between the 1970s and the 1990s. Once viewed as a savior to the local economy, especially for destinations on the periphery, cultural tourism eventually fell prey to its success. Richards and Raymond explained the basis for such an unwitting out-turn, "cultural tourism is seen as a growth market, and cultural tourists are viewed as desirable tourists ... there has been a massive growth in the number of cultural attractions" (2000, p. 16).

As supply began to outstrip the demand and competition intensified, some destinations with limited cultural offerings resorted to "McDonaldizing" their culture and heritage (Binkhorst & Den Dekker, 2009; Richards & Raymond, 2000). In their desperate attempts to appeal to the consumers, heritage and cultural destinations took a path of "creative destruction" through the provisioning of commodities and experiences that are "blatantly disconnected to or are only vaguely symbolic of, the heritage theme" (Mitchell & de Waal, 2009, p. 162). Consequently, these destinations jeopardize their cultural uniqueness, making it difficult to distinguish one from another (Binkhorst & Den Dekker, 2009).

An inability to demonstrate the unique culture of a place reveals cultural delinquency which undermines tourists' quests for originality and authenticity (Binkhorst & Den Dekker, 2009). A need for diversification and innovation was deemed obvious, and creative tourism began to transpire from such demand. On a similar note, the evolution of creative tourism can be interpreted as the interactive descendant of cultural tourism (Blapp & Mitas, 2020; Richards & Wilson, 2007), where the visitors and the visited co-create cultural experiences (Richards, 2011). Besides active participation over passive consumption, creative tourism also concerns tourists' interests in living culture over static heritage sites (Richards, 2011).

One of the ontological underpinnings of creative tourism, in Binkhorst and Den Dekker's words, is "tourists as human beings moving around in various environments" (2009, p. 322). Such a humanist approach enables tourism providers to extend their perspective of tourists as mere "clients" to "humans" with lives beyond tourism (Binkhorst & Den Dekker, 2009). This also has implications for travelers visiting communities. Recognizing tourists as human beings is important because some literature has dehumanized "the tourist," potentially perpetuating the inimical impacts generated in host communities.

Humanizing tourism may aid us in capturing the agency of travelers, as well as their capability to be responsible agents drawing on their ethical compasses while traveling. Fennell (2014) for example, uses the metaphor of a cloak, which some travelers choose to leave at home. If, however, we bring and wear our cloak of morality when we travel, we may, as Fennell (2014) argues, advance our moral practices and contribute to the protection of the places we visit. Packing and honing our (cloak) moral compass may serve as an impetus to purchase offerings or engage in experiences that align with one's ethics, values, self-image, or perceptions of what is authentic (Pine & Gilmore, 2013). In other words, Pine and Gilmore (2013) noted, "authenticity has become the new consumer sensibility" (p. 29).

Considerations regarding authenticity are a powerful source of information in developing creative tourism experiences and marketing (Binkhorst & Den Dekker, 2009) tailored to individuals' values and needs. Besides, as Poon (1993) claimed, the "new tourist" (though no longer a new phenomenon) likes being in charge and responding to their travel needs using various tools available—courtesy of the internet and information technology. Hence, a pre-determined, one-size-fits-all travel package or "staged entertainment and fun" (Binkhorst & Den Dekker, 2009, p. 314) is no longer palatable to new tourists. These emerging phenomena further justify the call for the co-creation of tourist experiences, with creative tourism as the most reliable response. Therefore, creative tourism must not be driven by a "product-centric point of view" (Binkhorst & Den Dekker, 2009) but by a human-centric perspective, involving humans (tourists and locals alike) in co-creating experiences.

While the scope and activities related to creative tourism have diversified since its emergence in the 1990s, the context is still the same. Providing authentic, active and engaging experiences remains a cornerstone of creative tourism (Richards, 2009; Baixinho et al., 2021; Duxbury et al., 2021). However, Baixinho et al. (2021) argue that the feature of creative tourism should not be limited to the co-creation of tourism experience for economic gains; instead, it must expand to the transference and development of "creative knowledge and skills in the course of shared and co-created activities" (p. 1). Special attention should be paid to the destinations' social, cultural and environmental sustainability, especially in the peripheral regions, i.e., rural and island communities (Baixinho et al., 2021) with waning cultures or fragile ecosystems. As revealed through the EUROTEX project, in addition to diversifying and innovating the tourist experience to stimulate the local economy, creative tourism can help the social and cultural sustenance of the host community (Richards, 2009). Such potential should be further investigated and implemented by adapting a regenerative agenda for creative tourism.

Duxbury et al. (2021) were the first to offer an alignment between creative tourism and regenerative tourism. For example, Duxbury et al. (2021, p. 1) offered this: "Creative tourism initiatives can inspire new ideas and avenues of activity and contribute to the cultural vitality and potential regeneration dynamics through reinforcing distinctive elements of local identity … and serving as platforms for regional collaboration, exchange, and development."

Given the unfavorable impacts tourism has generated, it seems creative tourism may serve as a tool to co-imagine with local stakeholders how to plan authentic experiences for visitors. In this vein, regenerative tourism reminds us that centering communities should be at the heart of tourism planning as advocated by scholars (i.e., Higgins-Desbiolles et al., 2019). We will now turn our attention to regenerative tourism, which may help us understand how to strengthen the anticipated outcomes of creative tourism.

Regenerative Tourism

Regenerative tourism focuses on resiliency, intergenerational equity, and awareness of planetary boundaries. Adopting a regenerative approach supports a mindset shift, pushing beyond linear thinking to adopting whole systems thinking approaches, with an eye on creating value for entire ecosystems and stakeholders (Hutchins & Storm, 2019). Importantly, regenerative travel seeks to integrate with local communities, attending to what already exists in a place, rather than building more. Accordingly, the intentionality in regenerative tourism is to recognize that tourism cannot be separated from the community where it takes place (Designing Tourism, 2022). Critically, tourism must be woven into the community fabric, necessitating care, intentionality, engagement, and authentic dialogue. If carried out with intention and care, regenerative tourism may leave ecosystems, communities, and economies better off than they were before the tourism encounter.

An important aspect of regenerative tourism is intentionally building connections with communities in places where tourism happens (Designing Tourism, 2022). Soliciting diverse views and experiences is essential to community engagement, as such centering of marginal voices draws attention to the importance of living in relationships with others, reciprocity, and the intricate web of caring (Giddens, 1982; Noddings, 1992). Prioritizing feminist and Indigenous perspectives was recently recognized as an omission in the articulation of the United Nations Sustainable Development Goals (e.g., Boluk et al., 2019); perhaps unsurprisingly given that some critiques of the goals note a reinforcement of capitalism (Hall et al., 2019).

Acting in resistance to neoliberal ideology, it is important to intentionally center and engage with marginalized and diverse voices. Such work responds to some concerns in the placemaking literature around complacency, inequities, settler colonialism, racism, and socio-economic elitism (Burns & Berbary, 2021). Accordingly, participatory democracy, nurturing open and authentic dialogue, and participatory decision-making must be prioritized when pursuing regenerative tourism (Designing Tourism, 2022). There is a recent and growing appetite for scholarship on regenerative tourism, with one recent special issue of the *Journal of Tourism Futures* focused on 'Transformation and the regenerative future of tourism' (Ateljevic & Sheldon, 2022). While the term is starting to be taken up in the mainstream, there are few in-depth examples of how regenerative tourism is making transformative impacts within communities undergoing destination and community development.

Recently, a few communities have started to share their destination development/ master plans, emphasizing how such communities may make tourism work for and regenerate their communities. Emerging examples include Belgium's Flanders (Ruiz, 2020), New Zealand's Bay of Plenty (Tourism Bay of Plenty, 2022), and Hawai'i's Kaua'i (Hawai'i Tourism Authority, 2021). Flinders Island in the Bass Strait of the Australian State of Tasmania has recently started to share their regenerative tourism journey (Designing Tourism, 2022). Missing from this list from both a theory and praxis lens is a Canadian context. Accordingly, to address this scholarly gap and respond to a practical need brought forward by a community partner to engage in a sense of place exercise, our work focuses on St. Jacobs Village, a community internationally recognized for tourism.

St. Jacobs Village

St. Jacobs Village is a quaint hamlet in southwestern Ontario located within the historic township of Woolwich, near the city of Waterloo (Explore Waterloo Region, 2022). Located in the traditional territories of the Attawandaron, Anishnaabe, and Haudenosaunee peoples, St. Jacobs, along with its sister village Elmira, is famed for its Mennonite community, which has emigrated mainly from Pennsylvania since 1806. Since its founding, St. Jacobs has experienced a period of emergence and decline, a detailed discussion of which is beyond this chapter's scope. Nevertheless, it is imperative to mention that the inception of tourism in St. Jacobs in the 1970s was not inimical; instead, it regenerated economic and social vitality (Dahms, 1991; Mitchell, 1998).

A creative product of the entrepreneurial spirit spearheaded by Milo Shantz (Mitchell, 1998), tourism in St. Jacobs was timely, cost-effective, and perceived to be a risk-free intervention, which was overwhelmingly welcomed by the traditional community of Pennsylvania Dutch and Germanic Mennonites. The eventual progression of tourism in St. Jacobs, however, followed Butler's (1980) "tourism areas lifecycle," generating incremental tension between the residents' community and the community of growth-minded, profit-centric entrepreneurs. One of the most comprehensive research series, led by Clare Mitchell and colleagues (Mitchell, 1998; 2013; Mitchell & de Waal, 2009; Mitchell & Randle, 2014), provided two innovative tools, creative destruction and creative enhancement, to trace such incidental, yet adversarial evolution of tourism in St. Jacobs. According to Mitchell (2013), tourism in St. Jacobs, post-1996, has been heading for the post-destruction stage.

Since the 1970s, the rural affinity of St. Jacobs, fueled by its Mennonite heritage and culture, has attracted tourists (Dahms, 1991; Getz, 1993; Mitchell, 1998). Understandably advertised as a heritage and cultural site, St. Jacobs evolved as a tourist shopping village (TSV) (Getz, 1993) and remained the same until recently. TSVs, as referred to by Getz (1993), are the towns and villages that primarily appeal to the tourists through retail opportunities in a pleasant setting marked by historical or natural amenities.

The analysis presented here is based on a multi-year collaboration with the St. Jacobs Village Business Improvement Area (BIA). The BIA reached out to the researchers for assistance in carrying out a re-envisioning and sense of place exercise, recognizing some of the changes to the community (i.e., increased diversity, young families moving to the village, and new entrepreneurs joining the community). The research partnership is a multi-year project. The analysis presented here was based on the initial data collection phase, resulting in 16 semi-structured interviews with key informants, including local businesses, residents, and the tourism development community working in collaboration to improve and diversify the tourism offerings in St. Jacobs Village. Our chapter now turns to a brief discussion of the methodology and methods used in gathering the data and the analysis carried out.

METHODOLOGICAL STATEMENT

Our study forms part of a larger, multiphased research project exploring the developing of a sense of place among community members in St. Jacobs Village. The current study reflects our initial phase of research, guided by an interest in co-identifying key stakeholders in the village to gain insights into the sense of place. Accordingly, 16 semi-structured interviews were carried out from September to November 2019. The data were analyzed following the strategies of grounded theory, which involved memo writing, constant comparative analysis, and open and intermediate (axial) coding (Charmaz, 2011). The informants represent diverse professional/occupational and voluntary contributions within the village. Here, our research involves the participating members (referred to by gender-specific pseudonyms) representing diverse communities whose collective interests served to drive our project. In the following sections, we present some of the key findings which emerged from our analysis.

KEY FINDINGS AND ANALYSIS

Sense of Place

St. Jacobs is a "*nice town,*" as succinctly described by Nora, a veteran business owner and operator. Most other participants portrayed St. Jacobs as a "quaint" or a "rural" community, which Adam, a restaurateur, articulated as a "cute little hamlet suburb of the greater Waterloo region." This idyllic appeal of St. Jacobs is rooted mainly in its cultural landscape, but is also attributed to, as most participants agreed, the welcoming, kind, friendly, and generous residents of St. Jacobs, both longstanding and newly settled.

Sense of Community

Though there is tangible evidence of Mennonite heritage within the village periphery, evidence suggests that St. Jacobs is no longer a predominantly Mennonite community. As revealed by Emma, a tourism marketer, "there is that big connection to the Mennonite ... they do have Mennonite Story." But she acknowledged, "Mennonites have moved to Elmira." Ruth, an Anabaptist cleric, attributed such a departure of intangible culture from its tangible heritage site to gentrification. As a result, Ruth claimed, "old families are seeing their way of life and neighbourhood, transition quickly and feeling alienated in their own community." Emma too echoed Ruth's concerns by saying, "I think, [the] struggle going forward is how do they make sure that they maintain that sense of community and not get swallowed up by Waterloo [nearby urban center] that [is] slowly encroaching on that area."

This contested sense of community, as represented through the disparate voices of St. Jacobs, seems to be impacting the longstanding families more. Ruth articulated, "this is a village that wants to welcome and be neighbours helping neighbours ... just don't diminish the core values of St. Jacobs ... don't diminish the way neighbours relate to each other as neighbours." Furthermore, Miller and Martin (self-identified as "invisible-Mennonites") expressed concerns regarding the presence of recent investors, who did not grow up in St. Jacobs and "view the village as a usual business opportunity and do not reflect the local community's needs." Amid such a contested sense of community, others shared optimism about the BIA leadership in the village. For example, Gavin, a heritage tourism conductor, referring to Milo Shantz's leadership in revitalizing St. Jacobs, shared, "I'm starting to get a sense of it going back that way." Nora enthusiastically noted, "we are trying ... to get everybody together." Nora's sentiment revealed that community dialogue is imperative for co-designing the future of St. Jacobs.

Support for Tourism

Many of our informants recognized the value of tourism in St. Jacobs. For instance, Adam declaimed, "Tourism is the anchor, or like the rock that everything else is built on." Tourism revenues, as others echoed, are essential for restoring and preserving spaces and facilities with heritage values. Beyond its economic justification, tourism's contribution to the community was recognized by Ruth, who underscored, "Tourism is an absolute opportunity ... I am all on board." Even though some tourists are not interested in the religious perspective, Ruth expressed "that if they come, they learn, feel included, feel part of the village."

Specific attention was paid to the importance of how tourism supports arts, culture and heritage within the community and "If the tourists don't come," Ruth said, "there's no theatre ... then there's no theatre for the residents." Without tourists, Ruth added, there will be no year-round income to keep the restaurants open that seniors drive to in their scooters "to have a $3.50 bowl of soup." Though coming from a very different background, Sam, a franchise developer and business consultant, agreed with Ruth, "If the village did not have the tourists coming … it would be less vibrant. It would not be viable for many businesses that are here now. They would close. It would be a quieter area." While our data analysis highlighted some of the value created by tourism, we heard tensions regarding the development and presence of tourism in the village.

Tensions Surrounding Tourism

One tension regarding the presence of tourism, Sam noted, is that "There are things that we have to tolerate, lots of traffic, lots of people on the sidewalks—doesn't bother me as long as there are more." Ruth challenged a sole emphasis on revenue generation guiding the village by recognizing that "money is not your most important asset to the community." Furthermore, to elaborate on her contestation, Ruth referred to some of the challenges stemming from over-tourism, i.e., clogged roads, parking, sewage, or other plumbing. Referring to the type of tourism developed "on the back of Old Order Mennonites," Ruth grappled, "It is hard to watch them … with eyes that aren't so gracious towards the … Mennonite people." Though "There's no boundary or barrier," she added, "you have to be part of the unwritten rules about wanting to be good neighbours, respectful neighbours to one another." It is not about being conservative, Ruth poignantly pointed out, "but it is about being respectful of those who are."

Miller and Martin were more forthright in expressing other follies of tourism. They suggested that most bus tourists in summer go to "whatever or wherever" guided tours take them. Incidentally, they stumble upon the Mennonite Story (an interpretive center) and say, in Martin's words, "Oh by golly, there's the Mennonites, so, uh, come on." Miller added, "Our product is information" and experience, not just merchandise. Unfortunately, most tourists, primarily affluent retirees, come to St. Jacobs, Miller claimed, "not to look at the falls … not to look at the river, it's to shop." In Martin's words, these tourists are "disrupting the everyday lives of the plain people in their communities."

These two self-identified "invisible-Mennonites" also laid bare some of the most revealing facts about St. Jacobs Farmers' Market, which product developers like Emma consider the biggest draw for tourists. According to Martin, the Farmers' Market does not consist of farmers or Mennonites, though some dress up as such. Miller echoed, "you go to any store that advertises Mennonite furniture … [it] hasn't been touched by a Mennonite." Miller claimed that lack of authenticity and unethical practices to generate revenues hurt the community and commerce. Stone Crock (a Mennonite-themed restaurant) used to be a draw, and now not so much, as Martin opined, "perhaps because it's competing with other non-traditional and less expensive eateries and cafes."

Though not as explicitly as the longstanding community members like Ruth, Miller and Martin, similar concerns also transpired from the other community members. Austin, a multinational company executive, summarized such concerns by arguing, "I don't think it [St. Jacobs] any longer knows … its brand or purposes."

Austin added, "I think St. Jacobs finds itself in the midst of confusion of … what are we all about?" Such confusion necessitates dialogue, continuing to draw on diverse voices to better understand a collective vision for St. Jacobs Village and how tourism may improve the community in which it takes place.

EMERGING TOURISM AGENDA

Now that we have presented a picture of the contested sense of community, including the perception and tension surrounding tourism, we focus on the emerging tourism agenda propagated through the voices of St. Jacobs' disparate communities. While agreeing St. Jacobs has earned its reputation as a heritage destination, Austin argued that heritage alone would not "bring sales in the absence of a brand." Austin emphasized, "create a brand … invest time to get that right, and then create the experience that will sustain it." Despite Austin's concerns around the lack of branding, our data analysis revealed there is significant interest within the participants regarding how to situate their operations within the overarching traditions of St. Jacobs. Gavin shared, "we must align ourselves with the Mennonite philosophy of people helping people and invest in being on the same page through better communication." In Adam's words, "we try as a restaurant to also be part of the cultural direction and development, like maintaining traditions or starting new ones." Tyler from the creative arts community proposed, " the more unique each business, the better it is for everybody." Diversifying the tourism product emerged as an important theme which we discuss next.

Diversification of Tourism Offerings

Apart from an interest in retaining the Mennonite culture and heritage, most participants emphasized the need to diversify tourism offerings that serve resident communities. Heinrich, an executive, working with the local government and a Mennonite himself, suggested, "St. Jacobs is going to stand the test of time to diversify the economy. It can't just be about tourism." As part of such a diversification strategy, he, along with other participants, strongly supports the development of new-age offices and creative start-ups. He also favored creating public spaces for the residents and the businesses to host events.

Adam also supports diversifying his restaurant offerings to appeal to millennials and young families from the local community and encourage them to spend more time in the village. Like other participants, Adam reiterated the importance of initiating compelling reasons to bring new tourists to St. Jacobs and keep the locals in the vicinity. Adam also favored creating multicultural experiences in the village, including concerts, tournaments, and nature-based tourism. Bill and Scarlett, both small business owners, admit that tourism is an opportunity to enhance community cohesion by developing meeting spaces with the capacity to host public and private events. Scarlett highlighted "the need for more diversity as far as opportunities for people are concerned between residents and visitors to get together and interact." As a continuation of the diversification agenda, some participants gave prominence to community-centered development.

Community Centered Development

Mia, who came of age in St. Jacobs and is currently managing projects for local businesses, including the BIA, proposed the idea of a more community-focused than tourism-focused development agenda. On the one hand, she attributed her interests to the rise of a creative economy. On the other hand, Mia justified the need for leisure-scapes for the influx of creative entrepreneurs coming from Waterloo and the young families settling in the new development areas. Mia recommended that the business offerings be spread over the weeks instead of weekends when St. Jacobs is typically overrun by tourists.

Mark, an educator, shared that St. Jacobs should be showcased as "a village that welcomes people of all walks of life." Honing the importance of intergenerational engagement and diversity, Mark suggested that St. Jacobs must celebrate the different cultures of communities, not only Mennonites, through events and activities. Nora echoed Mark's idea of "keeping that whole mentality of the community involvement" while being mindful of challenges in maintaining "a tough balance between … wanting to keep it as a small-town type thing and yet opening doors and welcoming everybody." She also shared Ruth's concern about the "longstanding members of the community feeling 'pushed out'" in the face of diversification, multiculturalism or gentrification.

Creative Turn in Tourism Development

Though not directly alluded to by the participants, there are indications that tourism in St. Jacobs is ready to take a creative turn. Heinrich, a proponent of diversification, believes in keeping the allure of Mennonite culture alive. Heinrich emphasized, "If St. Jacobs dies, God forbid, um, that wouldn't be good for the local economy." Thus, Heinrich offered a broader vision considering "Woolwich, not just St. Jacobs in the village." Instead of crowding the tourists inside the village parameters or the Farmers' Market, Heinrich shared a desire to expand tourism offerings to places of an active lifestyle, i.e., Mennonite farms. His rationale included favoring expansions such as, "in this day and age, more and more people are looking for getting out of the city and coming and experiencing what it's like to be out in the rural area." Heinrich's vision aligns with the tenets of creative tourism and could be regenerative in practice in terms of cultural appreciation, and learning about and supporting regenerative agricultural practices.

Emma, a product developer for a destination marketing organization, claimed that to get back to the "tourism game," St Jacobs needs to ramp up events highlighting history and traditions and placemaking through storytelling and the development of artifacts. Emma also favors creating a holistic experience for tourists, which would warrant the village and the market to collaborate and partner physically and digitally. Emma noted that people visiting the Farmers' Market should know, "three kilometres up the street, I can have a whole other experience, and it's great." Scarlett asserted something pertinent to the anticipated development of creative tourism with a regenerative agenda. She proffered: "It's really important that we as a village [are] able to change and adapt to what people want and give them reasons to keep coming back and not assume that they are just going to come because we are here."

Further to what Scarlett shared, it is imperative for residents to co-create meaningful and authentic experiences that will also return benefits to the local community over the long term.

BIA Leadership

Some of our informants revealed the challenges in leadership in St. Jacobs Village. Specifically, they talked about the absence of a vision and a loss of enthusiasm surrounding tourism development since Mercedes Corporation, a local corporation, was sold in 2018. As they noted, Mercedes was essentially a part of the community, which invested in and cared for the community. However, most informants, whether community members or business operators, expressed their hope about the ensuing role of the BIA. Reverberated in the voices of St. Jacobs, BIA leadership is gaining a favorable reputation in the community because of its willingness to lead a sustainable tourism redevelopment. Ruth asserted, "I think the BIA has their eye on [leading]." However, she stated advisedly, "The [BIA] is to improve businesses and help businesses make more profit. But they can't succeed without making the village a better place to live." Ruth emphasized that tourism redevelopment projects can "cross-fertilize" the people and help them see each other as "mutually-affirming leaders in the community." Tyler added, "we all have to pick up the slack for sure," where the BIA can play an important role, provided it becomes a full-time operation run by relatively permanent leadership.

DISCUSSION

As most of our research participants alluded to, St. Jacobs Village bears tangible emblems of rurality and "Mennonite-ness," which are built mainly on the facades of renovated buildings and structures, the murals of horses and buggies, or vintage signages and marble quilts, to name a few. Our analyses of the voices representing the longstanding communities with Anabaptist roots exposed a different reality: In the face of encroaching modernity, tourism's prosperity, and outmigration, St. Jacobs's intangible rurality, and cultural authenticity, and Mennonite philosophy of *people caring for people* have declined.

Participants whose businesses rely solely on the visitors' economy are understandably convinced that they would not survive without tourism. As some longstanding residents revealed, tourism revenues are essential for the restoration and preservation of spaces and facilities with heritage values and contribute to the continuation of daily lives. However, they understand the volatility associated with tourism, thereby actively investigating the diversification of their businesses and operations. As part of such a diversification strategy, businesses increasingly target domestic tourists and local communities, for example, through renting their spaces or facilities for private and corporate social events. Such rendering also shone the light on the alternative economic practices that aim not only for profit generation but also to strengthen the sense of community and place.

Despite such shared interests, as tourism proceeded through the stages of creative destruction, tension began to emerge, which resulted, as investigated in Clare Mitchell's longitudinal research, not only in the decline of the critical cultural components but tempered the awareness of such decline, at least up until the recent past. For example, during the surveys carried out in the 1990s, respondents frequently spoke about the exploitation of Mennonite culture. In contrast, the survey results from a decade later suggest that such concerns subsided (Mitchell, 2013). Mitchell (2013) claimed that the reason was not that the exploitation ceased or slowed down, but the voices raising such concerns did, as many local farm families sold their properties and relocated to more distant locales to practice their lifestyle.

Nearly a decade after Mitchell's (2013) revelation of post-destruction phenomena, our investigation discovered promising awareness of over-tourism and unsustainable growth and interest in creatively regenerating tourism. While most participants continue to herald the Mercedes Corporation's leadership in turning St. Jacobs from a rural task-scape to a heritage-scape, the eventual loss of "idyllic rural representation" (Mitchell, 2013, p. 376) has prompted interest in creating a leisure-scape, an emerging phenomenon, that Mitchell (2013) termed as "creative enhancement."

Unlike creative destruction, the creative enhancement does not displace an existing landscape or economic function entirely; instead, it leads to a multifunctional financial state (Mitchell, 2013). The ideas proposed by the participants towards the creation of leisure spaces comprising multifunctional avenues exemplify "creative enhancement," which not only will serve the community but also serve the tourists and generate revenues. With a Mennonite heritage on one side and on the other close proximity to a major urban center (i.e., Waterloo), as one of the interview participants expressed, St. Jacobs offers the "best of both worlds." Cultivating such a two-pronged offering is a recipe for creative enhancement (Mitchell, 2013), which is in motion in St. Jacobs. The gentrification of retail forward buildings in new office spaces, upscale condos, craft breweries, beer gardens, or trendy restaurants, favored by some participants, strongly suggests creative enhancement is in full play at St. Jacobs.

We, however, like to be cautiously optimistic. As uncovered through our analysis, it is not only tourism that concerns the remaining longstanding community members but also the increasing number of new residents settling within the community, some with little or no concern for the former. Expanding tourism from tangible heritage to intangible culture deep into the Mennonite farms and community seems encouraging from a creative tourism perspective. We wonder if such expansion would further jeopardize the Old-order Mennonite community's living and breathing culture and lifestyles. With a mode of tourism that facilitates only tourists' gaze and not the co-creation of experiences with regenerative terms set by the community, it is likely that St. Jacobs may end up repeating what one participant argued: "tourists' relentless urge to catch a glimpse of 'Mennoniteness' compelled Mennonites to move further and further away from St. Jacobs, where they could practice their lifestyle beyond the tourists' gaze."

While there are contradicting views, firstly driven by the conflicting sense of community and secondly, embedded in the continuing tensions about whether or not St. Jacobs should re-emerge as a heritage-scape to draw tourists or as a leisure-scape serving the community, the timing seems right for bringing the various propositions under one umbrella, with the BIA acting as the facilitator, mutually drawing on creative and regenerative tourism practices. It is also worth mentioning that the urge for multipurpose economic functioning offered through diversification, some more explicit than others, is ripe for further exploration of regenerative tourism (Cave & Dredge, 2020). Such a study, we believe, would help us to develop a comprehensive approach by proposing an alternative-capitalist, if not non-capitalist, framework alongside the capitalist economic framework for St. Jacobs that serves the diverse communities.

CONSIDERATIONS MOVING FORWARD

Destinations that have fallen victim to the unsustainable development of cultural tourism and eventual over-tourism need to seriously consider degrowth in line with regenerative practices. To reclaim their identity and authenticity, mislaid in the mix of commodification and McDonaldization, destinations should try re-envisioning tourism agendas by engaging in community dialogue. Peripheral destinations with limited capacity like St. Jacobs must limit the quantity of mass-produced inauthentic and inexpensive products, services and experiences that attract the masses. Instead, they should invest in developing the quality of the same in terms of authenticity, be it product or experience. Importantly, regenerative tourism does not propose building more; it offers contemplation and dialogue regarding how tourism may be woven into the fabric of a community.

Clearly, improved quality and authenticity may come with higher prices, but such costs may play an important role in regenerating the community. Some scholars such as Pine and Gilmore have drawn attention to the experience-seeking tourists, noting an interest in "buy[ing] real from the genuine, not fake from some phony" (2013, p. 29). Accordingly, destinations must not be afraid of staying real, and our analysis here relays a vital interest in creating authentic experiences in St. Jacobs. One of the main reasons we are focusing on creative tourism as part of the regenerative tourism development agenda in spatially constrained but culturally rich destinations, such as St. Jacobs, is because degrowth may facilitate regrowth. The potential for regeneration improves when tourists appreciate the work and skills involved in producing craft products, and they are usually willing to pay a comparatively higher price.

It is not a simple task to highlight ideal creative and regenerative tourism practices based on one destination. But the halt in tourism as a consequence of the pandemic has served as an important pause encouraging destination communities to reconsider how they want to plan and offer tourism. We believe that creative tourism with a regenerative agenda must be relevant to local contexts. Involvement of the local stakeholders is paramount to the management, including planning, organizing, coordinating, monitoring, and leading of local tourism. For creative–regenerative tourism to be realistic, it needs to be tailored to socio-cultural amenities, natural resources and expectations of communities within specific locations. As such, our collaborative work and iterative dialogue with the St. Jacobs community are important to better understand and co-create a collective vision. The next phase of data collection will be guided by participatory action research and organized around a series of community workshops. Such workshops will encourage dialogue across stakeholders, which we hope will get us closer to a community vision and regenerative destination plan.

Recognizing some of the conflicting interests from our first phase of data collection, we recognize the importance of co-creation at the core of creative and regenerative tourism. Accordingly, participative and consultative leadership should be the cornerstone of the development strategy in St. Jacobs Village. Such leadership should be left to the community whose regeneration, be it social, cultural, or economic, is the foremost goal of regenerative tourism, where creative tourism could play a significant role.

REFERENCES

Ateljevic, I. (2020). Transforming the (tourism) world for good and (re)generating the potential "new normal." *Tourism Geographies*, *22*(3), 467–475. https://doi.org/10.1080/14616688.2020.1759134.

Ateljevic, I., & Sheldon, P.J. (2022). Guest editorial: Transformation and the regenerative future of tourism. *Journal of Tourism Futures*, *8*(3), 266–268.

Baixinho, A., Santos, C., Couto, G., Albergaria, I.S., Silva, L.S., Medeiros, P.D., & Simas, R.M. (2021). Islandscapes and sustainable creative tourism: A conceptual framework and guidelines for best practices. *Land*, *10*(12), 1302. https://doi.org/10.3390/land10121302.

Becken, S. (2017). Decarbonising tourism: Mission impossible? *Tourism Recreation Research*, *44*(4), 419–433.

Binkhorst, E., & Den Dekker, T. (2009). Agenda for co-creation Tourism Experience Research. *Journal of Hospitality Marketing & Management*, *18*(2–3), 311–327. https://doi.org/10.1080/19368620802594193.

Blapp, M., & Mitas, O. (2020). Creative tourism in Balinese rural communities. *Current Issues in Asian Tourism*, 219–246. https://doi.org/10.4324/9780429291593-16.

Boluk, K., & Mottiar, Z. (2014). Motivations of social entrepreneurs: Blurring the social contribution and profits dichotomy. *Social Enterprise Journal*, *10*(1), 53–68.

Boluk, K., & Panse, G. (2022). Recognizing the regenerative impacts of Canadian women tourism entrepreneurs through a feminist ethic of care lens. *Journal of Tourism Futures*, *8*(3), 352–366.

Boluk, K., Cavaliere, C.T., & Higgins-Desbiolles, F. (2017). Critical thinking to realize sustainability in tourism systems: Reflecting on the 2030 Sustainable Development Goals. *Journal of Sustainable Tourism*, *25*(9), 1201–1204. https://doi.org/10.1080/09669582.2017.1333263.

Boluk, K., Cavaliere, C., & Higgins-Desbiolles, F. (2019). A critical framework for interrogating the United Nations Sustainable Development Goals 2030 Agenda in Tourism. *Journal of Sustainable Tourism*, *27*(7). doi:10.1080/09669582.2019.1619748.

Boluk, K., Krolikowski, C., Higgins-Desbiolles, F., Carnicelli, S., & Wijesinghe, G. (2021). Re-thinking tourism: Degrowth and equity rights in developing community-centric tourism. In C.M. Hall, L. Lundmark, & J.J. Zhang (eds), *Degrowth and Tourism: New Perspectives on Tourism Entrepreneurship, Destinations and Policy* (pp. 152–169). Abingdon: Routledge.

Burns, R., & Berbary, L.A. (2021). Placemaking as unmaking: Settler colonialism, gentrification, and the myth of "revitalized" urban spaces. *Leisure Sciences*, *43*(6), 644–660. https://doi.org/10.1080/01490400.2020.1870592.

Butler, R.W. (1980). The concept of a tourist area cycle of evolution: Implications for management of resources. *The Canadian Geographer/Le Géographe Canadien*, *24*(1), 5–12. https://doi.org/10.1111/j.1541-0064.1980.tb00970.x.

Butler, R.W. (2020). Overtourism in rural areas. In H. Séraphin, T. Gladkikh, & T. Vo Thanh (eds), *Overtourism* (pp. 27–43). Cham: Palgrave Macmillan. https://doi-org.proxy.lib.uwaterloo.ca/10.1007/978-3-030-42458-9_3.

Carnicelli, S., & Boluk, K. (2022). A theory of care to socialise tourism. In F. Higgins-Desbiolles, A. Doering, & B. Bigby (eds), *Socialising Tourism* (pp. 58–71). Abingdon: Routledge.

Cave, J., & Dredge, D. (2020). Regenerative tourism needs diverse economic practices. *Tourism Geographies*, *22*(3), 503–513. https://doi.org/10.1080/14616688.2020.1768434.

Charmaz, K. (2011). A constructivist grounded theory analysis of losing and regaining a valued self. In F.J. Wertz, K. Charmaz, L.M. McMullen, R. Josselson, R. Anderson, & E. McSpadden (eds), *Five Ways of Doing Qualitative Analysis* (pp. 165–204) New York: Guilford Press.

Dahms. F. (1991). St Jacobs, Ontario: From declining village to thriving tourist community. *Ontario Geography*, *36*, 1–13. https://doi.org/info:doi/.

Designing Tourism (2022). *The Islander Way: Flinders Island*. Designing Tourism. Retrieved May 16, 2022, from https://www.islanderway.co/flinders-island.

Dredge, D. (2017). Foreword. In P.J. Sheldon and R. Daniele (eds), *Social Entrepreneurship and Tourism Philosophy and Practice* (pp. v–vii). Cham: Springer.

Duxbury, N., Bakas, F.E., Vinagre de Castro, T., & Silva, S. (2021). Creative tourism development models towards sustainable and regenerative tourism. *Sustainability*, *13*(1), 2.

Everingham, P. & Chassagne, N. (2020). Post COVID-19 ecological and social reset: Moving away from capitalist growth models towards tourism as Buen Vivir. *Tourism Geographies*, *22*(3), 555–566.

Explore Waterloo Region (2022, September 19). *The Village of St. Jacobs*. https://explorewaterloo.ca/st-jacobs/.

Fennell, D. (2014). Foreword. In C. Weeden & K. Boluk (eds), *Managing Ethical Consumption in Tourism* (pp. xv–xvii). London: Routledge.

Getz, D. (1993). Tourist shopping villages: Development and planning strategies. *Tourism Management*, *14*(1), 15–26. https://doi.org/10.1016/0261-5177(93)90078-Y.

Giddens, A. (1982). Power, the dialectic of control and class structuration. In A. Giddens, *Profiles and Critiques in Social Theory* (pp. 197–214). London: Macmillan.

Hall, C.M., Lundmark, L., & Zhang, J.J. (eds) (2020). *Degrowth and Tourism: New Perspectives on Tourism Entrepreneurship, Destinations and Policy*. Abingdon: Routledge.

Hall, R.P., Ashford, R., Ashford, N.A., & Arango-Quiroga, J. (2019). Universal basic income and inclusive capitalism: Consequences for sustainability. *Sustainability*, *11*(16), 4481. https://doi.org/10.3390/su11164481.

Hawai'i Tourism Authority (2021). *Kaua'i Destination Management Action Plan: 2021–2023.* HTA Kauai Action Plan. Retrieved May 16, 2022, from https://www.hawaiitourismauthority.org/media/6771/hta-kauai-dmap.pdf.

Higgins-Desbiolles, F., Carnicelli, S., Krolikowski, C., Wijesinghe, G., & Boluk, K. (2019). Degrowing tourism: Rethinking tourism. *Journal of Sustainable Tourism, 27*(12), 1926–1944. https://doi.org/10.1080/09669582.2019.1601732.

Higgins-Desbiolles, F., Doering, A., & Bigby, B.C. (2021). Introduction: Socialising tourism: reimagining tourism's purpose. In F. Higgins-Desbiolles, A. Doering, & B.C. Bigby (eds), *Socialising Tourism* (pp. 1–22). Abingdon: Routledge.

Hutchins, G., & Storm, L. (2019). *Regenerative Leadership: The DNA of Life-Affirming 21st Century Organizations*. Royal Tonbridge Wells, UK: Wordzworth Publishing.

Mitchell, C.J. (1998). Entrepreneurialism, commodification and creative destruction: A model of post-modern community development. *Journal of Rural Studies, 14*(3), 273–286. https://doi.org/10.1016/S0743-0167(98)00013-8.

Mitchell, C.J. (2013). Creative destruction or creative enhancement? Understanding the transformation of rural spaces. *Journal of Rural Studies, 32*, 375–387. https://doi.org/10.1016/j.jrurstud.2013.09.005.

Mitchell, C.J., & de Waal, S.B. (2009). Revisiting the model of creative destruction: St. Jacobs, Ontario, a decade later. *Journal of Rural Studies, 25*(1), 156–167. https://doi.org/10.1016/j.jrurstud.2008.09.003.

Mitchell, C.J., & Randle, K. (2014). Heritage preservation and the "differentiated countryside": Evidence from Southern Ontario. *The Canadian Geographer/Le Géographe Canadien, 58*(4), 429–442. https://doi.org/10.1111/cag.12131.

Mottiar, Z. & Boluk, K. (2017). Understanding how social entrepreneurs fit into the tourism discourse. In P. Sheldon & R. Daniele (eds), *Social Entrepreneurship and Tourism* (pp. 117–131). Cham: Springer.

Mottiar, Z., Boluk, K., & Kline, C., (2018). The roles of social entrepreneurs in rural destination development. *Annals of Tourism Research, 68*, 77–88.

Noddings, N. (1992). In defense of caring. *Journal of Clinical Ethics, 3*(1), 15–18.

O'Reilly, A.M. (1986). Tourism carrying capacity: Concept and issues. *Tourism Management, 7*(4), 254–258.

Phi, G.T. (2019). Framing overtourism: A critical news media analysis. *Current Issues in Tourism, 23*(17), 2093–2097. https://doi.org/10.1080/13683500.2019.1618249.

Pine, B.J., & Gilmore, J.H. (2013). The experience economy: Past, present and future. In J. Sundbo & F. Sørensen, *Handbook on the Experience Economy* (pp. 21–44). Cheltenham, UK, & Northampton, MA, USA: Edward Elgar Publishing. https://doi.org/10.4337/9781781004227.00007.

Pollock, A. (2019). *Regenerative Tourism: The Natural Maturation of Sustainability.* Retrieved July 2, 2020, from https://medium.com/activate-the-future/regenerative-tourism-the-natural-maturation-of-sustainability-26e6507d0fcb.

Poon, A. (1993). *Tourism, Technology and Competitive Strategies*. Wallingford, UK: CAB International.

Richards, G. (2005). Textile tourists in the European periphery: New markets for disadvantaged areas? *Tourism Review International, 8*(4), 323–338. https://doi.org/10.3727/154427205774791519.

Richards, G. (2009) Creative tourism and local development. In R. Wurzburger, A. Pattakos, & S. Pratt (eds), *Creative Tourism: A Global Conversation* (pp. 78–90). Santa Fe: Sunstone Press.

Richards, G. (2011). Creativity and tourism: The state of the art. *Annals of Tourism Research, 38*(4), 1225–1253. https://doi.org/10.1016/j.annals.2011.07.008.

Richards, G. & Raymond, C. (2000). Creative tourism. *ATLAS News, 23*, 16–20.

Richards, G., & Wilson, J. (eds) (2007). *Tourism, Creativity and Development*. London: Routledge.

Ruiz, M. (2020). *A paradigm shift from quantity to quality tourism: Interview with Visit Flanders' marketing director Elke Dens*. Travel Tomorrow. Retrieved May 16, 2022, from https://traveltomorrow.com/a-paradigm-shift-from-quantity-to-quality-tourism-interview-with-visit-flanders-marketing-director-elke-dens/.

Sheldon, P.J. (2022). The coming-of-age of tourism: Embracing new economic models. *Journal of Tourism Futures*, *8*(2), 200–207. https://doi.org/10.1108/jtf-03-2021-0057.

Sheldon, P.J., Pollock, A., & Daniele, R. (2017). Social entrepreneurship and tourism: Setting the stage. In P.J. Sheldon and R. Daniele. *Social Entrepreneurship and Tourism Philosophy and Practice* (pp. 1–17). Cham: Springer.

Tourism Bay of Plenty (2022). *Official Tourism & Travel Guide: Tauranga: Bay of Plenty, New Zealand*. Retrieved May 13, 2022, from https://www.bayofplentynz.com/.

Weeden, C. & Boluk, K. (eds) (2014). *Managing Ethical Consumption in Tourism*. London: Routledge.

Wheeller, B. (1993). Sustaining the ego. *Journal of Sustainable Tourism*, *1*(2), 121–129. https://doi.org/10.1080/09669589309450710.

PART IV

PROCESSES AND PRACTICES

19. Ethical considerations for rural community tourism

Tazim Jamal and Blanca Camargo

INTRODUCTION

The protracted COVID-19 pandemic has reminded us painfully of our common humanity and our societal vulnerabilities, revealing deep inequalities in income and little social support for high-risk groups. Extreme global poverty is estimated to have risen by 150 million people,[1] while adversities faced by refugees displaced from war-torn areas and migrants seeking safety and sustainable livelihoods away from regions affected by extreme weather, geopolitical problems, and terrorism have risen immensely.[2] COVID-19 has increased the burden of villagers migrating between rural and urban places as lockdowns and other restrictions have gravely affected precarious workers and families surviving from meal to meal to meal, day to day. Vaccines offer hope for new beginnings (despite vaccine nationalism and inequitable global distribution to the severely affected countries) and a chance to rethink and prepare for sobering new futures and the new "normals" that await. Old ways of seeing and "measuring" success predominantly in terms of economic growth (quantifiable visitor numbers and revenues) are clearly insufficient for a sustainable planet and the well-being of hundreds of millions of rural inhabitants in subtropical regions that face the brunt of climate change impacts.[3] Of course, low-lying islands and coastal areas, polar regions, and other regions will also be gravely affected and issues of food security, food sovereignty, exploitation and depletion of precious resources in the local to global commons are just a few of the issues to be addressed.

Valuable discussion has been arising on notions related to *regenerative tourism* (Dredge, 2021) and restorative tourism[4] as communities and destinations struggle with the recovery process and seek more just and equitable futures and resilience to present and future global threats (pandemics included). But what ethical principles guide tourism and the new normals that await? What does it mean to put purpose before profit, and seek value beyond the obsessive compulsion of capital and commodification? For economist and former Bank of Canada and Bank of England governor Mark Carney, radical foundational change is required to redress misplaced values in financial markets, tackle climate change and move towards a just, sustainable future as we emerge from the pandemic (Carney, 2021). It is a future that builds equitable human values (value for all), not market values, and is deeply interrelated with nature being a part of it, not apart from it. The seven values Carney (2021) puts forward are solidarity, fairness, responsibility, resilience, sustainability, dynamism and humility – all laced with compassion – and resonate well with the pandemic. But let us take this 'god's eye view' (Haraway, 1988) of the former Bank of Canada and Bank of England governor to the situated, on-the-ground rural communities and their inhabitants around the world. What matters to them as they take up tourism development (if it's not already foisted on them)? What values guide them as they embark on the road to recovery and rebuilding communities and livelihoods ravaged by the pandemic worldwide? Carney raises compassion as a key value. But values and principles require careful scrutiny of how they are constituted and recognition that they vary from place to place, group to group, person to person. There may be some general or universal values in addition to local and particular ones. See, for instance, the quote below, that is commonly attributed to Albert Einstein. Your understanding of compassion may differ from Einstein's and Carney's, but the overall sense of the term remains, and you might agree that this is a universal human value.

> A human being is a part of the whole called by us universe, a part limited in time and space. He experiences himself, his thoughts and feeling as something separated from the rest, a kind of optical delusion of his consciousness. This delusion is a kind of prison for us, restricting us to our personal desires and to affection for a few persons nearest to us. Our task must be to free ourselves from this prison by widening our circle of compassion to embrace all living creatures and the whole of nature in its beauty … (Nguyen, 2017)

So, we might all agree, then, that the COVID-19 pandemic has not only revealed our vulnerabilities, but also shown the importance of values like compassion and empathy to help endure the many months of isolation, uncertainty and fear, and the death and disruption experienced by so many.

Perhaps "widening our circle of compassion" as per the above quote is one of the most important principles to also bear in mind as we embark in the next few sections to understand the places and spaces of rural tourism, and the practices and values that its inhabitants cherish. Though we engage with them academically, it is always with critical change and praxis in mind. It means engaging at the local level with the situated practices of the place, i.e., the community residents' experiences and interactions with rural tourism, their agency, initiative and resistance to injustice, and their efforts to enact "good" tourism.[5] Praxis involves critical thinking, action and change, seeking transformation towards good tourism with empathy, inclusiveness and openness to "other" worldviews and other values and ways of being and relating (Jamal et al., 2021). It celebrates *situated knowledges* (Haraway, 1988), including local and traditional knowledge, in addition to more general scientific and technical knowledge.

The next section discusses some perspectives and examples of rural community tourism development. This helps identify some ethical priorities and principles of justice that are discussed in the subsequent section. A case example follows, on the Palenqueros, a diasporic Afro-Colombian group; we explore some of the issues and challenges faced by the Palenqueros, how they exercise agency and strive to resist exploitation and domination, and how they engage in change towards greater resilience and sustainability. The chapter closes with some concluding thoughts on the task that awaits (post)pandemic futures and *regenerative tourism* (see Dredge, 2021).

While there are clearly a range of ethical principles and "universal" human values that all forms of good tourism share, it is important to note that there may also be particular group values and principles (Young, 2011) that are related to the specific local context, history and cultural heritage of the place and its people.

RURAL COMMUNITY TOURISM DEVELOPMENT

Many community-oriented approaches to rural tourism development emphasize the development of local, small-scale tourism initiatives, local control, resident involvement in planning and decision making, gender equity, environmental conservation, capacity building and social benefits (e.g., George et al., 2009; Hwang et al., 2012; see also Mair & Reid, 2007). Popular examples include pro-poor tourism (PPT), fair trade tourism and volunteer tourism, much of which occurs in small rural communities, though not exclusively so (Wearing & McGehee, 2013). Pro-poor approaches focus on grassroots strategies that enhance benefits and opportunities for the poor, aiming for poverty alleviation, capacity building, gender equality and income equity, as well as individual and community empowerment through direct participation in development and decision making (e.g., Manyara & Jones, 2007). A *sustainable livelihoods* perspective, such as informs community-based natural resource management (CBNRM), aims for environmental conservation through providing diversified opportunities, e.g., through wildlife viewing and ecologically sustainable tourism that promotes local participation in wildlife management and resource use rights of the local communities. The general principle relied on here is that rural communities will exercise greater stewardship and support conservation if they derive economic benefits from the ecological resources that form part of their physical and social environments (Mbaiwa & Stronza, 2010). Community-based natural resource management, however, can be significantly affected by external stakeholders and by market mechanisms that do not coincide with local values and practices related to land and nature (Mostafanezhad et al., 2016). Different worldviews and conflicting values clash over socially and culturally appropriate approaches to community development, poverty alleviation and governance (see, for example, Gohori & van der Merwe, 2020, on rural tourism in Manicaland Province, Zimbabwe).

Furthermore, challenges arise in rural regions where financing, management and marketing knowledge, and capacity building are in short supply (e.g., Victurine, 2000). However, attempts to address these through international development aid projects and structural adjustment programs funded by the International Monetary Fund and World Bank (among others), have often resulted in underdevelopment and dependency (Telfer, 2009). Many have been closely linked to issues of economic leakage, dependency, and domination by external stakeholders historically through colonialism and subsequently through neocolonialism and neoliberal globalization. Neoliberal approaches emphasize the efficiency of large-scale free market capitalism and economic globalization. But the costs of "development" are high in terms of destination vulnerability, social and economic inequalities, high leakage of tourist expenditures back to the core generating regions, and continued dependency and underdevelopment (Britton, 1982; see also Mosedale, 2016).

The popular destination of Bali is illustrative of places where "development" has created a conduit for rapid tourism growth with over 6 million annual visitors in 2019. Rural Bali and the ancestral *subak* water irrigation system have come under pressure and threat due to tourism growth and urbanization. Climate change and over-consumption of water by the tourism industry (65% of the island's water reserves are estimated to be used to supply tourist facilities) have put increasing stress on its water resources, disproportionately affecting its rural agricultural communities (Colorni, 2018). It is not surprising, says Colorni, that a significant number of Balinese people, especially the farmers, are skeptical of the solutions proposed by various UN agencies and their own government, including tourism development that has contributed significantly to land grabbing, loss of over 25% of agricultural land over 25 years and rapid depletion of farming households and individuals (see also Wardana, 2018; 2019). As Colorni illustrates, Green Revolution policies that started in 1969 have further undermined farmers' ability to retain and effectively manage their landholdings, affecting their health as well as their economic, ecological, and cultural resilience. As such, the "food security and sovereignty of the island as a whole is under threat" and cultural–spiritual traditions woven into communal *subak* irrigation and rice cultivation systems are also under threat from the Green Revolution, in addition to invasive tourism development and rampant commodification (Colorni, 2018, p. 4).

Fortunately, local Balinese communities are resisting global and political pressures, engaging the communitarian and collective decision making structure of the *subak* to engage in small-scale, ecologically sensitive tourism to offer visitor experiences and conserve rural agrarian livelihoods in historical rice-terraced landscapes that are deeply interrelated with the *subak* in their cultural heritage.[6] Nyambu Ecological Tourism Village, for instance, is a community ecotourism project in Bali where visitors can learn about traditional Balinese eco-agricultural methods and the rich cultural heritage of Balinese rice. As they learn during their walking tour and afterwards, over freshly cooked traditional food in the community, rice in Bali is a cultural tradition where ancestral *subak* water management practices interweave the human world, the spirit world and nature in the Balinese philosophical principle of Tri Hita Karan (also designated in the UNESCO World Heritage List[7]). "What makes this system even more unique is its strong spiritual element: life in the *subak* revolves around the *pura*, a temple specially built by farmers to worship the Goddess Sri. Every rice field has a shrine, every *subak* a temple, and up to 15 blessing ceremonies are held per season to ensure a good harvest".[8]

As a communal eco-village, Desa Nyambu exemplifies the *subak* principles of democratic and egalitarian participation in sustainable agriculture. Direct resident involvement fosters ecological and cultural conservation as well as sustainable livelihoods. Local community initiatives developed through direct resident participation are oriented to minimizing economic leakage and sharing benefits equitably (Dianasari, 2018). Direct resident involvement in interpreting and sharing with visitors the stories, rituals, foods and other traditions that shape the rice terraces, also enables social flourishing and sustains a dynamic, spiritual, cultural heritage that is deeply interwoven in lived experience and everyday life.

Of critical importance, then, is *direct resident participation* in the development, planning and management of community-based tourism (CBT), *organizational forms* and *institutional structures* that facilitate local ownership, conservation and stewardship, as well as equity and fairness in the distribution of the costs and benefits of tourism development. Local approaches to economic development can include collaborative networking, community-based enterprise (CBE), social enterprise and social entrepreneurship, micro-credit NGOs and cooperatives, as well as individual business enterprises situated within the community. Such locally based, resident-driven enterprises enable the potential for capacity building and community development, where "community" here refers broadly to local residents living within a specific geographical or politically defined area.

Confusion over concepts like "community-driven" tourism versus "community-based" tourism continue to occupy scholars, and various frameworks for community participatory processes in rural conservation and tourism have been proposed (e.g., Matarrita-Cascante et al., 2019; see also Manyara and Jones, 2007). A defining and vital parameter here is the degree of resident involvement for local control, i.e., resident-driven decision making in tourism development and planning, as well as resident involvement in the management and marketing of tourism, are crucial, along with associated principles of self-empowerment, autonomy, and agency. As ecofeminist Vandana Shiva describes, "economic globalization as we are seeing it unfold is not a process of ever widening circles of inclusion. It is a process of ascending hierarchies that concentrate power and exclude people from participating in the political and economic life of their societies" (Shiva, 1997, p. 2). She goes on to state that ecological feminism "sees in the current trend the ultimate concentration of capitalist patriarchy and its violence against nature and women" (p. 4).

As the example of Desa Nyambu described above illustrates, place-based, resident-driven tourism development and ethical principles and values are crucial to resist exploitation and domination, and to reclaim the local from unsustainable paradigms of economic growth and "mass" tourism facilitated by neocolonialism and neoliberal globalization. Postcolonial and ecofeminist critiques have challenged the dominant Eurocentric, modernist values that have driven "development" and emphasized the need to redress economic inequalities, loss of intangible human–environmental and cultural–heritage relationships with land and nature, as well as the vital importance of action reclaiming the local (see Shiva, 1988; Salleh, 1984). They also offer valuable insights for re-situating the missing gendered, diverse body in tourism, re-appropriating feminist values and emotional aspects like love, care, friendship, and empathy.

Santafe-Troncoso and Loring's 2021 study of Indigenous food sovereignty in Ecuador is illustrative of such a place-based, situated approach. They draw upon a framework for food sovereignty known as the Six Pillars Framework provided by the Nyéléni Forum for Food Sovereignty, which emphasizes: (1) a focus on food for people; (2) valuing food providers; (3) localizing food systems; (4) placing control locally; (5) building knowledge and skills; and (6) working with nature (Nyéléni, 2007). As their research demonstrated, "food sovereignty is not only about food production and farmers' struggles, but also about justice and connections among all the varied actors and elements in a food system" (p. 4). And, as the Nyéléni report described, "people's right to healthy and culturally appropriate food produced through ecologically sound and sustainable methods, and their right to define their own food and agriculture systems" (Nyéléni, 2007, p. 9, cited in Santafe-Troncoso and Loring, 2021, p. 4). The Kichwa and non-Kichwa Indigenous communities on the Chakra Route (Cacao Route) visited by tourists use traditional agroforestry methods, enabling cultural flourishing, food security and biodiversity conservation. The Kichwa women farmers (also known as Chakra Mamas) draw on traditional knowledge passed down over generations to cultivate and manage *chakra* gardens using agro-ecological practices. Four Kichwa cacao farming cooperatives and ten Kichwa community-based tourism associations are among the key stakeholders involved in a collaborative governance system of the Chakra Route. Working in a communitarian ethos, the Kichwa women play a leadership role in various entrepreneurial activities, providing tours in the *chakra* gardens and sharing culturally rich food experiences. The *chakra* is a restorative and healing place, a source of diverse foods and food security, and a "platform for agency, building community, teaching traditional knowledge, expressing cultural identity, empowering women, stewarding the environment, and maintaining spiritual wellness" (Santafe-Troncoso and Loring, 2021, p. 10). As their research showed, there is a rich spiritual and ancestral relationship between the women, *chakra* and Kichwa culture.

Examples such as the above offer hope and direction for recovery and restorative healing from the immense inequalities generated by COVID-19, further exacerbated by the ongoing climate crisis. The values described above resonate well with rural agricultural and tourism practices elsewhere, for example, regenerative agriculture and slow food tourism, where small-scale producers and alternative economies counter hegemonic neoliberal pressures through a paradigm of degrowth and provisioning, prioritizing healthy ecosystems, cultural flourishing and sustainable livelihoods. Like the *chakra* gardens, such rural endeavors (e.g., small-scale farmers, vineyards and wineries aiming to be ecologically and socially sustainable), are praxis oriented. They foster civil society, democracy and social action through active consumer–producer interactions, joint participation in food acts and policy actions to address the carbon and environmental costs of large-scale agriculture (and monoculture), including issues of deforestation, environmental degradation, biodiversity loss and related losses of food security plus adverse impacts on human health and community resilience (Shiva, 2000; Wise, 2019). Sustainability principles for alternative rural tourism and practices like slow food tourism must therefore include reflexive praxis and experiences that "facilitate conscious reflection on the nature and ethics of what we eat – both as part of our daily habits and as part of our travel experiences" (Fusté-Forné & Jamal, 2020).

The deep inequalities revealed by the COVID-19 pandemic and the sobering realities of climate change must be addressed in *regenerative* tourism and *just* transitions that enable nurturing alternative local economies for provisioning and food security, reducing food miles and working towards carbon zero, long-term sustainable futures, resilient communities and the well-being of humans and non-human others. What other principles, then, guide just, equitable rural tourism development in the 21st century?

SOME ETHICAL PRIORITIES IN A (POST)PANDEMIC WORLD

The tourism development paradigms that have driven mass tourism and neoliberal forms of development have typically been based on modernist principles that have privileged individualism and self-interest, economic growth (and reliance on quantifiable measures of "success", like visitor numbers), applying science and technology towards instrumental use of natural and cultural resources for economic growth, and expanding capital markets that require fueling by constant consumer consumption. Too little attention has been paid in such endeavors to the fair and equitable distribution of the costs and benefits of development. Moreover, there is far too little regulation of transnational corporations that wield immense influence and power, but have little accountability to the local places that provide the "attractions" to serve large visitor markets (Telfer, 2009).

By contrast, "good" rural tourism endeavors lay claim to some common principles such as *local ownership* of tourism-related enterprises, *direct resident participation* and *local control* over tourism development and the local commons (natural and cultural), *local economic and social benefits* including capacity building, enabling gender equity and women's empowerment, facilitating social capital (social relationships, social bonds, reciprocity) and individual as well as communal agency and empowerment (see Borlido & Coromina, 2018; Scheyvens, 2002). It entails a collaborative relationship between scientific, technical, traditional and local knowledges towards individual and communal well-being and healthy, flourishing ecosystems. It also involves holistic, gender-equitable, non-patriarchal, culturally situated, *pluralistic* approaches to development and CBT, welcoming a *pluriverse* (Escobar, 2018) of diverse worldviews, values and knowledge systems, rather than being driven by a monoverse of dominant modernist/colonial narratives. It is a relational, non-dualistic approach where cultural relationships with living and other than living things de-center anthropocentric human dominance over nature and the biophysical world – cosmos, emotions and embodied relations weave human beings in rich, interwoven relations with the "other" in the places and spaces of tourism.

Such a posthumanist paradigm of experience and belonging eschews dualistic binaries such as ecology "versus" culture, and welcomes diverse agencies and hybrid identities, countering stereotypes and enabling generative and transformative ways of being and becoming. Just and equitable futures call for such a relational ethic and principles of justice and well-being to guide regenerative tourism as places begin to embark on restorative strategies to revitalize battered economies and socio-economic recovery, and engage in biophysical and cultural conservation to enable resilient communities and flourishing ecosystems. The principles of justice are accompanied, too, by an ethic of care, and would include equity, recognition and fair representation, inclusiveness and good governance, as well as dignity, respect and care for human and non-human others (Jamal, 2019; Jamal & Higham, 2021).

Social isolation and the slowing of mobilities and economies offer a pause to contemplate new opportunities for rural tourism to address such hopeful, posthumanist new futures. As described below, good rural tourism development requires addressing identity politics, colonial histories (including settler colonialism), and ways of being and caring for land, engaging in degrowth and establishing alternative economies like community social enterprise (i.e., other than market capitalism), and human–environmental/human–animal relationships and nature's rights. Post-development advocates like Gustavo Esteva and Arturo Escobar call for a paradigm shift to a holistic, post-development approach to CBT that incorporates a cultural, relational, autonomous approach for communal well-being (see Escobar, 2018; Esteva, 2018). Others such as Karst (2017) have drawn on Indigenous worldviews like *Buen Vivir* (good living) to examine culture–spiritual–nature relationships and notions of well-being in CBT and ecotourism. *Buen Vivir* recognizes the rights of nature, and its rich concept encompasses social–cultural and spiritual relationships, reciprocity and communality, self-determination and autonomy. Additionally, it translates into a political ontology, resisting modernist notions of development to celebrate Indigenous well-being and ways of being (Escobar, 2018).

The *capabilities approach* put forward by development theorists Sen and Nussbaum in the context of international development also offers useful insights to advancing social justice and well-being in the rural development context. Following a perspective of human flourishing or well-being, they argue that justice should be about helping people live lives that they consider worthwhile rather than focusing on the appropriate distribution of primary goods, as John Rawls does (Sen, 1999; Nussbaum, 2011). Moral values of equal dignity and equal respect for people's dignity guide the concept of well-being in Nussbaum's approach. The basic idea is that "some living conditions deliver to people a life that is worthy of the human dignity that they possess, and others do not … A focus on dignity is quite different, for example, from a focus on satisfaction" (Nussbaum, 2011, p. 30, cited in Jamal, 2019, p. 68). But some basic capabilities (a minimum threshold of capability) are needed to enable a dignified and minimally flourishing life. As she asks: "Among the many things that human beings might develop the capacity to do, which ones are the really valuable ones, which are the ones that a minimally just society will endeavor to nurture and support?" (Nussbaum, 2011, p. 28).

Nussbaum responds that ten central capabilities are needed to enable a minimally flourishing life of self-worth and human dignity: bodily health; bodily integrity; senses, imagination, and thought; emotions; practical reason; affiliation; other species; play; and control over one's environment. Human dignity, respect and self-worth are especially important to bear in mind and incorporate in community development as destinations and communities begin to emerge from a protracted and traumatic pandemic that has left many rural inhabitants destitute with few options other than migrating to urban centers or attempting dangerous crossings over borders for a better life elsewhere (World Bank, 2020). We therefore discuss these principles further below, followed by a case example that illustrates their importance in the context of community tourism development.

Dignity and Respect

Dignity refers to the inalienable, unconditional worth of every human being (Kant, 1998). As Kant explains, individuals need to be acknowledged as unique, valued and rational beings who can make their own decisions based on their desires, interests and priorities. Being treated instrumentally as a means to an end, i.e., treated as an object to further someone else's interests, reduces an individual's notion of self-worth and dignity. Kantian moral values of equal dignity and equal respect for people's dignity guide the notion of well-being in Nussbaum's (2011, pp. 32–33) approach to a "dignified and minimally flourishing life" noted above.

Tourism plays an important role in promoting human dignity and well-being, recognizing and affirming the cultural identity of cultural groups, contributing to strengthening self-esteem, cultural pride, dignity and self-worth. It can also help to redress economic as well as social and cultural inequities. But it can also contribute to injustices such as worker exploitation and lack of a living wage, pollution affecting human and ecological health, habitat degradation and exploitation of the local commons, exclusion and marginalization of locals in tourism planning and development, as well as various forms of racism and discrimination, including demeaning and stereotypical portrayal of local people in tourism marketing, that affect the well-being of those who live and work in the tourism destination.

It is especially important to be mindful of the above in the context of vulnerable populations and marginalized groups who may be subject to racism, exploitation and domination, currently and historically (e.g., Schachter, 1983). The United Nations Declaration of Rights of Indigenous People (United Nations, 2007) outlines some valuable principles that facilitate Indigenous well-being and dignity. These include the rights to self-determination; to autonomy; to self-government; to participation in public life and decision making affecting their lives; to maintain and develop their institutions and means of subsistence; to practice, revitalize, and teach their language, cultural traditions, customs, and ceremonies; to strengthen their spiritual relationship with their lands and water bodies; to conservation and protection of their environment and productive capacity of their land; and to determine and develop priorities and strategies or use of their lands and other resources.

Tourism researchers have also identified a range of principles to guide Indigenous rural tourism (e.g., Whyte, 2010; Harbor and Hunt, 2021). Drawing on their research on the Maya Train project in Mexico, Camargo and Vazquez-Maguirre (2021) identify five guiding principles for human dignity, respect and the communal good of the Indigenous people exposed to mass tourism development: self-determination and self-control of tourism development; decent and meaningful tourism work; prioritization of entrepreneurial opportunities for Indigenous people; positive recognition; and oneness with the natural environment. They argue that the Maya Train, a 1,500-kilometer train route mega-tourism project proposed by the 2018–2024 federal government of Mexico, and designed to improve the quality of life, conserve the environment and trigger sustainable development in Southern Mexico (Tren Maya, 2021b), has reproduced postcolonial patterns of discrimination and lack of respect for Indigenous communities' rights and self-determination. For instance, there was no evidence that the federal government undertook meaningful consultation or obtained informed consent prior to launching the project to private investors. Also, international organizations such as the United Nations, universities, and research centers, not local Indigenous councils, have been invited to oversee the project (Tren Maya, 2021a). The voice and concerns of non-indigenous stakeholders seem to be privileged over those of the Indigenous people, who will bear the direct impacts of such development in their ancestral land and who will probably lose control over their ecocultural environment to mass tourism. As the authors say, Maya ecocultural resources will again be used as means to further the tourism industry of Mexico.

The case example below also illustrates the need to incorporate principles for dignity, respect and well-being in rural tourism development, as well as approaches to justice, such as distributive justice and recognition justice.

CASE EXAMPLE: PALENQUE, THE PALENQUERA AND INTANGIBLE CULTURAL CONSERVATION

San Basilio de Palenque, known as Palenque, is a rural community in the state of Bolivar, on the Caribbean coast of Colombia. Located 50 kilometers from Cartagena, the most popular tourist destination in Colombia, it is inhabited by descendants of African slaves who, led by the slave Benkos Biohó, escaped from their Spanish masters during colonial times (16th to early 19th century) and settled in inaccessible areas to avoid being captured. The rural isolation has helped the Palenqueros maintain their creole language, rituals, oral, religious and medical traditions, and socio-cultural structures to this day, and their cultural space was proclaimed a UNESCO Masterpiece of the Oral and Intangible Heritage of Humanity in 2005. The UNESCO declaration brought visibility to Afro-Colombian heritage, but at the same time highlighted the poverty, abandonment, marginalization, dispossession and racial discrimination that the Palenqueros have been suffering since colonial times (Rodríguez Morales, 2010; Zavala Guillen, 2018). The 2018 Census revealed that 50% of Afro-Colombians that live in rural areas suffer from multidimensional poverty; 27.9% are illiterate; 92.4% have informal jobs; and 51.2% do not have access to potable water, among other indicators (DANE, 2019). These conditions caused the migration of many Palenqueros, particularly women, to the nearby cities of Cartagena and Barranquilla, where they work as domestic workers, selling fruits and traditional sweets[9] made with their ancestral recipes, giving massages or braiding tourists' hair at the local beaches.

Palenque has a subsistence economy based on small-scale agriculture (yam, cassava, plantains, corn and pumpkin), cattle, and the production of traditional sweets and candies that are sold in bigger cities (Alcaldía Municipal de Mahates, 2021). Since its proclamation as an intangible UNESCO heritage of humanity, the rural village started to receive visitors – students and anthropologists at first – which sparked the interest of the community to develop tourism, specifically ethnic rural tourism, as a vehicle of sustainable development, cultural preservation and cultural pride. Palenque opened a restaurant, a few traditional homestays, a cultural center and developed tours to cater to the 200–300 people that started visiting them every month. But these efforts to develop tourism are isolated and fragmented, according to a community leader (J. Pérez, personal communication). The Palenqueros have not received support from the government to develop tourism nor to safeguard their culture, besides the minimal obligations stipulated by the UNESCO proclamation (see also Obezo Casseres, 2013). Moreover, new profiles of tourists are creating impacts that are affecting the community dynamics and threatening Palenquero culture. For instance, many tourists arrive in the remote community uninvited, uninformed, and expecting to see a primitive culture. Other tourists are using the village as a camping site where they party and engage in behavior that is not acceptable among Palenqueros, like walking around the streets in bikinis or underwear, smoking marijuana, or trying to engage in sexual relations with locals. Obezo Casseres (2013) reports internal competition to host and guide tourists that is affecting the relationships among community members and the quality of the cultural interpretation provided to visitors.

Meanwhile, Palenqueras have become an icon in Cartagena's tourist landscape; they can be seen wearing colorful dresses and selling fruit in the main tourist areas, as souvenirs in tourist shops (Figure 19.1) or in tourist marketing (Figure 19.2). Their image has been not only commodified, but also reduced to a hyper-sexualized representation (that evokes their colonial past) for the benefit of the tourism industry with little, if any, compensation to the Palenqueras (Camargo and Lawo-Sukan, 2015). Tourists take photos of them, on many occasions without permission, but get upset when Palenqueras request money for the right to take a photograph. The government invites two or three known Palenqueras to national and international events to promote Cartagena with no tangible benefit to the Palenquero community being "represented." On the contrary, they have been harassed by the police and restricted from working on public beaches and tourist areas where they sell their products and services to tourists. But Palenqueras are resisting and exerting their agency through organizations and cooperatives[10] that are fighting for their right to access the public spaces of Cartagena on equal terms with tourists and other tourist providers (Lamus Canavate, 2010). In 2019, they took to the main streets of the historic center of Cartagena to protest against the new Police Code that prohibits them from selling fruit on stands, an economic activity that many of them have been doing for more than 30 years (El Tiempo, 2019).

Source: Photograph by Blanca Camargo.

Figure 19.1 *Commodification of Palenqueras as souvenirs*

Source: https://procolombia.co/.

Figure 19.2 *Palenqueras image in Colombia travel portal*

Palenque has the potential to become one of the most important ethnic and rural destinations in the Caribbean region of Colombia. Tourism can play an important part in the much-needed economic and cultural well-being of its people and in reducing racial discrimination against them within the wider Colombian society. But, according to community leaders, there is an urgent need to strengthen the Palenquero culture and identity, in particular among the children, and build internal capacities to manage tourism and its impact before continuing with any further development. There are community-driven initiatives to support ethnic rural tourism, like the Annual Festival of Drums and Cultural Expressions (Festival de Tambores y Expresiones Culturales), living history tours, music and culinary workshops, and to produce more competitive and marketable sweets using the Palenquera brand.[11] However, they struggle to obtain financial and marketing resources to increase the benefit of these initiatives (Asopraduce, n.d). Moreover, in Colombia the COVID-19 pandemic has caused a plunge in tourism, which was the livelihood of many Palenqueras and their families, deepening their vulnerability and the racial inequities they have been experiencing for many years.

Palenqueros have used their cultural and territorial practices, their *marronage*, to survive slavery, colonialism, discrimination, racial violence, and, more recently, COVID-19 (Pérez Palomino & Zavala Guillen, 2020; Zavala Guillen, 2021). They will protect themselves from predatory private tourism companies that will, in their own words, "convert us into workers in our own territory that we positioned with our community efforts" (Asopraduce, n.d.). But much more is needed to guide good rural tourism development in Palenque and the respectful and equitable treatment of the Palenqueras by the tourism industry. The government needs to facilitate financial, technical and marketing support for Palenqueros' social enterprises and integrate them fairly in the tourism value chain. Their enterprises represent Palenque's community vision and their values, and are a vehicle for community empowerment, agency and cultural preservation. It also needs to ensure Palenqueras are given equal access to tourism's economic opportunities and be respectfully and fairly compensated for the use of their cultural goods and images for tourism purposes (and not stereotyped).

RURAL TOURISM TO WHAT END?

The interrelatedness of ecological, political, and social-cultural domains is evident in the rural tourism examples described above and they call for integrated, collaborative, and inclusive approaches to tourism development and planning that embrace diverse worldviews and cosmologies (e.g., Espeso-Molinero et al., 2016; Whyte, 2010). As the chapter indicates, much greater thought has to be given to the ontological and epistemological assumptions, values and principles driving tourism development and community-oriented forms of tourism. The old development paradigms that are embedded in modernist values of progress and anthropocentric worldviews have entrenched deep inequities in the accumulation of wealth and distribution of costs and benefits globally, and unsustainable resource use and exploitation of global and local commons. Moving forward, the protracted pause in the global economic machinery of tourism due to the COVID-19 pandemic will hopefully enable an opportunity to revisit the purpose of tourism development. Certainly, one good purpose is to better address structural injustices, human rights issues and disparities in fair and equitable distribution of the costs and benefits of tourism between the Global North and the Global South, and between core tourism-generating regions and the rural peripheries (Bianchi & de Man, 2021; Bianchi and Stephenson, 2014; Britton, 1982). Another is the hope for *regenerative tourism* (see Dredge, 2021) that incorporates principles of justice and ethics, including communality and social well-being, building capabilities and sustainable livelihoods that enable lives of dignity and self-worth, contributing to food security, healthy environments and biodiversity conservation (Fennell, 2017; Jamal, 2019; 2021: Smith & Duffy, 2003).

Regenerative tourism should also be seen as opportunities for *restorative tourism*, for healing troubled pasts and present conflicts, healing damaged ecosystems, enabling food security and food sovereignty, facilitating self-determination, autonomy, cultural survival and heritage conservation. As the above examples of the Balinese and Kichwa communities show, it is inclusive of diverse worldviews, spiritual cosmologies and ways of being, as well as diverse knowledge systems (traditional knowledge, local knowledge, scientific knowledge, gendered and diverse knowledges, etc.). It is situated not just in an ethic of justice but also an ethic of care for human and non-human others (Jamal, 2019). Digital technologies and digital marketing will play an important role here as disruptive technologies continue to transform urban and rural landscapes. But they too need to be used carefully and thoughtfully, as tools to collaboratively enable good tourism experiences, help redress environmental damage and related social injustices, and contribute to community resilience and social well-being, as well as long-term sustainability and conservation. The example above of the rural Afro-Colombian Palenqueros illustrates the challenges of commodification and exploitation of the Palenquera women, and the need for greater inclusiveness of marginalized groups and minority populations in decision making, as well as greater control over how they wish to share and market their cultural goods, both on site and "virtually" too.

Challenging futures lie ahead with future pandemics, climate change, and other global threats. Empowering and strengthening rural communities as stewards of the natural and cultural environments, facilitating food security and conservation of the local and global commons, reclaiming the local through community empowerment, capacity building and effective, community-driven policy, governance and democracy, are critical priorities in which rural tourism has a valuable role to play.

NOTES

1. https://www.worldbank.org/en/news/press-release/2020/10/07/covid-19-to-add-as-many-as-150-million-extreme-poor-by-2021#:~:text=The%20COVID%2D19%20pandemic%20is,severity%20of%20the%20economic%20contraction.
2. https://www.who.int/news-room/feature-stories/detail/a-virus-that-respects-no-borders-protecting-refugees-and-migrants-during-covid-19.
3. https://19january2017snapshot.epa.gov/climate-impacts/international-climate-impacts_.html.
4. https://goodtourismblog.com/2021/06/towards-a-new-paradigm-for-regenerative-tourism-and-just-futures/.
5. Community here is defined very simply as a bounded geographical area with a network of things and inhabitants, both human and non-human).
6. For more on the *subak*, see MacRae and Arthawigual (2011), and also https://nowjakarta.co.id/travel/archipelago-diaries/subak-the-spirit-behind-rice-cultivation.
7. https://whc.unesco.org/en/list/1194/documents/.

8. https://i-discoverasia.com/bali-subak/.
9. The 2018 Colombian census estimated that seven out of ten Palenqueros currently live in urban cities (DANE, 2019).
10. An example of such organizations is "Graciela Cha Inés", which is a group of 80 Palenqueras who sell food on Cartagena's beaches.
11. https://www.facebook.com/PalenqueraColombia/?ref=page_internal.

REFERENCES

Alcaldía Municipal de Mahates (2021). *Nuestro municipio.* http://www.mahates-bolivar.gov.co/municipio/nuestro-municipio.

Asopraduce (n.d). *Palenquera: la marca de San Basilio de Palenque.* https://www.mincit.gov.co/CMSPages/GetFile.aspx?guid=c4d5a7e7-416d-4c99-b5e9-e9c1be6a44e4.

Bianchi, R.V., & de Man, F. (2021). Tourism, inclusive growth and decent work: A political economy critique. *Journal of Sustainable Tourism, 29*(2–3), 353–371.

Bianchi, R.B., & Stephenson, M.L. (2014). *Tourism and Citizenship: Rights, Freedoms and Responsibilities in the Global Order.* New York: Routledge.

Borlido, T., & Coromina, L. (2018). Social capital as cause and consequence of rural touristic development: The case of Lindoso (Portugal). *European Journal of Tourism Research, 19,* 71–85.

Britton, S. (1982). The political economy of tourism in the Third World. *Annals of Tourism Research, 9*(3), 331–358.

Camargo, B., & Lawo-Sukam, A. (2015). San Basilio de Palenque (re)visited: African heritage, tourism, and development in Colombia. *Afro-Hispanic Review, 34*(1), 25–45.

Camargo, B.A., & Vázquez-Maguirre, M. (2021). Humanism, dignity and indigenous justice: The Mayan train megaproject, Mexico. *Journal of Sustainable Tourism, 29*(2–3), 371–390.

Carney, M. (2021). *Value(s): Building a Better World for All.* Toronto: Penguin Random House.

Colorni, R.R. (2018). Tourism and land grabbing in Bali: A Research Brief. Transnational Institute. https://www.tni.org/files/publication-downloads/tourism_and_land_grabbing_in_bali.pdf.

DANE (2019). Población negra, Afrocolombiana, Raizal y Palenquera. Resultados del Censo Nacional de Población y Vivienda 2018. https://www.dane.gov.co/files/investigaciones/boletines/grupos-etnicos/presentacion-grupos-etnicos-poblacion-NARP-2019.pdf.

Dianasari, D.A.M.L. (2018). Ecological tourism village as a tool for sustainability (Case study Nyambu Tourism Village, Bali). *Journal of Advanced Management Science, 6*(4), 197–202. doi:10.18178/joams.6.4.197-202.

Dredge, D. (2021, May 25). *Regenerative tourism rising ... and why it can't be unseen.* Medium. https://medium.com/the-tourism-colab.

El Tiempo (2019, March 27). *Palenqueras en Cartagena protestaron contra el Código de Policía* [video]. YouTube. https://www.youtube.com/watch?v=xhO9GQ0Nkmk.

Escobar, A. (2018). *Designs for the Pluriverse.* Durham, NC: Duke University Press.

Espeso-Molinero, P., Carlisle. S., & Pastor-Alfonso, M.J. (2016). Knowledge dialogue through Indigenous tourism product design: A collaborative research process with the Lacandon of Chiapas, Mexico. *Journal of Sustainable Tourism, 24*(8–9), 1331–1349, doi:10.1080/09669582.2016.1193188.

Esteva, G. (2018). Development. In W. Sachs (ed.), *The Development Dictionary: A Guide to Knowledge as Power* (2nd edn, pp. 1–23). London: Zed Books.

Fennell, D.A. (2017). *Tourism Ethics* (2nd edn). Bristol: Channel View Publications.

Fusté-Forné, F., & Jamal, T. (2020). Slow food tourism: An ethical microtrend for the Anthropocene. *Journal of Tourism Futures, 6(*3), 227–232. https://doi.org/10.1108/JTF-10-2019-0120.

George, E.W., Mair, H., & Reid, D.G. (2009). *Rural Tourism Development: Localism and Cultural Change*. Bristol: Channel View Publications.

Gohori, O., & van der Merwe, P. (2020). Towards a tourism and community-development framework: An African perspective. *Sustainability, 12*(13), 5305. https://doi.org/10.3390/su12135305.

Haraway, D. (1988). Situated knowledges: The science question in feminism and the privilege of partial perspective. *Feminist Studies, 14*(3), 575–599.

Harbor, L.C., & Hunt, C.A. (2021). Indigenous tourism and cultural justice in a Tz'utujil Maya community, Guatemala. *Journal of Sustainable Tourism, 29*(2–3), 214–233, doi:10.1080/09669582.2020.1770771.

Hwang D., Stewart, W.P., & Ko, D. (2012). Community behavior and sustainable rural tourism development. *Journal of Travel Research, 51*(3), 328–341. doi:10.1177/0047287511410350.

Jamal, T. (2019). *Justice and Ethics in Tourism*. Abingdon: Routledge.

Jamal, T. (2021, June 8). Towards a new paradigm for regenerative tourism and just futures. *Good Tourism Blog*. https://goodtourismblog.com/2021/06/towards-a-new-paradigm-for-regenerative-tourism-and-just-futures/.

Jamal, T., & Higham, J. (2021). Justice and ethics: Towards a new platform for tourism and sustainability. *Journal of Sustainable Tourism, 29*(2–3), 143–157.

Jamal, T., Kircher, J., & Donaldson, J.P. (2021). Re-visiting design thinking for learning and practice: Critical pedagogy, conative empathy. *Sustainability, 13*(2), 964. Special issue on "Tourism: A Fecund Frontier for Cross and Multidisciplinary Enquiries in Sustainability." Guest editors: J.M Cheer, S. Graci, & C. Dolezal. https://doi.org/10.3390/su13020964.

Kant, I. (1998). *Groundwork for the Metaphysics of Morals*. Cambridge: Cambridge University Press.

Karst, H. (2017). "This is a holy place of Ama Jomo": Buen vivir, indigenous voices and ecotourism development in a protected area of Bhutan. *Journal of Sustainable Tourism, 25*(6), 746–762. doi:10.1080/09669582.2016.1236802.

Lamus Canavate, D. (2010). Negras, palenqueras y afrocartageneras. Construyendo un lugar contra la exclusión y la discriminación. *Reflexión Política, 12*(23), 153–166.

MacRae, G.S., & Arthawiguna, I.W.A. (2011). Sustainable agricultural development in Bali: Is the subak an obstacle, an agent or subject? *Human Ecology, 39*, 11–20. https://doi.org/10.1007/s10745-011-9386-y.

Mair, H., & Reid, D.G. (2007). Tourism and community development vs. tourism for community development: Conceptualizing planning as power, knowledge, and control. *Leisure/Loisir, 31*(2), 403–425, doi:10.1080/14927713.2007.9651389.

Manyara, G., & Jones, E. (2007). Community-based tourism enterprises development in Kenya: An exploration of their potential as avenues of poverty reduction. *Journal of Sustainable Tourism, 15*(6), 628–644. doi:10.2167/jost723.0.

Matarrita-Cascante, D., Sene-Harper, A., & Ruyle, L. (2019). A holistic framework for participatory conservation approaches. *International Journal of Sustainable Development & World Ecology, 26*(6), 484–494. doi:10.1080/13504509.2019.1619105.

Mbaiwa, J.E., & Stronza, A.L. (2010). The effects of tourism development on rural livelihoods in the Okavango Delta, Botswana. *Journal of Sustainable Tourism, 18*(5), 635– 656. https://doi.org/10.1080/09669581003653500.

Mosedale, J. (ed.) (2016). *Neoliberalism and the Political Economy of Tourism*. Abingdon: Routledge.

Mostafanezhad, M., Norum, R. Shelton, E. J., & Thompson-Carr, A. (eds) (2016). *Political ecology of tourism: Community, power and the environment*. Abingdon: Routledge.

Nguyen, T. (2017, November 2). *Einstein on compassion*. Medium. Retrieved June 5, 2021, from https://medium.com/noblevator/einstein-on-compassion-3f90dc737185.

Nussbaum, M. (2011). *Creating Capabilities: The Human Development Approach*. Cambridge, MA: The Belknap Press of Harvard University Press.

Nyéléni (2007). *Declaration of Nyéléni*. http://www.nyeleni.org/spip.php?article290.

Obezo Casseres, K.G. (2013). *Turismo, Patrimonio y etnicidad: Dinámicas del turismo étnico-cultural en San Basilio de Palenque* [Master's dissertation, Universidad Tecnológica de Bolivar, Cartagena, Colombia].

Pérez Palomino, J.N., & Zavala Guillen, A.L. (2020, June 18). The loss of territory and the impact of Covid-19 on Afro-descendant communities in Latin America. *Geography Directions*. https://blog.geographydirections.com/2020/06/18/the-loss-of-territory-and-the-impact-of-covid-19-on-afro-descendant-communities-in-latin-america/.

Rodríguez Morales, M.M. (2010). The statistical invisibility of black, Afro-Colombian and Palenque-rooted populations in Colombia. *Trabajo Social*, *12*, 89–99.

Salleh, A.K. (1984). Deeper than deep ecology: The eco-feminist connection. *Environmental Ethics*, *6*, 339–345.

Santafe-Troncoso, V., & Loring, P.A. (2021). Indigenous food sovereignty and tourism: The Chakra Route in the Amazon region of Ecuador. *Journal of Sustainable Tourism*, *29*(2–3), 392–341.

Schachter, O. (1983). Human dignity as a normative concept. *American Journal of International Law*, *77*(4), 848–854. https://doi.org/10.2307/2202536.

Scheyvens, R. (2002). *Tourism for Development: Empowering Communities*. Harlow, UK: Prentice Hall. http://doi.org/10.1016/j.tourman.2005.07.013.

Sen, A. (1999). *Development as Freedom*. Oxford: Oxford University Press.

Shiva, V. (1988). *Staying Alive: Women, Ecology, and Development*. London: Zed Books.

Shiva, V. (1997). Economic globalization, ecological feminism and sustainable development. *Canadian Woman Studies*, *17*(2), 22–27.

Shiva, V. (2000). *Stolen Harvest: The Hijacking of the Global Food Supply*. London: Zed Books.

Smith, M., & Duffy, R. (2003). *The Ethics of Tourism Development*. Abingdon: Routledge.

Telfer, D.J. (2009). Development studies and tourism. In T. Jamal & M. Robinson (eds), *The SAGE Handbook of Tourism Studies* (pp. 146–165). London: SAGE Publications.

Tren Maya (2021a). *Aliados*. https://www.trenmaya.gob.mx/aliados/.

Tren Maya (2021b). *El Tren Maya nos une*. https://www.trenmaya.gob.mx/.

United Nations (2007). *United Nations Declaration on the Rights of Indigenous People*. https://www.un.org/development/desa/indigenouspeoples/wp-content/uploads/sites/19/2018/11/UNDRIP_E_web.pdf.

Victurine R. (2000). Building tourism excellence at the community level: Capacity building for community-based entrepreneurs in Uganda. *Journal of Travel Research*, *38*(3), 221–229. doi:10.1177/004728750003800303.

Wardana, A. (2018). Legal engineering over space in Bali. *Australian Journal of Asian Law*, *19*(1), 1–12.

Wardana, A. (2019). *Contemporary Bali: Contested Space and Governance*. Singapore: Palgrave Macmillan.

Wearing, S., & McGehee, N. (2013). Volunteer tourism: A review. *Tourism Management*, *38*, 120–130. https://doi.org/10.1016/j.tourman.2013.03.002.

Whyte, K.P. (2010). An environmental justice framework for indigenous tourism. *Environmental Philosophy*, *7*(2), 75–92.

Wise, T.A. (2019). *Eating Tomorrow: Agribusiness, Family Farmers, and the Battle for the Future of Food*. New York: The New Press.

World Bank (2020). *COVID-19 to add as many as 150 million extreme poor by 2021* [Press release 2021/024/DEC-GPV]. https://www.worldbank.org/en/news/press-release/2020/10/07/covid-19-to-add-as-many-as-150-million-extreme-poor-by-2021.

Young, I.M. (2011). *Justice and the Politics of Difference*. Princeton, NJ: Princeton University Press.

Zavala Guillen, A.L. (2018). *Geographies of marronage: Dispossession and resistance in the last Palenque* [Doctoral dissertation, University of Sheffield].

Zavala Guillen, A.L. (2021). Afro-Latin American geographies of in-betweenness: Colonial marronage in Colombia. *Journal of Historical Geography*, *72*, 13–22. https://doi.org/10.1016/j.jhg.2021.02.002.

20. Sustainable tourism development in rural areas: an alternative paradigm using the appreciative inquiry approach

Roslizawati Che Aziz

INTRODUCTION

Sustainable development is now widely promoted as a holistic concept that aims to integrate social, economic, and cultural policies to ensure high-quality growth in the context to which it is applied (Gronau & Kaufmann, 2009). Rather than putting emphasis on the question of whether the tourism industry can be sustainable or how sustainable can it be, it is better to focus on the question of whether tourism can support sustainable development or not. Sustainable development must improve the economic standard of living and people's quality of life, while at the same time protecting and enhancing the natural and cultural environment (Reid & Schwab, 2006). Reid and Schwab (2006) added that although sustainable development has been widely accepted as an international standard, for many countries it remains a concept that is seen by citizens as artificially superimposed on their domestic political environment.

The term "sustainable" can be defined in various ways according to people's perceptions and beliefs. According to the World Commission on Environment and Development (1987), sustainable tourism is "tourism that respects both local people and traveller, cultural heritage and the environment", which seeks to provide people with an exciting and educational holiday that is also beneficial to the nation's economy. In contrast, George et al. (2009) and Reid and Schwab (2006) asserted that there is no single definition for the term "sustainable development" in tourism as they believed that a great deal of ambiguity is associated with the concept because it is value-laden and interpreted differently by different groups. However, Gronau and Kaufmann (2009) claimed that a clear contradiction between sustainable principles and tourism development can be identified in the field of spatial inequality of development.

Sustainable development is a visionary development paradigm. Over the past 20 years, government and private sector businesses have accepted it as a guiding principle (Tsonis & Cheuk, 2000). More sustainable development directions are needed, which require new discussions, cooperation and most importantly, dedication from tourism-related businesses (Castellani & Sala, 2010; Scott, 2011). Although rural tourism in Malaysia has been under development for a long time, major gaps still exist in this industry, especially in the field of sustainable rural tourism. Many efforts have been made to enhance the demand for rural tourism as well as to improve the planning, management and development approaches for rural tourism. Nevertheless, efforts have been insufficient and there are still disturbing issues, which do not reflect sustainable rural tourism practices (Jaini et al., 2012). The development of rural tourism has been seen as a tool to improve accessibility to remote areas, where it provides more opportunities between tourists and local communities, such as entrepreneurial elements, engagement and empowerment of communities (Mathew, 2005).

The tourism industry has tremendous potential for sustainable tourism development, particularly at environmentally and culturally sensitive sites (Bramwell & Lane, 2012). Along with rapid growth, a diversification of tourism products and destinations is taking place, increasing demand for alternative types of tourism. Thus, sustainable development for community tourism should improve quality of life by optimizing local economic benefits, protecting the natural and built environment, and providing a high-quality experience for visitors. They should aim to provide long-term economic linkages between communities and industries (Nunkoo & Ramkissoon, 2010), and minimize the negative effects of tourism on the natural environment—and thus maybe improve the socio-cultural well-being of the communities (Egbali et al., 2011). Tourism activities should also be developed in adherence to the principles of sustainable development and in collaboration with various stakeholders (Mtapuri & Giampiccoli, 2019), for example, tourism companies and organizations, both at local and national level.

Community stakeholders, including the state government, tourists, hosts, as well as tour operators and other tourist-related businesses should implement ethical responsibilities and codes of conduct to ensure the sustainability of tourism development. To cater to changing demands, there is also a need for greater ties between environmentally oriented sustainable development and the role of entrepreneurship in economic development in underdeveloped communities (Hall et al., 2010). In addition, Calabrese et al. (2010) suggested five mechanisms to encourage these changes: (1) institutions should develop sustainable tourism civic education for communities; (2) cohesion and insight for consciousness-raising events can promote sustainable tourism development attitudes; (3) lively public hearings should be encouraged to turn passive into active participation; (4) mechanisms for reporting environmental damage should be created; and (5) community environmental clean-up days should be held.

Sustainable Development Strategies for Rural Tourism Development

One of the fundamental assumptions underpinning the concept of sustainable tourism development is that there has been an increase in environmental concern on the part of tourists and hence, growing demand for "responsible" travel experiences. Indeed, other researchers have consistently demonstrated that tourism is relatively immune to environmental concerns (or that "responsible" tourist behaviour is motivated by factors other than environmental concern), such as the cost of travelling and climate change (Hunter & Shaw, 2007). Thus, a significant gap remains between the idealism of the concept of sustainable tourism development as explored at length within academic circles and the reality of tourism development in practice. Despite the large amount of literature on this topic, a consistent framework of so-called sustainable tourism has not yet been developed (Gronau & Kaufmann, 2009) and the general assumption that there can be a sustainable form of tourism is often questioned. For instance, objections were raised about the various contradictions between the holistic concept of sustainability and the more product-centred perspective of the global industry. Illuminating the holistic perspective, Lane (1994; 2005) viewed sustainable tourism as a balanced triangular relationship between "host areas and their habitats and people, holiday makers and the tourism industry" (Lane, 1994, p. 10) with no stakeholder upsetting the equilibrium. This focus on a balanced relationship is supported by Sharpley (2000, p. 232), who stated that "the potential for sustainable tourism development exists if no single factor or stakeholder predominates".

The careful balance of economic and environmental concerns is truly important as the tourism industry also provides an important source of income, employment and wealth to the country. Thus, there is a need to ensure that the tourism industry remains both environmentally and economically sustainable (Siti-Nabiha et al., 2009). In ensuring the sustainability of the industry, protection of the environment is very important. Indeed, sustainable development has become an important criterion in attracting tourists and it has been proven that sustainability in tourism can only be achieved if all the stakeholders are really applying the concept and practices of sustainability (Jaini et al., 2012). Moreover, tourism can be an important module while introducing the concept of sustainability in rural areas. Designated forms of tourism based on authenticity and local culture as well as entrepreneurship can also connect the objectives of sustainability and the interest of the locals to improve their standard of living (Gronau & Kaufmann, 2009).

For the tourism industry to be sustainable, it needs to be integrated with a broader sustainable development strategy, particularly with effective multi-stakeholder engagement (Siti-Nabiha et al., 2009). It should also be integrated with other management functions and actions in line with any strategy that is implemented. More importantly, sustainable tourism development requires a process of planning and management that brings together the interests and concerns of a diverse group of stakeholders in a sustainable and strategic way (Sharpley, 2009). To ensure the sustainability of rural tourism development, a framework should also be built to understand the relationship between tourism impacts and community identity. This has been addressed by Hwang et al. (2011), where they found that community-based action in rural tourism development relates to a capacity to protect one's community from outside threats and foster development that aligns with its sense of community. Focusing on the community's involvement in rural tourism development is of utmost importance, as community members are the industry's key players in tourism activities. Local communities generally need to ensure that a sustainable approach is taken regarding the development of other economic activities in the area and that such development does not erode fragile tourism resources. If development is to be sustainable, it must improve the standard of living and quality of life while protecting and enhancing the natural and cultural environment (Reid & Schwab, 2006).

Community Capacity in Sustainable Tourism

Community capacity, however, is widely addressed as an essential strategy for improving the process of tourism development that can benefit local communities (Moscardo, 2008; Walker, 2008). Community leaders play a fundamental role in addressing the issues of tourism development, particularly in rural communities. Tourism activities should preserve the culture of the local communities and provide adequate economic opportunities for local communities whilst guarding them against exploitation. Due to the nature of the tourism industry, sustainable tourism development should be based on coordinated actions between the different sectors involved (Siti-Nabiha et al., 2009). Indeed, integrated and meaningful multi-stakeholder participation from a broad spectrum of host communities is needed, while partnerships should be at the federal, state and local level and should be backed by strong political leadership (Haven Tang & Jones, 2012; Ranđelović et al., 2012; Siti-Nabiha et al., 2009). It has been noted that little attention has been given to purposefully investigating the roles and responsibilities of local governments in addressing sustainable development within tourism contexts (Ruhanen, 2012). Studies also argue that effective management systems for sustainable tourism development ultimately require intervention and regulations from the public sector. The government's involvement and direction in addressing or attempting to meet the objectives of sustainable development are widely supported by local communities (Liu, 2010; Weaver, 2006).

Besides ensuring the preservation of the environment, the role of tourism as an agent of socio-economic growth and development has become more important. Within the developed world, peripheral or poor regions are increasingly focusing on tourism as a means of stimulating economic and social regeneration, while for many less developed countries, tourism has come to represent a vital ingredient in their development policies. Although tourism has undoubtedly contributed to the socio-economic development in a number of less developed countries, such development is less evident, challenging the widespread belief in tourism's potential development contribution (Sharpley, 2009). The concepts and principles of sustainable rural tourism development are illustrated in Figure 20.1, which consists of three important elements, namely, social, environmental and economic. The figure shows a basic understanding of the concept of sustainable development from the context of tourism, which represents a sustainable approach to economic, social and environmental development. Hence, it plays an important role in the spatial development of these aspects and the spreading of benefits to undeveloped regions. Therefore, researchers believed that there would be powerful incentives to conserve these resources if the right stakeholders were involved.

Previous studies (e.g. Park et al., 2012; Pearce, 2008; Richard et al., 2013) highlighted the significance of involving the local community in decision-making processes pertaining to sustainable tourism development. They believed that all stakeholders involved in the tourism industry need to implement "inclusive strategies", not only within the context of sustainable tourism presentation and interpretation, but also local communities in the tourism planning and development process. Other than that, partnerships from a sustainable tourism perspective imply the need to collaborate with various stakeholders to aim for sustainable use of resources. Partners within the sustainable tourism environment have been broadly classified into the following categories: institutions (heritage sites, attractions, events and shops), the public sector (chamber of commerce, city or government, who decide on tax deductions), the ancillary sector (such as gasoline stations, restaurants, lodgings, gift shops, etc.), tourism organizations (such as the convention and visitor bureaus, state tourism offices, national and regional tourism organizations), the local community, and the academic community (the educational institutions).

Based on the figure, it is believed that all stakeholders involved in the tourism industry need to implement "inclusive strategies", not only within the context of sustainable tourism presentation and interpretation, but also involving local communities in the tourism planning and development process.

APPRECIATIVE INQUIRY (AI) FOR SUSTAINABLE TOURISM DEVELOPMENT

Appreciative inquiry (AI) has been applied in the field of community development, especially in developing countries. Organizers used AI to help create and strengthen community development organizations in India (Lewis et al., 2008). Even after the successful implementation of the AI approach in the development field, however, it has been irregularly used for research purposes, especially in the field of tourism (Nyaupane & Poudel, 2012; Raymond & Hall, 2008b). In 1986, AI was first established by David Cooperrider in his doctoral thesis titled "Appreciative inquiry: Toward a methodology for understanding and enhancing organizational innovation" (Cooperrider et al., 2008; Cooperrider & Whitney, 2005). It has been practised around the world for more than a decade by non-profit organizations, businesses, families, health-care organizations, schools and the government. Since the late 1980s, AI has been promoted in a wide variety of organizations and for many different purposes and has been applied to strategic planning, culture transformation, increasing customer satisfaction, organization redesign and leadership development. Recently, AI has also been applied to integrate organizations after a merger, to build alliances and union–management partnerships for peacebuilding and for implementing educational reform and economic development efforts (Whitney & Trosten-Bloom, 2010).

Appreciative inquiry is an increasingly important area, which offers a positive and strength-based approach to rural community development (Cooperrider & Whitney, 2005). Although AI's origins are in academia and there is a considerable body of literature covering its organizational applications and impacts, AI has its roots in organizational development and strategies that help organizations transform themselves. It has traditionally been used as an approach to organizational analysis, learn-

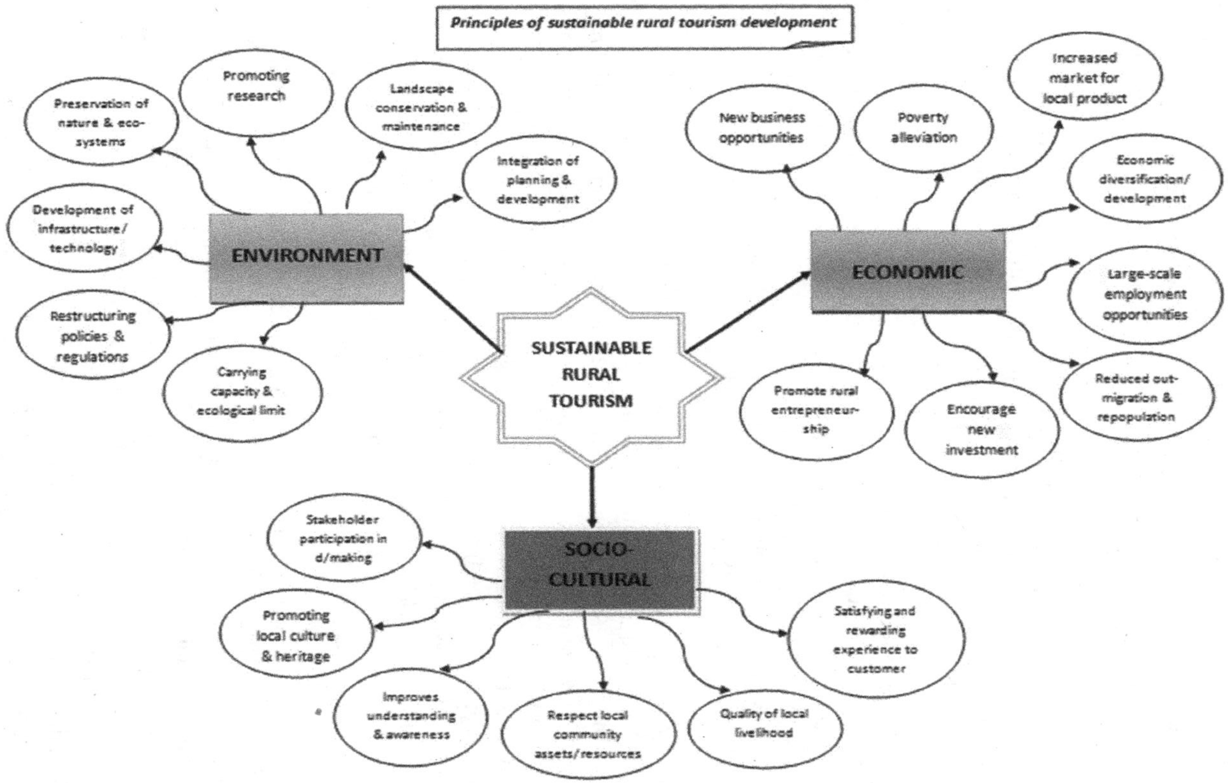

Sources: Che Aziz (2019); Dwyer & Edwards (2010); Hwang et al. (2011).

Figure 20.1 Principles of sustainable rural tourism development

ing and development that is uniquely intended for discovering, understanding and fostering innovations and transformation in the human/social system (The Mountain Institute, 2000). One of the most significant discussions about AI is a strength-based participatory action research method that is based on the constructivist paradigm and follows a grounded theory procedure (Cooperrider & Whitney, 2005). It serves as a framework for personal development or coaching, partnership or alliance-building and large-scale communities or organizations (Finegold et al., 2002; Raymond & Hall, 2008b).

The AI approach is typically used to facilitate positive development within organizations, and it could be employed in this way within the tourism industry (Raymond & Hall, 2008b). This approach has been suggested by previous studies (Koster & Lemelin, 2009; Nyaupane & Poudel, 2012; Raymond & Hall, 2008b), which illustrate that the tasks of tourism development in rural areas, the conservation of the environment and livelihood improvement, are only made possible through the joint efforts of a range of stakeholders, including governments, tourism entrepreneurs, non-profit organizations and the local people themselves. It is an applicable and interesting approach as AI treats people as human beings, not machines. It creates identity and knowledge in relation to one another such as through storytelling, and sharing the same values, beliefs and wisdom. They recommended applying AI to tourism research for various reasons (e.g. for problem-centric issues in tourism research) because by understanding pressing problems and their causes, researchers can devise and implement appropriate solutions.

More importantly, the AI approach could be an appropriate method that may help to empower rural communities, since they are often alienated by technology and technical jargon (Nyaupane & Poudel, 2011; 2012), as this method does not require respondents to read text-based instruments. Alternative methods like interviews and focus groups, as well as brainstorming, may encourage participants to speak more honestly about any problems they may have experienced (Raymond & Hall, 2008b). Efforts to understand the experiences of people living in poverty, Indigenous people, and women, are becoming ever more important, which compels a need for non-traditional methods of data gathering and verification (Koster & Lemelin, 2009; Michael, 2005; Nyaupane & Poudel, 2012). Thus, the AI approach has been identified as an appropriate method to gain an in-depth understanding of local community knowledge (Nyaupane & Poudel, 2012; Raymond & Hall, 2008a). Nyaupane and Poudel (2012) and Raymond and Hall (2008a) demonstrated that AI may offer researchers another worldwide methodology for framing and conducting tourism research. It does not mean that AI will replace all research approaches or can overcome all the challenges of conducting research. However, AI can be considered as a new method, approach, or strategy for tourism research, which initiates positive changes, especially in rural communities.

AI has the potential to describe how focusing on local community involvement serves to create a compelling vision of tourism development in their areas where it has been classified as a strength-based participatory action research method (Nyaupane & Poudel, 2012). The power of appreciating and valuing others through

AI brings people together and AI has shown success when applied in both public and private educational settings (Judy & Hammond, 2006). Similarly, in the field of tourism research, AI processes also provide opportunities for local communities and tourism stakeholders to focus on positive developments, and to collaboratively and generatively discover new possibilities not previously considered (Laszlo & Cooperrider, 2010). It also emphasizes the generation of positive ideas among individuals within an organization, and at the same time, these ideas provide the structure for creating positive change and demonstrate what the people value about themselves and their community or organization (Raymond & Hall, 2008b).

AI has been developed to gain and build enduring relationships between communities and the tourism industry based on the simple assumption that every organization or community has something that works well, and that strength can be the starting point for creating positive change (Raymond & Hall, 2008a; 2008b). The application of AI in planning and managing conservation and development programmes and activities in the tourism field provides an additional approach that helps motivate people to plan and manage a collective vision of the best possible future (Koster & Lemelin, 2009; The Mountain Institute, 2000). Inviting people to participate in dialogues and share stories about their past and present achievements, assets, unexplored potentials, innovations, strengths, opportunities, benchmarks, high-point moments, traditions, lived values, core and distinctive competencies, expressions of wisdom and possible futures can be identified as a "positive core" (Cooperrider et al., 2008; Cooperrider & Whitney, 2005). From this, AI links the energy of the positive core directly to any change agenda and this links to energy and excitement and a desire to move toward a shared dream.

The AI process engages entire communities and their stakeholders to create a future that works for everyone and can be used to guide conservation, large group meetings or whole-system efforts (Finegold et al., 2002; Shariff et al., 2009). No matter the purpose, the appreciative inquiry 4-D cycle is the foundation for change. The essence of all these models is based on a set of principles which generally follow the framework of the 4-D cycle (see Figure 20.2). Each AI process is home-grown and designed to meet the unique challenges of the organization and industry involved (Cooperrider & Whitney, 2005). The "four D's" of AI take many forms.

1. *Discovery:* At this stage, participants will reflect on and discuss the best of *what is* concerning the object of inquiry. Personal and organizational highpoints are discovered, as well as what people value and how they hope and wish to enhance their organization's social, economic and environmental vitality.
2. *Dream:* At this stage, participants are asked to picture their group's ideal image and try to figure out the common aspirations of all stakeholders. They should then attempt to create a symbolic representation of that image. This stage also requires participants to carefully listen to stories about their group at its best and to share their hopes and dreams for their collective future (Cooperrider et al., 2008). This helps create a clear results-oriented vision and uncover potential while asking higher-purpose questions, such as "What might be?" When the best of "what is"

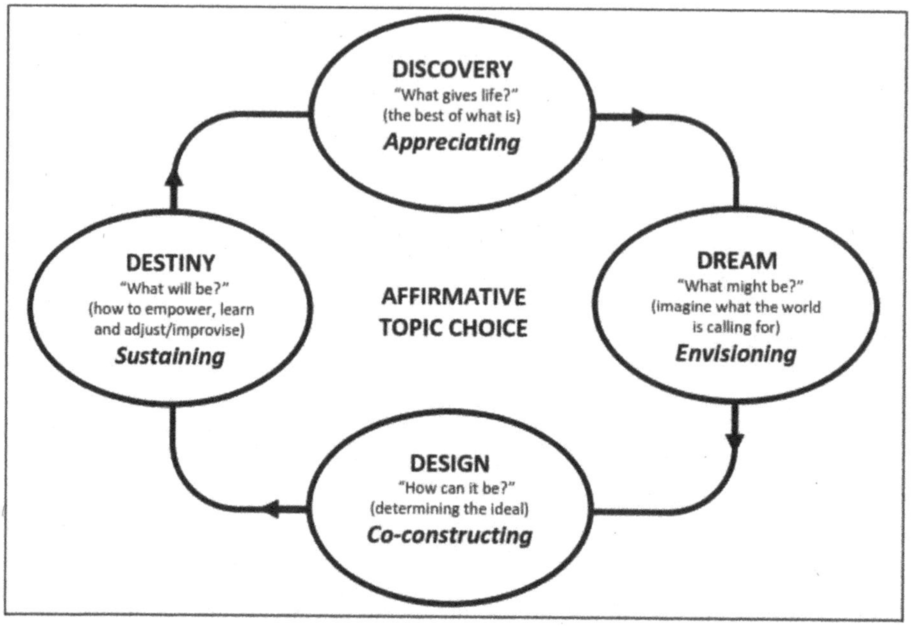

Source: Whitney & Trosten-Bloom (2010, p. 6).

Figure 20.2 The appreciative inquiry 4-D cycle

has been identified, the mind naturally begins to search further and to envision new possibilities (Cooperrider et al., 2008).

3. *Design:* With a common vision of the future established, participants are asked to develop real-world proposals for the future. This involves more than just speculation; it requires proactive, inspiring statements of intent grounded in the successes of the past combined with new ideas for the future (Cooperrider et al., 2008). This stage will help enhance the organization by leveraging both its own past success and successes elsewhere with "strategic intent", signalling what the organization wants more of and recognizing that the future is built around what can be and what is (Laszlo & Cooperrider, 2010). These activities are conducted in large group forums or within small teams (Whitney & Trosten-Bloom, 2010).
4. *Destiny:* The destiny phase involves a series of actions inspired by the previous stages, which support a continuous learning and innovation process; this causes "what will be" to emerge (Cooperrider et al., 2008). This phase specifically applies to personal and organizational commitments and the paths forward (Whitney & Trosten-Bloom, 2010), delivers new images of the future, and is sustained by nurturing a collective sense of purpose. The destiny phase is ongoing and brings the organization back full circle to the discovery phase, in which continued application of the method may result in new topic choices, dialogues, and continued learning.

Table 20.1 *Proposed questions based on appreciative inquiry 4-D cycle*

Focus Group Phase	Questions
1. DISCOVERY	What tourism assets do you have in and around your community? What type of tourism is working in your community? What kinds of activities do tourists undertake? Provide the best examples. What have you done to improve your community's livelihood through tourism? What positive linkages exist now between tourism and the resources you have?
2. DREAM	Please close your eyes if you feel like it. How do you envision your community 25 years from now? Think about what "ideal tourism" means in your community and for your children and grandchildren/for future generations.
3. DESIGN	Putting dreams into practice. What actions and strategies do you feel are needed to achieve these dreams? Where/what? How?
4. DESTINY	We have achieved or learned about tourism and its potential contribution to your community. What comes next? How can the outcomes and what we have learned be sustained? How and where might these ideas be used in the future?

Source: Moore (2019).

AI as a research tool offers novel ways of thinking about asking positive questions and hence building trust and rapport with participants. In turn, power relations and hierarchies between researcher and participants can be broken down. AI, as a field research tool, can be applied to a range of different leisure settings and with different communities to recognize what is important to them (within their current culture and values). Rather than merely emphasizing a problem-solving approach, it can also be used to identify organizational best practices.

It is very clear from the above statement that AI helps communities to improve their standard of living. During the literature search, it was found that very few studies were published showing AI's effectiveness in sustainable tourism development. Stavros et al. (2015), for example, explained that AI is a co-operative and co-evolutionary search for the best in people, their organizations, and the environment in which they operate. So, by carrying forward this argument, it could be inferred that in AI the focus is on "what is working well". However, a few years after the establishment of AI approach, several criticisms of the appreciative inquiry model have emerged. The most salient and widely discussed tend to focus on (1) the lack of strong evidence supporting the model's efficacy and (2) the model's emphasis on positivity.

When theoretical, conceptual, or procedural models are applied in community, organization, or team processes, the efficacy of a given model will depend on the quality of implementation, which can encompass a wide range of complex factors that can positively or negatively impact outcomes (for example: *Was the facilitation strong or weak? Did the facilitators understand the model and did they maintain fidelity to the model? Was a sufficient amount of time allocated for the process? Or was the process rushed? And so on.*). Consequently, it is often difficult to determine what may have gone right or wrong with the application of a given model or process. Perhaps the most potentially problematic dimension of appreciative inquiry's positive framing is that an AI process may be used in ways that reinforce and perpetuate racial or cultural bias, prejudice and discrimination. By insisting that an inquiry, dialogue, or engagement process focuses exclusively on positive questions, comments, and ideas, for example, the AI process can potentially be used—intentionally or unintentionally—in ways that silence legitimate concerns and criticisms raised by the victims of bias, prejudice and discrimination.

CONCLUSION

The appreciative inquiry (AI) approach has been identified as an alternative tourism research tool for evaluating the impacts of rural tourism for sustainable development. AI is one of the recent approaches that have been used in tourism research, particularly at the community level. The approach is typically used to facilitate positive developments and may find use within the tourism industry. Nyaupane and Poudel (2012) stated that AI acts as an alternative research method to conventional action research in the organizational development field and is applicable to tourism studies. In addition, Raymond and Hall (2008) suggested that AI might be conducted in volunteer tourism processes similarly to how it might be applied to commercial tourism. Recent AI development has found that the process not only helps an organization create images of the future, it also creates and nurtures energy, a renewed commitment to change, and a sense of hope among people working to achieve the future (Calabrese et al., 2010; Michael, 2005). The AI approach has emerged to counter traditional problem-solving approaches because it focuses on the strengths and successes of individuals, organizations and communities (Cooperrider et al., 2008). In a shift from the traditional approach of identifying, analysing and solving problems, AI builds on the strengths of organizations or communities to develop a positive approach in addressing issues (Raymond & Hall, 2008b). This method seeks out the strengths of individuals, organizations, communities, and societies, giving life, health and excellence to the development of those human systems (Whitney & Trosten-Bloom, 2010).

AI may find use as an alternative tourism research tool for community development in rural tourism. It is believed that there is significant potential for the adoption of the AI approach within the tourism discipline (Nyaupane & Poudel, 2012; Raymond & Hall, 2008b). These researchers believe that understanding the rural population's knowledge, needs and priorities, without alienating them from research, is an important goal. Existing studies have found that AI application is perceived to be a significant recent innovation in action research and is an innovative research instrument tool. The process has been shown to be enjoyable in that it allows individuals to frame problems in a more positive light. A major focus of the AI approach is learning the strengths and successes of people. AI is recognized as an important tool for empowering local people to take advantage of tourism development opportunities as it offers a positive, strengths-based approach to community development and was developed to establish enduring relationships between communities and the tourism industry. Past studies claimed that there is significant potential for the adoption of the AI approach as it is rarely used currently in the field of tourism and it provides an opportunity to close the gap between theory and practice.

REFERENCES

Bramwell, B., & Lane, B. (2012). Towards innovation in sustainable tourism research? *Journal of Sustainable Tourism*, *20*(1), 1–7.

Calabrese, R., Hester, M., Friesen, S., & Burkhalter, K. (2010). Using appreciative inquiry to create a sustainable rural school district and community. *International Journal of Educational Management*, *24*(3), 250–265.

Castellani, V., & Sala, S. (2010). Sustainable performance index for tourism policy development. *Tourism Management*, *31*(6), 871–880.

Che Aziz, R. (2019). S.O.A.R Analysis: A new model for the strategic sustainable tourism development and special interest tourism. In J.H. Costa (ed.), *Special Interest Tourism in Southeast Asia* (Vol. 8, pp. 107–131). Hershey, PA: IGI Global.

Cooperrider, D.L., & Whitney, D. (2005). *Appreciative Inquiry: A Positive Revolution in Change*. Oakland, CA: Berrett-Koehler Publishers, Inc.

Cooperrider, D.L., Whitney, D., & Stavros, J.M. (2008). *Appreciative Inquiry Handbook – For Leaders of Change* (2nd edn). Brunswick, OH: Crown Custom Publishing, Inc. & San Francisco, CA: Berrett-Koehler Publishers, Inc.

Dwyer, L., & Edwards, D. (2010). Sustainable tourism planning. In J.J. Liburd & D.U. Edwards (eds), *Understanding the Sustainable Development of Tourism* (pp. 19–43). Oxford: Goodfellow Publishers Ltd.

Egbali, N., Nosrat, A.B., & Ali, S. (2011). Effects of positive and negative rural tourism (Case study: Rural Semnan Province). *Journal of Geography and Regional Planning*, *4*(2), 63–76.

Finegold, M.A., Holland, B.M., & Lingham, T. (2002). Appreciative inquiry and public dialogue: An approach to community change. *Public Organization Review*, *2*, 235–252.

George, E.W., Mair, H., & Reid, D.G. (2009). *Rural Tourism Development. Localism and Cultural Change*. Bristol: Channel View Publications.

Gronau, W., & Kaufmann, R. (2009). Tourism as a stimulus for sustainable development in rural areas: A Cypriot perspective. *An International Multidisciplinary Journal of Tourism*, *4*(1), 83–95.

Hall, J.K., Daneke, G.A., & Lenox, M.J. (2010). Sustainable development and entrepreneurship: Past contributions and future directions. *Journal of Business Venturing*, *25*(5), 439–448.

Haven Tang, C., & Jones, E. (2012). Local leadership for rural tourism development: A case study of Adventa, Monmouthshire, UK. *Tourism Management Perspectives*, *4*, 28–35.

Hunter, C., & Shaw, J. (2007). The ecological footprint as a key indicator of sustainable tourism. *Tourism Management*, *28*(1), 46–57. http://linkinghub.elsevier.com/retrieve/pii/S0261517705001019.

Hwang, D., Stewart, W.P., & Ko, D. (2011). Community behavior and sustainable rural tourism development. *Journal of Travel Research*, *51*(3), 328–341.

Jaini, N., Aris Anuar, A.N., & Daim, M.S. (2012). The practice of sustainable tourism in ecotourism sites among ecotourism providers. *Asian Social Science*, *8*(4), 175–179.

Joyner, L., Qwynne Lackey, N., & Bricker, K.S. (2019). Community engagement: An appreciative inquiry case study with Theodore Roosevelt National Park gateway communities. *Sustainability*, *11*(24). https://doi.org/10.3390/su11247147.

Judy, S., & Hammond, S. (2006). *An Introduction to Appreciative Inquiry*. Slocan Park, BC: Silva Forest Foundation.

Koster, R.L.P., & Lemelin, R.H. (2009). Appreciative inquiry and rural tourism: A case study from Canada. *Tourism Geographies*, *11*(2), 256–269.

Lane, B. (1994). What is rural tourism? *Journal of Sustainable Tourism*, *2*(1–2), 7–21.

Lane, B. (2005). Sustainable rural tourism strategies: A tool for development and conservation. *Interamerican Journal of Environment and Tourism*, *1*(1), 12–18.

Laszlo, C., & Cooperrider, D.L. (2010). Creating sustainable value: A strength-based whole system approach. In T. Thatchenkery, D.L. Cooperrider, & M. Avital (eds), *Positive Design and Appreciative Construction: From Sustainable Development to Sustainable Value* (1st edn, pp. 17–33). Bingley: Emerald Group Publishing Limited.

Lewis, S., Passmore, J., & Cantore, S. (2008). *Appreciative Inquiry for Change Management: Using AI to Facilitate Organizational Development*. London: Kogan Page Limited.

Liu, Z. (2010). Sustainable tourism development: A critique. *Journal of Sustainable Sustainable Tourism*, *11*(6), 459–475.

Mathew, L. (2005). Eco-rural tourism and sustainable development. *Integral Liberation*, *9*(3), 243–251.

Michael, S. (2005). The promise of appreciative inquiry as an interview tool for field research. *Development in Practice*, *15*(2), 222–230.

Moore, C. (2019). 119+ appreciative inquiry interview questions and examples. https://positivepsychology.com/appreciative-inquiry-questions/.

Moscardo, G. (2008). Community capacity building: An emerging challenge for tourism development. In G. Moscardo (ed.), *Building Community Capacity for Tourism Development* (pp. 1–15). Wallingford: CABI.

Mtapuri, O., & Giampiccoli, A. (2019). Tourism, community-based tourism and ecotourism: A definitional problematic. *South African Geographical Journal*, *101*(1), 22–35. https://doi.org/10.1080/03736245.2018.1522598.

Nunkoo, R., & Ramkissoon, H. (2010). Modeling community support for a proposed integrated resort project. *Journal of Sustainable Tourism*, *18*(2), 257–277.

Nyaupane, G.P., & Poudel, S. (2011). Linkages among biodiversity, livelihood, and tourism. *Annals of Tourism Research*, *38*(4), 1344–1366.

Nyaupane, G.P., & Poudel, S. (2012). Application of appreciative inquiry in tourism research in rural communities. *Tourism Management*, *33*(4), 978–987.

Park, D.B., Lee, K.W., Choi, H.S., & Yoon, Y. (2012). Factors influencing social capital in rural tourism communities in South Korea. *Tourism Management*, *33*(6), 1511–1520. https://doi.org/10.1016/j.tourman.2012.02.005.

Pearce, P.L. (2008). Sociocultural benefits of tourism to destinations. In G. Moscardo (ed.), *Building Community Capacity for Tourism Development* (pp. 29–40). Wallingford: CABI.

Ranđelović, N., Stefanović, V., & Azemović, N. (2012). Suva Planina as development area of rural tourism. *Economics of Agriculture*, *44*(59), 333–345.

Raymond, E.M., & Hall, C.M. (2008a). The development of cross-cultural (mis)understanding through volunteer tourism. *Journal of Sustainable Tourism*, *16*(5), 530–543.

Raymond, E.M., & Hall, C.M. (2008b). The potential for appreciative inquiry in tourism research. *Current Issues in Tourism*, *11*(3), 281–292.

Reid, M., & Schwab, W. (2006). Barriers to sustainable development: Jordan's sustainable tourism strategy. *Journal of Asian and African Studies*, *41*(5/6), 439–457.

Richard, T., Chandrarathne, B., & Chandana, J. (2013). Engaging the rural communities of Sri Lanka in sustainable tourism. *Worldwide Hospitality and Tourism Themes*, *5*(5), 464–476.

Ruhanen, L. (2012). Local government: Facilitator or inhibitor of sustainable tourism development? *Journal of Sustainable Tourism*, *21*(1), 80–98.

Scott, D. (2011). Why sustainable tourism must address climate change. *Journal of Sustainable Tourism*, *19*(1), 17–34.

Shariff, Z., Gramberg, B. Van, & Foley, P. (2009). The usefulness of appreciative inquiry as a method to identify mass sports program success. *Transylvanian Review of Administrative Sciences*, *30*, 118–131.

Sharpley, R. (2000). Tourism and sustainable development: Exploring the theoretical divide. *Journal of Sustainable Tourism*, *8*(1), 1–19.

Sharpley, R. (2009). Sustainable tourism development: A critique. In R. Sharpley (ed.), *Tourism Development and the Environment: Beyond Sustainability?* (pp. 57–86). Abingdon: Earthscan.

Siti-Nabiha, A.K., Abdul Wahid, N., Amran, A., Che Haat, H., & Abustan, I. (2009). Towards a sustainable tourism management in Malaysia. *Journal of Business*, *4*, 301–312.

Stavros, J., Cole, M.L., & Hitchcock, J.A. (2014). Appreciative inquiry research review & notes: A research review of SOAR. *AI Practitioner*, *16*(3), 1–7.

The Mountain Institute (2000). *Community-Based Tourism for Conservation and Development: A Resources Kit*. Washington, DC: The Mountain Institute.

Tsonis, J.L., & Cheuk, S. (2000). Sustainability challenges: Changing attitudes and a demand for better management of the tourism industry in Malaysia. In C. Ghenai (ed.), *Sustainable Development – Policy and Urban Development – Tourism, Life Science, Management and Environment* (pp. 177–200). Winchester: InTech.

Walker, K. (2008). Tools to enhance community capacity to critically evaluate tourism activities. In G. Moscardo (ed.), *Building Community Capacity for Tourism Development* (pp. 86–100). Wallingford: CABI.

World Commission on Environment and Development (1987). *Report of the World Commission on Environment and Development: Our Common Future*. Oxford: Oxford University Press.

Weaver, D. (2006). Issues in sustainable tourism. In D. Weaver, *Sustainable Tourism: Theory and Practice* (pp. 18–38). London: Butterworth Heinemann.

Whitney, D., & Trosten-Bloom, A. (2010). *The Power of Appreciative Inquiry: A Practical Guide to Positive Change* (2nd edn). San Francisco: Berrett-Koehler Publishers, Inc.

21. Tourism governance for rural community well-being: challenges and creative opportunities

Gianna Moscardo

INTRODUCTION

The contrast between the core and the periphery is an important one in many social science disciplines, including economics (Althouse et al., 2020), geography (Eder, 2018), and sociology (Nordlund, 2018). Peripheral regions are defined by their distance, including geographical distance, physical access, social distance and political distance, from some centre. Peripheral regions are those that are further away from the mainstream, with less power, and fewer resources and opportunities. Rural communities are often located in such peripheral places. A common theme in discussions of this core–periphery model is the need to find different pathways to economic development and to improve the well-being of the communities in peripheral regions. Rural areas are, by definition, peripheral regions with the term rural commonly defined as being non-urban, with low population densities (Hawley et al., 2016). Tourism has long been used as a policy tool or strategy in these contexts (Lun et al., 2016). Tourism in these areas is typically based on access to natural or cultural heritage and scenery, and traditional agricultural lifestyles, or focused on relaxation and escape, with links to community-based tourism (Zielinski et al., 2020).

Despite a long history of using tourism for rural community development, there is little clear and consistent evidence that tourism is a successful development strategy for these places (Thees et al., 2020). This chapter argues that there is considerable research describing the barriers and challenges to the effective use of tourism as a rural development strategy, but there have been only limited attempts to explore the underlying conditions that create these barriers and challenges and only limited suggestions for ways to tackle these problems. This chapter further argues that there are two related underlying conditions that need to be considered. The first is a tendency to use business strategy planning rather than community development planning as the default approach to tourism governance, resulting in a consistent perception of rural communities as resources for global tourism rather than tourism as a global resource for specific rural communities. The second is a failure to empower and engage rural community residents in the planning, management and practice of tourism. Awareness of these two underlying conditions leads to new directions for tourism governance based on focusing on destination community well-being (DCW) and harnessing the power of stories. The chapter will describe these new directions for improved governance in rural tourism using a variety of case studies of tourism in rural destinations.

CHALLENGES FOR USING TOURISM AS A DEVELOPMENT TOOL

In 2011 Moscardo (2011a) reviewed 100 case studies of tourism used as a rural community development strategy and identified 12 barriers to the effective development of tourism in peripheral regions. These barriers have been confirmed in subsequent reviews (Ghaderi & Henderson, 2012; Lenao, 2015; Lun et al., 2016; Keyim, 2018; N'Drower et al., 2021). These barriers have also been identified for community-based tourism (CBT), regardless of the nature of the destination (Dodds et al., 2018; N'Drower et al., 2021). Box 21.1 lists both Moscardo's (2011a) barriers and others identified in subsequent papers. Moscardo (2011a) argued that these barriers were interconnected and offered a framework describing these interconnections. In this framework she argued that two core barriers combine to create the other barriers that ultimately lead to either failed tourism or tourism that is not sustainable. The two core barriers were a lack of destination community stakeholder knowledge of tourism as a complex system and the dominance of external agents in the governance of tourism. Together these core barriers resulted in a lack of local tourism leadership and poor coordination of community stakeholders, which in turn contributed to community conflict over tourism, limited or no formal tourism planning, and limited or no local destination community involvement in tourism development. The lack of destination community involvement in tourism then ultimately resulted in poor outcomes. Moscardo (2011a) concluded the review by suggesting that the answer lies in developing the tourism knowledge of destination communities.

While this chapter agrees that supporting improved tourism knowledge amongst destination community stakeholders is an important element of more successful sustainable tourism development in rural regions, it proposes that the problem is deeper than these two core barriers. More specifically, this chapter argues that tourism governance is typically driven by business strategic planning approaches, which assume that financially successful tourism businesses are the ultimate goal of tourism governance. This approach disempowers destination community residents who are not seen as having expertise relevant to tourism business success. This chapter further argues that for tourism to be an effective rural community development strategy it must instead be based on a community development planning approach that explicitly seeks to empower rural communities and assumes that the ultimate goal of tourism governance is the improvement of DCW.

BOX 21.1 COMMONLY IDENTIFIED BARRIERS TO THE EFFECTIVE DEVELOPMENT OF TOURISM IN PERIPHERAL REGIONS

- Reliance on external agents for tourist market information and access.
- Dominance of external agents in tourism planning and development decisions.
- Limited or no formal tourism planning.
- Limited community awareness of tourism processes, especially negative tourism impacts.
- Unrealistic expectation of tourism benefits.
- Lack of local tourism leadership.
- Lack of local community coordination.
- Limited local community participation/empowerment in tourism governance.
- Limited connections to tourism distribution systems.
- Poor infrastructure development.
- Lack of skills and capital within the destination community.
- Conflict within the community over tourism development.
- Limited innovation in tourism products.
- Poor integration of tourism with other sectors.
- Limited local experience with ICT-enabled communication.
- Lack of local entrepreneurial ability.
- Use of culturally inappropriate methods for gaining local knowledge and participation.

Sources: Ghaderi & Henderson (2012); Moscardo (2011a); Lenao (2015); Lun et al. (2016); Keyim (2018); N'Drower et al. (2021).

ALTERNATIVE APPROACHES TO TOURISM PLANNING AND DEVELOPMENT IN RURAL COMMUNITIES

The lack of destination community or resident empowerment in tourism governance is a common theme in the literature (cf. Carter et al., 2015; Farmaki, 2015; Moscardo, 2019; Thees et al., 2020). While much of this discussion then focuses on mechanisms for enhancing destination community or resident involvement in tourism governance, few have considered why this might be case. Using a review of social representations of tourism in planning, Moscardo (2011b) argues that it is because tourism planning is driven by ideas borrowed from business strategy rather than from community development planning or the broader planning literature in general. This results in what Saarinen et al. (2017) refer to as the boosterism approach. In this approach planning is driven by the need to increase tourist numbers in a destination and to ensure that tourism businesses are profitable. Such a focus ignores whether these tourists or businesses provide any net benefits to the destination residents. The alternative is to examine the community development and broader planning literature for better approaches.

An examination of this broader literature on planning confirms the importance of citizen empowerment in and through planning processes (Green & Haines, 2015; Morse, 2014; Moscardo, 2019; Oliver & Pitt, 2013). This examination also reveals three major themes in contemporary planning approaches:

1. more explicit recognition of community well-being beyond economic factors as the primary goal of planning (Connelly et al., 2016; Green & Haines, 2015; Healey, 2016; Moscardo, 2019; Scott, 2019);
2. recognition of the need to move from technical, positivist, planner as expert and leader approaches to more innovative, creative, post-positivist, planner as facilitator approaches (Drescher et al., 2019; Green & Haines, 2015; Groves et al., 2013; Morse, 2014; Moscardo, 2019); and
3. greater awareness of the cultural dimensions of planning (Green & Haines, 2015; Heiner et al., 2019; Hibbard, 2016; McMillan et al., 2016; Moscardo, 2019).

Community planning frameworks that explicitly use community well-being or quality of life (QoL) include the Community Capitals Framework (Emery & Flora, 2006; Flora et al., 2004; Flora et al., 2016) and various approaches to sustainable livelihoods (Dent et al., 2013; Serrat, 2017). These frameworks argue that community well-being depends upon the stock of different forms of capital (natural, social, human, financial, built, cultural, and political) that exists within each community. The Community Capitals Framework for development and planning argues that planners need to identify strategies to enhance all these different capitals (Emery & Flora, 2006; Flora et al., 2004; Flora et al., 2016). Sustainable livelihoods frameworks take this further, arguing that the best development strategies are those make use of existing capital to generate more capital in areas that communities aspire to improve (Dent et al., 2013; Serrat, 2017).

The concept of destination community well-being (DCW) or destination community quality of life (QoL) has emerged in recent years as an important element in discussions of tourism impacts on destinations (cf. reviews by Hartwell et al., 2018; Uysal et al., 2016). Hartwell and colleagues (2018) note that research into resident perceptions of tourism impacts has begun to either replace overall attitudes towards tourism with some measure of well-being or QoL as the overall indicator of resident acceptance of tourism, or to place well-being/QoL as the mediator between perceived impacts and overall attitudes towards tourism. Most of these studies are, however, very descriptive, listing tourism impacts rather than analysing how tourism results in those impacts. Three recent studies have begun to apply systems thinking to move towards understanding the processes that link tourism to its impacts of DCW and thus offer some guidance for tourism managers (Konovalov et al., 2019; Moscardo et al., 2017; Ridderstaat et al., 2014).

Despite this growing awareness of DCW in discussions of tourism impacts and the examples of planning frameworks based on community well-being, there is little discussion of DCW in papers on destination governance. One exception has been the work of Moscardo and Murphy (2014), which builds upon the Community Capitals Framework to suggest a tourism planning process focused on finding forms of tourism that can directly enhance the various capitals that support DCW. Figure 21.1 sets out this DCW tourism planning process and presents planning as a circular rather than a linear process, centred on the destination community rather than on objectives linked to tourism. The process does not include any components linked to tourism business-related goals or to audits of the destination resources available for tourism, which are typically included in traditional tourism planning processes. This DCW tourism planning approach incorporates elements from Flora's capitals framework for community development (Emery & Flora, 2006; Flora et al., 2004; Flora et al., 2016) and ideas about Living Labs, where destination residents develop, implement and then evaluate tourism opportunities (Thees et al., 2020). In Figure 21.1, after developing resident capacity to control and manage tourism planning, auditing

Source: Adapted from Moscardo & Murphy (2014).

Figure 21.1 *Destination community well-being approach to tourism planning*

current stocks of the various types of community capital and identifying types of capital that residents see as either in deficit or as aspirational goals, this tourism planning approach then seeks to co-create ideas for new tourism ventures or strategies. The resulting future tourism scenarios generated by the destination residents are assessed for viability including market acceptance and sustainability. Those that are both desired by the destination residents and judged to be viable and sustainable are then the ones that are included in the tourism planning processes for implementation.

The development of future tourism scenarios is an important element of this DCW tourism planning process. Future scenarios are types of stories that can be used to express both fears and aspirations and to encourage creativity and innovation (Burnam-Fink, 2015). Future scenarios are especially useful in helping people to think beyond their traditional comfort zone (Amer et al., 2012). The use of these stories of future tourism is also consistent with the use of more creative approaches in planning in general, including story creation, future scenarios development and drama approaches to planning (Drescher et al., 2019; Moscardo, 2019).

This move away from assumptions that technical positivist approaches are the most appropriate to approaches that are more flexible and based on post-positivist thinking is not unique to planning. It is both part of a renewed focus on empowering citizens in more ethical frameworks and recognizes the importance of stories in all aspects of human thought and cognition. This is part of the story turn in social sciences (Gottschall, 2012). Moscardo (2020, p. 1) defines a story as

> a very specific type of narrative … designed explicitly to entertain, engage emotions and change their audience, … that … must include descriptions not just of a set of causal chains of events, but also … of characters, how they respond to these events and the consequences of these reactions.

This story turn, both in and beyond planning, recognizes the importance of stories across cultures and their central role in how many traditional and First Nations communities communicate and collectively make decisions (cf. Fox et al., 2017). The use of stories in tourism planning recognizes the importance of finding ways for all destination communities to be able to engage in tourism governance.

Engaging in tourism governance involves two main processes – engagement in planning processes and decisions and in ongoing tourism management through direct participation in tourism businesses, practices and/or experiences. The following sections demonstrate the use and benefits of a DCW approach to tourism planning and how story-based tourist opportunity design can include destination residents in ongoing tourism management and practices. These sections present case studies of different forms of tourism in a variety of rural communities.

PLANNING TOURISM FOCUSED ON DESTINATION COMMUNITY WELL-BEING

A number of papers have attempted to link tourism to both negative and positive impacts on rural destination communities. Naidoo and Sharpley (2016), for example, found that rural residents in Mauritius saw agritourism as offering more for community well-being than enclave coastal resort tourism. This was especially the case for supporting local culture, improving local educational opportunities and enhancing local livelihood diversification. While this study in Mauritius compared two different forms of tourism, another, focused on rural regions in northern Australia, compared different types of tourists and their impacts on DCW (Moscardo et al., 2013). This paper found a variety of links between different patterns of tourist mobility in the regions and impacts on the different capitals that make up DCW. Essentially, longer-stay slow tourists made more positive contributions across all the capitals than did short-stay fast tourists. While these studies provide some insights into the links between different characteristics of tourism and impacts on the various capitals that make up DCW, there have been almost no published attempts to connect DCW to tourism planning. Table 21.1 provides a summary overview of the five rural community cases of taking a DCW and community capitals approach to tourism planning, as set out Figure 21.1.

Although four of the five rural regions included in these cases are in Australia, they cover a range of different types of rural community with different current stocks of community capitals, and with different levels of experience with tourism and different existing styles and scales of tourism. In all the cases, levels of government beyond the local or regional saw tourism as an important rural community development strategy. These government agencies had a consistent reliance on external experts to develop tourism plans and a strong focus on dominant international markets and large-scale, externally funded and managed resorts as the most commonly suggested strategy for attracting tourists. Across all the locations, rural residents displayed a cautious optimism about tourism in general but lacked enthusiasm for the large-scale resort options. In all cases the outcomes of the first major steps in the DCW approach to tourism planning were presented to the formal government agencies responsible for tourism development policy and management. In some cases, the scenarios were taken further for implementation both by relevant government agencies but also by the residents themselves, but the onset of COVID-19 has limited both this process and any moves to conduct evaluations. In other cases, government agencies were reluctant to abandon existing traditional tourism planning approaches and to pursue more resident-centric options. The cases with slower change tended to be those where local governments were less engaged in tourism governance, a trend noted in other aspects of government planning and policy processes (Amundsen et al., 2018).

Compared to the traditional tourism planning approach, outcomes taking a DCW approach centred on destination residents resulted in both a wider range of more innovative tourism options and much more detailed and clearer pathways linking the tourism to specific benefits to the capitals that make up DCW. The Magnetic Island

Table 21.1 *Summary of three cases of a DCW approach to tourism planning*

Case	Location	Main tourism forms	Steps in DCW approach	Further information
Magnetic Island, Australia	Small island community off north-east Australian Coast	Long-standing, mostly small-scale tourism Marine tourism Access to protected natural environments Wildlife tourism Relaxation/escape	DCW audit Tourism scenarios generated Scenarios assessed Some implemented	Moscardo & Murphy, 2014; 2015
Can Tho, Vietnam	Rural villages in the southern Mekong Delta	Emerging tourism Village homestays Agritourism Cultural experiences	DCW audit Tourism scenarios generated	Murphy et al., 2014
Atherton Tablelands, Australia	Agricultural hinterland region in Northern Australia adjacent to major coastal tourism resort	Long-standing but small-scale tourism Food tourism Access to protected natural environments Wildlife tourism Relaxation/escape	DCW audit Tourism scenarios generated Some implemented	Konovalov et al., 2019; Konovalov et al., 2021
Bowen, Australia	Agricultural and mining coastal region in Northern Australia	Emerging tourism focused on marine activities		
Whitsundays, Australia	Agricultural region that includes a major coastal resort in Northern Australia	Established large-scale coastal resort tourism focused on one location with limited tourism elsewhere		

community, for example, focused on educational tourism, which provided visitors with access to the various protected natural environments linked to the destination rather than the larger-scale coastal resort style of tourism advocated in traditional plans. The community in this destination provided considerable detail on how this type of tourism would contribute to multiple forms of capital through the use of locally owned, smaller-scale tour operators and accommodation and other service providers. The community specifically chose a type of tourism that relied on smaller businesses as they recognized that this would provide more direct economic and local employment benefits with minimal leakage. It would also enhance actions linked to improving the sustainability of the residents' lifestyles by creating positive support for local initiatives. It would also keep the level of tourism within the capacity of the local infrastructure and limit negative impacts on residents' lifestyles.

The community participants in the Can Tho DCW planning approach did not have the choice of style of tourism as the national government had already determined that village-based agritourism was the preferred development option. Within this limitation the residents were, however, still able to articulate ways in which the tourist experience opportunities could be adapted to enhance the well-being of local communities. Of particular importance to them was the inclusion of cultural demonstrations for visitors, engagement of visitors in agricultural activities and a programme for developing cooperative linkages between villages. The arguments for these strategies included avoiding direct competition between villages, making the best use of local social and cultural capital, and offering more varied packages to encourage visitors to stay in the region longer. In the Atherton Tablelands the residents focused on family activities and wellness tourism, which again offered a very different approach to the larger-scale resorts identified in the traditional planning approaches. These two options were chosen because they simultaneously made the most of the environmental and social resources already within the community and they potentially contributed to increasing the number of amenity migrants, especially families, who would both support the maintenance of services such as schools and medical facilities and increase human and financial capital in the region. The Bowen group, with the lowest level of tourism, smallest population and least amount of existing infrastructure, saw food-themed tourism and events as the best way forward as both options could be started without significant investment, based on day trips and short stays from the adjacent established tourism destination. They were especially focused on identifying alternative sources for funding, such as corporate sponsorship from mining and resources businesses and crowdsourced funding from tourists, to develop local events that could attract tourists to stay longer and offer opportunities for local social capital building.

Common to all the case study rural communities was an informal sustainable livelihoods approach. The residents involved in the DCW planning were able to identify more and different resources to support development than are typically investigated in traditional tourism planning approaches. They built very clear, direct, and sophisticated linkages between the types of tourism they saw in their tourism futures and how these could be managed to enhance multiple forms of DCW capital. They identified tourism success as its contributions to their community well-being needs and aspirations rather than in terms of tourism numbers or profits. None of them favoured tourism becoming a dominant economic activity. These cases challenge dominant narratives about rural communities as being in some sort of deficit state, tourism success being about numbers of tourists, and rural communities being less able than "external experts" to plan for and manage tourism.

CASE STUDIES OF STORY-BASED RURAL COMMUNITY ENGAGEMENT IN TOURISM PRACTICE

Two core features of the suggested tourism future stories generated in the DCW approach to tourism planning were the focus on tourist experience opportunities that used or were supported by smaller locally owned businesses or by cooperative local networks, and the importance of connecting tourists to local stories. These stories could be those of the locals involved in the tourism venture, the stories of the places being visited, or co-created stories. The DCW planning cases generated a wide range of different types of tourism but three were common to all the different rural regions in the cases – providing access to natural heritage in the rural region, food tourism and cultural tourism. The following sections provide more specific tourism examples from the cases in the previous sections.

Providing Access to Natural Heritage

Aquascene (see https://www.aquascenecharters.com.au/the-aquascene-story.html for more information) is an example of a small, family owned, locally based tour operator on Magnetic Island that offers tourists unique access to natural heritage in the region, with a strong educational component. The business runs a range of different boat-based tours around the Island and the immediate vicinity, giving visitors access to bays that tourists could not otherwise visit. The business focuses on telling tourists two types of story – the story of the owners and their family and the stories of the natural environments they visit. The stories of the natural environment are a traditional use of stories in tourist experiences, used in tourist interpretation in many places. Telling stories about the tour operator is not uncommon, but it is increasingly important as tourists seek to engage in the local lifestyle. The business has won multiple awards for customer service, environmental sustainability and conservation education and action. Tourist reviews highlight the importance of being on a small tour with knowledgeable locals telling stories of the place:

> We had an amazing day out on the water with Aquascene. Adam and the team took us on a great trip around the island and told us great tales about the island and history. The visibility for snorkelling wasn't fabulous (nothing to do with the tour) but because we weren't with a big company, Adam was able to plan the trip to get to the best part of the island based on the conditions.

Cultural Tourism

In Can Tho, the Fabulous Mekong Eco-Tours (https://fabulousmekong.com/) is another small, locally owned tour business that provides tourists with access to rural village experiences. In this case their three-day cultural tour is one of the most highly rated on various online booking sites. It includes visits to the typical attractions such as floating markets and a bird sanctuary, as well as a homestay. Visitors are encouraged to engage in local cultural activities such as weaving, dancing and food preparation and they share a meal with their homestay hosts. This tour, while focused on telling tourists the stories of the local people and their culture, also offers an opportunity for tourists to develop their own story. Tourist comments highlighted the importance of personal involvement and achievement:

> Having found these girls on social media it was great to meet them! We so enjoyed this complete experience – combining the tours with staying at Lily's Homestay, helped us feel immersed in the Mekong and we learned so much. Lily, Kim and Tran make the day out fun. You feel involved with the activities and genuinely part of the local cultural experience.

Food Tourism

Food-based tours have always been a central part of tourism to the Atherton Tablelands, reflecting its long history of agriculture in general and its recent history of specialist or niche agricultural ventures. The region offers both a self-drive food trail (see https://www.athertontablelands.com.au/travel-directory/taste-paradise-food-trail/), which links tourists to specific locally owned food producers, and several small, organized food-themed tours. As with previous cases, the focus is on telling the stories of the rural community and local food producers – "the owner, has been running foodie tours for 13 years so he knows the best places to visit and has a wealth of stories to tell about the region as you eat and drink your way around the Tablelands".

CO-CREATING STORIES AND MORE EXPLICIT STORY-BASED TOURIST EXPERIENCE OPPORTUNITIES

Each of the existing cases of successful rural community tourism described in the previous sections was focused on storytelling to the tourists, with stories being mostly about the destination and/or the tourism businesses or tour guides. Evidence from other rural regions suggests that more explicitly building tourist experiences around local stories can be effective at providing residents with greater involvement in tourism governance and enhances the spread of benefits in the rural community (Hartman et al., 2019; Idziak et al., 2015; Moscardo, 2018). Hartman and colleagues (2019) describe the development of story-themed tourist experiences, including themed tours to different locations in rural regions in the Netherlands. Idziak and colleagues (2015) considered the development of themed villages in Poland that have been created for tourists and often focused on a specific story, such as Tolkien's novels. Moscardo (2018) described several story-based tourist experiences built on mobile dramas offered in parts of Scandinavia. In all cases tourists get to experience and co-create a story by moving through the rural region and interacting with locals in different capacities. Locals in turn get to engage directly with tourists and are more involved in the provision of tourism services. This builds connections, and local residents gain direct benefits from tourism. In all cases tourism was more directly linked to positive impacts on different forms of DCW capital, and locals played a much greater role in tourism governance. Other examples of this more explicit story-based tourism products can be found in Croatia (https://www.istrainspirit.hr/en/project/) and in a series of family holidays inspired by children's books (https://www.blacktomato.com/story/).

CONCLUSIONS AND FUTURE DIRECTIONS

In 2011, the author conducted a review of 100 studies of rural communities using tourism as a development strategy and concluded that the key to improving rural community outcomes was to develop residents' knowledge of tourism as a complex system (Moscardo, 2011a). Subsequent research and engagement with tourism in different rural communities have moved me to a different position. Possibly I was wrong in my original conclusion or possibly a further decade of government usage of tourism as a rural community development strategy, massive growth in tourism and the rise of the internet and social media have changed residents' knowledge about tourism. Regardless, this chapter argues that barriers to the effectiveness of tourism as a rural community development option stem from the lack of effective participation by rural destination residents in tourism governance. This chapter has argued that the continued use of business strategy planning in tourism, which assumes rural communities lack expertise in tourism and which focuses on the success of tourism businesses and increases in tourism numbers as the goal of planning, has failed to overcome many of the barriers to sustainable tourism in rural communities. The chapter offers an alternative approach to tourism planning based on recent themes in the community development planning literature that highlight the need to think about the various capitals that make up DCW and that focus on the stories that destination residents can generate as to what they would like future tourism to be. The cases studies of this DCW approach to tourism planning revealed that residents have complex and sophisticated models of tourism and generate more innovative tourism proposals than external experts following more traditional tourism planning processes. Placing local stories at the centre of tourism also offers opportunities for greater involvement of residents in tourism, giving them both more direct benefits and also enhancing their opportunities to engage in ongoing tourism management governance.

REFERENCES

Althouse, J., Guarini, G., & Porcile, J.G. (2020). Ecological macroeconomics in the open economy. *Ecological Economics, 172*. https://doi.org/10.1016/j.ecolecon.2020.106628.
Amer, M., Daim, T., & Jetter, A. (2013). A review of scenario planning. *Futures, 46*, 23–40.
Amundsen, H., Hovelsrud, G.K., Aall, C., Karlsson, M., & Westskog, H. (2018). Local governments as drivers for societal transformation. *Current Opinion in Environmental Sustainability, 31*, 23–29.
Burnam-Fink, M. (2015). Creating narrative scenarios. *Futures, 70*, 48–55.
Carter, R.W., Thok, S., O'Rourke, V., & Pearce, T. (2015). Sustainable tourism and its use as a development strategy in Cambodia. *Journal of Sustainable Tourism, 23*(5), 797–818.
Connelly, S., Green, K., Markey, S., & Roseland, M. (2016). Peaks and valleys on the prairies. In L. Hallstrom, M. Beckie, G. Hvenegaard, & M. Mundel (eds), *Sustainability Planning and Collaboration in Rural Canada* (pp. 64–78). Edmonton: University of Alberta Press.
Dent, D., Dubois, O., & Dalal-Clayton, B. (2013). *Rural Planning in Developing Countries: Supporting Natural Resource Management and Sustainable Livelihoods*. Abingdon: Routledge.
Dodds, R., Ali, A., & Galaski, K. (2018). Mobilizing knowledge. *Current Issues in Tourism, 21*(13), 1547–1568.
Drescher, M., Feick, R., DeGeer, C., & Shipley, R. (2019). Participatory methods for identifying cultural heritage landscapes. In M. Scott, N. Gallent, & M. Gkartzios (eds), *The Routledge Companion to Rural Planning* (pp. 446–457). Abingdon: Routledge.
Eder, J. (2019). Innovation in the periphery. *International Regional Science Review, 42*(2), 119–146.
Emery, M., & Flora, C. (2006), Spiraling-up: Mapping community transformation with community capitals framework, *Community Development, 37*(1), 19–35.
Farmaki, A. (2015). Regional network governance and sustainable tourism. *Tourism Geographies, 17*(3), 385–407.
Flora, C.B., Flora, J.L., & Fey, S. (2004). *Rural Communities: Legacy and Change*. Boulder, CO: Westview Press.
Flora, C.B., Flora, J.L., & Gasteyer, S. (2016). *Rural Communities* (5th edn). Boulder, CO: Westview Press.
Fox, C.A., Reo, N.J., Turner, D.A., Cook, J., Dituri, F., Fessell, B., et al. (2017). "The river is us; the river is in our veins": Re-defining river restoration in three Indigenous communities. *Sustainability Science, 12*(4), 521–533.
Ghaderi, Z., & Henderson, J.C. (2012). Sustainable rural tourism in Iran. *Tourism Management Perspectives, 2*, 47–54.
Gottschall, J. (2012). *The Storytelling Animal*. Boston, MA: Houghton Mifflin Harcourt.
Green, G.P., & Haines, A. (2015). *Asset Building and Community Development* (4th edn). Los Angeles, CA: Sage.
Groves, C., Munday, M., & Yakovleva, N. (2013). Fighting the pipe. *Environment & Planning C, 31*, 340–356.
Hartman, S., Parra, C., & de Roo, G. (2019). Framing strategic storytelling in the context of transition management to stimulate tourism destination development. *Tourism Management, 75*, 90–98.
Hartwell, H., Fyall, A., Willis, C., Page, S., Ladkin, A., & Hemingway, A. (2018). Progress in tourism and destination wellbeing research. *Current Issues in Tourism, 21*(16), 1830–1892.
Hawley, L., Koziol, N., Bovaird, J., McCormick, C., Welch, G.W., Arthur, A.M., & Bash, K. (2016). Defining and describing rural. *Rural Special Education Quarterly, 35*(3), 3–11.
Healey, P. (2016). The planning project. In S. Fainstein & J. DeFilippis (eds), *Readings in Planning Theory* (4th edn, pp. 139–155). Oxford: Blackwell.

Heiner, M., Hinchley, D., Fitzsimons, J., Weisenberger, F., Bergmann, W., McMahon, T., et al. (2019). Moving from reactive to proactive development planning to conserve Indigenous community and biodiversity values. *Environmental Impact Assessment Review*, *74*, 1–13.

Hibbard, M. (2016). Development planning with cultural integrity: Self-determination. *Journal of Planning Education and Research*, *36*(2), 158–166.

Idziak, W., Majewski, J., & Zmyślony, P. (2015). Community participation in sustainable rural tourism experience creation: A long-term appraisal and lessons from a thematic villages project in Poland. *Journal of Sustainable Tourism*, *23*(8–9), 1341–1362.

Keyim, P. (2018). Tourism collaborative governance and rural community development in Finland. *Journal of Travel Research*, *57*(4), 483–494.

Konovalov, E., Murphy, L., & Moscardo, G. (2019). An exploration of links between levels of tourism development and impacts on the social facet of residents' quality of life. In A.M. Campón-Cerro, J.M. Hernández-Mogollón, & J.A. Folgado-Fernández (eds), *Best Practices in Hospitality and Tourism Marketing and Management* (pp. 77–107). Cham: Springer.

Konovalov, E., Moscardo, G., & Murphy, L. (2021). Transforming tourism governance. In N. Pappas & A. Farmaki (eds), *Tourism Dynamics* (pp. 96–114). Oxford: Goodfellow Publishers.

Lenao, M. (2015). Challenges facing community-based cultural tourism development at Lekhubu Island, Botswana. *Current Issues in Tourism*, *18*(6), 579–594.

Lun, L.M., Pechlaner, H., & Volgger, M. (2016). Rural tourism development in mountain regions. *Journal of Quality Assurance in Hospitality & Tourism*, *17*(4), 389–411.

McMillan, L., Prosper, K., Moffitt, M., & Davis, A. (2016). Netukulimk narratives. In L. Hallstrom, M. Beckie, G. Hvenegaard, & M. Mundel (eds), *Sustainability Planning and Collaboration in Rural Canada* (pp. 188-207). Edmonton: University of Alberta Press.

Morse, S. (2014). *Smart Communities*, 2nd edn. San Francisco: Jossey-Bass.

Moscardo, G. (2011a). The role of knowledge in good governance for tourism. In E. Laws, H. Richins, J. Agrusa, & N. Scott (eds), *Tourist Destination Governance* (pp. 67–80). Wallingford: CABI.

Moscardo, G. (2011b). Exploring social representations of tourism planning. *Journal of Sustainable Tourism*, *19*(4–5), 423–436.

Moscardo, G. (2018). Tourist experience design: A storytelling framework. In L.P. Cai & P. Alaedini (eds), *Quality Services and Experiences in Hospitality and Tourism* (pp. 93–107). Bingley, UK: Emerald.

Moscardo, G. (2019). Rethinking the role and practice of destination community involvement in tourism planning. In K. Andriotis, D. Stylidis, & A. Weidenfeld (eds), *Tourism Policy and Planning Implementation* (pp. 36–52). Abingdon: Routledge.

Moscardo, G. (2020). The story turn in tourism. *Journal of Tourism Futures*, *7*(2), 168–173. https://doi.org/10.1108/JTF-11-2019-0131.

Moscardo, G., & Murphy, L. (2014). There is no such thing as sustainable tourism. *Sustainability*, *6*(5), 2538–2561.

Moscardo, Gianna, & Murphy, Laurie. (2015). Educating destination communities for sustainability in tourism. In G. Moscardo & P. Benckendorff (eds), *Education for Sustainability in Tourism: A Handbook of Processes, Resources, and Strategies* (pp. 135–154). Berlin: Springer-Verlag.

Moscardo, G., Konovalov, E., Murphy, L., & McGehee, N. (2013). Mobilities, community well-being and sustainable tourism. *Journal of Sustainable Tourism*, *21*(4), 532–556.

Moscardo, G., Konovalov, E., Murphy, L., McGehee, N.G., & Schurmann, A. (2017). Linking tourism to social capital in destination communities. *Journal of Destination Marketing & Management*, *6*(4), 286–295.

Murphy, L., Moscardo, G., & Blackman, A. (2014). Tropical communities as resources for tourism or tourism as a resource for tropical communities. In T.R. Tiller (ed.), *Proceedings of BEST EN Think Tank XIV: Politics, Policy and Governance in Sustainable Tourism* (pp. 232–250). Sydney: University of Technology Sydney.

Naidoo, P., & Sharpley, R. (2016). Local perceptions of the relative contributions of enclave tourism and agritourism to community well-being. *Journal of Destination Marketing & Management*, *5*(1), 16–25.

N'Drower, F., Moscardo, G., & Murphy, L. (2021). "Tourism brings good things": Tourism and community development in rural Papua New Guinea. *Tourism Review International*, *25*, 229–245.

Nordlund, C. (2018). Power-relational core–periphery structures. *Network Science*, *6*(3), 348–369.

Oliver, B., & Pitt, B. (2013). *Engaging Communities and Service Users*. Basingstoke: Palgrave Macmillan.

Ridderstaat, J., Croes, R., & Nijkamp, P. (2016). The tourism development–quality of life nexus in a small island destination. *Journal of Travel Research*, *55*(1), 79–94.

Saarinen, J., Rogerson, C.M., & Hall, C.M. (2017). Geographies of tourism development and planning. *Tourism Geographies*, *19*(3), 307–317.

Scott, M. (2019). Spatial planning and the rural economy. In M. Scott, N. Gallent & M. Gkartzios (eds), *The Routledge Companion to Rural Planning* (pp. 219–236). Abingdon: Routledge.

Serrat, O. (2017). *Knowledge Solutions*. Singapore: Springer.

Thees, H., Pechlaner, H., Olbrich, N., & Schuhbert, A. (2020). The Living Lab as a tool to promote residents' participation in destination governance. *Sustainability*, *12*(3), 1120. https://doi:10.3390/su12031120.

Uysal, M., Sirgy, M.J., Woo, E., & Kim, H.L. (2016). Quality of life (QOL) and well-being research in tourism. *Tourism Management*, *53*, 244–261.

Zielinski, S., Jeong, Y., Kim, S.I., & Milanés, C.B. (2020). Why community-based tourism and rural tourism in developing and developed nations are treated differently? *Sustainability*, *12*(15), 5938. https://doi.org/10.3390/su12155938.

22. "Salvaging" presence: tourism as regional development strategies for rural communities

Dominic Lapointe

There is a strong advocacy for tourism as a regional development strategy. According to its champions, tourism, with its low entry cost, its high labor needs and its flow of "new" currencies, can turn a depressed restructuring economy into a vibrant striving community. While tourism development still maintains that mantra in a growth-oriented strategy (Brouder, 2018), the criticism and doubts are plentiful (Müller & Jansson, 2007; Marcotte et al., 2020) even when it is considered successful (Lebon & Lapointe, 2018). This contradiction is mostly evident due to an ideological perspective of development, an industrial view of tourism in which tourism is a product on sale, a service to consume, an economic strategy to grow. This industrial view of tourism shadows the understanding of how tourism works as a regional development strategy. This chapter aims to deconstruct this ideological perspective and will do so by positioning regional development and tourism within the structure of geographically unequal development (Harvey, 2006) and exploring how it is part of the mechanism of salvage capitalism (Tsing, 2015). Afterwards, moving away from the industrial understanding of tourism, I will explore tourism as an economy of mobility through the *économie résidentielle* (Terrier, 2009; Pecqueur & Tallandier, 2011). This strand of research is based on the transformation of part of the regional economy from a value creation structure to a value salvaging structure, through the consumption activities of the mobile population. I will conclude with a discussion of what this change of lens involves regarding understanding the place of tourism for rural communities and the reweaving of tourism in a world shaped by COVID-19.

REGIONAL DEVELOPMENT OR THE UNEQUAL DISTRIBUTION OF DEVELOPMENT

Regional development, as a concept and a practice, is an acknowledgment that development is unequally distributed across space, but also across class, race, gender, and so forth. In the 1970s, the emergence of theories of dependency (Cardoso, 1970; 2009) and unequal development (S. Amin, 1973; Emmanuel, 1969) exposed these contradictions. Thus, according to these theories, the economic growth of peripheral areas first and foremost benefits the center, and, as such the growth of developing countries benefits developed countries (S. Amin, 1973). Dependency theory postulates that peripheral economies are dependent on the capital and markets of the center (Cardoso, 1970). It is this dependence, and not local culture, tradition and institutions, that hinders development (Frank, 1969). The periphery plays the roles of a supplier of raw materials, as well as the market for the manufactured products of the center (S. Amin, 1973). In turn, manufactured products sold to the periphery undermine the local economy, which is not able to compete against the products of the center (Cardoso, 1970; Frank, 1969). It is in this circular production process that a so-called unequal exchange is created, where surplus value is appropriated by the economies of the center, which thereby find themselves maintaining their technological and economic advantages over the periphery. This can therefore create a cultural and economic structure characterized by an unequal relationship between developed and dependent areas. As Cardoso makes clear,

> the structuring model that is emerging is one which is made up of highly integrated and structured sectors capable of defining themselves and defending their interests, and of a large periphery which is becoming more numerous, but which is socially and politically amorphous. (Cardoso, 1970, p. 314)

The economic, political and technological disparities created by this unequal structure are also the source of social inequalities: differences in income, standard of living, levels of investment, access to education, health care, etc. (S. Amin, 1973). If this so-called dependent or unequal development is mainly expressed in terms of differences between developed and underdeveloped countries, countries of the Global South and countries of the Global North, similar center/periphery relationships exist within nation states, whether they are in developed countries or not. Paradoxically, development creates these gaps and inequalities by concentrating wealth and productive assets in certain poles (Polèse, 1994).

These inequalities are constitutive of the capitalist mode of accumulation and at the source of a territorial differentiation between center and periphery (Klein, 2008; Harvey, 2006). In his "Notes towards a theory of uneven geographical development", David Harvey (2006) illustrates these spatial inequalities of development through four conditionalities: (1) the material embedding of capital accumulation processes in the web of socio-ecological life; (2) the accumulation by dispossession; (3) the law-like character of capital accumulation in space and time; and finally, (4) political, social and class struggles at a variety of scales (Harvey, 2006, p. 60). In this theory, Harvey (2006) exposes how capitalism engages non-capitalistic processes, like socio-environmental processes, in the merchant exchange through commodification and privatization. Development, in this theory, is associated with the capacity of capital accumulation in some areas versus others, and the process of resistance by different social groups.

DESTINATION AS A REGIONALITY: UNEQUAL DEVELOPMENT IN TOURISM

As a general theory, Harvey's text, and its predecessors, doesn't account much for tourism in its formulation. The text only refers to tourism as an example of the process of accumulation by dispossession in the commodification of histories and culture for tourists' consumption. Indeed, tourism is part of the accumulation by dispossession conditionality. As such, it can close space and leave aside previous users (Jeannite & Lapointe, 2016; Manuel-Navarrete, 2016), redefine space and place through new symbolic orders (Fitchett et al., 2021; Morin & Lapointe, 2020), and even displace populations (Young & Markham, 2019).

The other point of Harvey's theory I want to stress as relevant for tourism studies, in the context of regional development, is the production of regionalities. It refers to regional class alliances resulting from investments in the built environment and the structure of local capital accumulation. The structure of these alliances may vary: they can become exclusive and therefore hostile to outside influence; they can widen or fracture and be undermined by conflict; or be in a comfortable, hegemonic position, depending on the circumstances. In short, the theoretical stage of the production of regionality can be characterized by identity and loyalty as well as by the assumption of responsibility for infrastructure and services associated with the accumulation of capital by local protagonists (Harvey, 2006). This production of regionality can be associated with the all-encompassing concept of destination, where a concentrated, spatial fix of surplus capital in one place creates the condition of accumulation (Yrigoy, 2014; 2021).

While the notion of destinations can refer to many dimensions of tourism (Kadri et al., 2011), it is mostly used to refer to the place sold to tourists and the organization of services for tourists in that particular place. However, a destination can refer to lived places, to political or administrative space, to a nation state, or even be created exclusively for tourism. Destinations have two dimensions: static, the physical space and infrastructure, and dynamic, the complex web of interactions of the tourists in place (Framke, 2002). As Dredge and Jenkins stipulate, destinations vary in terms of scales, coordination and policies (Dredge & Jenkins, 2003). Using A. Amin and Thrift's (1994) institutional thickness concept, they show how tourism destinations were produced on top of previous state-defined, regional boundaries and created a complex (dis)equilibrium between scalar responsibilities of local organizations and regional organizations. In short, tourism creates a regionality around a space of desires and fantasies of potential visitors, with investments in symbolic values, through representations, and economic values, and through infrastructure to facilitate capital accumulation (Lapointe et al., 2015; Yrigoy, 2014; 2021) by private and public stakeholders (Dredge & Jenkins, 2003). Tourism's regional development strategies are built around setting up regionality under the name of the destination, creating a system of capital accumulation (Lapointe et al., 2018; Yrigoy, 2014) through a marketing narration that links the destination, the regionality, with spatial processes of leisure mobility at different scales (Framke, 2002). This regionality structures an accumulation system at the local and regional scale that is reproduced in time, and networked with other regionalities through complementary dynamics, like tours (Wang, 2006) and competition through the marketing of experience (Lapointe et al., 2018).

Nature tourism, associated with conservation, is a good example of this process. Through the neoliberal turn shaping the development of ecotourism and conservation, nature and the environment are in the market as commodities (Fletcher, 2019; Fletcher & Neves, 2012; Tardif & Sarrasin, 2018), justifying conservation through its capacities to attract tourism revenue to finance its operations. It becomes a whole reframing of the environment that competes, even replaces, previous entanglements and place-based ways of life in and around protected areas (Lapointe, 2011; Lapointe & Gagnon, 2009). It also creates competition between species, where some emblematic species, like pandas or elephants, are deemed more valuable to protect than other species, as their capacity to attract tourists' money is higher. Nature tourism will then structure a regionality to be able to accumulate capital through tourists' consumption of nature, with the construction of hotels and restaurants, the provision of specialized services like guiding services, but also all the linkages necessary for tourism to happen at the periphery of conservation areas (transport, construction materials, sports equipment, etc.).

While a lot of ink on pages and pixels on the screen have been spent talking about destination management, marketing and development, there are two elements of this form of regionality called destination that are unexplored. First, what is brought to the market and how is it made available for the market, and second, what is the "productive challenge" (Nadou & Pecqueur, 2020) around which the destination might strive or collapse? To answer those two questions, I will first go back to Harvey's work on the material embedding of capital accumulation processes in the web of socio-ecological life.

THE SALVAGE ECONOMY, OR HARVESTING NON-CAPITALIST PROCESSES IN CAPITALISM

As noted above, Harvey's theory of uneven geographical development stresses the capacity of capitalism to transform socio-ecological processes to commodities in the market economy. We can think about how agriculture brings a complex socio-ecological process that includes photosynthesis, biotic life in the soil, human labor and so forth, to the market, as we can buy the end product of agriculture: food. But you also can buy seeds and soil amendments, like fertilizers, minerals, compost, etc. As for tourism, we can include landscapes, history, culture, and more, as examples of this transformation of the material embeddedness of capitalism into the socio-ecological web of life. In this vision, elements and processes of the socio-ecological web of life are brought into capitalism, and they produce and reproduce as part of a growth-oriented capitalistic process. Once again, we can think of theme parks or closed resorts where nature, sociability, people and place are turned into products (Minca, 2010). This accounts for only a fragment of tourism's development. Subtler and more complex integration of the socio-ecological web of life is brought into the market exchange, but without transforming the whole process through capitalist production, and not always with a process of accumulation by dispossession, for example, the higher price of a hotel room with a view of a particular landscape. The landscape is produced by a socio-ecological process but it is transformed by capitalist values, the higher price or what could be called a position rent, without a full integration to capitalist production of every interaction producing the landscape.

In her book *The Mushroom at the End of the World*, Anna Lowenhaupt Tsing (2015) describes *salvage economy* as the processes that allow non-capitalist interactions and goods to be absorbed into capitalist commodity trade systems. Thus, value creation and exchange on the periphery of capitalism are found to be integrated into the value chain of capitalism. She postulates four key ideas at the heart of the salvage economy: (1) the translation of non-capitalist interactions and goods into capitalist goods allowing accumulation; (2) the spaces on the periphery of capitalism see non-capitalist and capitalist processes flourish simultaneously; (3) the supply chain works to link these heterogeneous processes together; and (4) economic diversity is at the heart of capitalism even if it allows the emergence of a space of contestation or even rejection of capitalism (Tsing, 2015, p. 301).

Tsing (2015) focuses on the notion of the patch, which she defines as a relatively stable assembly of worldmaking projects around boundary objects. This assembly is made of entanglements of human and non-human relationships that make life possible (Tsing, 2015; 2018). Following the sociology of translation (Callon, 2006), these patches are spaces where the translation is articulated between regimes of meaning and values. These translations make it possible to set up complex value chains in which these boundary objects, like Tsing's matsutake mushroom, change roles and meanings as they move from one patch to another, from a gift to commodities, back to gift and sociability. These passages are the moments of translations.

Following Tsing (2015), Mostafanezhad & Suryanata (2018) look at a new agro-food initiative in Hawai'i and note that:

> We draw on this conception to analyze AFIs [agro-food initiatives] to highlight exchanges that build communities, social ties, and citizenship while producing, yet not always profiting from, agricultural commodities. … For many farmers, the value of farming is constituted through the circulation in and between commodity and social systems of exchange. (Mostafanezhad & Suryanata, 2018)

They stress that it is through circulation that agro-food initiatives bring non-capitalist values, like environmental concerns, or community building, into the market through new forms of value embedded in the narratives of a new agriculture for values (Mostafanezhad & Suryanata, 2018). Tourism is built on a similar process. A basket of heterogeneous goods and services, tourism is both tangible and intangible consumption that brings into the market economy processes that have been created outside capitalism, on the periphery of capitalism, and/or as hybrids with capitalistic productions. The example of the itinerary as a consumer product is a good illustration of this dual nature of tourism consumption (Wang, 2006). Wang examines the different paradoxes of the itinerary within the tourist phenomenon. One of the paradoxes identified is that of materiality/immateriality in the commodification of the route where, on the one hand, the route involves material consumption of food, transport, memories, and therefore physical objects produced to meet the needs of the tourist. These goods, produced mainly through capitalist processes with cycles of investment, return on capital, profits, etc., are offered and consumed by tourists. However, part of this can come from the periphery of capitalism, including "authentic" food consumption, and integrates capitalistic market exchange through tourism, which adds a symbolic value that can be turned into profit and capital. The moment when the symbolic is turning into profit and capital is a moment of translation.

While translation can be done at the frontier of capitalist/non-capitalist processes, capitalism can also alienate the complex entanglement in trying to reproduce the moment of translation solely within capitalism, to deterritorialize it. Tsing (2018) gives the example of plant nurseries, which produce plants totally within the capitalist exchange system and ships them wherever those plants are needed to replace locally embedded plant production. While we will not extend here to the ecological catastrophe that travels with and through this production system (for this see Tsing et al., 2021), going back to tourism, I can expose cruise tourism as a similar alienation of natural, social and cultural entanglements into a placeless mobile capitalist process, reproducing on the boat and in staged experiences in ports of call for large cruise ships, to maximize profits for shareholders while having the fewest links to the territorialized tourism economy, the destination as regionality (Renaud, 2020).

This moment of translation has also been addressed through the use of actor–network theory efforts to understand the relationship between human and non-human interactions in tourism (Ren, 2011; van der Duim et al., 2013) especially the notion of ordering (Ren, 2011), which entails the different ways of relating between human and non-human. Ren (2011) exposed four ways of ordering associated with the oscypek cheese: the traditional cheese, the tourism cheese, the modern cheese, and the unique cheese. What she calls the "tourism" cheese is a translation of the traditional cheese to translate in market exchange, through tourism, the immaterial dimensions of the cheese, mostly those entangled in the traditional cheese, like identity, in the experience economy.

Different amenities are brought into the market for tourists' consumption. They are brought into the market through translation of non-market, non-capitalistic goods and processes (like landscape, identity) and production. This happens mainly through the constitution of a regionality called a destination, which allows regional stakeholders to accumulate capital through tourism. But in a perspective of regional development, destination, and place in general, are not self-contained with clear boundaries as they are crossed by different forms of mobilities, which need our scrutiny to understand the place tourism takes and to not equate development solely as numbers of tourists or expenditure.

IN-PLACE ECONOMY, OR SALVAGING TOURISTS AS AN ECONOMIC INPUT

Tourism, as a regional development strategy, usually conceptualizes tourists as consumers, and this is exemplified in the usual performance indicators like spending at the destination, the number of nights' stay, and the amount of paid entrance fees on site. This is not to be challenged as tourists are, among other things, consumers, but at the same time, without tourists, some market exchange would not be touristic; we need tourists for tourism. I suggest reconsidering tourists not solely as consumers, even if they consume, but as a mobile resource salvaged by tourism, principally with imaginaries and desire. This turnaround involves designing tourism's regional development strategies as a process of salvaging the ebb and flow of non-residents and turning their presence into economic gains while keeping in mind the quality of life of resident populations. Instead of seeing the production of tourism services, we can see it as a development of a living place where tourists are one of its components, but we can also see tourism as regional development within a possible interdependent spatial system. To make this flip, we look to Davezies and Talandier (2014) and Talandier (2013) and the concept of *économie résidentielle*, or in-place economy.

The residential economy is challenging the economic base theory in which the productive economy, the creation of added value through production and exportation, is considered the base of the economy and of local growth (Hoyt, 1954). The base is dissociated from the residential economy, which is considered as a passive economy that caters to the consumption of the workers of the base sector (Hoyt, 1954). This theory concentrates regional development efforts on export strategies in which jobs in the "base" are considered multipliers in the residential economy, which is dependent on the "base" for its growth. Recent work on residential economy (Davezies & Talandier, 2014; Pecqueur & Talandier, 2011) breaks with this vision of the residential economy as a secondary and passive part of the economy and stipulates that some "non-productive areas" can base their economic development on consumption by different mobile individuals such as commuters, retirees, and tourists, instead of on export-based economic activities. This type of regional strategy is creating all sorts of issues, such as residential gentrification, growing social inequalities between new residents and "locals", tensions between migrant retirees and local service economy workers, exclusion and expropriation through speculative land and house markets, and volatility caused by territorial competitions (Davezies & Talandier, 2014). Indeed, the residential economy has also been labeled as a parasite economy. While not denying the impacts of the residential transformation of the economy, Davezies and Talandier (2014) positioned the productive and residential economies as possibly interdependent and complementary.

> So, a regional economy does not only depend on incoming streams, circulating locally (as posited by the export base theory), but also on minimizing outgoing streams. The presence of in-place activities encourages the capture of new economic bases (derived from residents and tourists), determines local consumption of goods and services, enables wealth to cir-

> culate and reduces outgoing streams. To maximize the wealth of a territory thus involves maximizing created and captured wealth, minimizing wealth evasion and fostering its circulation. (Talandier, 2013, p. 5)

One interesting aspect of the residential economy is that it integrates time. The residential economy is linked to the productive economy through a network of mobility where values created somewhere are captured somewhere else through movement in daily time (commuter), yearly time (tourists) and lifetime (retirees). This breaks with a functional understanding of regional development, in which development is not only about specializing through niches within a geographical division of production, but also about residents' different time of life. Regional development becomes a combination of productive and in-place economy, trying to answer productive issues, such as quality of life, access to leisure, and socio-environmental amenities throughout a lifetime.

This theoretical angle also allows us to resituate tourism development at the regional level within the fabric of societies and its inscription in different regimes of mobility (Salazar & Schiller, 2016). Where dominant export-based theories are mostly concerned with the mobility of goods and tourism becomes export goods as foreigners spend money at the destination, the in-place economy is giving attention to the local circulation of capital and the needs of the resident population while acknowledging the entanglement of local consumption within multi-scalar and multi-temporal regimes of mobility that influence regional development through presence and absence at different places and different times. Tourism becomes one of these mobilities that can be salvaged, like a resource, in the local economy through market exchange.

TOURISM AS REGIONAL DEVELOPMENT STRATEGY

Tourism, as a regional development strategy, as noted above, is traditionally focused on growth and capital accumulation. Therefore, it is not surprising to witness policies, initiatives and strategies relying on the creation of the condition of accumulation in space and time (Harvey, 2006) through a regional system of accumulation, which is called a destination and reproduces center–periphery inequalities. Thinking about tourism's regional development is commonly restricted to the implementation of the conditions of accumulation, and the inscriptions of the destination in global and national flows of mobility, creating a tourism supply chain, per se. While there are many directions that can be taken along this rhizomatic understanding of tourism development (Deleuze & Guattari, 1980), we can stress how the states and local authorities act in tourism development through a combination of asset protection, like national parks, heritage sites, traditional practices, and, in the name of competition and free markets, self-financing through tourism activities, support of for-profit organizations, and investor security in the market, real estate, and land tenure regulations. These actions concentrate on the condition of accumulation but also on defining the points of contact to salvage non-capitalist processes within tourism capitalist exchange. In the discussion to this point, I have reviewed three moments, the creation of regionality, salvage capitalism, and the residential economy, to offer elements of an alternative framework, which can help to consider the role of tourism in regional development beyond growth and accumulation.

This effort implies that I frame tourism as the consumption of idiosyncratic and unique ways of life in place through movement and encounters. I elaborate a framework to acknowledge this assumption, as living in place is built through time and space with an array of relationships between social agents but also more-than-human life and materialities. Those relationships are salvaged within the tourism exchange. Indeed, as described earlier, this framework is elaborated through a threefold embeddedness in the capitalism process, through salvaging in capitalism a non-capitalist process, the constitution of regionality (destination) to allow accumulation, and the consumption of mobility flow across time and space.

Our framework, while allowing for the analysis of local and regional development, embraces a larger scope, in acknowledging a diversity of the process in the dynamic of the regionality called a destination, from the socio-environmental processes of non-capitalist actions, to the dichotomy of presence and absence of the mobile subjects interacting with tourism's regionality and value chain. So, let's quickly revisit each of them to ponder how they can work in thinking about, or analyzing, tourism as a regional development strategy.

The first moment would be to identify the point of translation between territorial resources and life in place. What are the actual non-capitalist processes that are translated into tourism goods and how it is done? Here we are not looking at general principles but at the actual entanglement that makes it possible to translate it into the market. While capital accumulation-based strategies would focus on identifying resources to salvage into the experience economy, and even resources to alienate from their entanglement to reproduce it in predictable and controlled ways to maximize accumulation, I suggest we look in a different direction. Resources to bring in tourism practices could be looked at with the lens of life-sustaining entanglement to be shared and stewarded. This calls for documenting how they have become entangled with the particular way of life of the area, how they might, and probably are, partially disentangled, or even alienated, in capitalist production and then how, through tourism, they can be re-entangled with a place-based way of life; a place-based way of life that has intrinsic values as a testimony of complex socio-ecological entanglement.

The second moment is the constitution of regionality, which in growth-based capitalist regional development would be the stabilization of the condition of accumulation in time and space through tourism. But, region, territories, places, are not just about capital accumulation as they are also the intersection of the locus of socio-ecological relationships and entanglements. As tourism in peripheral economies is often a substitute for declining production (Marcotte et al., 2020; Müller & Jansson, 2007), this substitution doesn't have to be within the same accumulation parameters. The passage from natural resources exploitation towards a tourism economy can move beyond extractivism and high-profit-oriented activities to dwell on the particularities of living in such a place, especially given that the particular inscription of life in that place is what attracts tourists.

The third moment would be to look at mobility in space and time. As I have noted, the residential economy ties complex mobility in a territorial knot, weaving three forms of mobilities in time: time of the day (commute), time of the year (work and leisure), and lifetime. As I have discussed earlier, the residential economy is focused on the consumption of mobile subjects in place. Here we can understand how mobility in place unfolds, and particularly how the three different mobilities are entangled with the particular inscription of life in place (formerly known as destination) and the needs of members of the three mobilities. Here I advocate expanding what is taken into account in tourism development as it is usually centered on tourists' needs and desires, therefore creating an alienating alterity disentangled from the needs and desires of other mobility, other times of life. Tourism as regional development shouldn't create alterity between tourism and the needs and desires of other users of place (formerly known as destination), but instead focus on how it brings the alterity of particular ways of life in place. These target tourists as a "resource", a resource that can foster the complex socio-ecological entanglements that create places instead of disentangling them to create a somewhat set-apart uniform space to be consumed by tourists.

For example, let's say that the building of a winter ski resort necessitates the logging of ten hectares of hillside maple groves. From a traditional growth-oriented regional development perspective, this loss of habitat, and history, could be justified. Investment in the built environment, job creation and capital accumulation for shareholders of the resort, would all be strong arguments to allow to such a development. Local authorities would deliver permits and authorization, they might even change the zoning scheme, as they will capture their share of resources, especially through land taxes, usually the main source of income for local communities. Furthermore, the resort will offer to its international clients a locally sourced and produced "terroire", which gives a sense of the place and its food craft, but will have to ask for scaled-up production for their suppliers to be able to cater for the throughput of such a resort. This will generate disentangled reproduction of the original products issuing from a history of living in place, in a uniformization process of offering according to deterritorialized standards of international demand.

On the other hand, the vision I offer would consider the entanglement of those hillside maple groves, their role in the watershed, how they make the land what it is, and how this entanglement allows people to live in the lower area of the watershed. Here the accumulation could be understood as the relationships that humans and more-than-humans build together. Skiing would need to be reconsidered as the main tourist activity. Lower impact, regenerative activities would need to be considered and would include the density and diversity of particular cultural linking with place and living in this place. History and food practices would be central to this regional development of tourism. Tourists would be salvaged within this process in order to densify the entanglement, as a shared encounter with other human and more-than-humans, and not the other way around (as in the previous paragraph). The seasonality of tourism would also be entangled within regenerative cycles of the patch that would stabilize under such entanglement.

While this idealist vision of tourism as regional development might seem to come out of a dreamy academic haze, we have to consider the fact that in the periphery, what are turned into tourist attractions are places that have been left out of productive capitalism, under an unequal development, as the forest or the mine is not profitable anymore under the diktat of capital. In this context, and the context of multiple crises facing our world, and tourism (Cheer et al., 2021), such a different way of conceptualizing development and creating regionality (destination) in tourism could be done according to different standards than capital accumulation. In this regard, I do follow Gibson-Graham's (2011) vision that the capitalist economy is the tip of the iceberg of what regions are. In their vision of (post)development, the structural imperative dictated by capital shall give way to a situated ethico-political imaginary, which considers alternative paths to development anchored in well-being and complex interdependencies. Their vision of (post)development is based on a relational understanding of place and the life in such places, therefore recognizing the diversity of transaction, relationship, and translations, that build interdependence and can foster well-being for human and more-than-human populations (Gibson-Graham, 2011).

This construction of tourism as a regional development strategy is not building a regionality (destination), which would mostly allow capital accumulation, but would look at accumulating links and encounters to foster and steward the human and more-than-human entanglements that make places unique through time and space. Those processes can be engulfed by capitalism and lose their non-capitalistic forms, but the identification of those processes, their role in the more than capitalist worldview, can also be the locus of creating diversity, fostering their relevance, expanding the recognition of their intrinsic values and therefore sustaining idiosyncratic ways of life in place at the core of the tourism encounters and consumption. This would call for a multisectoral policy framing involving land planning, arts and culture, and economic development, while reconsidering land planning and cultural support beyond the market, including mobility in place and time, to foster interaction with the in-place economy that gives space to the different time of life of residents, the productive needs of the regional system, and the non-capitalistic processes that produce this place as a place of value to live in and visit.

CODA – COVID-19 SHUFFLING OF TOURISM

As these lines are written, the COVID-19 pandemic is still raging, travel restrictions are many, and situations are shifting fast. As the ballet of sanitary restrictions is still ongoing, tourism and mobility are transformed and hard to predict. A reflection on how this might affect the connectivity of peripheral tourism areas is already relevant, even if, in hindsight, some of the conclusions might be off. While tourism has been deemed a growth sector since the 1950s, COVID-19 shrank it, and transformed it into a restructuring sector. In terms of regional development, tourism was a sector invested in supporting a transition from a production economy to a residential economy, based on salvaging non-capitalist practices and values within the capitalist exchange but also salvaging mobile presence for consumption. In some instances, it reproduces a functional geographical division of labor instead of a community needs approach, creating enclave patterns (Ek & Tesfahuney, 2019; Manuel-Navarrete, 2016; Saarinen, 2016) and alterity (Équipe MIT, 2002; Lapointe, 2020). Within the restructuration, or reassemblage, of tourism as it interplays with the COVID-19 pandemic and its health consequences, looking at tourism's regional dynamics at the intersection of salvaging economy and residential economy appears to be of relevance. As the COVID-19 pandemic unfolds in surging waves, and as mobility resumes slowly, there is some research for local particularism as an exotified experience, where tourism feeding on domestic mobility is looking for a change of scenery in the neighborhood, therefore salvaging non-capitalist values in tourism. The work from home orders, with their many inequalities and injustices (Rose, 2020), blurred the home–work–leisure spatial triad, concentrating them all at the same place or in smaller circuit than before the pandemic. This crystalizes the residential dimension of the economy while evacuating part of the mobility side of it, for now, as new patterns and new scales of commuting might reappear once immunity is attained. It might create a sensitivity in tourism for resident populations, but it might also create a larger alterity with digital nomadism in which value production and consumption will be deterritorialized from the traditional localization criteria and reterritorialized along digital bandwidth infrastructure.

Looking at tourism's regional development through this lens open new questions for rural and peripheral tourism development. First it creates a discursive space for the acknowledgment of the non-market value practices and relations that are salvaged into capitalism through consumption of tourism. How are these complex assemblages salvaged? Who operates the translations and how they translate into tourism? How do those tourism "patches" get stabilized? What is the meaning of these practices before the point of salvaging? How can the presence and absence of tourists play into the salvaging process? How has the presence and absence of tourists stabilized or enfeebled the pre-existing, non-capitalist patch and capitalist regionalities?

To close and open the conversation we raised with this chapter, I would like to go back to Anna Tsing, and particularly to her warning about the damage done to our world. Contemporary tourism is part of the major collapse of the Holocene entanglements that allow life on Earth as we know it. Tourism as a growth strategy, especially in its monstrous last-chance version, is not only accelerating the process but acknowledging implicitly its own role in the collapse in salvaging through capitalism, and even arguing that we can rely on capital to save those last patches of entanglements, fragile ecosystems and ghostlike memories of past entanglement described on plywood information boards in national parks. Our humble contribution to try to step away from this view of development, to advocate for a form of tourism regionality that echoes patch and entanglement, based on fostering linkages, adding to local aspirations in the stewardship of complex entanglements that makes life in a particular place unique, instead of replacing it through a generic simplified reproduction in which all steps and links aim at accumulating capital.

REFERENCES

Amin, A. and Thrift, N. (1994). *Globalisation: Institutions, and Regional Development in Europe*. Oxford: Oxford University Press.

Amin, S. (1973). *Le Développement inégal: essai sur les formations sociales du capitalisme périphérique*. Paris: Éditions de Minuit.

Brouder, P. (2018). The end of tourism? A Gibson-Graham inspired reflection on the tourism economy. *Tourism Geographies*, *20*(5), 916–918.

Callon, M. (2006). Sociologie de l'acteur réseau. In M. Akrich, M. Callon, & B. Latour (eds), *Sociologie de la traduction. Textes fondateurs* (pp. 267–276). Paris: Presses des Mines.

Cardoso, F.H. (1970). Les obstacles structurels et institutionnels au développement. *Sociologie et sociétés*, *2*(2), 297–316.

Cardoso, F.H. (2009). Beyond the international crisis: Economy, politics, and culture. *Americas*, *61*(4), 8.

Cheer, J.M., Lapointe, D., Mostafanezhad, M., & Jamal, T. (2021). Global tourism in crisis: Conceptual frameworks for research and practice. *Journal of Tourism Futures*, *7*(3), 278–294.

Davezies, L., & Talandier, M. (2014). *L'Émergence des systèmes productivo-résidentiels. Territoires productifs – territoires résidentiels: quelles interactions?* Paris: La documentation française.

Deleuze, G., & Guattari, F. (1980). *Mille plateaux*. Paris: Editions de Minuit.

Dredge, D., & Jenkins, J. (2003). Destination place identity and regional tourism policy. *Tourism Geographies*, *5*(4), 383–407. doi:10.1080/1461668032000129137.

Ek, R., & Tesfahuney, M. (2019). Topologies of tourism enclaves. *Tourism Geographies*, *21*(5), 864–880. doi:10.1080/14616688.2019.1663910.

Emmanuel, A. (1969). *L'échange inégal: Essai sur les antagonismes dans les rapports économiques internationaux*. Paris: F. Maspero.

Équipe MIT. (2002). *Tourisme 1: Lieux communs*. Paris: Belin.

Fitchett, J., Lindberg, F., & Martin, D.M. (2021). Accumulation by symbolic dispossession: Tourism development in advanced capitalism. *Annals of Tourism Research*, *86*(1), 103072. doi:10.1016/j.annals.2020.103072.

Fletcher, R. (2019). Ecotourism after nature: Anthropocene tourism as a new capitalist "fix". *Journal of Sustainable Tourism*, *27*(4), 522–535.

Fletcher, R., & Neves, K. (2012). Contradictions in tourism: The promise and pitfalls of ecotourism as a manifold capitalist fix. *Environment and Society*, *3*(1), 60–77.

Framke, W. (2002). The destination as a concept: A discussion of the business-related perspective versus the socio-cultural approach in tourism theory. *Scandinavian Journal of Hospitality and Tourism*, *2*(2), 92–108. doi:10.1080/15022250216287.

Frank, A.G. (1969). *Capitalism and Underdevelopment in Latin America: Historical Studies of Chile and Brazil*. New York: Monthly Review Press.

Gibson-Graham, J.-K. (2011). Forging post-development partnerships: Possibilities for local and regional development. In A. Pike, A. Rodriguez-Pose, & J. Tomaney (eds), *Handbook of Local and Regional Development* (pp. 226–236). London: Routledge.

Harvey, D. (2006). Notes towards a theory of uneven geographical development. In D. Harvey, *Spaces of Global Capitalism* (pp. 69–116). London: Verso.

Hoyt, H. (1954). Homer Hoyt on development of economic base concept. *Land Economics*, *30*(2), 182–186.

Jeannite, S., & Lapointe, D. (2016). La production de l'espace touristique de l'Île-à-Vache (Haïti): illustration du processus de développement géographique inégal. *Études caribéennes*, *33–34*. doi:10.4000/etudescaribeennes.8810.

Kadri, B., Khomsi, M.R., & Bondarenko, M. (2011). Le concept de destination. Diversité sémantique et réalité organisationnelle. *Téoros. Revue de recherche en tourisme*, *30*(1), 12–24. http://journals.openedition.org/teoros/1229.

Klein, J.L. (2008). Territoire et développement : du local à la solidarité territoriale. In G. Massicotte (ed.), *Sciences du territoire: perspectives québécoises* (pp. 315–334). Québec: Presses de l'Université du Québec.

Lapointe, D. (2011). *Conservation, aires protégées et écotourisme des enjeux de justice environnementale pour les communautés voisines des parcs?* [Doctoral dissertation, Université du Québec à Rimouski].

Lapointe, D. (2020). Reconnecting tourism after COVID-19: The paradox of alterity in tourism areas. *Tourism Geographies*, *22*(3), 633–638.

Lapointe, D., & Gagnon, C. (2009). Conservation et écotourisme: une lecture par la justice environnementale du cas des communautés voisines du Parc national de la Guadeloupe. *Études caribéennes*, *12*. doi:10.4000/etudescaribeennes.3577.

Lapointe, D., Sarrasin, B., & Guillemard, A. (2015). Changements climatiques et mise en tourisme du fleuve St-Laurent au Québec. Analyse critique des représentations. *VertigO – la revue électronique en sciences de l'environnement, Special issue 23*. doi:10.4000/vertigo.16575.

Lapointe, D., Sarrasin, B., & Benjamin, C. (2018). Tourism in the sustained hegemonic neoliberal order. *Revista Latino-Americana de Turismologia*, *4*(1), 16–33. doi:10.34019/2448-198X.2018.v4.13915.

Lebon, C., & Lapointe, D. (2018). Community well-being between climate risk and tourism development. In B.S.R. Grimwood, H. Mair, K. Caton, & M. Muldoon (eds), *Tourism and Wellness: Travel for the Good of All?* (pp. 173–190). Lanham, MD: Lexington Books.

Manuel-Navarrete, D. (2016). Boundary-work and sustainability in tourism enclaves. *Journal of Sustainable Tourism*, *24*(4), 507–526.

Marcotte, P., Lapointe, D., Sarrasin, B., & Bourdeau, L. (2020). Nulle part ailleurs. La périphérie touristique en question. *Téoros. Revue de recherche en tourisme*, *39*(2).

Minca, C. (2010). The island: Work, tourism and the biopolitical. *Tourist Studies*, *9*(2), 88–108. doi:10.1177/1468797609360599.

Morin, A., & Lapointe, D. (2020). S'approprier la nature: le développement géographique inégal par le tourisme et l'extraction sur Anticosti. *Téoros. Revue de recherche en tourisme*, *39*(2).

Mostafanezhad, M., & Suryanata, K. (2018). Is farming sexy? Agro-food initiatives and the contested value of agriculture in post-plantation Hawai'i. *Geoforum*, *91*, 227–234. doi:10.1016/j.geoforum.2018.03.011.

Müller, D.K., & Jansson, B. (2007). The difficult business of making pleasure peripheries prosperous: Perspectives on space, place and environment. In B. Jansson & D.K. Müller (eds), *Tourism in Peripheries: Perspectives from the Far North and South* (pp. 1–18). Wallingford: CABI.

Nadou, F., & Pecqueur, B. (2020). Pour une socioéconomie de l'intermédiation territoriale. *Géographie, Économie, Société*, *22*(3), 245–263.

Pecqueur, B., & Talandier, M. (2011). *Les espaces de développement résidentiel et touristique– état des lieux et problématiques*. https://hal.archives-ouvertes.fr/hal-01353484.

Polèse, M. (1994). *Économie urbaine et régionale: logique spatiale des mutations économiques*. Paris: Economica.

Ren, C. (2011). Non-human agency, radical ontology and tourism realities. *Annals of Tourism Research*, *38*(3), 858–881.

Renaud, L. (2020). Reconsidering global mobility – distancing from mass cruise tourism in the aftermath of COVID-19. *Tourism Geographies*, *22*(3), 679–689. doi:10.1080/14616688.2020.1762116.

Rose, J. (2020). Biopolitics, essential labor, and the political-economic crises of COVID-19. *Leisure Sciences*, *43*(1–2), 211–217. doi:10.1080/01490400.2020.1774004.

Saarinen, J. (2016). Enclavic tourism spaces: territorialization and bordering in tourism destination development and planning. *Tourism Geographies*, *19*(3), 425–437. doi:10.1080/14616688.2016.1258433.

Salazar, N.B., & Schiller, N.G. (2016). *Regimes of Mobility: Imaginaries and Relationalities of Power*. Abingdon: Routledge.

Talandier, M. (2013). Redefining the in-place economy and women's role in the local economy of highland areas. *Journal of Alpine Research/Revue de géographie alpine*, *101*(1). doi:10.4000/rga.2028.

Tardif, J., & Sarrasin, B. (2018). Biodiversité et conservation néolibérale au Québec: la place de l'écotourisme. *Études caribéennes*, *41*.

Terrier, C. (2009). Distinguer la population présente de la population résidente. *Insee, Courrier des Statistiques*, *128*, 63–70.

Tsing, A.L. (2015). *The Mushroom at the End of the World*. Princeton, NJ: Princeton University Press.

Tsing, A.L. (2018). Getting by in terrifying times. *Dialogues in Human Geography*, *8*(1), 73–76. https://doi.org/10.1177/2043820617738836.

Tsing, A.L., Deger, J., Keleman Saxena, A., & Zhou, F. (2021). *Feral Atlas: The More-Than-Human Anthropocene*. Stanford, CA: Stanford University Press. https://feralatlas.org/.

Van der Duim, R., Ren, C., & Johannesson, G.T. (2013). Ordering, materiality, and multiplicity: Enacting Actor–Network Theory in tourism. *Tourist Studies*, *13*(1), 3–20. doi:10.1177/1468797613476397.

Wang, N. (2006). Itinaries and the tourist experience. In C. Minca & T. Oakes (eds), *Travels in Paradox: Remapping Tourism* (pp. 65–76). Lanham, MD: Rowman & Littlefield Publishing.

Young, M., & Markham, F. (2019). Tourism, capital, and the commodification of place. *Progress in Human Geography*, *44*(2), 276–296. doi:10.1177/0309132519826679.

Yrigoy, I. (2014). The production of tourist spaces as a spatial fix. *Tourism Geographies*, *16*(4), 636–652. doi:10.1080/14616688.2014.915876.

Yrigoy, I. (2021). Strengthening the political economy of tourism: Profits, rents and finance. *Tourism Geographies*, 1–20. doi:10.1080/14616688.2021.1894227.

23. Rural communities, tourism, and the roles of academics

Bernard Lane

INTRODUCTION

Rural community tourism issues are very special. Rural communities have complex collections of built and human heritage and are typically extremely aware of their vulnerability to change, loss of services, and income reductions and they are especially aware and worried about new people – tourists are incomers. Communities often need help to understand their situations, the dangers and, importantly, the opportunities that tourism can bring to them, and how they, as communities, can seek to manage tourism and their own futures, individually and collectively.

This chapter looks at when, how, and why academics have learned to understand and work with communities, the techniques that can be used, the timescales involved, funding and the "project method", the creation and functioning of local focus groups, and a range of product development, marketing, and conservation techniques. The chapter focuses on working with rural communities in the developed world, and on trying to create sustainable tourism outcomes, including governance, partnership creation with the public and private sectors, and with other communities. It concentrates on rural tourism, which includes small towns in rural areas (Lane, 1994a).

THE BACKGROUND

Bob McKercher (1993) explained the relationship between tourism and communities in the first-ever issue of the *Journal of Sustainable Tourism*. It was a succinct, useful, detailed, and far-reaching assessment:

> Tourism enjoys a love–hate relationship with its host community. It is both a much sought after and much reviled activity. On one hand, it has been identified as an economic saviour, generating employment, income and tax revenue, and acting as a catalyst for regional development. On the other hand, it has been described as a pariah that destroys host societies and cultures and cuts a swathe of environmental destruction wherever it goes. (McKercher, 1993, p. 6)

One of the most obvious, but frequently forgotten, features of tourism is that it cannot easily be hidden within a community. New factories can be located at the edge of a town, and their traffic routed in and out via dedicated roads, and then away from the community via ring roads. Landscaping can hide those factories. Tourism brings new people, the tourists, openly into the community, usually into the town or village centre, and they take up car parking spaces there, spaces that local people have often used for many years. And in many cases those tourists look, sound, and behave differently to local people. Mammals (including humans) tend to be herd animals, and highly territorial. Communities are, therefore, normal when they notice outsiders, complain about them, and despite the extra spending they might bring, see them as problems. As noted above, tourists do take local people's parking spaces. Perhaps the ultimate horror, some tourists like the places they visit so much that they buy properties in those places and add to the often-cited second home problem (Bjørn et al., 2008; Müller et al., 2004).

McKercher (1993, p. 7) found eight other "tourism versus community" problems:

1. As an industrial activity, tourism consumes resources, creates waste, and has specific infrastructure needs.
2. As a consumer of resources, tourism has the ability to over-consume resources.
3. Tourism, as a resource-dependent industry, must compete for scarce resources to ensure its survival.
4. Tourism is a private-sector-dominated industry, with investment decisions being based predominantly on profit maximisation.
5. Tourism is a multifaceted industry, and as such, it is almost impossible to control.
6. Tourists are consumers, not anthropologists, even though they often show great interest in an area and its people.
7. Tourism is entertainment. And it is tourists who are entertained, not local communities.
8. Unlike other industrial activities, tourism generates income by importing clients rather than exporting its product.

The last four of the above points are especially difficult for communities to cope with.

WHEN DID COMMUNITIES FIRST BECOME CLOSELY INVOLVED IN TOURISM?

Modern tourism dates back to the nineteenth century, and particularly to the building of railway systems, which allowed the mass movement of tourists from industrial towns and cities to many of the then newly built resorts, usually in seaside locations, and in scenic areas. As Walton (2009), points out, the history of tourism is much more complex than the above sentence implies, but the early history of tourism generally did not involve local communities in any major way. There were, however, early objections to tourism's expansion and to its potential for bringing change. The views of the English poet William Wordsworth, and his fears of the railways bringing large numbers of damaging tourists to the English Lake District, are well known. Wordsworth was followed by the writer John Ruskin (1819–1900), whose often-quoted words from 1876 described visitors to what is now the Lake District

National Park as "the stupid herds of modern tourists … leading to the opening of taverns and skittle grounds round Grasmere, which will soon, then, be nothing but a pool of drainage, with a beach of broken ginger beer bottles" (see https://www.hdrawnsley.com/index.php/conservation/railways-and-the-lake-district-1840-1880). But these are the views of individual, thoughtful, literary commentators, not the considered views of local communities.

Four great changes allowed and strongly encouraged community-based comments on tourism from the 1960s onwards.

Changes in Society

Prior to the 1960s, communities were relatively passive, leaving economic change and development to business. In most Western countries, however, local groups interested in their environments began to grow from the 1960s onwards, as education levels rose, and public concern about change and the environment grew. Hopkins (2008) provides an example of how broadly based community activities can be: his book describes the changing driving forces at work in a small English market town, and their relevance to many other similar situations. Cultural change has led to the formation of pressure groups on many issues, including tourism. Those groups have been international, national, and local.

Changes in Tourism

At the same time, in the 1960s, tourism began to change. Mass, resort-based tourism lost its monopoly. Special interest tourism grew, based on growing niche markets, which were also linked to rising levels of education. Michael Hall and Betty Weiler recorded that change in their ground-breaking book, *Special Interest Tourism* (Hall & Weiler, 1992) and chapters covered educational travel, art and heritage tourism, ethnic tourism, nature-based tourism, adventure tourism, scuba diving and yachting tourism. Special interest tourism/niche market tourism/rural tourism allowed small, often remote rural communities to become involved in tourism, to use tourism to help regenerate local economies, support local service provision, and provide jobs (see Lane, 1995; Lane & Clemenson, 1995).

The Growth of Lifestyle Entrepreneurship

In 2000, Ateljevic and Doorne published their now well-known paper about lifestyle entrepreneurs, who start businesses, often tourism-related businesses, to support their lifestyles in specific, often scenic, rural places. These businesses may also involve growing numbers of women working in tourism for the first time (see Scheyvens, 2000; Acharya & Halpenny, 2013). The movement towards increasing self-employment is not peculiar to rural communities or to tourism. It occurs in many other places and fields.

The Growing Interest in Community Projects and Public–Private Partnerships as Ways to Regenerate Local Economies and Support Conservation

As early as 1994, sustainable rural tourism, if skilfully managed, was shown to have strong potential for both conservation and economic development (Lane, 1994a; 1994b). Eagles (2002) expanded on those ideas, and they were made real by the work of the European Union's LEADER Groups (see https://enrd.ec.europa.eu/leader-clld/leader-toolkit/leaderclld-explained_en).

WHY AND HOW CAN COMMUNITIES BECOME INVOLVED IN TOURISM?

- By making statements about tourism – for and against – to the media and/or to local politicians, establishing community interest in tourism in their locality.
- By working with public sector/private sector tourism partnerships to influence local tourism management.
- By organising petitions to seek to change policies.
- By working for changes in tourism management as a group endeavour.
- By working with tourism academics to understand the often complex and special ways in which tourism as a phenomenon works, as an activity, as a business and as an agent for conserving and/or damaging local life, prosperity and biodiversity.

HOW, WHEN, AND WHERE DID ACADEMICS BECOME INVOLVED WITH COMMUNITIES AND TOURISM?

This is an interesting question. The answer dates back to the mid-nineteenth century, if one considers the evolution of community interest in adult education for all, covering economics, lifestyles, and related events. In that period, a Danish Lutheran Church Minister, Nikolai Grundtvig (1783–1872) developed the idea of folk high schools, offering practical education to adult rural people to help them better understand their worlds, their possibilities and especially their futures. The first folk high school opened in 1844. There are now 70 folk high schools in Denmark. They are credited with strong roles in helping create modern dairy farming in Denmark, capable of successfully exporting Danish dairy (and pig meat) products around the world.

The folk high school idea spread throughout Scandinavia, and to Germany, Austria, and Switzerland, and they form a vital part of society in those regions to this day, and to a more limited extent in the UK, USA, France, and parts of Africa (see https://danishfolkhighschools.com/about-folk-high-schools/history). There is no requirement for a folk high school to exist before academics can work with communities; their ideas and methods can be useful everywhere.

Interestingly, the USA, and then Canada, led the way in terms of encouraging academics to get involved with community tourism. In the USA, many of the oldest and strongest public universities, the so-called "land grant universities", had historic links to agricultural lands and derived much of their income from local land granted by central government to selected state universities in the nineteenth century. In return, they worked on local agricultural and community education issues. From the 1960s onwards, they produced guidance notes and worked with rural communities via their "extension departments", offering adult education and training in rural tourism.[1] Michigan, for example, is just one classic and well-researched example (see https://www.canr.msu.edu/search?siteContext=&siteTitle=Agritourism&searchbox-type=&q=rural+tourism).

Many extension departments helped produce guides and handbooks helping communities to understand rural tourism (see, for example, Koth et al., 1991, and for material from Canada, see Alberta Tourism, 1988; 1991; Jamal & Getz, 1999). Local communities obtained research knowledge about tourism, its issues, opportunities, and threats, from this work. Academics gained knowledge too, and living laboratories in which to conduct research work (see, as an example, Idziak et al., 2015). In the early days of the processes outlined above, tutors travelled to meet the communities and people involved, often covering long distances. They chaired and taught in person local courses and focus groups. But more recently sessions have become possible using the web, and now the use of Zoom and Microsoft Teams has opened many new possibilities.

What Do Academics Bring to This Process?

Knowledge about tourism

The average "normal" person knows little about why people travel, how they travel, and their expenditure on the various segments of the travel and tourism process. They know little about marketing. And often, they even know little about what their own area has to offer in terms of tourism "product". Tourism academics normally know about such things, or, if not, they know how to find answers.

Network knowledge and contacts

Local people rarely understand which parts of the public sector do what, nor do they always know who to contact in the many governing councils and non-elected statutory commissions which exist in most developed countries. Academics usually do know, can explain these things, and can help communities to make contact with relevant organisations.

Skills

Academics know how to bid to obtain money for local projects. Some tourism academics know about markets and marketing. Most know how to set up, administer and analyse questionnaires for visitors.

Importance

Universities still have some status in the minds of the public, and in the minds of local politicians. Academics need to know how to use that status to attract people to take part in their meetings/groups.

Examples of good practice

Many academics know – or can find out about – examples of communities which have successfully worked on community tourism development and management.

Discussion/focus group management

The process of teaching in universities often involves small discussion groups or tutorials. Local community focus groups led by academics can profit from the experience that academics have of managing such tutorials, ensuring that the group does more than just talk, that all members of the group take part, and that they begin to work together in a long-term way (Scott et al., 1989). Unlike working with undergraduate students, community groups have, however, to take decisions: Keske and Smutko (2010) discuss this issue and ways forward.

Media links

Groups need to be seen to be valuable if their findings are to be taken seriously. Academics need to know how to issue regular press releases to local print media, and how to arrange local radio and television sessions. Media links can attract more visitors and can build local credibility politically.

The community course or tourism discussion group

The coming together, or creation of a local group, perhaps led by an academic, especially one with experience of, or linked to, adult education or to an extension studies department, is typical of the early stages of the emergence of community-based tourism development. There are a number of requirements:

- The academic involved must know the community or area concerned and must see achievable community-led tourism potential there in terms of its scale, future economic good, and avoidance of future problems.
- That person must also get to know and understand local politicians and the businesses, local, regional and in some cases, even national businesses, who could be involved.
- The meeting place must be suitable and accessible to as many people as possible. The timing of meetings is also a central issue, including which days, and times. Traditionally evening sessions from 1930 to 2100 are the norm, but some communities and groups within communities meet at weekends – often on Saturday afternoons.
- Sustainable tourism knowledge is a must – not only to avoid overdevelopment or damaging developments, but also to enable group members to defend themselves from anti-tourism people in the area.

- The group should, therefore, conduct an overall SWOT analysis of the community (strengths, weaknesses, opportunities and threats) at an early stage in order to help guide not just their tourism aspirations, but the wider questions their communities face.
- The academic must understand that the group are not students, but people who want to explore the possibilities and problems of tourism activities. Lectures can be included, but discussion sessions are extremely important for both the leader of the group and for the group. The concepts of learning by doing and doing by learning are especially useful (Scott et al., 1989).
- Academics have to realise the stresses involved for the members of the group, many of whom have to assess their personal risks, in tourism work, financially, domestically, and socially, a very different situation to "normal" student groups. Many academics have to learn to understand how to take risks carefully.
- And academics also have a second responsibility. The group may need their help for a number of years. That responsibility can be fulfilled more easily now that Zoom meetings can be held, but it remains a long-term responsibility.
- And finally, academics doing this type of work need to realise how much they take for granted about what the general public know. Thus, for example, this author recently worked with a rural community in the UK that urgently needed the economic help that tourism might bring. But they did not want overnight visitors, only day visitors, which they believed would be better because they would disturb them less in the evening. They would not accept that they would take their daytime parking places, could not be regulated by the number of beds available, and would not believe that the net gain from overnight accommodation would be far greater than from day visitors. A small town wanted more visitors to help maintain its small independent shops. It had good rail services to several large cities, rail companies keen to work with them, few car parking spaces, and excellent local heritage walks. The idea of Slow Tourism was suggested to them. They were shocked. They did not want anything that was slow! And have, therefore, done nothing (for a full discussion about the issues of slow tourism see Dickinson et al., 2011).

WHAT CAN COMMUNITY TOURISM GROUPS DO?

The creation and operation of community tourism development groups is an essential part of sustainable rural tourism development. They bring local ownership, help to get people to work together, present a more unified product to the visitor, build entrepreneurial capacity in the community and, not least, make the job of any future development agent easier by providing one group to work with instead of many different people. They also provide a forum for training and other purposes.

A Sample Menu for Community Tourism Groups

A "portmanteau" list of 50 things that they could do is given below, which has been developed and tested by this author over 30 years for use in European, Japanese, and North American situations. In practice, the list can be tailored to suit each situation. It is a list of ideas rather than a definitive checklist.

1. Bring together business people, administrators, potential business people, and other interested parties in the area to discuss issues and find ways forward for their community's future.
2. Examine what other rural communities have achieved in tourism development, note which things went wrong, and why, and which were successful, and why. Explain how to work in partnership with other businesses and with the public sector (Getz & Jamal, 1994; Hall, 1999; Bramwell & Lane, 1999; 2000).
3. Organise local discussions, seminars, and other working groups, invite outside "experts", and practitioners, using these occasions to generate interest, ideas, and new forum members.
4. Begin to understand the regional, national, and international tourism markets, and discuss their strengths, weaknesses, opportunities and threats (SWOT) for tourism in the locality.
5. Draw up both a long-term and an annual strategy to build on strengths, eliminate weaknesses and threats, and seize opportunities. These strategies should follow the concept of long-term sustainability, conserving tourism's environmental and cultural assets, and using tourism where possible as a tool for the overall improvement of the socio-economic and political life of the area.
6. Decide upon a marketing campaign for use outside the area, and a linked campaign to persuade those who do visit the area to stay longer. This could include researching, writing, funding, printing, launching, and distributing a local brochure, and also setting up and maintaining a website. Although print media may be deemed outdated by many, it can have value, and is both tangible and easy to use in some circumstances. Groups can either join an existing website or set up a website on the local area. Look at other areas' sites to find suitable styles to work from.
7. Having set up a site, arrange links from other regional or local sites with things in common.
8. Set up an ongoing group to continue the work on a long-term basis when the opportunity seems ripe.
9. Act as a lobbying group to get "the authorities" to do things. Ensure the group has active relationships with councillors, their Member of Parliament and, in some cases, their Member of the European Parliament (Long & Nuckolls, 1994).
10. Discuss with your district or county tourism officer the need for a region-wide tourism management strategy encompassing infrastructure improvements, marketing, conservation, and planning guidelines (Lane, 1994b).

11. Invite organisations such as the regional tourist board, countryside agency, public transport authorities and others to talk to the group to find out what they have to offer to the group – and what the group can offer to them.
12. Learn how to handle and use the press and other media. Appoint a press officer from amongst the group.
13. Develop, waymark, publicise and maintain circular walks of various lengths in such a way that visitors are encouraged to use local services or transport facilities at the beginning or end of their walks – or en route.
14. Develop cycle routes on the lines discussed above, and encourage the development of local cycle hire facilities, usually as an additional part of an existing business.
15. Join in discussions with public transport operators and planners to encourage the development of a traffic planning system that will protect communities and habitats, encourage walking and cycling, and the use of local businesses – and be attractive to visitors. Car parking provision will be an important part of any system.
16. Encourage the production, use, and marketing of local food and drink products.
17. Encourage the production and marketing of local craft products.
18. Encourage local accommodation providers and restaurants to serve local food products, and to make their use clear when they are serving them.
19. Encourage local businesses and events generally to showcase local crafts etc.
20. Set up weekend breaks, offering accommodation, meals, and activities in a package.
21. If possible, join rural communities in other places in marketing these weekend breaks, in order to save expense and increase their effectiveness by joint marketing.
22. Consider developing tourist information outlets of various types and sizes, and site these, when possible, in shops, pubs, stations or other businesses, many of which need additional visits and activity to help them survive.
23. Persuade existing museums to modernise their displays to involve and animate visitors of all ages: where necessary, consider new heritage centres, which may also be designed to act as tourism information and specialised retail and/or hospitality places.
24. Make contact with local historians and/or the local history society. Use their knowledge to inform your members about the history of the area and capitalise on the interest in heritage by providing simple, accurate, interesting, and relevant historical information for visitors. See points 26, 27 and 28 below.
25. Learn the techniques of the eco-museum concept: begin to valorise the many small assets of the area: set up, for example, nature trails, craft trails, church trails, tomb trails, architectural trails, farm trails, geological/mineral trails, industrial archaeology trails. Integrate into any printed material on these trails, information on how to find food, accommodation, and retail opportunities.
26. Set up a guide training programme. Offer evening and/or afternoon guided walks.

27. Encourage older residents to act as guides: tape the recollections of some older residents to use in local information centres, museums, etc. They can also be very effective working in tourism information centres, often having encyclopaedic knowledge of the locality.
28. The web and the smart phone have together now allowed heritage interpretation and self-guided walks, which make a very valuable contribution to rural tourism's palette of products. Websites can be downloaded and heritage information listened to on smart phones while walking. Further, the websites can guide walkers, because they can be geographically aware, removing the need for fixed signposts and information boards (try https://soundcloud.com/culturenight/stream-of-consciousness-by-tom-lane-with-olwen-fouere-self-guided-walking-tour as an example – albeit an urban example). Walking is a classic slow tourism act, creating little environmental damage, healthy for the participants, and good for the community because it is slow, encouraging more overnight stays, bringing more net revenue into the area.
29. Encourage shopkeepers to come together to discuss how best they could improve their efficiency and effectiveness at selling to both the local and the visitor market.
30. Encourage new entrants to the accommodation sector by offering introductory training afternoons; encourage existing accommodation owners to improve their operations by attending further training, or quality circle discussions.
31. Encourage farmers, foresters, and naturalists to discuss how they could benefit from tourism; invite successful operators from other regions to talk about their experiences.
32. Encourage at least one farm in your area to operate as an open farm, as a visitor attraction and/or with a farm shop.
33. Involve local schools and colleges in visitor surveys, translation work, and career discussions. Try to form a permanent link with the staff involved.
34. Translate some information into German, French or other relevant languages, and make it available for businesses or guests in the area. Multilingual compliment slips with key responses in up to three languages would be a start.
35. Involve the church, by devising church and tomb trails, using church buildings for events, and by offering a "welcome ministry" in churches in popular places.
36. Work with local pubs or other hospitality enterprises to set up a pub trail, try special brews, events, etc.
37. Explore the possibility of hosting regular farmers' markets.
38. Set up local festivals and cultural events. Use these to generate shoulder season trade, publicity, and additional facilities for local people.
39. Develop the area's image by devising promotional material, postcards, shopping bags, unified letterheads, etc.
40. Consider the idea of a local area brand.
41. Hold a competition for new postcards to boost local thinking about the area and to increase pride of place.
42. Encourage press visits to obtain press coverage from outside the area.

43. Ensure that detailed, clear, and attractive local area material is available at every bedside to enthuse visitors to stay longer and return. These collections are often known as bedside browsers.
44. Set up databases of past visitors and set up direct marketing procedures.
45. Encourage visits by the area's Member of Parliament to ensure political support.
46. Organise study trips for forum members to other areas, or even abroad.
47. Discuss possibilities for working with tour operators who might bring additional business. Understand how tour operators work and choose operators suitable for and sympathetic to the area.
48. Use both bedside information (see point 43) and the database to advertise the potential of the area as a base for incoming businesses of all kinds.
49. Audit facilities for people with disabilities of various kinds – access, sight, hearing – and seek to both improve and publicise those facilities.
50. Remember that in most areas there are limits to growth, and that too much tourism development could both make the area unattractive in the long term and distort the economy of the area.

THE PROJECT METHOD

The ideas described above are typical of the relatively new method of introducing effective and, if possible, swift and positive change to the fortunes of rural areas. The "project method" uses a short-term ideas intervention, often funded by a public sector agency and staffed by outsiders to the area, well presented to the media, to the business community, and to the public. The project is typically of three years' duration, although often extensions can be sought. The EU's LEADER concept uses the project method to bring urgent, popular, and effective change to rural areas suffering population and/or economic decline.

THE LONG TERM

Only a few local tourism groups survive in the long term (i.e. over ten years or so). If the groups succeed, the need for them, as working groups, declines. The local economy and businesses have diversified. Often local government takes over key activities such as training, information, and marketing. Much depends on the leadership of the group. Good leaders are central to success (see Clarke, 1999). And group members can disagree strongly and disastrously about policies (see Idziak et al., 2015). But if the group succeeds only for a few years, it usually has a lasting influence on the area.

Some of the longer-term tourism governance schemes can be provided by protected area managements. Thirty years ago, protected areas tried hard to oppose tourism enterprises (Sharpley & Pearce, 2007). Then, gradually, they came to realise that sustainable rural tourism could legitimise and assist their work, providing political and economic support for their conservation work (Eagles, 2002; Eagles, et al., 2013). In Austria, the management of the large Grossglockner National Park (for background, see https://www.tripadvisor.co.uk/Attraction_Review-g608655-d4939753-Reviews-Grossglockner_National_Park-Heiligenblut_Carinthia_Austrian_Alps.html) has joined with their local communities to create marketing, infrastructure building, new tourism products, and training opportunities (Getzner et al., 2014). Lane and Kastenholz (2015) have noted that rural tourism is entering a new, second-generation era where it is taking on the qualities of a decentralised sustainable destination, with new professional governance possibilities. Community partnership governance could be part of that process.

There are growing opportunities and needs for academics to work in the new world of rural tourism. They will need to develop new skills of trust creation, partnership making, and management (Bramwell & Lane, 1999; 2000) in addition to business skills (Higuchi & Yamanaka, 2017). Rural tourism has seen many changes in its short life (Lane & Long, 2000; Gössling & Lane, 2015); it is likely to see many more. Even urban tourism can learn much from its countryside cousins (Long, 2000).

CONCLUSIONS (OR THE FUTURE RURAL WORLD)

This chapter has ranged over a long period of time and raised many demanding issues. Many "conventional" academics may find some of the issues discussed difficult or doubtful. Personal progress in the academic world has increasingly been based on the publication of research papers. Success in publishing has increasingly been determined by journal editors and reviewers, few of whom have been directly involved with rural communities, and/or with the many types of tourism businesses that have grown in recent years. But the financial demands now being put on many academic institutions may bring change. Success in raising money for rural regeneration work – including rural tourism management and marketing – could be important. Training public sector staff about tourism management and marketing could be a growth area. That type of work presents opportunities and challenges quite different from working with young undergraduates and postgraduates. It could bring many changes to the university sector.

"May you live in interesting times" is an English expression (https://en.wikipedia.org/wiki/Proverb) that is claimed – perhaps falsely – to be a translation of a traditional Chinese curse (https://en.wikipedia.org/wiki/Chinese_culture). While seemingly a blessing, the expression is normally used ironically; life is better in "uninteresting times" of peace and tranquillity than in "interesting" ones, which are usually times of trouble. But the trend towards practical, "real world", teaching and research has already begun in many subject areas.

That the academic world of tourism might now change is, in some ways, a form of "poetic justice". Rural communities have already had to change radically. Agriculture has seen intensification, new crops, and the rise of vegetarianism (reducing the value of stock farming), and there has been growing interest in organic farming methods.

Rural communities have faced depopulation, reduced service levels of many kinds, and the arrival of tourists. Along with tourists, property prices have risen, and townspeople have arrived, with once-quiet communities now crowded. New economic and people skills have become necessary. And life in many rural places is now dominated by conservation issues (Becken & Job, 2014), often-controversial issues such as rewilding, and the challenges of protected area politics (see Pettorelli et al., 2019). There will be plenty of demand for future researchers and academics engaged in community tourism!

NOTE

1. For a readable, recent, and open access paper about this activity, try Jin et al. (2021).

REFERENCES

Acharya, B.P., & Halpenny, E.A. (2013). Homestays as an alternative tourism product for sustainable community development: A case study of women-managed tourism product in rural Nepal. *Tourism Planning & Development, 10*(4), 367–387.

Alberta Tourism (1988). *Community Tourism Action Plan Manual*. Edmonton: Alberta Tourism.

Alberta Tourism (1991). *Market Planning Skills Program Manual for Communities*. Edmonton: Alberta Tourism.

Ateljevic, I., & Doorne , S. (2000). "Staying within the fence": Lifestyle entrepreneurship in tourism. *Journal of Sustainable Tourism, 8*(5), 378–392.

Becken, S., & Job, H. (2014). Protected areas in an era of global–local change. *Journal of Sustainable Tourism, 22*(4), 507–527.

Bjørn P., Kaltenborn, B., Andersen, O., Nellemann, C, Bjerke T., & Thrane, C. (2008). Resident attitudes towards mountain second-home tourism development in Norway: The effects of environmental attitudes. *Journal of Sustainable Tourism,16*(6), 664–680.

Bramwell, B., & Lane, B. (1999). Collaboration and partnerships for sustainable tourism. *Journal of Sustainable Tourism, 7(*3–4), 179–181.

Bramwell, B., & Lane, B. (eds) (2000) *Tourism Partnerships and Collaboration: Politics, Practice and Sustainability*. Bristol: Channel View Publications.

Clarke, J. (1999). Marketing structures for farm tourism: Beyond the individual provider of rural tourism. *Journal of Sustainable Tourism, 7*(1), 26–47.

Dickinson, J.E., Lumsdon L.M., & Robbins, D. (2011). Slow travel: Issues for tourism and climate change. *Journal of Sustainable Tourism, 19*(3), 281–300.

Eagles, P.F.J. (2002). Trends in park tourism: Economics, finance and management. *Journal of Sustainable Tourism, 10*(2), 132–153.

Eagles, P.F.J., Romagosa, F., Buteau-Duitschaever, W.C., Havitz, M., Glover, T.D., & McCutcheon, B. (2013). Good governance in protected areas: An evaluation of stakeholders' perceptions in British Columbia and Ontario Provincial Parks. *Journal of Sustainable Tourism, 21*(1), 60–79.

Getz, D., & Jamal, T. (1994). The environment–community symbiosis: A case for collaborative tourism planning. *Journal of Sustainable Tourism, 2*(3), 152–173.

Getzner, M., Lange Vik, M., Brendehaug, E., & Lane B. (2014). Governance and management strategies in national parks: Implications for sustainable regional development. *International Journal of Sustainable Society*, *6*(1–2), 82–101.
Gössling, S., & Lane, B. (2015). Rural tourism and the development of Internet-based accommodation booking platforms: A study in the advantages, dangers and implications of innovation. *Journal of Sustainable Tourism*, *23*(8–9), 1386–1403.
Hall, C.M. (1999). Rethinking collaboration and partnership: A public policy perspective. *Journal of Sustainable Tourism*, *7*(3–4), 274–289.
Hall, C.M., & Weiler, B. (1992). *Special Interest Tourism*. London: Belhaven Press.
Higuchi, Y., & Yamanaka, Y. (2017). Knowledge sharing between academic researchers and tourism practitioners: A Japanese study of the practical value of embeddedness, trust and co-creation. *Journal of Sustainable Tourism*, *25*(10), 1456–1473.
Hopkins, R. (2008). *The Transition Handbook*. Dartington, UK: Green Books.
Idziak, W., Majewski, J., & Zmyślony, P. (2015). Community participation in sustainable rural tourism experience creation: A long-term appraisal and lessons from a thematic villages project in Poland. *Journal of Sustainable Tourism*, *23*(8–9), 1341–1362.
Jamal, T., & Getz, D. (1999). Community roundtables for tourism-related conflicts: The dialectics of consensus and process structures. *Journal of Sustainable Tourism*, *7*(3–4), 290–313.
Jin, X., Wu, H., Zhang, J., & He, G. (2021). Agritourism development in the USA: The strategy of the State of Michigan. *Sustainability*, *13*(20), 11360. doi.org/10.3390/su132011360.
Keske, C., & Smutko, S. (2010). Consulting communities: Using audience response system (ARS) technology to assess community preferences for sustainable recreation and tourism development. *Journal of Sustainable Tourism*, *18*(8), 951–970.
Koth, B., Kreag, G., & Sem, J. (1991). *The Rural Tourism Development Project Training Guide*. St. Paul: Minnesota Extension Service, University of Minnesota.
Lane, B. (1994a). What is rural tourism? *Journal of Sustainable Tourism*, *2*(1–2), 7–21.
Lane, B (1994b). Sustainable rural tourism strategies: A tool for development and conservation. *Journal of Sustainable Tourism*, *2*(1–2), 102–111.
Lane, B. (1995). Niche marketing and rural tourism. In OECD (eds), *Niche Markets and Rural Development* (pp. 81–109). Paris: OECD.
Lane, B., & Clemenson, H. (1995). Niche marketing: Impacts and policy issues for rural development. In OECD (eds), *Niche Markets and Rural Development* (pp. 15–26). Paris: OECD.
Lane, B., & Kastenholz, E. (2015). Rural tourism: The evolution of practice and research approaches – towards a new generation concept? *Journal of Sustainable Tourism*, *23*(8–9), 1133–1156.
Lane, B., & Long, P. (2000). Rural tourism development. In W.C. Gartner & D.W. Lime (eds), *Trends in Outdoor Recreation, Leisure and Tourism* (pp. 299 –308). Wallingford: CABI.
Long, P. (2000). Development regimes in the city fringe: The case of *Discover Islington*, London. *Journal of Sustainable Tourism*, *8*(3), 190–206.
Long, P.T., & Nuckolls, J.S. (1994). Organising resources for rural tourism development. *Tourism Recreation Research*, *19*(2), 19–34.
McKercher, B. (1993). Some fundamental truths about tourism: Understanding tourism's social and environmental impacts. *Journal of Sustainable Tourism*, *1*(1), 6–16.
Müller, D.K., Hall, C.M., & Keen, D. (2004). Second home tourism impact, planning and management. In C.M. Hall & D.K Müller (eds), *Tourism, Mobility and Second Homes. Between Elite Landscape and Common Ground* (pp. 15–32). Bristol: Channel View Publications.
Pettorelli, N., Durant, S.M., & du Toit, J.T. (eds) (2019). *Rewilding*. Cambridge: Cambridge University Press.
Scheyvens, R. (2000). Promoting women's empowerment through involvement in ecotourism: Experiences from the Third World. *Journal of Sustainable Tourism*, *8*(3), 232–249.

Scott, I., Denman, J., & Lane, B. (1989). *Doing by Learning: A Handbook for Organisers and Tutors of Village-Based Community Development Courses*. Cirencester: ACRE.

Sharpley, R., & Pearce, T. (2007). Tourism, marketing and sustainable development in the English national parks: The role of national park authorities. *Journal of Sustainable Tourism*, *15*(5), 557–573.

Walton, J.K. (2009.) Histories of tourism. In T. Jamal & M. Robinson (eds), *The Sage Handbook of Tourism Studies* (pp. 115–129). London: Sage Publications.

PART V

EMERGING MATTERS AND NEW DIRECTIONS

24. Community events, rural–urban interdependencies, and rural community development

Kyle Rich

INTRODUCTION

Events and festivals have emerged as a common feature in the landscape of rural tourism. Particularly in Western/developed contexts, events are often leveraged for their potential to attract tourists, generate income, and build community. Events range in size and scope from small/local agricultural fairs and markets that generate a handful of visitors from within a host region, to large-scale spectator events that bring in thousands of international attendees. Indeed, some marquee events rely on rural and remote places. For example, the infamous Burning Man Project takes places annually in the Black Rock Desert, just outside Gerlach, Nevada (Garrett, 2010) – a town of less than 100 people. More commonly, however, organizers in rural areas seek to use events and festivals as a means of deriving a variety of benefits for municipalities, regions, and communities more broadly. These events and festivals often seek to embody or capitalize on unique elements of these local areas by staging or portraying them for visitors (Mannon & Glass-Coffin, 2019). While the elements of these rural places used in tourism strategies are often related to the land or natural environment (e.g., natural resources) they are also – sometimes – more whimsical or lighthearted. For example, Mair (2009a) described a Star Trek Festival developed in the rural town of Vulcan, Alberta, which sought to capitalize on the name of the town as part of economic development efforts. Events and festivals are therefore important elements of the rural tourism landscape and implicated in broader economic and community development practice.

Much public policy and academic attention has been devoted to the strategic use of events and festivals to promote rural tourism and development. For example, rural tourism scholars have discussed the use of event portfolios (Ziakas, 2010), the role of events within participatory approaches to planning (D.G. Reid et al., 2004), and issues related to capacity and succession planning of events and festivals (Frost & Laing, 2015). Although events and festivals are often implicated in economic development strategies and activities (Chhabra et al., 2003), event experiences and many of the impacts they produce for attendees and host communities are firmly rooted in leisure. Chalip (2006) suggested that events create liminal spaces in which typical social norms and hierarchies are suspended, temporarily allowing communities to engage with new ideas, discussions, and ultimately activities that influence social and cultural development. These transformations – such as converting busy streets into pedestrian markets or agricultural fields into festival spaces and camping venues – impact the way people see, experience, and understand communities and the places they are associated with. These impacts are experienced by both tourists and residents of host communities. As such, unpacking the social impacts of events and festivals involves dissecting a complex nexus of factors that are related to tourists and hosts, insiders and outsiders, costs and benefits, as well as transformations and continuities.

In this chapter, I consider some of the social and cultural implications of rural events and festivals. Grounded in the theoretical perspective that leisure affords tourists the freedom to attend events, which in turn offer liminal spaces and transformative potential to change how we see and experience communities, I focus on the movement of people, capital, and ideas between rural and urban spaces. Specifically, I consider the idea of rural–urban interdependencies (Caffyn & Dahlström, 2005) and how these connections (rather than divisions between urban and rural places) are implicated in both the hosting of events and festivals and the outcomes they may manifest for rural communities. Drawing on my own research and experiences participating in ice hockey tournaments in various rural parts of Canada, I illustrate the importance of considering interdependencies for both academics and policymakers interested in rural tourism and development more broadly. Importantly, I highlight the social and cultural implications of these interdependencies and how power, leisure, and social change may all be implicated in rural events and tourism. The chapter proceeds with an overview of rural–urban interdependencies and then an analysis of the aforementioned events before concluding with some reflections on the potential of this theoretical approach and future avenues for research.

RURAL–URBAN INTERDEPENDENCIES

Interdependencies have emerged in the literature in the context of changing ideas about rural and regional development. In many contexts, the rise of neoliberal policy agendas and the reduction in the role of the state in providing resources and services for communities led to the decentralization and downloading of responsibilities to the local level (Harvey, 2007). These trends, coupled with growing skepticism about top-down policy and government models, fueled a growing sense of localism and self-reliance within regions and municipalities (Daniels et al., 2019). The way of thinking that emerged about policy, planning, and development practice has been called "new regionalism" (Scott, 2007; Wheeler, 2002). New regionalism involves a focus on specific territories or places, more holistic (e.g., considering social, economic, political, and environmental factors) and incremental development approaches, as well as more fluid and emergent ways of thinking about how a region is defined – and by extension, how it operates (Daniels et al., 2019). New regionalism offered a more bottom-up and place-specific way of thinking about communities and how they change and develop.

Within this shift towards new regionalism, clearer thinking about the interdependencies of places emerged. Partially in response to the shortcomings of thinking about regions as discrete and finite, interdependencies also resist a tendency (in public discourse and media representation) to rely on urban–rural divides or distinctions. Rural and urban places are connected and depend on each other in many ways. Rural areas provide food, primary resources for construction, and pollution processing for densely populated urban areas, and conversely urban areas provide economic centers, technology, and workforces required to sustain rural economies (Reimer et al., 2019). Importantly, rural and urban places are also linked through social and cultural mechanisms. Rural areas are home to (and stewards of) natural environments where many urban residents and international tourists alike access recreational activities and leisure opportunities. Some of these rural environmental characteristics, such as mountain ranges, rainforests, deserts, and waterways are also key cultural symbols for urban centers and entire nations. For example, in Canada, the Rocky Mountains and the Great Lakes waterways are quintessential symbols of Canadian heritage, which draw visitors each year from both within the country and from around the world. Therefore, the relationships between rural and urban places are complex and intersect across spheres of economic, cultural, political, and social influence (Caffyn & Dahlström, 2005) and recreational, leisure, and tourist activities are implicated in rural–urban interdependencies in various ways (Rich, 2021a).

At its core, the idea of interdependence suggests that "changes in one place affect the other" (Reimer et al., 2019, p. 150). Although they provide a broad and sweeping framework for thinking about rural–urban relationships, interdependencies can be broken down further for the purpose of analysis. Reimer and colleagues (2019) suggested that interdependencies can be analyzed across four spheres: exchanges (economies), institutions, environments, and identities. Exchanges refers to the idea of economic interdependence, which is often the most common topic of discussion among academics and practitioners. Exchanges here refers to both goods and labor and can occur directly between individuals or can occur in complex ways that are intermediated by supply chains or layers of corporate transactions. Given the increasingly globalized state of most economies, these transactions may often involve stakeholders from outside the region(s) in question or indeed in another country altogether (Woods, 2013). Institutional interdependence refers to the ways in which policies and decision-making of public (government), private, and nonprofit organizations have implications for the relationship between rural and urban places. For example, changes in policy regarding transportation and land use may have implications for how citizens are able to flow to and from regions to access work or where a company decides to locate their facility. Ultimately, these kinds of decisions impact where and how both urban and rural folks are able to access employment, goods, and services.

The two spheres of interdependence which have received less attention in the literature are those of the environment and identities. Interdependencies related to the environment are becoming an increasingly prevalent topic of discussion in both academic and public discourse. As noted above, rural environments provide the natural resources for construction in urban areas, and they play the essential role of processing the high levels of pollution generated in urban centers. Urban areas also depend on rural environments as sources of clean water and food production as well as spaces for recreation and leisure (Reimer et al., 2019). As issues related to climate change are finding their way onto political agendas, the environmental aspects of rural–urban interdependencies are becoming increasingly clear, as are their potential to impact other spheres of influence (e.g., economic development). Interdependence based on identity refers to the notion that people form attachments to places and their attachments inform the way they understand these places and themselves (Reimer et al., 2019). These place-based identities play an important role in flows of people and ideas within and between regions. For example, there are a variety of complex processes and experiences surrounding the motivations of young people who migrate out of rural areas in search of specific educational and economic opportunities (Looker & Naylor, 2009), while others choose to remain or migrate into rural areas for various reasons (MacMichael et al., 2015). As alluded to above, identities are also implicated in how urban folks understand and engage with rural places – as visitors, tourists, or temporary residents.

In the remainder of this chapter, I explore the various ways that events and festivals are implicated in and may be analyzed through a lens of rural–urban interdependencies. In doing so, I highlight the social and cultural implications of the theoretical approach as an area which – I believe – may be an important future area of research. While academics and policymakers often discuss the real and potential economic impacts of events and festivals (e.g., Bernthal & Regan, 2004; Chhabra et al., 2003; Long & Perdue, 1990; Ziakas, 2010), much less attention has been paid to their social impacts in rural areas (Alves et al., 2010; S. Reid, 2007). Similarly, in the literature on new regionalism and interdependencies, less work has focused on social and environmental spheres of influence (Reimer et al., 2019). Therefore, in this chapter, while I discuss economic and institutional implications, I also aim to map out a preliminary analysis and to stimulate further discussion and research into the social and cultural aspects of rural–urban interdependencies and how they relate to sport, recreation, leisure, and tourism. I illustrate the potential of this theoretical framework by considering examples of rural ice hockey tournaments, which have recently been the focus of my own research as well as the context for some of my own leisure activities.

RURAL SPORTS EVENTS

Sport is a common feature of the recreation and leisure landscapes of many communities and as such can be used as a lens through which to examine social issues and gain a better understanding of community life. In rural places, sport is commonly discussed as a platform upon which local residents develop social capital and place-based identities (Spaaij, 2009; Tonts, 2005; Tonts & Atherly, 2010). As important volunteer and community organizations, sport clubs and facilities also play a variety of roles in rural communities. For example, Mair (2009b) described how curling clubs in Canada act as third places and fulfill various roles beyond the typical scope of sport organizations, including that of a banquet hall, bar, kitchen, and social space for intergenerational relationship building. Sport and recreation have also been used to explore the impacts of economic restructuring (Oncescu & Robertson, 2010; Tonts & Atherly, 2005) and municipal amalgamations (Rich et al., 2017), as well as tourism development (Costa & Chalip, 2005) in rural areas, and the impacts that changes and developments in these fields have on the lived experiences of rural citizens.

My research focuses broadly on sport, community, and social change (Rich et al., 2021; Rich & Misener, 2020a; Misener et al., 2022) and has recently examined rural ice hockey tournaments and the various ways in which they are implicated in local culture in rural Canada. For example, I have looked at tournaments in Powassan, Ontario, organized around kin-based competition structures (colloquially called "family hockey"), and how these tournaments subvert typical ethics of sport/athleticism to allow for new and sometimes unexpected ways of constructing and expressing ideas of community (Rich et al., 2014). I have also looked at an annual event in rural Saskatchewan where team members from the local First Nation and carded Rodeo Cowboys play against each other in a spectacular performance of sport and local culture aptly (and for some controversially) named "The Battle of Little Big Puck." This investigation explored how narratives about historical and contemporary relationships and community membership impacted how the event was politicized and communicated within and outside the region (Rich et al., 2020). Most recently, I looked at the "Heritage Cup" in Barry's Bay, Ontario. This tournament is organized around cultural groups in the region (i.e., Irish, German, Kashubian/Polish, and Algonquin) and reflects a strong network of local organizations and activists who are engaged in promoting an awareness of cultural heritage and settlement in the area (Rich, 2021b). While all of these sports events have their own unique elements, they share common threads in that they reflect specific elements of local rural cultures (e.g., family structure, heritage, social practices) and capitalize on the liminal spaces created by these events to (re)construct ideas about what it means to be rural and a member (or to not be a member) of the host community (Rich, 2021b). Although the focus in much of this work has, thus far, been to examine ideas of rurality in and through these events, the interconnection between ideas of urban and rural has also become apparent. In what follows, I draw from examples related to these events as well as some related literature to sketch out how the notion of interdependencies may be a useful tool for understanding the role of these events in their respective host communities and how they can help us understand the social processes and impacts of tourism based around these events and others like them.

Economic Interdependence

The idea of interdependencies between rural and urban places can be used to examine economic aspects of these events. As people migrate in and out of rural areas for many reasons, events (unlike regular sport or recreation programming) present a unique opportunity to attract visitors to a rural area for a short period of time. Along with the flow of people to rural areas, we can expect there to be an economic impact of these visitors. While economic impacts may be felt in the usual ways (e.g., hotel stays, restaurant and store sales, travel expenses, etc.), these events also provide substantial opportunities for volunteer organizations to raise funds for a variety of community purposes. For example, the family hockey tournaments are hosted by the local minor hockey association and a community center board and all funds raised are reinvested into the programming of those organizations (Rich et al., 2014); the responsibilities associated with the Heritage Cup (e.g., snack bar, catering, lottery sales, etc.) are shared each year between volunteers of the local heritage societies and the local minor hockey association, who benefit financially from the funds raised at the event; and each year the Battle of Little Big Puck raises funds for a local charity, organization, or cause (Rich et al., 2020). In each of these cases, these sports events draw people to these rural areas and support local causes and community organizations. Through a lens of interdependence, we see how these events allow local organizers to convert the human capital of volunteers (as workers and organizers) into substantial financial capital for community organizations. These examples illustrate how economic interdependence related to tourism is not only applicable or relevant to the private sector, but rather that rural–urban interdependencies also support the flow of financial capital to rural communities through civil society and volunteer organizations.

Institutional Interdependence

Institutionally, these events may be implicated in the policy- and decision-making of a variety of organizations.[1] As noted above, civil society organizations such as sport clubs and heritage groups actively leverage the opportunity provided by sports events. While they benefit financially, the hosting of these events is also implicated in a variety of outcomes for community members. The ways community organizations mobilize around these events are often framed by policy and (in)action of governments at various levels. For example, forced municipal amalgamation and the construction of a highway bypass in Powassan (or more specifically, the former town of Trout Creek) led to a sense of urgency and collective response of community members, which was ultimately acted out through the community center board, who reinvigorated their programming – some of which was centered around the family hockey tournament (Rich et al., 2017). In Barry's Bay, the strong connection to various heritage groups can be traced back to historical government practices of settlement in the region (Murray, 2018). These institutional forces that traverse rural and urban space ultimately shape the way these events are organized and enacted in these communities.

Various institutional structures related to community development are implicated in these events, including the sport system in Canada. The Canadian sport system is a federal system that operates through a network of national and provincial nonprofit organizations, which are largely organized around individual sports (Thibault &

Harvey, 2013). Although these organizations are generally responsible for supporting both recreational sport participation as well as development pathways for elite athletes to progress to the highest levels of competition, the support provided is not always effective or appropriate for community sport organizations in rural areas. These rural clubs may seek out diverse and innovative ways of creating and maintaining participation opportunities (Rich & Misener, 2019; Rich et al., 2022) and simultaneously supporting their communities outside sport (Mair, 2009b) – many of which involve hosting events and tournaments. As such, the institutional changes enacted by (governments and nongovernmental) organizations within and outside sport impact the way rural folks organize and engage with sports events. Interdependence in this case is helpful in understanding how institutional practices shape the organization and conceptualization of the events themselves as well as how the events are leveraged for various outcomes in rural communities. The flow of people, capital, and ideas to and from rural areas is therefore deeply implicated in these institutional structures.

Environmental Interdependence

Although the environment may not be the first sphere of interdependence that comes to mind when considering sports events and rural tourism, there are certainly implications to consider. Environmental sustainability is an issue that permeates all sectors and is particularly relevant for tourism. Whereas major concerns have been raised about the viability of large-scale and mega-events such as the Olympic Games, scholars have suggested that small-scale sports events may be part of a sustainable tourism development strategy (Gibson et al., 2012).

In the context of rural Canada, sports events generally, and ice hockey tournaments specifically, are implicated in the environment in various ways. Firstly, the imagery of rural sporting landscapes is often intertwined with narratives of Canadian identity. For example, the imagery of "pond hockey" is a prevalent signifier that is visible in many contexts from professional sports events like the National Hockey League's Heritage Classic (Ramshaw & Hinch, 2006) to the back of the Canadian five-dollar bill (Edwards & Kulczycki, 2018). Ironically, scholars have predicted that current climate change trajectories will lead to substantially fewer outdoor skating days in major Canadian cities (Robertson et al., 2015). Hockey tournaments are often accompanied by a variety of outdoor winter activities (e.g., a family hockey tournament hosted as part of a winter carnival), and therefore a changing climate poses a real threat to the viability of these events. Second, given that rural sports events often rely on many participants traveling to host communities to participate (Rich, 2021b), it is reasonable to expect that these tourists will have a greater carbon footprint than if they were to access participation opportunities in their own communities. With growing awareness of the environmental impacts of various sport and recreation activities, we are developing more nuanced understandings of how heterogeneous patterns of day-to-day sport participation have varying levels of environmental impact, which can be measured by carbon footprinting (Wicker, 2019). While certainly not an exhaustive list, these examples demonstrate how – through events – rural and urban areas are connected and implicated in issues related to the environment in complex and interdependent ways.

Identity Interdependence

In many cases, events offer people an opportunity to return to a place where they feel some sort of attachment or affective connection. The liminal spaces created around events and festivals offer both a reason to visit rural places as well as an opportunity for host communities engage in a process of (re)creating identities and affective connections related to rural places (Rich, 2021b). Given that these sports events are implicated in the movement of people to and from rural places, they are involved in identity interdependence in various ways.

At all of the sports events that I observed and participated in (for research or leisure), community members recognized that they served as a sort of "homecoming" (Rich, 2021b, p. 8) which drew people back to host communities. Presumably, this reflects an attachment to the place and, by extension, the idea that people maintain a sense of identity associated with rural places despite having moved on or away. As many rural youths migrate to more urban areas for employment and education, these events draw them back – reflective of the identity sphere of interdependence (Reimer et al., 2019). The structure of these sports events is also explicitly about elements of identities: families, heritage, and cultural or ethnic backgrounds. Layered on top of these identity markers are ideas about what it means to be, identify with, or see the world from a rural perspective (Rich & Misener, 2020b; Rich, 2021b). For example, the mural on the wall of the rink in Barry's Bay reads "Welcome *back* to the Madawaska valley" (emphasis added, Rich, 2021b, p. 8). I read this mural to infer that people come and go regularly but are often drawn back for the sport and recreational activities that take place in the arena. The politicization of elements of the Battle of Little Big Puck was also firmly grounded in the idea that folks from urban areas or other parts of the country "don't understand" (Rich et al., 2020, p. 57). Although these ideas were firmly grounded in a narrative of differences between urban and rural places, these differences rely on shared understandings of what it means to identify as a member of a community or being from a place. Effectively, these identities rely on each other in interdependent ways. That is, rural ways of knowing cannot exist as such without an urban equivalent to compare to. Importantly, these ideas are not static and fixed, but fluid and developing, similar to the communities with which they are associated.

IMPLICATIONS, APPLICATIONS, AND LIMITATIONS

In this chapter I have attempted to sketch out how the idea of rural–urban interdependencies may be a useful tool for examining the impacts of events and festivals in and for rural communities. While the analysis is certainly not an exhaustive or systematic account, it is my hope that I have provided an insightful and useful theoretical approach to examining some issues related to tourism and rural development. In conclusion, I offer a few thoughts on implications and avenues for future inquiry as well as some extant limitations of this approach.

While the interdependencies of rural and urban places are often discussed in relation to economic and institutional spheres, less attention has been focused on environmental and identity interdependencies (Reimer et al., 2019). Using case studies of sports events, I have attempted to highlight some of the ways that we might examine the social and cultural elements of tourism related to these events, and how interdependencies might be a fruitful way to conceptualize these processes. This approach has implications for both theory and practice of rural tourism planning and development as it highlights the complex and interdependent ways in which the movement of people, capital, and ideas may impact host communities. While planning for rural tourism may explicitly consider how to attract people and capital from urban areas, without considering the social and cultural ramifications of these processes, our understanding of rural tourism may remain partial and incomplete. Interdependencies may provide a useful framework to consider in this regard.

Theoretically, the notion of interdependencies also provides a helpful framework for thinking through power relationships involved in sport, leisure, and tourism. Sports activities such as the tournaments and events described in this chapter may (in some cases) be overlooked as components of a municipal or regional tourism strategy or plan. In particular, if they are not organized or developed specifically for the purpose of generating tourism or economic impacts, they may not be considered seriously. The various spheres of interdependence outlined above highlight the social, cultural, and environmental implications of these events that may be of interest to both scholars and practitioners precisely because of the impacts that they may engender in and for communities. Furthermore, the framework of interdependencies highlights the complex and messy ways that the different spheres of interdependence may intersect and intertwine. In line with the theoretical framing of new regionalism, considering these rural–urban interdependencies challenges us to think about place-based or regional analyses and to consider the problematics of development and planning processes that are entirely top-down or bottom-up in nature. Interdependencies highlight how changes in economic, social, cultural, and environmental spheres of both urban and rural places may have ramifications for citizens and visitors from a much larger geographical region than we might expect.

It is also important to highlight how the ideas and examples discussed in this chapter are quite narrow in scope. Indeed, both the spheres of interdependence and examples discussed here come from rural Canada, and the research described was undertaken from settler/Eurocentric perspectives. While I hope that the ideas advanced here are useful and thought provoking for a variety of readers, additional work is needed to examine how these ideas may be taken up, understood, and refuted in diverse social and cultural contexts as well as epistemological and ontological perspectives. I hope that the discussion provided here might stimulate some of this work. Moving forward, the idea of rural–urban interdependencies can help us think through some of the complex and intersecting ways that tourism, leisure, sports, and events are implicated in community development in rural areas.

NOTE

1. While beyond the scope of this chapter to explore in depth, in Canada (and elsewhere) the institutional interdependencies of urban and rural regions were starkly highlighted through public health measures that were put in place in 2020 in response to the COVID-19 pandemic (e.g., see Rich, 2021).

REFERENCES

Alves, H.M.B., Cerro, A.M.C., & Martins, A.V.F. (2010). Impacts of small tourism events on rural places. *Journal of Place Management and Development*, *3*(1), 22–37.

Bernthal, M.J., & Regan, T.H. (2004). The economic impact of a NASCAR racetrack on a rural community and region. *Sport Marketing Quarterly*, *13*(1), 26–34.

Caffyn, A., & Dahlström, M. (2005). Urban–rural interdependencies: Joining up policy in practice. *Regional Studies*, *39*(3), 283–296.

Chalip, L. (2006). Towards social leverage of sport events. *Journal of Sport & Tourism*, *11*(2), 109–127.

Chhabra, D., Sills, E., & Cubbage, F.W. (2003). The significance of festivals to rural economies: Estimating the economic impacts of Scottish Highland Games in North Carolina. *Journal of Travel Research*, *41*(4), 421–427.

Costa, C.A., & Chalip, L. (2005). Adventure sport tourism in rural revitalisation—An ethnographic evaluation. *European Sport Management Quarterly*, *5*(3), 257–279.

Daniels, J., Douglas, D.J., Vodden, K., & Markey, S. (2019). What is New Regionalism? In K. Vodden, D.J. Douglas, S. Markey, S. Minnes, & B. Reimer (eds), *The Theory, Practice and Potential of Regional Development: The Case of Canada* (pp. 30–55). Abingdon: Routledge.

Edwards, J., & Kulczycki, C. (2018). "Who wants to go the outdoor rink?": A media analysis of Canada's outdoor rinks and the connection to Canada's national identity. *Leisure/Loisir*, *42*(4), 389–424.

Frost, W., & Laing, J. (2015). Avoiding burnout: The succession planning, governance and resourcing of rural tourism festivals. *Journal of Sustainable Tourism*, *23*(8–9), 1298–1317.

Garrett, R. (2010, April 20). Designing Black Rock City. *The Burning Man Journal*. Retrieved March 15, 2021, from https://journal.burningman.org/2010/04/black-rock-city/building-brc/designing-black-rock-city/.

Gibson, H.J., Kaplanidou, K., & Kang, S.J. (2012). Small-scale event sport tourism: A case study in sustainable tourism. *Sport Management Review*, *15*(2), 160–170.

Harvey, D. (2007). *A Brief History of Neoliberalism*. Oxford: Oxford University Press.

Long, P.T., & Perdue, R.R. (1990). The economic impact of rural festivals and special events: Assessing the spatial distribution of expenditures. *Journal of Travel Research*, *28*(4), 10–14.

Looker, D.E., & Naylor, T.D. (2009). "At risk" of being rural? The experience of rural youth in a risk society. *Journal of Rural and Community Development*, *4*(2), 39–64.

MacMichael, M.F., Beazley, K., Kevany, K., Looker, D., & Stiles, D. (2015). Motivations, experiences, and community contributions of young in-migrants in the Maitland area, Nova Scotia. *Journal of Rural and Community Development*, *10*(4), 36–53.

Mair, H. (2009a). Searching for a new enterprise: Themed tourism and the re-making of one small Canadian community. *Tourism Geographies*, *11*(4), 462–483.

Mair, H. (2009b). Club life: Third place and shared leisure in rural Canada. *Leisure Sciences*, *31*(5), 450–465.

Mannon, S.E., & Glass-Coffin, B. (2019). Will the real rural community please stand up? Staging rural community-based tourism in Costa Rica. *Journal of Rural and Community Development*, *14*(4), 71–93.

Misener, L., Rich, K., & Pearson, E. (2022). Tensions and opportunities in researching social change in sport management. *Sport Management Review*, *25*(2), 323–340.

Murray, D. (2018). *A "Colony of Unrequited Dreams"? Settler colonialism and the failed-settlement narrative in the Ottawa-Huron Tract, 1850–1910* [Doctoral dissertation, University of Victoria]. http://dspace.library.uvic.ca.

Oncescu, J., & Robertson, B. (2010). Recreation in remote communities: A case study of a Nova Scotian village. *Journal of Rural and Community Development*, *5*(1), 221–237.

Ramshaw, G., & Hinch, T. (2006). Place identity and sport tourism: The case of the heritage classic ice hockey event. *Current Issues in Tourism*, *9*(4–5), 399–418.

Reid, D.G., Mair, H., & George, W. (2004). Community tourism planning: A self-assessment instrument. *Annals of Tourism Research*, *31*(3), 623–639.

Reid, S. (2007). Identifying social consequences of rural events. *Event Management*, *11*(1–2), 89–98.

Reimer, B., Barrett, J., Vodden, K., & Bisson, L. (2019). Rural–urban interactions and interdependence. In K. Vodden, D.J. Douglas, S. Markey, S. Minnes, & B. Reimer (eds), *The Theory, Practice and Potential of Regional Development: The Case of Canada* (pp. 149–182). Abingdon: Routledge.

Rich, K.A. (2021a). Rural–urban interdependencies: Thinking through the implications of space, leisure, politics, and health. *Leisure Sciences*, *43*(1–2), 190–196. https://doi.org/10.1080/01490400.2020.1774001.

Rich, K.A. (2021b). Rural sport spectacles: Ice hockey, mythologies, and meaning-making in rural Canada. *Leisure Sciences*, *43*(6), 617–629. https://doi.org/10.1080/01490400.2020.1870591.

Rich, K.A., & Misener, L. (2019). Playing on the periphery: Troubling sport policy, systemic exclusion, and the place of rural sport in Canada. *Sport in Society*, *22*(6), 1005–1024.

Rich, K.A., & Misener, L. (2020a). Get active Powassan: developing sport and recreation programs and policies through participatory action research in a rural community context. *Qualitative Research in Sport, Exercise and Health*, *12*(2), 272–288.

Rich, K.A., & Misener, L. (2020b). Resisting, reproducing, and recreating rurality. In T.D. Glover & E.K. Sharpe (eds), *Leisure Communities: Rethinking Mutuality, Collective Identity and Belonging in the New Century* (pp. 192–201). Abingdon: Routledge.

Rich, K.A., Bean, C., & Apramian, Z. (2014). Boozing, brawling, and community building: Sport-facilitated community development in a rural Ontario community. *Leisure/Loisir*, *38*(1), 73–92.

Rich, K.A., Misener, L., & Trout Creek Community Centre Board (2017). Community centres as spaces for sport, recreation, and resiliency: The case of the Trout Creek Community Centre. In L. Brinklow & R. Gibson (eds), *Building Community Resilience: From Black Horses to White Steeds* (pp. 109–131). Charlottetown, PE: Island Studies Press.

Rich, K.A., Hoeber, L., & Weisgerber, A. (2020). The Battle of Little Big Puck: Narratives of community, sport, and relationships in rural Canada. *Journal of Rural and Community Development*, *15*(3), 45–64.

Rich, K.A., Spaaij, R., & Misener, L. (2021). Theorizing community for sport management research and practice. *Frontiers in Sports and Active Living*, *3*. https://doi.org/10.3389/fspor.2021.774366.

Rich, K., Nicholson, M., Randle, E., Donaldson, A., O'Halloran, P., Staley, K., Nelson, R., & Belski, R. (2022). Participant-centered sport development: A case study using the leisure constraints of women in regional communities. *Leisure Sciences*, *44*(3), 323–342. https://doi.org/10.1080/01490400.2018.1553124.

Robertson, C., McLeman, R., & Lawrence, H. (2015). Winters too warm to skate? Citizen-science reported variability in availability of outdoor skating in Canada. *The Canadian Geographer/Le Géographe Canadien*, *59*(4), 383–390.

Scott, J.W. (2007). Smart growth as urban reform: A pragmatic "recoding" of the new regionalism. *Urban Studies*, *44*(1), 15–35.

Spaaij, R. (2009). The glue that holds the community together? Sport and sustainability in rural Australia. *Sport in Society*, *12*(9), 1132–1146.

Thibault, L., & Harvey, J. (2013). *Sport Policy in Canada*. Ottawa, ON: University of Ottawa Press.

Tonts, M. (2005). Competitive sport and social capital in rural Australia. *Journal of Rural Studies*, *21*(2), 137–149.

Tonts, M., & Atherley, K. (2005). Rural restructuring and the changing geography of competitive sport. *Australian Geographer*, *36*(2), 125–144.

Tonts, M., & Atherley, K. (2010). Competitive sport and the construction of place identity in rural Australia. *Sport in Society*, *13*(3), 381–398.

Wheeler, S. (2002). The new regionalism: Key characteristics of an emerging movement. *Journal of the American Planning Association*, *68*(3), 267–278.

Wicker, P. (2019). The carbon footprint of active sport participants. *Sport Management Review*, *22*(4), 513–526.

Woods, M. (2013). Regions engaging globalization: A typology of regional responses in rural Europe. *Journal of Rural and Community Development*, *8*(3), 113–126.

Ziakas, V. (2010). Understanding an event portfolio: The uncovering of interrelationships, synergies, and leveraging opportunities. *Journal of Policy Research in Tourism, Leisure and Events*, *2*(2), 144–164.

25. Centering animals within rural tourism

Carol Kline

INTRODUCTION

With the rise in interest about animals in tourism and a resurgence of activity around tourism's role in rural resilience (Fountain et al., 2021) set upon the stage of staggering statistics of biodiversity and population loss (World Wildlife Fund & the Zoological Society of London, 2020), it seems appropriate to examine the intersection of animals and rural community development to identify areas for significant improvement. Indeed, many "animal welfare problems" are also environmental problems, entangled within established cultural and heritage norms that are often propped up by capitalist economic systems. The aim of this chapter is to "illuminate tourism's blind spots" (Tribe, 2008, p. 251) relative to animals in rural tourism development and to "engage with normative questions of values and desirable ends" (ibid., p. 247). I contextualize the chapter using critical theory to rebuke forms of animal welfare violations in rural tourism, and then quickly move to concepts of regenerative and transformative tourism so as to suggest a new path.

Most of the academic work on the intersection of animal-based tourism and rural tourism development has historically been associated with hunting and fishing, wildlife tourism, ecotourism, and agritourism (Clamen & Rossier, 1991; Dashper, 2015). In a 2000 report published on the Rural Tourism Management: Sustainable Options Conference (Hall, 2000), there was scant mention of animals. The introduction to the report noted that "the role of, and rural tourism's relationship with animals, was … only addressed by one or two papers" (Hall, 2000, p. 296) and "the wider role of animals in the consumption of rural tourism and questions of animal welfare may raise important ethical questions" (ibid., p. 298). Indeed, we have made progress since that time, but because humans concerned for the welfare of animals are often deemed overly sentimental, or anthropomorphizing, or ignorant of the natural hierarchy of the world (Deckers, 2009; Haynes, 2001; Singer, 1975), it is only recently that voicing this concern has become more acceptable in the mainstream. Indeed there continues to be an "anxiety attached to femininity and animality" (Probyn-Rapsey, 2018, p. 175).

Increasingly, we are seeing research about additional and nuanced contexts emerge, injured animal sanctuaries, tours of production farms for alligator hides or civet coffee, working animals such as donkeys or elephants, and racing sports involving horses, greyhounds, and camels. However, the lines blur between what is a rural tourism experience with animals and what is urban, as sloths and camels are brought into the city for selfies and increased land development for residential, tourist, and agricultural use continues to pervade what were once wild(er) spaces. Indeed, when the focus of rural tourism analysis is via animals, the context of animal encounters must first be delineated.

Rural tourism activities may include animals as integral to the direct experience, for example, riding seaside donkeys, eating Tibetan pig in a local restaurant, visiting a wildlife sanctuary, or taking a self-guided birdwatching tour. Alternatively, animals may appear as part of the backdrop, as in the case of grazing sheep, breaching whales, or roaming reindeer. While we know that animal encounters within rural communities occur in wild, semi-natural, and captive settings (Orams, 2002), their omnipresence is often unnoticed or taken for granted. Dashper (2021) outlines four factors to consider relative to the context of animals who are working/performing within tourism, namely the animal's degree of wildness–habituation–domestication; level of captivity and restriction; degree of agency; and the type of working expectations humans have placed upon them. As an example of these expectations, in their discussion of a well-accepted practice within agritourism – sheep-shearing – Tully and Carr (2021, p. 83) elucidate the ethical shortcomings of producing farm animal experiences that ignore the "complex physical, mental, emotional, and social interests" of the animals as active participants, because the sheep are viewed as objects for commodification and utility. The farm and farmed animals discussed in their work (sheep, llamas, cows, pigs, goats, horses) are deemed "tourism animals" who are there solely for entertainment, to generate an "awww" factor, or are leveraged as symbols to legitimize the farm backdrop. However, despite the important role they play in generating tourism revenue, the tourism animals are captive, domesticated animals, with little agency, and are expected to perform or interact upon demand. At times the "performance" might include a rough handling by farm workers while other times the animal is subject to continuous stimulation by visitors touching or hugging them, talking to them, and posing for selfies with them.

Historic writings and practices around rural community tourism development took it for granted that cultural and heritage employment of animals was intrinsic to the rural context, rarely questioning the welfare and agency of animals involved, much less viewing activities involving animals through an ethical lens. Küçükaltan and Dilek (2019) call this the meta-production of animals in tourism and, because animal-based tourism happens in theoretical realms as well as practical, they assert that as long as the moral status of animals in tourism is ignored, new ideas about the moral responsibility of tourism will not be developed. Fennell and Thomsen (2021) situate the moral consideration of animals around the pragmatic concept of animal suffering. Below is a framework delineating the various settings and related elements for rural community and tourism development that incorporates considerations of the animal tourism context. While the fragility of the ecosystem is an inherent contextual element of rural tourism, the number of ways to estimate/establish a measurement for fragility is complex, and discussion of them is beyond the scope of this chapter. They are therefore not included in the framework below.

Table 25.1 Contextual factors of animal tourism within rural tourism

Contextual element	Contextual range
Setting for animal	Wild, semi-wild, captive
Role/expectations of animal	Attraction, commodity, threat, companions, shared encounters, destination setting, destination identity or narrative
Level of contact/habituation	No contact, low contact, familiarity but independence, moderate contact or habituation, high habituation
Level of embodiment	Animal part is ingested by or worn on one's person; animal and humans touch frequently; animal is touched infrequently; animal is present but not touched; animal is remote
Tourism's impact on animal's behavior/life	High, moderate, low, seemingly non-existent
Agency of animal/restrictions on animal	Four types of agency: behaviorally passive or purely reactive; behaviorally pursuing desirable outcomes; engaging with the environment to gain skills and information for future use; and pursuing long-term goals through planning and reflection
Welfare of animal	Freedom from: thirst, hunger and malnutrition; discomfort and exposure; pain, injury, and disease; fear; freedom to express normal behavior
Fulfillment of animal	Elements of setting include nutrition and pleasurable eating, comfortable environment such as suitable shade/shelter and comfortable resting areas, fresh air, well-coordinated physical activity, engagement in rewarding and stimulating activities, social pleasure, confidence and a sense of control
Animal's level of conflict with human activity	High, moderate, low, seemingly non-existent
Impact of COVID-19 on animal	Shortage of food, in danger of losing home, having an enriched/relaxed environment, spatial expansion, disease transmission, poaching, unusual sightings
Conservation status of animal	Least concern, near threatened, vulnerable, endangered, critically endangered, extinct in the wild, and extinct

Sources: Adapted from Dashper (2021); IUCN (2021); Markwell (2015); Mellor (2016); Orams (2002); Shani (2013); Špinka (2019); Usui et al. (2021).

CRITICAL THEORY: A DEPARTURE POINT

Applying a critical theory lens to animals within rural tourism development, it is quickly evident that issues of power come into play (Tribe, 2008). For instance, one can begin to appreciate issues such as physical power over individual animals, the power to give or take agency, the power to create and determine the outcome of life-or-death situations, the power to provide a living context that offers or deprives intellectual and emotional stimulation, the power to disrupt natural social structures and behaviors, and the power to nurture or to desecrate environmental ecosystems. Additionally, human power is exerted in our language about animals and in our hegemonic assumptions of their place in a societal hierarchy where the human animals are always at the top. Broadly speaking, human power is displayed in the cultural practices that we create and deem as "norms," mostly because there is advantage to us to do so. Oppressive aspects of culture include the unquestioned nature of "heritage" as it relates to using animals in recreation or sport or entertainment, dominant discourses stoked by industrial interests regarding traditional animal meat-centric meals, and the narrative around animals who "like to work" and are thereby provided tasks to help human economic gains.

Critical theories can be employed to dismantle the archetype of "rural" and challenge omnipresent notions of a peaceful, unpretentious, and charismatic scene where wholesome activities and charming folkways are practiced daily. Tribe (2008, p. 248) notes that "mass contemporary culture" as well as tourism promotion paint a rosy picture of tourism's function in the world. For example, Tribe writes, "the plane as a symbol of escape and empowerment rather than of pollution; the five-star hotel as the ultimate in luxury rather than a symbol of social class and division; and the developing world as the playground of the exotic other rather than a site of development and self-determination." So too should the application of critical theories help us understand the role of animals in rural tourism and to elevate their status to *stakeholder*, freeing them from, "an ideological framing and interpretation" of dominant (human) actors (p. 249).

While Tribe likely was not focused on animals at the time of writing "Tourism: A critical business" (2008), his work relates powerfully to this topic. Here is a passage with my insertions in brackets: "critical tourism aims for understanding [of animals], belonging [of animals], being [of animals], emancipation [of animals], and accommodation [of animals] in and with the world. But if tourism does not become a fully critical business it will seek only technical solutions to issues of [animal] management and governance" (Tribe, 2008, p. 254). As academics continue to engage in the animal turn, multispecies turn, and justice turn in tourism, it is time to take the consideration of animals out of the *technical* domain. We turn to Ateljevic's work (2011; 2013; 2020) on the transmodern paradigm to find a hopeful and regenerative form of rural tourism as it relates to animals.

TRANSMODERNITY IN TOURISM: OUR DESTINATION

Ateljevic (2011) positions transmodernity as a framework that "gives us an enormous political weight to point to the agency and authority of tourism to possibly change the world for the better" thus changing the "dominant interpretations of tourism as being nothing more than a frivolous leisure activity or yet another form of economic development" (p. 497). She tells us that tourism can be a change agent when attached to a transmodern paradigm. Characteristics of transmodernity include active tolerance of one another as we recognize interdependency, resist values of domination or control, and lean toward flatter organizations and decision-making by consensus. Stewardship of the natural world, being open to the sacred and spiritual realms of life, and dismantling of hierarchies among cultures and people (and I would argue, species) are embodied in this paradigm (Ateljevic, 2013). Transmodernity opposes the "endless economic progress and obsession with material wealth" (ibid., p. 39) and rather ascribes to a quality-of-life measure of global advancement. Notably,

Ateljevic draws on the work of Riane Eisler to explain that we are witnessing the "world-wide movement towards a partnership system of caring economics" (ibid., p. 44). As an example of transmodernity in action, Garrido (2011) analyzes a case of rural tourism within Cádiz, Spain, where tourists protest a mass tourism development in order to protect a small town and rural way of life.

In 2020, Ateljevic proposed that a transformational direction for tourism would be to link itself to regenerative agriculture because of the sheer enormity of food embedded in the tourism system; it is a prominent feature of the tourism experience and is tied to considerable geophysical resources, financial gain, and significant employment. Other authors have also focused on food as a means to shape the trajectory of tourism to a more sustainable and meaningful one. However, it is only recently that the ethical issue of animals as food within the tourism context has even been raised (Bertella, 2020; Fennell & Markwell, 2015; Kline, 2018a; 2018b). Animals as food remains a problem because market pressures – sometimes fabricated by the tourism industry – can increase production of farmed animals needlessly, creating massive environmental impacts and presenting a false face of what and how often locals eat animals. Bertella (2020) spotlights the animals-as-food issue relative to the Sustainable Development Goals (SDG), namely individual physical health (SDG #3) and climate action (SDG #13). Mkono (2015) and Burns et al. (2018) discuss the paradox of tourists eating the same types of wildlife that they travel to view in a natural or semi-natural setting.

Bricker and Joyner (2018) question the use of meat from confined animal feed operations in "sustainable" tourism operations in the US, while Fennell and Markwell (2015) explored ethical dimensions of food in Australian ecotourism. Schmalz et al. (2020) demonstrate how food and leisure can be used for resistance and social equity; a notion that Adams and Messina (2018) set forth in their part-manifesto, part-cook book *Protest Kitchen*. Duffy et al. (2017) analyzed Cuban permaculture tourism through the Community Capitals framework, explaining the ripple effect of positive impacts that a sustainable food system embedded within tourism has made in a rural area outside Havana. Food tourism overlaps distinctly with animals in that they are often what is on the plate – sometimes an iconic dish representing a destination. As food, animals are squarely at the center of daily rural tourism practices, as well as in the multiple other roles animals play, including symbolizing the destination through marketing images, holding a central role within an area's heritage and stories, or figuring prominently as part of the backdrop or naturescape of a rural place. As such, Ateljevic's (2020) supposition for tourism to embrace regenerative agriculture would no doubt have a direct effect on the volume of animals "produced" in the name of food.

TOURISM AS A RURAL CHANGE AGENT

Just as Ateljevic outlines the transformational power that tourists can wield, Sheldon (2020) offers a similar view, citing touristic circumstances that can elicit selfless acts from tourists. She also observed that within the appropriate context, tourists can transition from consumer to change agent; tourism experiences that offer deep connectivity with the environment or with others, provide a chance for self-inquiry, and include engaged contribution to something beyond oneself, can shift human (i.e. tourist) consciousness. When this happens, it can initiate a transformational moment for tourists that may trigger a desire to contribute to (or regenerate or restore) a local community's assets, such as a streambed or forest habitat. Sheldon (2020) provides ample examples of established transformative settings and experiences, focusing on how we might facilitate those as providers. With a little imagination, we can apply Sheldon's suggestions to rural tourism experiences that would benefit animals directly or indirectly via environmental improvements.

Where Ateljevic focuses on the promise of the enlightened tourist as an empowered force for change in visitor activity and accountability, and Sheldon attends to experience design in order to facilitate inner transformation, Cave and Dredge (2018; 2020) describe alternative or regenerative economic operating systems to guide us to a new way of thinking about how tourism might be organized and carried out. They outline how enterprises, labor, property, transactions and finance can be shaped alternatively to capitalist models – and provide examples of where new models are being enacted. They warn us, however, that a shift won't be easy because "seven decades of growth have thwarted any appetite to imagine new and alternative economic models in tourism" (Cave & Dredge, 2020, p. 504). The World Wildlife Fund and the Zoological Society of London (2020, p. 4) agreed with this assessment in their 2020 *Living Planet Report*, stating that while we have failed to seek it thus far, we have to transform to "a society and economic system that values nature, stops taking it for granted and recognises that we depend on nature more than nature depends on us." Reports issued from the November 2021 Conference of Parties 26 held in Glasgow were equally glum, declaring that there is no other choice but for world leaders to recommit to net zero emissions by 2050 (as well as big reductions by 2030) in order to keep the 1.5 °C goal achievable. Obviously, rural environments – and the animals that live there – are most susceptible to the climate crisis as we have seen in the increasing number of natural disasters globally as well as the unequal effects of climate change on less developed nations.

The alternative economic examples provided by Cave and Dredge (2020) and the varied economic genres outlined by Sheldon (2021) (that is, collaborative, circular, creative, generosity, sacred, and regenerative), demonstrate that new approaches to address old problems are not just abstract or academic in nature. Cave and Dredge, Sheldon, and Ateljevic each demonstrate how their conceptual models of hopeful economies are currently a reality by sharing examples from around the globe. We are at tourism's coming-of-age moment where, just as in the human development process, maturation "requires learning, inner reflection, taking responsibility for consequences of choices made and a commitment to wiser choices" Sheldon (2021, second paragraph).

Rural economies, environmental health, and animal life are inextricably intertwined. We should thus continue to use critical theory to subjugate various forms of animal welfare violations in all tourism. However, I am basing the ideas in this chapter within concepts of regenerative and transformative tourism and trying to stay within an arena of hopeful and transmodern views. Ateljevic (2020, p. 3) tells us that we are currently seeing a "emerging global shift in human consciousness rather than just 'special interest' market segments." She suggests that hopeful tourism scholarship "strives for the transformation of our way of seeing, being, doing and relating in tourism worlds [and promotes] the creation of a less unequal, more sustainable planet through action-oriented and participant-driven learnings and acts."

Milking Cows

It is within this space of considering a transformative and regenerative future for animals that I wish to reflect upon two rural tourism concerns that have recently been in the news. They have not been couched as tourism issues, but they are, nonetheless. The first involves dairy cows and the dairy milk industry in general. The reputation of milk as a health food has recently begun to unravel as longitudinal trends have shown severe diseases attributed to a dairy-rich diet (Willett & Ludwig, 2020) and as the science behind the original health claims is dismantled (Goodman, 2020; Jacobs et al., 2020). There is significant medical evidence that milk and dairy can cause bone loss, bone fractures, and, when compared to plant-based sources of protein, higher mortality (Willett & Ludwig, 2020), contributing to heart disease, type 2 diabetes, Alzheimer's disease, and an increased risk of breast, ovarian, and prostate cancers (Physician's Committee for Responsible Medicine, n.d.). Additionally, traces of pesticides, growth hormones, and antibiotics have been found to be prevalent in non-organic samples of milk (Welsh et al., 2019). Further, speaking strictly to a US context, many of the studies, as well as the policies of distributing cows' milk to the public, were not carried out with cultural differences (i.e. eating preferences) and physiological differences (i.e. lactose tolerance or vitamin D absorption) in mind, pertaining to African American, Asian, and Native American populations. In a recent meta-examination of beneficial health claims by the milk industry published in the *New England Journal of Medicine*, Willett and Ludwig (2020, p. 651) conclude "the current recommendation to greatly increase consumption of dairy foods ... does not appear to be justified."

These studies about milk's effects on our health are the foundation for addressing the welfare violations that are becoming more commonly understood: if cows' milk is not beneficial from a health standpoint, and the dairy industry is hard on the environment, then it is even less justifiable to cause the life of dairy cows to be difficult. As with many environmental and animal welfare-focused documentaries, the 2017 release of *The Milk System* (Pichler, 2017) highlighted what has been known for some time – many dairy cows lead miserable lives. Despite the idyllic symbolism of cows, green pastures, and the honest livelihood of living off the land, the dairy industry is pushing cows and farmers past their limits. Cows are artificially inseminated often and typically have one calf per year; the newly born calves are immediately separated from the cow after birth – typically not allowed any of the milk – and are either killed or sold off as veal cows (males) or used to perpetuate the milk cycle (females). Cows mourn when their offspring are taken away (King, 2017). The udders of cows often become infected with bovine mastitis, which is spread easily through milking machines or human handling (Food and Agriculture Organization, n.d.). Mastitis is the most common disease of dairy cows (Cheng & Han, 2020); however, the overuse of antibiotics for mastitis has caused the industry to fear the emergence of antibiotic-resistant pathogens and to reduce the use of antibiotics. Additionally, udder edema is common in high-producing dairy cattle before and after giving birth (Ruegg, 2015). When the treatment for these conditions causes expense beyond what the farmer can bear, culling is recommended (Food and Agriculture Organization, n.d.). On some farms, the horns of cows are de-budded (to prevent a cow using them on a handler) and tails of cows are removed because they get in the way of milking; however, pain medication is not used. If and once a cow cannot get pregnant, they are killed.

Heritage Hunting with Dogs

The second issue related to rural tourism is the heritage sport of hunting with dogs. While the pastime has been long criticized because of the welfare of prey being chased, the practice of culling hunting dogs has been more elusive to the public. Recent footage has exposed what is known to be a relatively common occurrence in hunting circles. On Friday October 9, 2021, ITV Media in the UK broadcast footage of four separate incidents spanning April to September where the dogs at the kennels of the Duke of Beaufort's Hunt in Badminton, Gloucestershire were shot and killed. While this is not an illegal practice, there is a growing understanding in the public that hunting dogs are killed (by shooting) in several circumstances, as noted by Hounds Off (n.d., main webpage), an activist group within the UK that teaches "homeowners, landowners and tenants to protect their properties, livestock and pets from hunt trespass." Hounds Off, along with other anti-hunting and pro-conservation groups, bring attention not only to the animal welfare violations regarding the wildlife hunted but also in respect of the dogs doing the hunting.

It is estimated that 3,000–4,000 dogs are killed each year in the UK because they have reached the end of their hunting days (Burns et al., 2000), although this figure has been questioned as an underestimate (Hashman, 2016). However, there are many other factors behind the killing of these dogs, namely: (1) foxhounds are continually bred to improve the characteristics desired in the hunt; excess puppies may be put down; (2) dogs may not be able to adapt to the training; (3) dogs may be too vocal during a hunt; (4) dogs may become sick or injured, and (5) dogs may become ill from being kept in unhygienic conditions (thereby contracting disease) or fed meat from infected cattle (infected, for example, with bovine tuberculosis) (Burns et al., 2000; Griffin, 2007; Harris & Dorning, 2018; Hashman, 2016). Similar culls of sled dogs have been reported in the United States and Canada within the context of dogs who have ended their use for human profit, who do not bear the desired characteristics for the sport, or who are injured (Fennell, 2011). In addition to culling, in a report titled *Hunting with hounds and the spread of disease* by Harris and Dorning (2018, p. 3) the authors note that "all aspects of hunting with hounds pose a significant risk of disease transmission to both livestock and humans," citing biosecurity lapses, fouling by dogs while they are hunting, traveling with show hounds, and using fallen farm stock or dead foxes as feed for dogs.

Even though hunting has been officially banned in the UK since 2004, there is considerable evidence that it still exists. In 2021 a high-profile judgment was given against Mark Hankinson, director of the Master of Foxhounds Association, who was found guilty of giving advice "on how to hunt illegally by making it look like a hunt was in fact, trail hunting" (UK Courts and Tribunals Judiciary, p. 1), and was noted as "clearly encouraging the mirage of trail laying to act as cover for old fashioned illegal hunting" (p. 6). Oddly enough, despite the recent exposure of dog shooting and illegal hunting activity, the *Action Plan for Animal Welfare* published in the UK in May 2021 (UK Department of Environment, Food and Rural Affairs, 2021) only mentions hunting in reference to hare coursing (chasing rabbits with dogs), noting that it is illegal under the Hunting Act of 2004, but admitting the practice still exists. There is no mention of fox, deer, or bird hunting and the plan omits any reference to hunting dogs, much less the welfare of them.

OUR RESPONSIBILITY TO RURAL ANIMALS

It is not moral, much less responsible, to promote forms of rural tourism development that hide behind labels of heritage and tradition, such as commercial dairy farming (and the resulting local dairy products sold) and hunting with dogs. Current praxis of rural tourism involving animals must be examined and separated from nostalgia and convenience. How can a transformative and regenerative agenda be applied here? How can experiences that are not explicitly tourism-oriented, but part of the supply chain of tourism, be examined? How can the elements of the destination that make up the backdrop also be considered? Rural tourism development, done in a sustainable and ethical manner, extends beyond the direct tourist experience to the implicit, seemingly detached elements of rural communities and environments.

As we are learning and caring more about animals in tourism, it is neglectful to take for granted the fact that animals (such as milking cows and hunting dogs) are part of the rural cultural context without considering how, why, and for what purpose. The "five freedoms" (freedom from hunger and thirst; discomfort; pain, injury and disease; fear and distress; and freedom to express normal behaviors) are a great foundation for addressing animal welfare violations, but do not go far enough (Mellor, 2016) in regard to the questions we should ask relative to animals' role in rural tourism development. Indeed, Carr and Broom (2018) expect us to not only notice animals' situations, but to employ an animal-centric position in our research and practice. It seems short-sighted, however, to think that tourism scholars can do this alone. It has been heartening to see researchers partner with veterinarians, conservation scientists, sociologists and anthropologists, and philosophers, to advance the animal-based tourism literature. As described in more detail below, a positive step for transmodern tourism practice and research would be to partner with other disciplines, with practitioners, and with volunteer tourists to co-create a better form of rural tourism development. In the spirit of action-oriented and participant-driven learnings and acts (Ateljevic, 2020), it seems a natural point in our research trajectory to identify local, regional, and national/global animal-focused entities to work with and to drive forward a hopeful and transformative agenda.

MOVING FORWARD IN COLLABORATION WITH ANIMAL STUDIES

Using a community-based participatory research (CBPR) model to address tourism problems/inquiries is certainly not new (Koster et al., 2012). Further, scholars have long worked with local communities on research involving wildlife–human conflict or wildlife-based tourism (Stone & Nyaupane, 2018). What might indeed be a bit more novel is for tourism academics to engage with organizations working in the space of animal welfare, ethics, and/or advocacy. By partnering with carefully chosen animal studies programs and/or animal welfare organizations, we can learn nuanced physiological and emotional information about animals that will contribute to our own scholarship.

There are many examples already in existence within tourism of efforts that interlace ethics, advocacy, and practice. For example, research, advocacy, and teaching are combined in the Tourism Education Futures Initiative, the Critical Tourism Studies community, and Tourism RESET (where RESET stands for Race, Ethnicity, and Social Equity in Tourism). Academics and practitioners routinely come together within the Travel & Tourism Research Association as well as the aforementioned groups. However, within tourism, there is no formal "animal tourism" subgroup of academics such as you will find within disciplines like geography (e.g. American Association of Geographers: Animals Geography Specialty Group) or philosophy (e.g. Society for the Study of Ethics and Animals) or law.

Additionally, within the animal studies world, there are many conferences and organized groups where academics, practitioners, and activists merge. Some of the host organizations include Minding Animals, the Oxford Centre for Animal Studies, the Culture & Animal Foundation, the Aquatic Life Institute, and the European Association for Critical Animal Studies. And finally, there are independent scholars who focus on animal welfare and/or animal rights who would bring new insight to touristic systems, behavior, and animal-tourism theory. As one example, Kim Stallwood is an animal studies historian who recently has chronicled the story of Topsy, a female Asian elephant who was executed by electrocution at Coney Island, New York, in 1903 (Stallwood, 2018).

Engaging in collaborations with animal studies experts, as well as forming and institutionalizing our own collaborative, would bring a collective strength to efforts to improve animal welfare, educate tourists as well as suppliers, and make policy changes in the field. A brief agenda of critical needs in rural tourism development follows.

How Tourism Researchers Can Aid Animal Advocacy and Activism in Rural Communities

- Increasingly acknowledge the presence and contribution of animals within rural tourism; pursue opportunities to incorporate the welfare of animal "others."
- Form partnerships with animal welfare organizations at the fringe of tourism to further their research needs.
- Offer research assistance to organizations performing valuable service work for ecosystems, i.e. local habitat conservation restoration.
- Develop research agendas regarding animal-focused volunteer tourism in rural or fringe settings.
- Document solidarity tourism and other types of activism tourism; solidarity/activism tourism often occurs around areas of intensive agriculture or aquaculture.
- Explore the entanglement of heritage and culture relative to animals, including how visitors react to potential animal welfare concerns.
- Highlight pockets of restorative, regenerative, transmodern tourism systems and economies found all over the world related to ecosystem and animals; document "alternative and diverse tourism practices" (Cave & Dredge, 2020, p. 513).
- Coalesce academic communities of like-minded scholars revolving around animals, including tourism and non-tourism scholars.
- Work with local communities, attractions, trade associations and NGOs to develop transformational animal-focused experiences based on the recommendations found in Soulard et al. (2019) and Sheldon (2020).

How Animal Advocacy Organizations Can Advance Animal Tourism Research

- Bring issues to bear that may not be overt or readily known to tourism scholars. Approach local researchers to link universities' resources (research time, student projects) with organizational need.
- For larger organizations, allocate line-item funding for tourism-related research.
- Allow researchers consideration for theoretical work to align with pragmatic needs.
- Work with university faculty and independent scholars to design examples of animal-focused (or environmentally focused) transformative and regenerative experiences for tourists.

To achieve some real measure of improvement for animals, efforts couched within a transmodern space that strive to create transformative experiences for visitors are the ultimate goal. Alternative economies (Sheldon, 2021) can help shepherd tourism into its maturity. As Cave and Dredge (2018, p. 477) describe,

> the traditional economic focus will give way to hybrid social–political–economic practices and diverse organisational forms. Some of these will take the shape of communitarian CBT [community-based tourism] initiatives, and others in the digital tourism space. Yet other forms of this diverse economy [or these diverse economies] will unfold in the [corporate social responsibility] space as businesses confront the need to generate non-monetary value to address sustainability issues.

Many of us echo the sentiments of Hales et al. (2018, p. 195) who are "strongly concerned that their [academic] service is directed to other communities of interest beyond industry, communities that are typically confronting the ecological, sociological and/or cultural impacts resulting from the consequences of tourism development but whose voices are marginalised or overlooked in development discourses." As such, we engage with the issues that move us the most, continually circling back or "doubling down" on our concerns, moving our work forward and hoping to contribute not only to the academic community but to the collective academic voice on those concerns. And if we have taken up the cause for animals or environment or social justice in our scholarship, the hope of providing more color and dimensionality to the somberness and grief we feel can propel our work (Persson, 2021). This chapter has applied the critical lens to animal-based tourism in rural communities, related the work of others engaged in "hopeful" tourism to this context, addressed two specific cases where animal welfare is grossly ignored, and offered productive ideas for moving forward.

REFERENCES

Adams, C.J., & Messina, V. (2018). *Protest Kitchen: Fight Injustice, Save the Planet, and Fuel Your Resistance One Meal at a Time*. Newburyport, MA: Conari Press.

Ateljevic, I. (2011). Transmodern critical tourism studies: A call for hope and transformation. *Turismo em análise*, *22*(3), 497–515.

Ateljevic, I. (2013). Transmodernity: Integrating perspectives on societal evolution. *Futures*, *47*, 38–48.

Ateljevic, I. (2020). Transforming the (tourism) world for good and (re)generating the potential "new normal." *Tourism Geographies*, *22*(3), 467–475.

Bertella, G. (2020). Re-thinking sustainability and food in tourism. *Annals of Tourism Research*, *84*, 103005.

Bricker, K., & Joyner, L. (2018). Feed thy tourist well: CAFOs or cooperatives? In C. Kline (ed.), *Animals, Food, and Tourism* (pp. 96–112). Abingdon: Routledge.

Burns, T., Edwards, V., Marsh, J., Soulsby, E.J.S., & Winter, M. (2000). *Report of Committee of Enquiry into Hunting with Dogs in England & Wales*. Home Office. Retrieved October 26, 2021, from https://www.gov.uk/government/publications/report-of-committee-of-inquiry-into-hunting-with-dogs-in-england-wales.

Burns, G.L., Öqvist, E.L., Angerbjörn, A., & Granquist, S. (2018). When the wildlife you watch becomes the food you eat: Exploring moral and ethical dilemmas when consumptive and non-consumptive tourism merge. In C. Kline (ed.), *Animals, Food, and Tourism* (pp. 22–35). Abingdon: Routledge.

Carr, N., & Broom, D.M. (2018). *Tourism and Animal Welfare*. Wallingford: CABI.

Cave, J., & Dredge, D. (2018). Reworking tourism: Diverse economies in a changing world. *Tourism Planning & Development, 15*(5), 473–477.
Cave, J., & Dredge, D. (2020). Regenerative tourism needs diverse economic practices. *Tourism Geographies, 22*(3), 503–513.
Cheng, W.N., & Han, S.G. (2020). Bovine mastitis: Risk factors, therapeutic strategies, and alternative treatments—A review. *Asian-Australasian Journal of Animal Sciences, 33*(11), 1699.
Clamen, E., & Rossier, E. (1991). Animal-tourism and leisure in rural areas. *Equ'idée-Bulletin d'Information sur les Equides, 2*, 42–49.
Dashper, K. (ed.) (2015). *Rural Tourism: An International Perspective*. Newcastle upon Tyne: Cambridge Scholars Publishing.
Dashper, K. (2021). Conceptualizing non-human animals as "workers" within the tourism industry. In J. Rickly & C. Kline (eds), *Exploring Non-Human Work in Tourism: From Beasts of Burden to Animal Ambassadors*. Berlin: De Gruyter Publishing.
Deckers, J. (2009). Vegetarianism, sentimental or ethical? *Journal of Agricultural and Environmental Ethics, 22*(6), 573–597.
Duffy, L.N., Kline, C., Swanson, J.R., Best, M., & McKinnon, H. (2017). Community development through agroecotourism in Cuba: An application of the community capitals framework. *Journal of Ecotourism, 16*(3), 203–221.
Fennell, D.A. (2011). *Tourism and Animal Ethics*. Abingdon: Routledge.
Fennell, D., & Markwell, K. (2015). Ethical and sustainability dimensions of foodservice in Australian ecotourism businesses. *Journal of Ecotourism, 14*(1), 48–63.
Fennell, D., & Thomsen, B. (2021). Tourism & animal suffering: Mapping the future. *Annals of Tourism Research*, 91. https://doi.org/10.1016/j.annals.2021.103317.
Food and Agriculture Organization (FAO) (n.d.). *Mastitis Control*. Retrieved November 6, 2021, from https://www.fao.org/3/t0218e/t0218e04.htm.
Fountain, J., Cradock-Henry, N., Buelow, F., & Rennie, H. (2021). Agrifood tourism, rural resilience, and recovery in a postdisaster context: Insights and evidence from Kaikōura-Hurunui, New Zealand. *Tourism Analysis, 26*(2–3), 135–149.
Garrido, J.F. (2011, April 6). *Transmodernity and activism in tourism: The case of El Palmar (Cádiz, Spain)* [Master's dissertation, Wageningen University].
Goodman, B. (2020, February 14). *Rethinking milk: Science takes on the dairy dilemma*. WEBMD. Retrieved October 22, 2021, from https://www.webmd.com/diet/news/20200214/rethinking-mik-science-takes-on-the-dairy-dilemma.
Griffin, E. (2007). *Blood Sport: Hunting in Britain since 1066*. New Haven, CT: Yale University Press.
Hales, R., Dredge, D., Higgins-Desbiolles, F., & Jamal, T. (2018). Academic activism in tourism studies: Critical narratives from four researchers. *Tourism Analysis, 23*(2), 189–199.
Hall, D. (2000). Rural tourism management: Sustainable options conference. *The International Journal of Tourism Research, 2*(4), 295.
Harris, Stephen, & Dorning, Jo. (2018). *Hunting with hounds and the spread of disease*. A report commissioned by the League Against Cruel Sports.
Hashman, J. (2016, August 11). *They Love Their Hounds, Don't They?* Hounds Off. Retrieved October 22, 2021, from https://www.houndsoff.co.uk/archives/they-love-their-hounds-dont-they/.
Haynes, R.P. (2011). Competing conceptions of animal welfare and their ethical implications for the treatment of non-human animals. *Acta Biotheoretica, 59*(2), 105–120.
Hounds Off (n.d.). Retrieved October 22, 2021, from https://www.houndsoff.co.uk/.
IUCN (2021). *Red List Index*. International Union for Conservation of Nature and Natural Resources. Retrieved November 6, 2021, from https://www.iucnredlist.org/assessment/red-list-index.

Jacobs, E.T., Foote, J.A., Kohler, L.N., Skiba, M.B., & Thomson, C.A. (2020). Re-examination of dairy as a single commodity in US dietary guidance. *Nutrition Reviews*, *78*(3), 225–234.
King, B. (2017). Evidence for cows' minds and hearts: Why cows are far more than biological machines. *Animal Behavior and Cognition*, *4*(4), 512–514.
Kline, C. (ed.) (2018a). *Animals, Food, and Tourism*. Abingdon: Routledge.
Kline, C. (ed.) (2018b). *Tourism Experiences and Animal Consumption: Contested Values, Morality and Ethics*. Abingdon: Routledge.
Koster, R., Baccar, K., & Lemelin, R.H. (2012). Moving from research ON, to research WITH and FOR Indigenous communities: A critical reflection on community-based participatory research. *The Canadian Geographer/Le Géographe canadien*, *56*(2), 195–210.
Küçükaltan, E.G., & Dilek, S.E. (2019). A philosophical approach to animal rights and welfare in the tourism sector. *Journal of Tourism Leisure and Hospitality*, *1*(1), 1–11.
Markwell, K. (ed.) (2015). *Animals and Tourism: Understanding Diverse Relationships*. Bristol: Channel View Publications.
Mellor, D.J. (2016). Moving beyond the "five freedoms" by updating the "five provisions" and introducing aligned "animal welfare aims." *Animals*, *6*(10), 59.
Mkono, M. (2015). "Eating the animals you come to see": Tourists' meat-eating discourses in online communicative texts. In K. Markwell (ed.), *Animals and Tourism: Understanding Diverse Relationships* (pp. 211–226). Bristol: Channel View Publications.
Orams, M.B. (2002). Feeding wildlife as a tourism attraction: A review of issues and impacts. *Tourism Management*, *23*(3), 281–293.
Persson, A. (2021). Heart cracked open: From personal loss to radical mourning. *Mortality*, 1–15.
Probyn-Rapsey, F. (2018). The "crazy cat lady." In L. Gruen & F. Probyn-Rapsey (eds), *Animaladies: Gender, Animals, and Madness* (pp. 175–185). New York: Bloomsbury.
Physician's Committee for Responsible Medicine (n.d.). Dairy. Retrieved October 21, 2021, from https://www.pcrm.org/term/dairy.
Ruegg, P. (2015). *Physiological disorders of the udder in cows*. Merck Vet Manual. Retrieved November 6, 2021, from https://www.merckvetmanual.com/reproductive-system/udder-diseases/physiologic-disorders-of-the-udder.
Schmalz, D.L., Joyner, L., Duffy, L.N., Bricker, K.S., & Blomquist, K.K. (2020). The cycle of food socialization: Leisure as resistance. *Annals of Leisure Research*, *23*(4), 510–529.
Shani, A. (2013). Differentiating settings of tourist–animal interactions: An anthrozoological perspective. *Tourism Recreation Research*, *38*(1), 104–107.
Sheldon, P.J. (2020). Designing tourism experiences for inner transformation. *Annals of Tourism Research*, *83*, 102935.
Sheldon, P.J. (2021). The coming-of-age of tourism: Embracing new economic models. *Journal of Tourism Futures*, *8*(2), 200–207. https://doi.org/10.1108/JTF-03-2021-0057.
Singer, P. (1975). *Animal Liberation*. New York: Random House.
Soulard, J., McGehee, N.G., & Stern, M. (2019). Transformative tourism organizations and glocalization. *Annals of Tourism Research*, *76*, 91–104.
Špinka, M. (2019). Animal agency, animal awareness and animal welfare. *Animal Welfare*, *28*(1), 11–20.
Stallwood, K. (2018). Topsy: The elephant we must never forget. In A. Krebber & M. Roscher (eds), *Animal Biography* (pp. 227–242). Cham: Palgrave Macmillan.
Stone, M.T., & Nyaupane, G.P. (2018). Protected areas, wildlife-based community tourism and community livelihoods dynamics: Spiraling up and down of community capitals. *Journal of Sustainable Tourism*, *26*(2), 307–324.
Pichler, A. (2017). *The Milk System* [film]. Retrieved October 29, 2021, from https://www.dassystemmilch.de/.
Tribe, J. (2008). Tourism: A critical business. *Journal of Travel Research*, *46*(3), 245–255.

Tully, P.A., & Carr, N. (2021). Farm animals' participation in tourism experiences: A time for proper respect. In J. Rickly & C. Kline (eds), *Exploring Non-Human Work in Tourism* (pp. 83–100). Berlin: De Gruyter.

UK Department of Environment, Food and Rural Affairs (2021, May 12). *Action Plan for Animal Welfare* [Policy paper]. Retrieved October 30, 2021, from https://www.gov.uk/government/publications/action-plan-for-animal-welfare.

UK Courts and Tribunals Judiciary. *In the Westminster Magistrates' Court Regina v Mark Hankinson Decision*. Retrieved November 30, 2022, from https://www.judiciary.uk/judgments/r-v-hankinson/.

Usui, R., Sheeran, L.K., Asbury, A.M., & Blackson, M. (2021). Impacts of the COVID-19 pandemic on mammals at tourism destinations: A systematic review. *Mammal Review*, *51*(4), 492–507.

Welsh, J.A., Braun, H., Brown, N., Um, C., Ehret, K., Figueroa, J., & Barr, D.B. (2019). Production-related contaminants (pesticides, antibiotics and hormones) in organic and conventionally produced milk samples sold in the USA. *Public Health Nutrition*, *22*(16), 2972–2980.

Willett, W.C., & Ludwig, D.S. (2020). Milk and health. *New England Journal of Medicine*, *382*(7), 644–654.

World Wildlife Fund and the Zoological Society of London (2020). *Living Planet Report*. Retrieved October 22, 2021, from https://f.hubspotusercontent20.net/hubfs/4783129/LPR/PDFs/ENGLISH-FULL.pdf.

26. Shifting from benefiting to serving community: a case of regenerative tourism and building cultural capital through the Children's University Tasmania

Can-Seng Ooi and Becky Shelley

INTRODUCTION

It is uncontroversial to consider tourism an economic resource. Policymakers, industry players and the public acknowledge that tourism brings jobs, revenue, and other benefits. Unfortunately, the distribution of the benefits to the local community is frequently uneven. Tasmania – the southernmost island state of Australia – is no exception (Denny et al., 2019; Shelley et al., 2021). There are points of intersection and overlapping interests between tourism and community members. For example, tourism income is redistributed through infrastructure, services and welfare to residents. Tourism services are regarded by many as implicitly fostering the benefit and welfare of future generations (Wiltshier, 2020). Indeed, tourism offers opportunities that can be generative and can engage, intergenerationally, with children and families. Yet, this is an underdeveloped field. We identify one way tourism can support the community. There is a particular programme – Children's University Tasmania – that distributes benefits from tourism to the local community in a way that fosters positive, intergenerational change. Tourism can be used to increase educational aspirations. This chapter shows how it happens.

Children's University Tasmania mobilizes tourism resources for educational purposes within the community (Shelley et al., 2019). It also puts on the agenda the benefits of tourist attractions, focusing on residents as a primary 'market'. In Tasmania many tourism activities are freely available, such as going to museums or visiting parks. These are common 'trickle down' benefits. Some residents in regional and rural communities may not have the resources or confidence to participate in these free tourism experiences. This section of the community has largely been ignored in tourism policy discussions even though policymakers and the industry often make claims that tourism will benefit society-at-large. Yet tourism operators can engage with local children and their families to enhance cultural capital so that these children will have a better foundation to succeed. The case of the Children's University Tasmania draws lessons, and points to other opportunities for tourism to prioritize residents in their activities. Moreover, the programme enables tourism to boost its services in response to regional challenges.

Rural and regional communities in Australia have been exposed to the twin impacts of globalization and economic reforms. These impacts are spatially patterned. Rural and regional areas tend to have a narrowly based economy. The 'five sectors' that are dominant in Tasmania are agriculture, forestry and fishing; retail trade; accommodation and food services; public administration and safety; and health care and social assistance. The island state of Tasmania is regarded on a number of national measures as regional, for example, it has only one university, which is classed as a regional university, with its campuses dispersed across three subregions on the island. This reflects the regionally dispersed nature of the population. In addition to its regionally dispersed population, other defining characteristics of the Tasmanian community are lower income, employment and educational attainment levels compared to national counterparts. Nearly half of Tasmania's adult population lacks the literacy and numeracy skills they need for everyday life. As in other regional communities in Australia, national economic reforms have influenced economic disadvantages, and tourism has been regarded as a viable vehicle to generate economic growth. Local, state and federal governments have turned to tourism development as a development strategy to sustain regional communities (Ooi and Hardy, 2020). Tourism has grown to the point where it now accounts for a larger share of the Tasmanian economy than any other state in Australia.

Interest in sustainable and regenerative tourism highlights the challenge of using local resources, such as heritage, lived cultures, environment and physical spaces, for visitors, at the expense of residents. Community-led tourism has emerged as a development model that centres local people within the planning process and promotes equitable and sustainable practices (Jamal & Getz, 1995; Nair & Hamzah, 2015). A shift in emphasis to collaboration fosters a more regenerative model. As described above, policymakers and planners have tended to regard tourism sites and the communities in which they are located in terms of regional economic policy goals. The Children's University Tasmania connects tourism with broader social policy goals; for example, it aims to lift educational attainment levels of young people in the state by increasing children's participation in rich, experiential learning outside the curriculum (Ooi & Shelley, 2019). Learning occurs everywhere and informal learning influences formal learning. Tourist attractions in regional destinations become validated Children's University Tasmania 'learning destinations'.

This study contends that: educational equity matters; is critically important in regional Australia; and that tourism can play a role. In this context, the chapter addresses a social and community challenge in which tourism can be part of the solution. Consistent with our core argument, the chapter is structured to prioritize community before tourism. The first part of the chapter sets the scene for the analysis, outlining the connection between cultural capital and education outcomes in the context of informal learning. The second part of the chapter discusses how to centre the community within conceptions of regenerative tourism. In the third section, a deeper consideration of the Children's University Tasmania is undertaken.

To inform this analysis, the authors draw from research with tourism operators and from artefacts created by children who are members of the Children's University Tasmania. The final section refocuses on regenerative tourism and identifies four lessons.

TOURISM, EDUCATION, CULTURAL CAPITAL, AND THE CHILDREN'S UNIVERSITY

Many tourism studies have focused on how to enhance visitor experiences by transmitting local knowledge to them, such as through interpretation signage, guided tours and recommendations. The cultural knowledge provided to visitors is distilled and packaged, as attraction operators attempt a balance between deep knowledge and entertainment. Regardless, the visitors are engaged in informal learning, and are enhancing their cultural capital. Local residents, particularly children, would also benefit from the experiences. This is an opportunity for the community to enhance their cultural capital. Set in this context, this section concentrates on making the links between cultural capital and enriching young people's education outcomes in the community.

This chapter draws upon Bourdieu's sociological concept of cultural capital to accentuate the potential benefit of transmitted experiences curated and created for visitors to be shared with the local community. Cultural capital refers to cultural competencies, either in the embodied sense of valued lifestyles or in the institutionalized sense of educational credentials (Bourdieu, 1986). Bourdieu's concept of cultural capital has enabled us to view cultural capital as a resource – one that provides scarce rewards and under certain conditions may be transmitted from one generation to the next (Lareau & Weininger, 2003). Bourdieu and Passeron considered the relationship between children's stock of cultural capital and their accumulation of educational capital (Bourdieu & Passeron, 1977). While all modern societies espouse the values of equality and aim to offer every child the same chance to excel, the playing field is still not level. This unevenness, according to Bourdieu, is influenced by the family background of the children. Some children, he argues, have access to the types of cultural capital valued by education systems, while others do not. Bourdieu's work helps us understand why this is so.

Bourdieu observed that the educational system presupposes the possession of cultural competence, which only some students possess.

> [Education] is in fact the most effective means of perpetuating the existing social pattern, as it both provides an apparent justification for social inequalities and gives recognition to the cultural heritage, that is, to a social gift treated as a natural one. (Bourdieu, 1974, p. 34)

It has been argued that the educational system may function to legitimize social inequalities because it wrongly assumes the same starting points for all children (Sullivan, 2002). Educational attainment is then viewed as the result of individual abilities and gifts. By assuming that everyone has the same cultural capital resources upon which to draw and observing that children from higher socio-economic status backgrounds tend to have higher levels of educational attainment, educational credentials serve to reproduce and legitimize social inequalities, as these children are made to think that they deserve their place in the social structure (Bamford, 2014, p. 32). It may seem that Bourdieu's work suggests a biased education system that is resistant to change (English & Bolton, 2016), but there is a less pessimistic approach to his work. Cultural capital is not fixed, it is transmitted. It is valued differently in different circumstances, settings, or fields.

Cultural Capital in Action

There are different ways in which cultural capital is acquired. One of them is parental engagement with and participation in schools, and with their children's learning more broadly. Lareau & Weininger's (2003) examination of the influence of class-related cultural factors on the parents' compliance with teachers' requests for parental participation in schools is relevant to our research given the key role that parents play within the Children's University Tasmania – as it is parents who take children to the 'learning destinations'. Lareau & Weininger conducted a qualitative study of two first-grade classrooms located in two different communities. The participants include a white working-class community and a professional, middle-class community. The study reveals that both sets of teachers at the different schools considered parental involvement as indicative of the value which parents placed on education. During the study, Lareau & Weininger observed that working-class parents primarily placed the responsibility of education upon the teachers whereas the middle-class parents consider the process of educating to be a shared, collective experience.

It is important to understand differences in parental involvement among different groups of parents because parental involvement is a powerful determinant of the educational success of students. Gottlob (2010, p. 98) concludes:

> There is no magic strategy bullet to increase student achievement. Instead, we have to focus on building on the cultural capital of the community and helping parents gain the cultural capital they need to navigate the educational system. Schools need to work with, and in the context of the community. There has to be give-and-take and a mutually respectful relationship between them. It is imperative that schools find the way to build on the assets of the community to create genuine partnerships that benefit students, parents, and the community at large.

Learning, of course, occurs outside formal educational institutions. Self-directed learning, learning that is voluntary and beyond the formal curriculum, is also part of a process of cultural transmission. The acquisition of cultural capital is thus an ongoing socialization process.

Kisida et al. (2014) suggest that children can be activated to acquire the type of cultural capital valued by educational systems and thus compensate for family background characteristics. Their study is based on a large-scale study of an art museum educational programme operating across a set of American schools. Exposure to the institutional culture of the museum was seen to produce 'cultural consumers' who were then 'motivated to acquire new cultural capital'.

Seow and Pan (2014), in their literature survey, identify three frameworks to explain the impact of extra-curricular activities (ECA).

> First, the zero-sum framework posited that ECA participation has a negative effect on academic performance because students were devoting more time for their ECA activities at the expense of their academic studies. Second, the developmental framework theorized that ECA participation has a positive effect on academic performance indirectly as a result of the non-academic and social benefits associated with ECA participation. Last, the threshold framework hypothesized that ECA participation has a positive effect on academic performance up to a certain point beyond which participation leads to negative academic outcomes.

It is important to observe that measures of cultural capital are imperfect since they are limited to describing behaviours (e.g., participating in activities) rather than other types of cultural 'signals', such as attitudes, preferences, or credentials. In sum, everyone has cultural capital; the education system may privilege some forms of cultural capital over others. This, according to Bourdieu, can account for persistently uneven educational outcomes. The ways in which informal learning, in the form of extra-curricular tourism activities, can support the twin goals of raising educational aspiration and attainment and increasing parental engagement in children's learning are key considerations of this study.

REGENERATIVE TOURISM AND TRANSFORMING THE COMMUNITY

In the last two decades, tourism scholarship has become increasingly critical of the eroding impact of tourism on social justice and environmental preservation (Budeanu et al., 2016; Cave & Dredge, 2020; Higgins-Desbiolles, 2020; Jamal, 2019). Scholars have advocated a more sustainable, if not regenerative, form of tourism development, rather than an extractive one. Like so many types in this genre, regenerative tourism gives equal consideration to the social and natural environment as to economic outcomes (Pollock, 2019). Regenerative tourism advocates practices that claim to be transformational (Ateljevic, 2020). It advances an approach that values and celebrates people, place and diversity, and that aims for quality over quantity (Higgins-Desbiolles, 2020). Such tourism strategies should be developed from the ground up and be community-driven (Muganda et al., 2013; Sofield, 1993). Local communities live their culture, heritage and environment. They have found opportunities to thrive, and solutions to the challenges they face. The community should thus be consulted extensively and should lead the development (Okazaki, 2008).

However, engaging with the community has its challenges. Local communities are also replete with agendas and inequalities that can be further perpetuated through collaborative processes (Ooi, 2022; Scoones et al., 2020; Sin & Minca, 2014). There are established social hierarchies and informal practices that benefit the local elites. Regardless, some examples are available that show specific community solutions that deliver tourism and are largely beneficial with limited detrimental impacts (Cave & Dredge, 2020). It remains the case that tourism is seen largely as an economic enterprise that brings benefits to the community, albeit these tourism activities are likely to be community-led, and the benefits of tourism should be distributed more equitably.

In this context, the case of the Children's University Tasmania points to an opportunity, and also to a need to rethink tourism. Regenerative tourism desires to be community-focused, but current so-called regenerative practices still predominantly let tourism drive the agenda. The programme drives the social policy agenda of lifting up the educational attainment of the community, and through that, tourism-relevant activities, services and products may emerge. It is thus intentional that we have decided to focus on education and Bourdieu's concepts here. By doing so, we then subsume the interests of tourism to serving the community through the educational programme. It is then that tourism can be seen as truly positive, transformative, and collaborative. The implementation of regenerative practices requires the collaboration of multiple stakeholders, which commonly meets with obstacles and challenges (Higgins-Desbiolles, 2020). Tourism operators and businesses frequently set the agenda, as they seek out local cultural and natural resources for the industry. But in the Children's University Tasmania programme, the need to improve educational attainment in the community takes centre stage, and tourism activities are selected and shaped towards serving that goal. This is idealistic and may not be realized in actual practice, as our experience in Tasmania shows. There are still many tourism businesses and operators that do not participate in the programme. Their products and services may not be relevant (e.g. bars and casinos) or they have not bought into a community-led mindset. Regardless, a truly regenerative tourism process has started, in a rural development context.

TASMANIA AND THE CHILDREN'S UNIVERSITY

This chapter draws from a small number of research projects related to the Children's University Tasmania. The first compared survey responses from participants and non-participants from the same class groups to reveal a self-concept as a learner. The second involved in-depth interviews with learning destinations (involving the authors of this chapter). The final (ongoing) research project assesses the influence of the Children's University Tasmania on educational outcomes of students aged 7 to 14 years and continuing after they have exited the programme, until they leave school at 18 years old. It also contains several child-friendly research methods, such as workshops where children are invited to write a letter to the 'boss' of the Children's University Tasmania, or to draw a map detailing their experiences.

The Children's University is a social franchise and operates in the UK, Malaysia, New Zealand and Australia. A literature is developing, focused on its influence. Evaluations of the Children's University programmes conducted in the United Kingdom and in Adelaide have shown an indicative relationship with children's aspiration, social skills, self-confidence, and well-being (Gorard et al., 2017; Hamshaw, 2015; Harrison & Skujins, 2017; MacBeath, 2013). The Children's University aims to promote social mobility by providing high quality out-of-school-hours learning activities to children aged 7 to 14. It targets children and young people facing socio-economic disadvantages to ensure that every child, irrespective of parental means, has access to quality extra-curricular learning activities. Studies have shown the significance of extra-curricular participation on educational outcomes, and the ways in which access and opportunity to engage in extra-curricular activity are impacted by socio-economic characteristics (Mullan, 2014; Skattebol et al., 2012; Skattebol & Redmond, 2019).

The Children's University identifies informal learning as a rich site to foster aspiration, parental engagement, and educational success – if the social, economic and cultural barriers to participation can be reduced (Moore, 2014). Using the narrative of travel, when a child becomes a member of the Children's University, they are given a 'Passport to Learning', in which they record their participation in activities at learning destinations (Figure 26.1). After completing 30+ hours of validated learning, their achievement is celebrated at a formal graduation ceremony, a significant cultural experience itself.

The Children's University Tasmania is a mechanism to support parents and guardians to participate in a variety of extra-curricular learning experiences. Tourism experiences are key within this. During 2021 there were over 83 learning destinations and 586 active members who collectively participated in over 50,000 hours of extra-curricular learning. Participants report that the best thing is spending time with family. One learning destination described: 'The adults absolutely loved the activities. It crossed all the boundaries' (LD5) (see Shelley et al., 2019).

DOING TOURISM AS EXTRA-CURRICULUM ACTIVITIES IN RURAL COMMUNITIES

There is a significant body of literature that identifies the significance of extra-curricular participation on educational outcomes, and the ways in which access and opportunity to engage in extra-curricular activity is impacted by socio-economic characteristics (Karlsen Baeck, 2005, p. 218). So, how does tourism fit into this discussion? The Children's University Tasmania has partnered with local services and attractions that serve residents and/or visitors. They include museums, historic sites, wildlife sanctuaries, regional galleries, and events such as festivals. The Children's University Tasmania encourages tourism services to become validated learning destinations by highlighting their potential to enrich the local community, following the Children's University Australia guidelines (n.d., p. 12):

Figure 26.1 *Every Children's University Tasmania member has a 'Passport to Learning'*

> Being a learning destination is a great way to promote your organization to the local community and to engage with young people in your service delivery. Learning destinations can enrich communities and build valuable networks that help support children and their families in their educational journey.

The authors conducted interviews with the Children's University Tasmania's employees and at public learning destinations in 2017. An employee of the Children's University Tasmania observed that 'it provides a structure for parents to understand how and what to engage in with kids outside of school … But I think that is important as parents struggle sometimes with knowing what to do (Employee, Children's University Tasmania).

Tourism destinations that actively seek to engage local children also engage other caregivers. In this way it contributes to an intergenerational enrichment. An employee at a small regional art gallery in Tasmania stated,

> It is either a mother or a father and quite often a grandparent who will bring the children and it is very important for us to have contact with the older generation. We are very open to have a wide range of ages participating … I really encourage adult and child participation here. It is a family thing if they can. (Learning Destination Officer 1)

They added: '[Parents] know that art and creativity is good for you (a bit like broccoli), but they haven't had the opportunities themselves, but they sense that it should be encouraged and rewarded and applauded and helped. [During activities] they also have a go and are as happy as their kids' (Learning Destination Officer 1).

Early evidence is indicative of positive impacts of the programme upon a sense of achievement, the self-concept as a learner, and awareness of further learning as a pathway to a 'dream' job in the future. The Children's University Tasmania also highlights an avenue for the local tourism industry to be more proactive in using their social licence. While many of them serve visitors, they should use similar resources to serve residents. They can become learning destinations, and not just tourist attractions.

Obstacles remain. It is important to observe that the economic and cultural barriers to participation in local tourism and informal learning opportunities need further analysis. It is evident that even where people have free access to tourist attractions, there are cultural barriers to visitation. Skattebol and Redmond (2019) reveal a tendency for young Australians living in disadvantaged locations to resist or opt out of out-of-school hours opportunities that are costly or located in areas of perceived higher advantage. Tourism has a latent and largely unrealized potential to encourage further participation of parents in their children's learning. Learning that takes place outside the classroom benefits and enhances learning outcomes within the classroom.

GENERATIVE AND TRANSFORMATIVE OPPORTUNITIES

In spite of good intentions by tourism researchers, policymakers and attraction operators to make sure that tourism development serves the needs of the community beyond economics, examples are still largely tourism-centric. And in the rural context, per capita resources allocated for community development tend to be more limited than more densely populated urban places. As a result, more effective and efficient solutions need to be found. As criticisms hold, resources for tourism development take away opportunities from other improvements. Can we then allocate more resources for community development that are tourism-friendly, rather than allocate resources for tourism development that are community-friendly? Yes, and the case of the Children's University Tasmania shows that possibility.

In building up young people's engagement with their surroundings, they were asked to map out the places they have been with the Children's University. Figure 26.2 is an example of a Grade 2 child in a regional school (seven years old). They started with school and listed the sites they visited, such as 'looking for cristls [crystals]', bee farms, gardens, nature parks, McDonald's and also playing basketball.

Figure 26.2 *A seven-year-old's map detailing their trip with the Children's University Tasmania*

They ended with an image of the abacus. The child is acquiring cultural capital at various levels, including getting to know the geography of their surroundings, and becoming familiar and knowing more about various sites. This travel 'itinerary' is from a child's perspective. Many of them are also tourist attractions, even though they are not presented in that manner.

Another child, also aged seven, wrote a review of their Children's University Tasmania journey: 'I really enjoy the way you've set up CU. I find it very enjoyable. All activities I've done have been awesome and find it quite unfortunate that COVID-19 prevented me from doing more of these activities. The graduations are my favourite part of it all.'

In the context of tourism and the authentic experience, these experiences are important. Would these be interesting for visitors? These stories can be told fondly, but more importantly, the children went looking for crystals, visited a geology museum and engaged with the region's mining heritage. The content knowledge and destination experience overlap and intersect with that of visiting tourists. Local children and visiting tourists are building up their cultural capital through informal learning.

Four Lessons

We identify at least four lessons of thinking about the Children's University programme as part of a more regenerative tourism approach. First, the 'learning destinations' that signed up with the Children's University Tasmania are not all tourism sites. From one perspective, the case suggests resources created for tourism can also be deployed to support broader societal goals, such as raising educational aspiration and attainment. During off-peak tourist seasons, more can be done to engage local children in extra-curricular, experiential learning through tourism-related services and attractions. And the tourism industry has benefited from skill development initiatives funded by government. Fostering skills in inclusive and child-friendly practices within the tourism industry is encouraged. The beneficiaries of this approach include the child participants, but also the industry, as they build connections with a new audience, or customer base. Further, and from another perspective, because the industry receives public funds for development, it can be argued that the attractions and their programming should put the community first and not the other way around. The Children's University Tasmania seems to be spurring visitor attractions in that direction. For instance, one employee in a 'learning destination' notes that they had carried out training of customer services officers. Learning Destination Officer 2 observed:

> [T]he staff are getting a really good basis in educational principles so that when they are actually teaching or showing students … they are able to give them the information that they need in a way that has meaning to them … it means that the kids are walking away with something a little bit more than just an experience which is fantastic … We want to engage with the Children's University. For one, in terms of pure numbers, it brings us a new audience.

Second, ideally from a regenerative tourism perspective, the community should be the primary focus, but that is still not the norm today. In moving towards that, steps can be taken. So, in some instances, rich and interesting learning content on site at tourism attractions can also be harnessed to create online learning opportunities. Higher priority should be given in this area, particularly for regionally dispersed communities. Tourism assets can be crafted into mobile 'destinations'. These can be physical, or digital, or a combination. For example, the Tasmanian Museum and Art Gallery has an outreach programme. They provide a box that contains items and information that schools can use, while a museum employee connects via video-link and discusses the contents of the box. The museum becomes a mobile supporter of learning. Currently, this is used to support formal learning within the curriculum. There are opportunities to develop content that supports non-formal learning.

Third, whilst key public institutions such as museums, art galleries, and historic sites often have an explicit educational purpose and resources to support engagement strategies, the Children's University Tasmania case suggests that similar principles can be extended more broadly across the tourism sector. The example here shows that the narrative of travel is a tool that can stimulate the imagination and provide creative, fun and engaging ways to enhance learning about culture, history, science, natural and social environments. As such, it supports broader social and economic goals of improved educational attainment for children and young people, as well as fostering an understanding of their place (self and community) in a globalized world. For instance, an 11-year-old member wrote: 'Learning in this environment has really helped me to explore different parts of myself, and learn to be a positive role model in my community.'

Fourth, the Children's University Tasmania is also about the adults. The role of parents, caregivers, and home environments as influential agents on levels of educational engagement and attainment is well documented. As noted above, as in many rural places around the world, Tasmanian families have higher levels of socio-economic disadvantage than the national average. All parents have aspirations for their children. Yet, the pathway to achieve those goals can be hard to navigate, especially for socially excluded families. They benefit from programmes like the one described here to understand what experiences and events are available in their local community. Tourist attractions should not be places that are expensive and exclusive. Instead tourist attractions should be part of the community and be there to lift the population and to improve children's educational attainment.

CONCLUSION

In sum, this chapter reminds researchers and the practice community that a more regenerative form of tourism has to put the community first. There remains a tendency for social programmes to fit in and find synergies with tourism development initiatives. The Children's University Tasmania illustrates how resources are being used for local and community development, enhancing individual, social and cultural well-being, and tourism players can try to fit into that wider social aim of serving the community before visitors. Ideally, tourism is then part of wider social policy strategies that regenerate, invigorate and can engage, intergenerationally, with children and families. This would be the spirit of community-led and regenerative tourism, but it remains only a partially realized goal even for the Children's University Tasmania. As for next steps, policy settings should move in the right direction to encourage – if not necessitate – new tourism developments to fit in and work with wider social programmes like the Children's University from the beginning, so that the community-first ethos can be realized. This is even more salient in a rural regional context.

REFERENCES

Ateljevic, I. (2020). Transforming the (tourism) world for good and (re)generating the potential 'new normal'. *Tourism Geographies, 22*(3), 467–475.

Bamford, M. (2014). *The influence of cultural capital on teachers' perceptions of native and immigrant students' academic and social behaviors* [Doctoral dissertation, Florida State University].

Bourdieu, P. (1974). The school as a conservative force: Scholastic and cultural inequalities. In J. Eggelston (ed.), *Contemporary Research in the Sociology of Education* (pp. 32–47). London: Methuen.

Bourdieu, P. (1986). The forms of capital. In J. Richardson (ed.), *Handbook of Theory and Research for the Sociology of Education* (pp. 241–258). Westport, CT: Greenwood Press.

Bourdieu, P., & Passeron, J. (1977). *Reproduction: In Education, Culture and Society*. Beverly Hills: Sage.

Budeanu, A., Miller, G., Moscardo, G., & Ooi, C.-S. (2016). Sustainable tourism, progress, challenges and opportunities: An introduction. *Journal of Cleaner Production, 111*, 285–294.

Cave, J., & Dredge, D. (2020). Regenerative tourism needs diverse economic practices. *Tourism Geographies, 22*(3), 503–513.

Children's University Australia (n.d.). *Children's University Australia Marketing Materials*. Adelaide: Children's University Australia. Retrieved March 29, 2022, from http://www.banksiar7.sa.edu.au/docs/ChildrensUni.pdf.

Denny, L., Shelley, B., & Ooi, C. (2019). Education, jobs and the political economy of tourism: Expectations and realities in the case of Tasmania. *Australasian Journal of Regional Studies, 25*(2), 282–305.

English, F.W., & Bolton, C.L. (2016). *Bourdieu for Educators: Policy and Practice*. Thousand Oaks, CA: Sage.

Gorard, S., Siddiqui, N., See, B.H., Smith, E., & White, P. (2017). *Children's University: Evaluation Report and Executive Summary*. Durham University.

Gottlob, J.E. (2010). Building on cultural capital of low-income Hispanic parents and families in elementary school. *Dissertation Abstracts International Section A: Humanities and Social Sciences, 70*(7-A), 2391.

Hamshaw, K. (2015). *Sheffield Children's University Participation and Performance Analysis*. Sheffield University.

Harrison, A., & Skujins, P. (2017). *Children's University Australia Evaluation: Summary Report*. Adelaide: National Centre for Vocational Education Research.

Higgins-Desbiolles, F. (2020). Socialising tourism for social and ecological justice after COVID-19, *Tourism Geographies, 22*(3), 610–623.

Jamal, T. (2019. *Justice and Ethics in Tourism*. Abingdon: Routledge.

Jamal, T.B., & Getz, D. (1995). Collaboration theory and community tourism planning. *Annals of Tourism Research, 22*(1), 186–204.

Karlsen Baeck, U.-D. (2005). School as an arena for activating cultural capital: Understanding differences in parental involvement in school. *Nordisk Pedagogik, 25*, 217–228.

Kisida, B., Greene, J.P., & Bowen, D.H. (2014). Creating cultural consumers. *Sociology of Education, 87*(4), 281–295.

Lareau, A., & Weininger, E.B. (2003). Cultural capital in educational research: A critical assessment. *Theory and Society, 32*(5/6), 567–606.

MacBeath, J. (2013). *Evaluating provision, progress and quality of learning in the Children's University 2012*. University of Cambridge.

Moore, R. (2014). Capital. In M. Grenfell (ed.), *Pierre Bourdieu: Key Concepts* (pp. 98–113). Utrecht: Acumen.

Muganda, M., Sirima, A. & Ezra, P.M. (2013). The role of local communities in tourism development: Grassroots perspectives from Tanzania. *Journal of Human Ecology*, *41*(1), 53–66.
Mullan, K. (2014). Time use and children's social and emotional wellbeing and temperament. In G. Daraganova (ed.), *The Longitudinal Study of Australian Children Annual Statistical Report 2013* (pp. 51–70). Melbourne: Australian Institute of Family Studies.
Nair, V., & Hamzah, A. (2015). Successful community-based tourism approaches for rural destinations: The Asia Pacific experience. *Worldwide Hospitality and Tourism Themes*, *7*(5), 429–439.
Okazaki, E. (2008). A community-based tourism model: Its conception and use. *Journal of Sustainable Tourism*, *16*(5), 511–529.
Ooi, C. (2022). Sustainable tourism and the moral limits of the market: Can Asia offer better alternatives? In A.S. Balasingam, & Y. Ma (eds), *Asian Tourism Sustainability* (pp. 177–197). Singapore: Springer Nature.
Ooi, C., & Hardy, A. (2020). *Tourism in Tasmania*. Hobart: Forty South Publishing Pty Ltd.
Ooi, C.-S., & Shelley, B. (2019). The Children's University Tasmania: The transformative power of tourism. In T. Jamal (ed.), *Justice and Ethics in Tourism* (pp. 222–228). Abingdon: Routledge.
Pollock, A. (2019). *Regenerative tourism: The natural maturation of sustainability*. Activate the Future. Retrieved March 11, 2021, from https://medium.com/activate-the-future/regenerative-tourism-the-natural-maturation-of-sustainability-26e6507d0fcb.
Scoones, I., Stirling, A., Abrol, D., Atela, J., Charli-Joseph, L., Eakin, et al. (2020). Transformations to sustainability: Combining structural, systemic and enabling approaches. *Current Opinion in Environmental Sustainability*, *42*, 65–75.
Seow, P.-S., & Pan, G. (2014). A literature review of the impact of extracurricular activities participation on students' academic performance. *Journal of Education for Business*, *89*(7), 361–366.
Shelley, B., Ooi, C.-S., & Brown, N. (2019). Playful learning? An extreme comparison of the Children's University in Malaysia and in Australia. *Journal of Applied Learning & Teaching*, *2*(1), 16–23.
Shelley, B., Ooi, C.S., & Denny, L. (2021). The dialogic negotiation of justice. *Journal of Sustainable Tourism*, *29*(2–3), 488–502.
Sin, H.L., & Minca, C. (2014). Touring responsibility: The trouble with 'going local' in community-based tourism in Thailand. *Geoforum*, *51*, 96–106.
Skattebol, J., & Redmond, G. (2019). Troubled kids? Locational disadvantage, opportunity structures and social exclusion. *Children's Geographies*, *17*(1), 76–89.
Skattebol, J., Saunders, P., Redmond, G., & Bedford, M. (2012). *Making a Difference: Building on Young People's Experiences of Economic Adversity: Final Report*. Sydney: Social Policy Research Centre.
Sofield, T.H.B. (1993). Indigenous tourism development. *Annals of Tourism Research*, *20*(4), 729–750.
Sullivan, A. (2002). Bourdieu and education: How useful is Bourdieu's theory for researchers? *The Netherlands' Journal of Social Sciences*, *38*(2), 144–166.
Wiltshier, P. (2020). Health and welfare at the boundaries: Community development through tourism. *Journal of Tourism Futures*, *6*(2), 153–164.

27. Exploring the regenerative practices of Canadian women tourism social entrepreneurs: a feminist ethic of care

Karla A. Boluk, Gaurav Panse, and Sung Eun Jeon

INTRODUCTION

Sustainability Concerns in Tourism

Precipitating the pandemic, several indicators signaled the necessity to respond to a much-needed radical reform in the way tourism is practiced. Of primary concern, evidenced in mainstream headlines and scholarly literature, were trepidations regarding overtourism from destination communities. Overtourism, that is, too many travelers visiting the same place at the same time, may contribute to a variety of negative impacts such as degradation, congestion, and inflated pricing (i.e., Phi, 2019). International urban centers (i.e., Barcelona and Rome), heritage and cultural sites (i.e., the Great Wall of China, Machu Picchu, Venice) and rural areas in both Scotland and Northern Ireland (motivated by Harry Potter and Game of Thrones fans) have felt the adverse impacts of overtourism.

The negative impacts generated by the tourism sector have drawn attention to when societies reject characteristics of care in favor of unlimited growth, which is clearly not a successful strategy over the long term (Groenhout, 2004). Scholars have recently argued that overtourism reveals our inability to care in tourism, which has consequences for all stakeholders involved in or affected by the tourism sector (Carnicelli & Boluk, 2021). Tourism scholars have shared grave concerns regarding what unlimited growth may lead to in tourism, including the demise of culture, heritage, and community ties, and to environmental disrepair. Furthermore, as sustainability concerns in the sector of tourism have escalated, many critiques of sustainability have emerged.

Wheeller (1993) was among the first to critique the take-up of sustainability in tourism, arguing that the seizure of sustainability was intentional for achieving public relations outcomes. Specifically, Wheeller signaled that projecting a sustainability discourse supported aims of appearing as though the sector was showing care; however, a continuation of a "bugger it up and pass it down" (p. 125) philosophy persisted, meaning that future generations are left to endure the consequences of the hedonistic desires and decision-making of travelers and other key decision-makers such as governments and large businesses (Weeden & Boluk, 2014).

Nearly two decades after Wheeller's (1993) sustainable tourism critique, Higgins-Desbiolles' (2010) work argued that sustainable tourism was as elusive as ever, because neoliberal ideology continued to promote a consumerist culture. A decade after Higgins-Desbiolles' (2010) concerns about the elusiveness of sustainable tourism, Boluk et al. (2019) challenged the articulation of the United Nations Sustainable Development Goals (SDGs). The authors expressed concerns regarding the UN identifying the tourism sector as a key driver of the SDGs. In particular, the scholars argued that the SDGs, which serve as a global and cross-sector framework, reinforce capitalist tenets, and ignore important voices who should be included in sustainable tourism considerations; including feminist and Indigenous perspectives (Boluk et al., 2019).

Recognizing the palpable frustrations and concerns expressed by scholars regarding the tourism sector's inability to effectively engage in sustainable practices, scholars have made calls for new approaches and models to support clear and responsible tourism practices and outcomes (i.e., Dredge, 2017). Such calls signal an imperative moral and caring turn (Dredge, 2017) in the tourism sector, to ensure we do not destroy the places and communities where tourism, especially rural tourism, takes place. We argue that there are three critical provisions that may advance our thinking regarding how to cultivate an improved tourism sector that may generate desirable benefits. Further, we argue such provisions require additional consideration in the literature.

The first provision is that while we have recognized the limitations of sustainability evidenced in theory and praxis reinforcing neoliberal ideology, we have not adequately progressed our thinking on alternative approaches. In response to calls for new responses, we recognize the value of regenerative tourism as a continuation of sustainability thinking in tourism. Secondly, we highlight how the work of some tourism social entrepreneurs (TSEs) in Canada contribute to regenerative tourism. As such, our informants serve as examples of key community disruptors and change agents improving the conditions where tourism takes place, which requires further attention in the literature. Thirdly, in combination with the initial two concepts, we join Pritchard (2018) in advocating for, and seriously engaging with, feminist theory in tourism. As we hope to illustrate here, engaging with a feminist ethic of care may propel our thinking, contributing to just, regenerative, inclusive, and caring pathways in tourism. The next section will provide an overview of the regenerative tourism and tourism social entrepreneurship literature.

LITERATURE REVIEW

Regenerative Tourism

Regeneration is well documented in agricultural practices, emphasizing both the need to move beyond mechanistic methods and that the use of pesticides and fertilizers in agriculture deteriorates soil health (e.g., Brown, 2018; LaCanne & Lundgren,

2018). Regenerative thinking and practices have infiltrated into the fields of business and leadership (e.g., Hutchins & Storm, 2019) and, more recently, tourism, stemming from the thought leadership of Anna Pollock (2019). Regenerative tourism offers a resolution to the current fragmented way tourism is practiced, encouraging interconnectivity and the practice of higher levels of care (Pollock, 2019). Adopting a regenerative approach requires a mindset shift and whole systems thinking, with a commitment to creating value for entire ecosystems and stakeholders (Hutchins & Storm, 2019). Regenerative thinking in tourism is a future-oriented philosophy aiming to enhance destination communities (Ateljevic & Sheldon, 2022), embracing the goal of leaving ecosystems, communities, and economies better off than they were before the tourism encounter.

Pollock (2019) argues that regenerative tourism focuses on delivering net benefits and is a natural maturation, responding to some concerns regarding the short-term thinking associated with sustainability. Accordingly, regenerative tourism is a radical reform that is necessary in tourism and attends to the importance of re-evaluating the role of communities and environments where tourism takes place (Cave & Dredge, 2020). In the case of regenerative tourism, the emphasis lies in a community engagement approach, focused on co-creating and co-designing tourism to bring improvements to tourism communities (Duxbury et al., 2021); this forms the central focus of this chapter. Regenerative tourism has been offered as a potential remedy to the current mechanistic and fragmented way tourism is practiced, encouraging interconnectivity and the practice of higher levels of care (Pollock, 2019; Boluk & Panse, 2022).

Building on contemporary sustainability definitions, regenerative tourism emphasizes giving back to communities, cultures, heritage, places, landscapes (Duxbury et al., 2021), thus improving the nature of communities through tourism. As such, engaging in regenerative tourism requires prioritizing the needs of host communities over the desires of travelers and the profit-oriented ambitions determined by governments and businesses (Higgins-Desbiolles et al., 2019). To put regenerative practices into action requires meaningful and authentic dialogue, transparency, co-creating partnerships with each other and other living systems, drawing on feminist perspectives and Indigenous ways of knowing, as well as being and interacting with the world, which have been largely ignored and disregarded in tourism.

Some scholars advocate that the tourism sector needs to be led by cultivating regenerative travels post-pandemic (e.g., Ateljevic, 2020), countering irresponsible tourism practices (Higgins-Desbiolles et al., 2019). Pursuing a regenerative approach in tourism requires a level of maturity and realism regarding the destructive pathway of tourism (Sheldon, 2021). There are limited analyses on regenerative tourism in the scholarly literature. An additional response to the calls for new models responding to the destructive pathways of tourism comes from tourism social entrepreneurs (TSEs), who will be discussed next.

Tourism Social Entrepreneurs

Tourism social entrepreneurs (TSEs) balance profit generation with making a social impact, defying, and transforming, the current economic order by foregrounding social justice (Mueller, 2011). TSEs are defined as "change agents in a destination's social entrepreneurship system; the people who bring their vision, characteristics and ideas to solve the social problem and bring about the transformation of the tourist destination" (Sheldon et al., 2017, p. 7). Importantly, TSEs demonstrate alternative values with their sights on inclusivity, justice, and regenerative futures (Boluk & Panse, 2022). Clearly documented in the literature, the tourism sector presents plenty of opportunities for social entrepreneurs to positively impact destinations and improve the sector considering its inimical impacts (Dredge, 2017; Sheldon et al., 2017). The crucial roles of social entrepreneurs in tourism are beginning to be realized in the literature, especially in relation to sustainable destination development in rural areas (i.e., Aquino et al., 2018; Boluk, 2011a; 2011b; Boluk & Mottiar, 2014; Mottiar et al., 2018); however, the scholarship on TSEs is still only emerging (Boluk & Aquino, 2022).

The work of TSEs serves as an essential response to calls for improving the way we practice tourism. In the first edited volume on *Social Entrepreneurship and Tourism*, the editors (Sheldon et al., 2017) offered a research agenda encouraging scholars to respond to several nascent areas in the literature. Among several avenues of potential inquiry, the editors called for conceptual and theoretical research on TSEs, relational characteristics between entrepreneurs and others including communities, governments, and characteristics, qualities, and behaviors of TSEs. Recently, Boluk & Aquino (2022) scanned the literature charting some of the impacts of TSEs on Gartner's (1985) dimensions of entrepreneurship, encompassing the individual, process, organization, and environment. The authors' analysis identified that the exploration of TSEs is still developing and has centered on the various motivations propelling individuals to engage, enact, and/or respond to pressing social problems, the various roles TSEs play in their communities, the opportunities for policymakers, and the need for governments to intervene in support of local community decision-making, and in implementation to maximize tourism impacts (Boluk & Aquino, 2022). As Boluk and Aquino (2022) noted in their analysis, additional attention is required in the scholarship on TSEs to explore diversity, including individuals of a variety of races, sexualities, abilities, and genders.

There are few studies in the tourism social entrepreneurship literature which recognize the agency of women and their role in progressing sustainability and regenerative tourism (see Boluk & Panse, 2022; Higgins-Desbiolles & Monga, 2021; Kimbu & Ngoasong, 2016). In response to this, Kimbu and Ngoasong (2016) made a call for more research to understand how women engage in social entrepreneurial activities and benefit their local communities. Responding to this call, Higgins-Desbiolles and Monga (2021) highlighted the work of women social entrepreneurs in an event business focused on mitigating social inequity by supporting vulnerable women in Australia. Such research is imperative, given that women are recognized as essential change agents in communities and "are generally more purpose-driven than men," more likely to initiate businesses with the intention of making a difference (Global Entrepreneurship Monitor, 2020, p. 16), and to place a high importance on community well-being (Tajeddini et al., 2017). Yet, women are significantly understudied in the tourism social entrepreneurship literature. Informed by this gap, our work aims to explore how women TSEs may lead change in their communities, fostering regenerative tourism futures. The next section outlines our theoretical framework.

Feminist Ethic of Care

Care is identified as an emotion and activity, which is sometimes invisible, shared, and/or reciprocal, impacting one's social relationships and self-perception (Armstrong & Armstrong, 2001). Tronto (1993) refers to care as a basic human activity that consists of being aware of and attending to one's own needs and the needs of others. Scholars have recognized care work is often women's work and afforded limited value (Armstrong & Armstrong, 2001; Gilligan, 1993). The way in which caring happens is often a result of societal expectations based on gender roles (Pearlin et al., 1990). Gilligan (1993) argues that women are often socialized from childhood to define themselves by their relationships and their ability to care for others, which is reinforced throughout their lives.

Gilligan (1993) argues that "in the different voice of women lies the truth of an ethic of care, the tie between relationship and responsibility, and the origins of aggression in the failure of connection" (p. 173). Importantly, Gilligan (1993) recognizes the ideal of care as "an activity of relationship, of seeing and responding to need, taking care of the world by sustaining the web of connection so that no one is left alone" (p. 62). Noddings' (2012) ethic of care work is premised on the mutual nature of care between the one who provides care and the one who receives care. The ethic of care is an approach to morality emphasizing a complex web of interdependence with others (Gilligan, 1993; Noddings, 1984; Tronto, 1993).

The relational aspect of care is particularly meaningful considering the current issues regarding tourism as a force for marginalization and environmental destruction. As such, to ensure that "no one is left alone" (Gilligan, 1993, p. 62) it is important to critically engage with a feminist ethic of care to consider how TSEs practice care in tourism, contributing to regenerative futures. Taking up a feminist ethic of care is important in tourism given that Pritchard (2018) recognized that feminist theory has not been adequately attended to in the tourism scholarship. Notably, few scholars studying the role and impacts of TSEs have drawn on care theory (see Higgins-Desbiolles & Monga, 2021; Boluk & Panse, 2022; Jeon & Boluk, forthcoming, as examples). A feminist ethic of care is used as a theoretical framework in our work to provide insights into understanding how TSEs practice care, supporting their regenerative tourism activities.

STUDY CONTEXT: CANADIAN WOMEN TSES

The research described in this chapter is part of a larger project exploring the role of Canadian TSEs in progressing the United Nations Sustainable Development Goals. The broad goal of our research is to understand the role of women TSEs in contributing to regenerative futures, drawing on a feminist ethic of care. The context is particularly important given that much of the literature on TSEs focuses on communities in developing countries; as such there is a paucity of scholarship on TSEs in Canada. Furthermore, as previously remarked, there are few studies in the TSE scholarship focusing on the social impacts of women. There is also a lack of explicit recognition as to how TSEs contribute to regenerative tourism outcomes, as well as a lack of reference to a feminist ethic of care.

We refer to two studies in this chapter, including our analysis of 20 interviews, which included 11 semi-structured interviews with food TSEs, mostly in rural contexts across Canada (Boluk & Panse, 2022) and nine semi-structured interviews with a variety of TSEs (winery, bar, cooking school, restaurant, hostel, inn, mechanic) in one rural community, Prince Edward County, two hours east of Toronto, Ontario (Jeon & Boluk, forthcoming). Data was collected between May 2021 and December 2021 and analyzed using the strategies of grounded theory (Charmaz, 2011).

RESULTS

Across both data sets, our analysis revealed the importance of adopting a regenerative mindset in the work of TSEs, enhancing personal and community health, well-being, and quality of life. Specifically, our analysis shone a light on the importance of the TSEs' beliefs and values, which translated into intentional business decisions. Accordingly, their values influenced their interactions with colleagues, staff, suppliers, and their environments, as well as specific decisions made in developing their products, sourcing products/ingredients, and buying raw materials. A regenerative mindset was intentional in the data, as all of our informants explicitly emphasized a desire to improve their community from an environmental perspective by supporting the circular economy (i.e., minimizing or reducing waste, procuring or growing organic products, showing care for animals), from a human relations perspective via providing training and development opportunities, centering equity, inclusion and diversity, and responding to the needs of marginal (diverse populations, girls and women) or vulnerable populations (i.e., homelessness, addictions, those with food sensitivities).

The Boluk and Panse (2022) study examined how TSEs' work supported the consumption of real food by partnering with non-GMO producers, organic or fair-trade producers; many sourced indigenous ingredients, and others partnered with local farmers, ensuring their animals were responsibly raised. Others aligned themselves with networks (i.e., Feast On, Fair Wage Employer) supporting some of these interests and/or made financial donations (i.e., the Canadian Biotechnology

Action Network) to support value-aligned causes. Based on their value-oriented decisions, many of our informants reflected on their involvement in educating their clients, staff, and broader communities (including younger generations) about their regenerative practices. Such education manifested in informal (i.e., in-store client interaction – to raise awareness regarding intentionality in sourcing ingredients and their health benefits, communication with suppliers) and formal ways (i.e., employee equity training, youth programing – teaching the skills of creating and growing food) (Boluk & Panse, 2022; Jeon & Boluk, forthcoming). An additional example of formal support for employees was noted in the setting up of a not-for-profit organization supporting mental health in the hospitality sector, signaling this TSE's commitment to regenerating practices in the overall sector (Boluk & Panse, 2022).

Another way TSEs educated visitors about their regenerative practices was through immersive experiences. While the importance of sharing practices was signaled in both studies, our second study (Jeon & Boluk, forthcoming) unearthed the meaning in showing and sharing their practices with visitors. For example, the TSEs operating an inn built a garden greenhouse for visitors and their neighbors to use. Other TSEs operating a bar, cooking school, and hostel, with affiliated on-property farms, have used their farms to educate visitors about food security and ingredient sourcing.

Some TSEs highlighted the importance of introducing visitors to their farmers to better understand the process of regenerative farming practices. Many TSEs provided opportunities to visitors to harvest their own produce (Jeon & Boluk, forthcoming).

A novel contribution in Jeon and Boluk's (forthcoming) analysis was realizing some of the opportunities in pivoting their businesses due to the pandemic. This was an important finding in the face of media attention that relayed the crippling effects the pandemic had on small businesses. Our analysis revealed the support and care received by TSEs from their communities, ensuring their success during the pandemic. Furthermore, our analysis drew attention to the economic potential of adopting a regenerative mindset and approach. The economic value of adopting regenerative practices was interesting, given that some of our informants downsized their businesses. For example, the pandemic pivot required restaurant owners to reduce seating capacity. Another informant made scale adjustments to their business, including reducing occupancy in their inn down to just one room. This TSE reflected that "bigger isn't always better," understanding the opportunities to enhance their connections with their visitors and consider their own enjoyment and quality of life. In a similar way, other informants reflected on aspects of their businesses that they could redesign. For example, a couple of TSEs operating a repair shop chose to cease the line of their business that rented equipment to party-goers who had previously contributed to community disturbances. Upon reflection, our informants came to terms with the fact that this part of their business diminished their quality of life and the quality of life of their community, which was incongruent with their core values (Jeon & Boluk, forthcoming).

DISCUSSION AND IMPLICATIONS FOR FUTURE RESEARCH

The literature reviewed at the outset of this chapter presented some responses to calls for alternative models to support just and caring pathways in tourism, including regenerative tourism and TSEs, which have not been explored together until now (Boluk & Panse, 2022; Jeon & Boluk, forthcoming). The theoretical framework underpinning our analysis is a feminist ethic of care. This section will discuss our analysis and offer a conceptual model to aid in the illustration of the web of care demonstrated by our informants. It is our goal, through the presentation of our conceptual model, to bring visibility to the regenerative activities of our women TSEs, given that we know care work is often invisible and undervalued (Armstrong & Armstrong, 2001; Gilligan, 1993).

Gilligan's (1993) care theory establishes that care is an activity of relationship responding to individual needs; this was verified in our data. The importance of living in a relationship with one another and one's natural environment "to ensure no one is left alone" (Gilligan, 1993, p. 62) was understood as important in our analysis. Specifically, our informants reported developing relationships with their visitors, and producers, and seeking certifications that supported their values (e.g., Fair Wage Employer, Feast On, Canadian Biotech Action Network (advocating for policy against GMOs)). Such examples built on the relationships our informants nurtured, in particular with their suppliers, and with both local and broader communities including provincial, national, and international networks.

Further to the importance of relationships, an ethic of care requires "responsiveness to particular others, and responsibility over the long-term" (Robinson, 2006, p. 337). Beyond an interest in attending to relationships with visitors, suppliers, and networks, it was evident in our analysis that our informants were attentive to the needs of their employees, and broader communities. Specifically, strong philosophies regarding sustainability and regenerative practices drew attention to the TSEs' commitment and the lengths they were willing to go to support just futures. Clearly, some of these values are in direct conflict with operating a contemporary business in a capitalistic system supported by neoliberal ideology, which underscores individualism and profit, and which acts in opposition to regenerative practices (Ateljevic, 2020; Sheldon, 2021). Instead, our informants practiced and demonstrated care in a variety of ways, which we illustrate in Figure 27.1.

At the core of our figure is care for oneself; such care was recognized by a few of our informants, particularly in the second study. Our analysis revealed some of our TSEs attended to their own needs specifically when pivoting their businesses during the onset of the COVID-19 pandemic. We noticed that for some, downsizing their businesses resulted in "right-sizing" their businesses, allowing for more meaningful experiences with their visitors, and enhanced quality of life. Given the goals set by TSEs to make social impacts within their communities, and the limited scholarly research on burnout, particularly in times of crises such as the ongoing health and

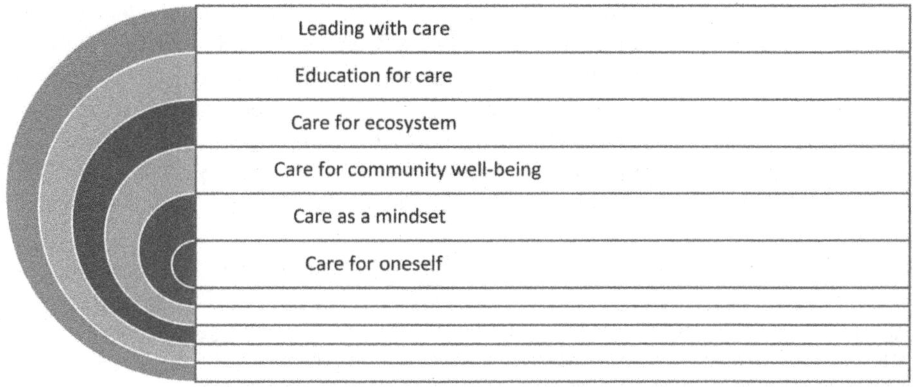

Figure 27.1 Tourism social entrepreneurs' web of care

climate crises, it is important for future research to explore how TSEs attend to their own personal care while engaging in important regenerative work.

The second component in our web of care reflects the importance of adopting a care mindset. There is limited scholarship on TSEs and mindsets (see Kline et al., 2014; 2017 as exceptions), despite our knowledge that TSEs set out to make positive impacts in their communities responding to governmental deficiencies and market gaps (Thompson, 2008). Thinking about mindsets is particularly important in engaging with regenerative tourism activities and opportunities to redesign tourism. Many examples in our data unveiled examples of care for the well-being of TSEs' communities. Generally, at the forefront of many of the food-related TSEs was support for healthy lifestyles through sharing gardens and allowing opportunities for locals and visitors to harvest produce. Further research exploring the mindsets of TSEs carrying out regenerative tourism practices could be valuable.

Our informants showed care for the whole ecosystems they are a part of when discussing intentionally sourcing ingredients, procuring local, organic, indigenous and/or fairly sourced produce, and ingredients, and prioritizing the conditions and treatment of animals in establishing partnerships with farmers. In one example, one of our informants traveled oversees to witness firsthand the conditions of farmers harvesting the tea she provided in her cafe. In another example, a TSE started a non-profit organization raising awareness about mental health in the hospitality sector and providing financial aid for those requiring support. Such care illuminates the importance of some of their relationships and their responsibility to nurture relationships with employees, farmers, and the broader hospitality sector.

An ethic of care was revealed in the intentional work TSEs engaged in around education. We noted informal, formal, and immersive education in the data. Informal education sometimes took place over the counter when interacting with customers about their intentional ingredient choices, whereas formal education was reflected in, for example, employee equity and diversity training, and youth programming teaching the skills of creating and growing food. Immersive education occurred when visitors to Prince Edward County, for instance, were supported in their experiences to wander around gardens, harvest their own produce, and interact with farmers. Our analysis elicited the role of altruism in some forms of educational experiences, particularly with young people who we thought the entrepreneurs might be unlikely to receive financial gains from. Altruism has been ignored in the literature on TSEs (Boluk & Aquino, 2022) and deserves further attention. This may also be an interesting line of inquiry in the regenerative tourism scholarship.

All of the aspects illustrated in our model revealed to us that our TSEs lead with a care orientation, which has guided their regenerative practices. Leadership has been attended to in the TSE literature. Importantly, regenerative leadership is one of the foundational fields that regenerative tourism emerges from, and as such, considering leadership in regenerative tourism is important, specifically, because, as we outlined at the beginning of our chapter, we believe leadership is precisely what is required in our sector to ensure "no one is left alone or hurt" (Gilligan, 1993, p. 63) in the communities where tourism takes place. We need more leaders to respond to the pressing issues of our time, and we need more research examining how we may support their work to bring about improvements to a sector that has caused inimical impacts.

CONCLUSIONS: CHARTING CARING PATHWAYS

Our findings from both studies reveal an array of examples of Canadian women TSEs, who have engaged in activities that have made regenerative impacts on their communities. Regenerative tourism has emerged as an important approach specifically in response to some scholars' calls for new models that may support an improved way of planning and designing tourism (Dredge, 2017). Furthermore, regenerative tourism has been sought as an approach for destinations to regain control, minimize harmful impacts, and create benefits emerging from the COVID-19 pandemic (Duxbury et al., 2021). TSEs are important stakeholders in our consideration of co-designing more responsible tourism destinations with communities (Mottiar et al., 2018; Higgins-Desbiolles et al., 2019). Particularly, as established in the literature, TSEs' work challenges the status quo and capitalist systems, with intentions to generate and return social impacts to their communities (Boluk, 2011a). However, the scholarship up until now has not explicitly considered the regenerative impacts of TSEs.

Our work has adopted a feminist ethic of care to help us better understand specifically how TSEs are contributing to regenerative practices in tourism. There are several reasons why engaging with an ethic of care is important in tourism including: (1) the urgent need to redesign the way we offer tourism; (2) there is a lack of scholarship attending to feminist theory in tourism (Pritchard, 2018); (3) to acknowledge women and other genders who are amplifying the voices of women, and (4) shedding light on care work that has been referred to as invisible work in the literature (Armstrong & Armstrong, 2001). Drawing attention to the regenerative and caring work of TSEs may provide inspiration across stakeholders and sectors.

Notably, pathways to understanding how the tourism sector may be more inclusive are imperative if we are to realize SDG 5, Gender Equality, and to support women in work progressing many of the SDGs. But, as the literature highlights, "gender equality is not possible with an *exclusive* focus on women and girls; after all, gender equality is a social ideal that necessarily involves people of all genders and changes in institutions that govern human opportunities, rights and behavior" (CARE International, 2016). As such, future research embracing a feminist ethic of care should seek to explore the networks, allies, and ambassadors supporting the regenerative practices of women TSEs.

Lastly, while an ethic of care is premised on the notion that everyone is mutually a giver and receiver of care, it is clear that some stakeholders in tourism are comfortable on the receiving end and are happy to take much more than they give. The literature paying attention to the negative impacts of tourism clearly draws attention to the takers and our tenuous ability to care in tourism. The work of social entrepreneurs in tourism may bring hope to our sector and if we take those entrepreneurs seriously, they may reveal a potential stakeholder who could help steer us in a less harmful direction. Attending to the regenerative practices of TSEs may provide us with some tangibles, which if cross-pollinated across the tourism sector could generate some important positive, beneficial, and potentially transformative impacts.

REFERENCES

Aquino, R.S., Luck, M., & Schanzel, H.A. (2018). A conceptual framework of tourism social entrepreneurship for sustainable community development. *Journal of Hospitality and Tourism Management*, *37*, 23–32.

Armstrong, P., & Armstrong, H. (2001). *Thinking it through: Women, work and caring in the new millennium*. Halifax, NS: Healthy Balance Research Program. ACEWH_hbrp_thinking_it_through_women_work_caring_new_millennium.pdf (dal.ca).

Ateljevic, I. (2020). Transforming the (tourism) world for good and (re)generating the potential "new normal." *Tourism Geographies*, *22*(3), 467–475. https://doi.org/10.1080/14616688.2020.1759134.

Ateljevic, I., & Sheldon, P.J. (2022). Guest editorial: Transformation and the regenerative future of tourism. *Journal of Tourism Futures*, *8*(3), 266–268.

Boluk, K. (2011a). Revealing the discourses: White entrepreneurial motivation in Black South Africa. *Tourism Planning and Development*, *8*(2), 199–213.

Boluk, K. (2011b). Fair trade tourism South Africa: A pragmatic poverty reduction mechanism? *Tourism Planning and Development*, *8*(3), 237–251.

Boluk, K., & Aquino, R. (2021). Assessing the impacts of social entrepreneurs and social enterprises in tourism. In A. Stoffelen & D. Ioannides (eds), *Handbook of Tourism Impacts* (pp. 117–133). Cheltenham, UK, & Northampton, MA, USA: Edward Elgar Publishing.

Boluk, K., & Mottiar, Z. (2014). Motivations of social entrepreneurs: Blurring the social contribution and profits dichotomy. *Social Enterprise Journal*, *10*(1), 53–68.

Boluk, K., & Panse, G. (2022). Recognizing the regenerative impacts of Canadian women tourism entrepreneurs through a feminist ethic of care lens. *Journal of Tourism Futures*, *8*(3), 352–366. https://doi.org /10.1108/JTF-11-2021-0253.

Boluk, K., Cavaliere, C., & Higgins-Desbiolles, F. (2019). A critical framework for interrogating the United Nations Sustainable Development Goals 2030 Agenda in Tourism. *Journal of Sustainable Tourism*, *27*(7), 847–864. doi:10.1080/09669582.2019.1619748.

Brown, G. (2018). *Dirt to Soil: One Family's Journey into Regenerative Agriculture*. White River Junction, VT: Chelsea Green Publishing.

CARE International (2016). Engaging men and boys for gender equality series: Man in the mirror—reflections on men and boys engaging gender work in development. Brief 3. Retrieved April 18, 2022, from CARE_EMB-Brief-3_2016 (1).pdf.

Carnicelli, S., & Boluk, K. (2022). A theory of care to socialise tourism. In F. Higgins-Desbiolles, A. Doering & B. Bigby (eds), *Socialising Tourism* (pp. 58–71). Abingdon: Routledge.

Cave, J., & Dredge, D. (2020). Regenerative tourism needs diverse economic practices. *Tourism Geographies*, *22*(3), 503–513.

Charmaz, K. (2011). A constructivist grounded theory analysis of losing and regaining a valued self. In F.J. Wertz, K. Charmaz, L.M. McMullen, R. Josselson, R. Anderson, & E. McSpadden (eds), *Five Ways of Doing Qualitative Analysis* (pp. 164–201). New York: The Guilford Press.

Dredge, D. (2017). Foreword. In P.J. Sheldon & R. Daniele (eds), *Social Entrepreneurship and Tourism: Philosophy and Practice* (pp. v–vii). Cham: Springer.

Duxbury, N., Bakas. F.E., Vinagre de Castro, T., & Silva, S. (2021). Creative tourism development models towards sustainable and regenerative tourism. *Sustainability*, *13*(1), 1–17.

Gartner, W.B. (1985). A conceptual framework for describing the phenomenon of new venture creation. *The Academy of Management Review*, *10*(4), 696–706.

Gilligan, C. (1993). *In a Different Voice*. Cambridge, MA: Harvard University Press.

Global Entrepreneurship Monitor (GEM) (2020). *Global Report*. Retrieved May 3, 2021, from https://www.gemconsortium.org/report/gem-2019-2020-global-report.

Groenhout, R.E. (2004). *Connected Lives: Human Nature and an Ethics of Care*. Oxford: Rowman & Littlefield Publishers.

Higgins-Desbiolles, F. (2010). The elusiveness of sustainability in tourism: The culture-ideology of consumerism and its implications. *Tourism and Hospitality Research*, *10*(2), 116–129.

Higgins-Desbiolles, F., & Monga, M. (2021). Transformative change through events business: A feminist ethic of care analysis of building the purpose economy. *Journal of Sustainable Tourism*, *29*(11–12), 1989–2007. https://doi.org/10.1080/09669582.2020.1856857.

Higgins-Desbiolles, F., Carnicelli, S., Krolikowski, C., Wijesinghe, G., & Boluk, K. (2019). Degrowing tourism: Rethinking tourism. *Journal of Sustainable Tourism*, *27*(12), 1926–1944. doi:10.1080/09669582.2019.1601732.

Hutchins, G., & Storm, L. (2019). *Regenerative Leadership: The DNA of Life-Affirming 21st Century Organizations*. Royal Tunbridge Wells: Wordzworth Publishing.

Jeon, S.E. & Boluk, K.A. (forthcoming). [Paper unpublished at time of writing, but a valuable contribution to this chapter.]

Kimbu, A.N., & Ngoasong, M.Z. (2016). Women as vectors of social entrepreneurship. *Annals of Tourism Research*, *60*, 63–79.

Kline, C., Shah, N., & Rudbright, H. (2014). Applying the positive theory of social entrepreneurship to understand food entrepreneurs and their operations. *Tourism Planning & Development*, 11(3), 330–342.

Kline, C., Boluk, K., & Shah, N. (2017). Exploring social entrepreneurship in food tourism. In P. Sheldon & R. Daniele (eds), *Social Entrepreneurship and Tourism: Philosophy and Practice* (pp. 135–154). Cham: Springer.

LaCanne, C.E., & Lundgren, J.G. (2018). Regenerative agriculture: Merging farming and natural resource conservation profitably. *PeerJ*, *6*, e4428. doi:10.7717/peerj.4428.

Mottiar, Z., Boluk, K., & Kline, C. (2018). The roles of social entrepreneurs in rural destination development. *Annals of Tourism Research*, *68*, 77–88.

Mueller, S. (2011). The transformative power of social entrepreneurship: How social entrepreneurs induce change on a firm, institutional and societal level. *Innovation & Entrepreneurship*, *1*(9/10), 1–8.

Noddings, N. (2012). The caring relation in teaching. *Oxford Review of Education*, *38*(6), 771–781. doi:10.1080/03054985.2012.745047.

Pearlin, L., Mullan, J., Semple, S., & Skaff, M. (1990). Caregiving and the stress process: An overview of concepts and their measures. *The Gerontologist*, *30*(5), 583–591.

Phi, G.T. (2019). Framing overtourism: A critical news media analysis. *Current Issues in Tourism*, *23*(17), 2093–2097. https://doi.org/10.1080/13683500.2019.1618249.

Pollock, A. (2019). *Regenerative Tourism: The Natural Maturation of Sustainability*. Retrieved April 29, 2022, from https://medium.com/activate-the-future/regenerative-tourism-the-natural-maturation-of-sustainability-26e6507d0fcb.

Pritchard, A. (2018). Predicting the next decade of tourism gender research. *Tourism Management Perspectives*, *25*, 144–146. https://doi.org/10.1016/j.tmp.2017.11.014.

Robinson, F. (2006). Beyond labour rights. *International Feminist Journal of Politics*, *8*(3), 321–342.

Sheldon, P.J. (2021). The coming-of-age of tourism: Embracing new economic models. *Journal of Tourism Futures*, *8*(2), 200–207. https://doi.org/10.1108/jtf-03-2021-0057.

Sheldon, P.J., Pollock, A., & Daniele, R. (2017). Social entrepreneurship and tourism: Setting the stage. In P.J. Sheldon and R. Daniele (eds), *Social Entrepreneurship and Tourism: Philosophy and Practice* (pp. 1–18). Cham: Springer.

Tajeddini, K., Walle, A.H., Denisa, M. (2017). Enterprising women, tourism, and development: The case of Bali. *International Journal of Hospitality & Tourism Administration*, *18*(2), 195–218.

Thompson, J.L. (2008). Social enterprise and social entrepreneurship: Where have we reached? A summary of issues and discussion points. *Social Enterprise Journal*, *4(*2), 149–161.

Tronto, J.C. (1993). *Moral Boundaries: A Political Argument for an Ethic of Care*. New York: New York University Press.

Weeden, C., & Boluk, K. (eds) (2014). *Managing Ethical Consumption in Tourism*. Abingdon: Routledge.

Wheeller, B. (1993). Sustaining the ego. *Journal of Sustainable Tourism*, *1*(2), 121–129. https://doi.org/10.1080/09669589309450710.

28. Transforming a dogsledding community: the 'Gafsele Open' and lifestyle migrants in sparsely populated northern Sweden

Marco Eimermann, Doris A. Carson and Linda Lundmark

INTRODUCTION

This chapter increases our understanding of how intra-European lifestyle migrants may transform communities in sparsely populated areas (SPAs) through their engagements in civil society, using the example of a dogsledding community in Arctic Sweden. In-depth narrative analysis of interviews with international migrant dogsledders and longer-term residents shows the heterogeneity of communities in sparsely populated settings and their diverse perspectives on community transformation and renewal in response to challenges of demographic shrinkage (Eimermann et al., 2022). The case study village of Gafsele in Åsele municipality (Figure 28.1) provides an interesting study context as it is home to a relatively large group of international migrants, many of whom were attracted by exceptional opportunities for dogsledding and an internationally renowned trail network. The local dogsledding club organizes an annual dogsledding event (the Gafsele Open) attracting Swedish and international participants. Balancing their incomes and lifestyles around dogsledding activities, migrants are engaging in the club and co-organizing the event as individuals and through their businesses (D.A. Carson et al., 2018; Eimermann & Singleton, 2021). This area is thus exploring its potential for local rural development through community- and nature-based tourism.

The chapter analyses four interviews to provide more in-depth qualitative insights into the impact of international migration and globalization processes on new rural community development related to tourism, leisure, and events. Previous studies have not paid due attention to small village communities in SPAs, such as those commonly found in the Swedish North and similar northern peripheries (D.B. Carson et al., 2020). Such areas are different from other rural contexts in that they are more distant and disconnected from general urban–rural and global core–periphery structures (D.B. Carson & Carson, 2014). They are too far from major urban centres and tourist markets to attract urban–rural migration and tourism flows to the same extent as rural areas located in the hinterland of larger cities, whilst they also lack the sorts of outstanding natural amenities and tourist attractions as a foundation for larger-scale tourism and resort developments (Koster & Carson, 2019). Quite often, these villages are also distant from other settlements within their vast and sparsely

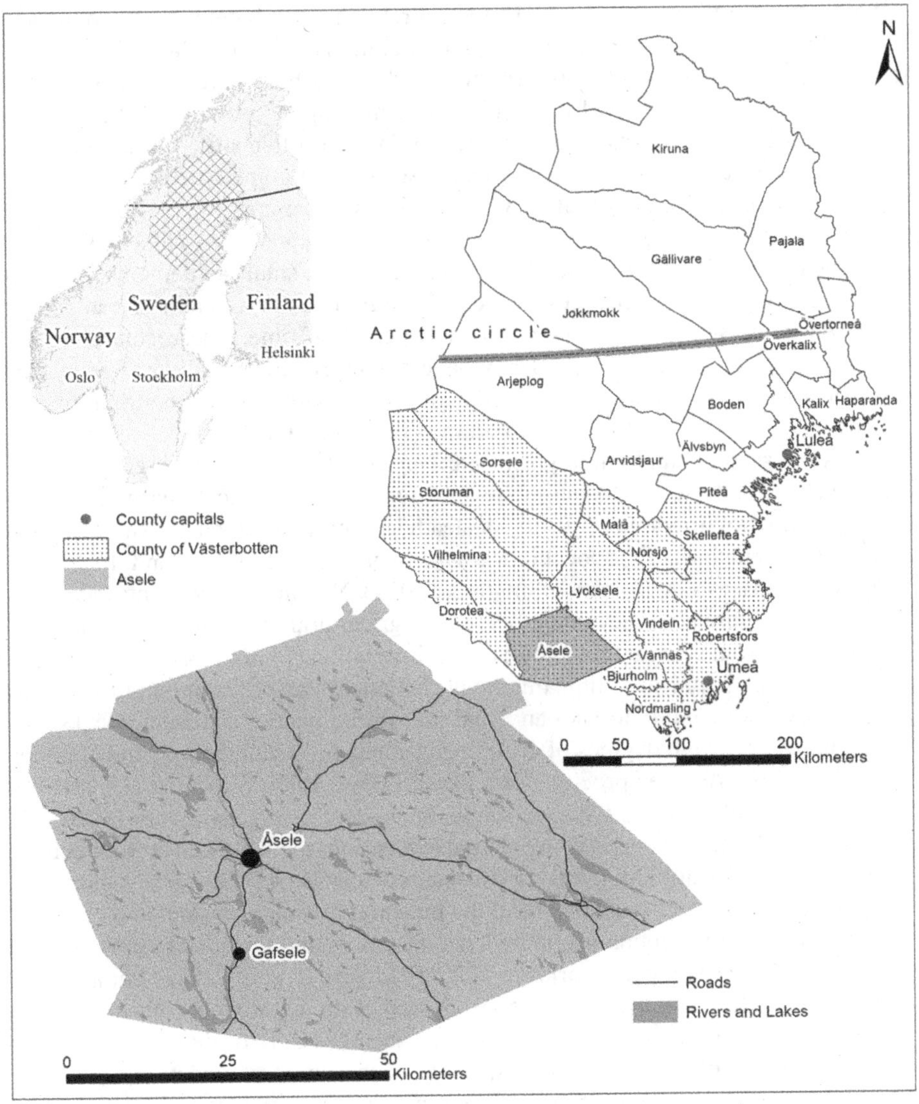

Source: Created by Linda Lundmark. Data: Lantmäteriet (the Swedish mapping, cadastral and land registration authority used with permission).

Figure 28.1 *The study area: Gafsele village in Åsele municipality (Arctic Sweden)*

populated peripheries and are only poorly connected as part of larger functional regions, particularly as different settlements have evolved around different land uses, economies, and socio-cultural characteristics (D.B. Carson & Carson, 2014).

Settlements in SPAs have a long history of relying on outsiders for economic and social development, with large resource corporations (e.g., mining and forestry), the military and other central government authorities being prominent drivers of infrastructure investment, and economic and demographic growth (Huskey, 2005). This dependence on externally driven interventions has often stifled local ambitions and community capacities for economic and socio-demographic diversification and renewal, and it is common that new initiatives, such as new tourism or lifestyle developments, are driven by in-migrants or external entrepreneurs (Tillberg Mattsson & Heldt Cassel, 2020). Another factor characterising small settlements in SPAs is that seemingly minor changes to local demographics or economies can trigger large and unforeseen changes within a short period of time. These conditions are all present in our case study area. The village of Gafsele, historically an agricultural settlement, shares few similarities and functional ties with the neighbouring town of Åsele, which was built around forestry and is the dominant municipal centre where most public services and employment are concentrated (Eimermann et al., 2022). Nevertheless, Gafsele has managed to transform itself on the basis of attracting a small number of migrants active in dogsledding, a process which has since brought about critical changes to the local business network, tourism flows, and the general image of the place (Eimermann & Singleton, 2021). Not only did they introduce new tourism and leisure activities into a declining agricultural community, but they also contributed to some extent to demographic rejuvenation (Eimermann et al., 2022).

Due to its geographical setting, this study challenges ideas of communities as characterized by traditional industries and land use, strong place attachment, local social 'bonding' capital, high degrees of civic engagement and volunteering, and common values based on trust, reciprocity and reliance (Tuan, 1986; Sharpe et al., 2016). It focuses on mobile people, who we see as an in-community subgroup (Skerrat & Steiner, 2013) within the overall local dogsledder community. Such mobile people, ranging from residential migrants to seasonal occupants, second-home owners and other frequent short-term visitors, have the potential to make considerable contributions to local village economies, knowledge networks and the social fabric of rural communities (Robertsson & Marjavaara, 2015; Moscardo et al., 2013) but are often treated as outsiders and, thus, excluded from studies on 'community' change. In this study, international dogsledders represent one such example of mobile people who enlarge the sense of community, as explained in the literature review below. Their stories are important sources of knowledge in human geography and related social sciences. Narrative analysis can contribute to our understanding of how space and social relations in the context of communities are reproduced (Riessman, 1993; Wiles et al., 2005).

Against this background, this chapter investigates community tourism that is not necessarily based on supporting market hegemony (Mair, 2018), but rather on less commercial activities that are meant to benefit members of the community from a social, cultural and emotional well-being perspective. The chapter illustrates lifestyle migrants' potentials and constraints in contributing to local community tourism in Åsele. It focuses on experiences in dogsledding as 'serious leisure', understood as a systematic and voluntary pursuit of an amateur activity that is sufficiently substantial, interesting, and fulfilling to strive for a leisure career, while building on a combination of special skills, knowledge, and experience (Stebbins, 2009, p. 764), often in an effort to improve conditions for the specific leisure activity for like-minded people. This focus on civic engagement provides insights into the meaning of community for the project participants. While events like the Gafsele Open can contribute to local rural development (Eimermann & Singleton, 2021; Eimermann et al., 2022), we also need to highlight obstacles in realizing this potential.

LITERATURE REVIEW: GEOGRAPHY, COMMUNITY AND LIFESTYLE MIGRATION

To understand the role of a community's physical and social environment for the development of community-based tourism (Sharpe et al., 2016), we emphasize human perceptions and spatial psychology through a focus on meanings, values, goals, and purposes with and in a place (Tuan, 1976). As a concept, place is 'combining the sense of position within society … with the sense of and identity with spatial location that comes from living in and associating with it' (Agnew et al., 2005, p. 444). As such, researchers have generally regarded 'traditional' communities as being limited in membership and geographical area, whose members' senses of community build on good interpersonal relationships (Tuan, 1986). The image of community life as village life has been applied to primitive communities in which people are strongly rooted and bound by locality in a confined space. In Western societies, such communities would consist of a tightly knit and stable group of family and close friends who are willing to help with minor and major issues (Shaffer & Anundsen, 1993; Sharpe et al., 2016).

However, bonds, social ties and stability can both be used in positive and negative terms. Most old-style communities were also 'homogen[e]ous, suspicious of outsiders, socially and economically stratified … and limited in opportunities (Shaffer & Anundsen, 1993, p. 6). Also, Tuan (1986) wonders what it means to know each other well, and how long it takes. Rather than habit, he suggests that intensity of personal contacts can be at least as important. He lifts the conceptualization of community far beyond considerations of the local geographical area and limited memberships (ibid.).

Thus, an enlarged sense of community also includes strangers – even those in another part of the world (ibid.) – who may become soulmates through a common interest like dogsledding. Such enlarged communities embrace heterogeneity and dynamism as their members are independent but committed to each other, participating 'in common practices' and identifying as part of something bigger than the sum of their parts (Shaffer & Anundsen, 1993). This builds a sense of solidarity, encompassing a shared identity and social norms through shared places, interests, affiliations, and ideologies (Bhattacharyya, 2004).

This can be called 'communities of practice', understood as a social group that comes together and must work out how to get along together. Such groups include partners, family, co-workers, club members, politicians, municipal officials, neighbours, and, in some studies, even dogs (Granås, 2018). The groups consist of diverse individuals that engage together in the 'doing' of social life and in the negotiation of meaning (O'Reilly, 2012). The concept of practice implies not just doing, 'it is doing in a historical and social context that gives structure and meaning to what we do' (Wenger, 1998, p. 47). Communities of practice provide the structures that enable and constrain agents in their social activities (O'Reilly, 2012).

In reality, communities can seldom be viewed as either exclusively bounded (homogeneous and exclusive) or enlarged (heterogeneous and inclusive). Therefore, this chapter views communities along a spectrum, ranging from functional communities (around a shared project like dogsledding races that brings members together) on the one side, to conscious communities (where members recognize their interdependence with each other and with others), to finally deep communities (where solidarity is solidified through internalized and natural everyday behaviour) (Sharpe et al., 2016, p. 4). This spectrum may further be filled with meaning by studying links between community and mobilities (Moscardo et al., 2013) or links with entrepreneurship (Steiner et al., 2018). In addition, a relatively novel field related to these research topics is lifestyle migration (Benson & O'Reilly, 2009; D.A. Carson et al., 2018).

As a lens to understand changing communities, lifestyle migration draws on anthropological material and social theories, focusing on 'the spatial mobility of relatively affluent individuals of all ages, moving either part-time or full-time to places that are meaningful because, for various reasons, they offer the potential of a better quality of life' (Benson & O'Reilly, 2009, pp. 2–3). This is motivated by social reasons, in particular, escaping from a place, a group or a job to recreate, restore or rediscover one's personal potential to realize one's 'true' desires (ibid.). Benson (2010, p. 54) suggests three groups of lifestyle migrants based on life course stages and other matters. Two of those are relevant in this chapter: young families, who migrate and choose a location influenced by their desire for a better, healthier life where their children can enjoy more freedom; and mid-life migrants; young, childless, ex-professionals who have left their jobs to emigrate and set up their own micro businesses in tourism or other sectors. There is a growing awareness that lifestyle migration cannot be separated from other forms of migration, such as labour or refugee migration, since social, economic, environmental and political factors play a role in all kinds of migration (Benson & O'Reilly, 2016).[1]

As is shown in more detail below, the lifestyle migrants interviewed in this study moved to Arctic Sweden in search of space, tranquillity and opportunities to engage in their nature-based special interests (D.B. Carson & Carson, 2018). They desire a better and healthier life related to dogsledding, and are attracted by snow-rich winters combined with more leisure opportunities, fewer restrictions on land use, affordable housing, less traffic congestion, less stress, and lower crime rates. Large variations in seasonal weather and landscape conditions do, however, present a challenge. The migrants deal with these in various ways, for example by working hard in their nature-based tourism firms during the winter to live a relaxed lifestyle during the summer (D.A. Carson et al., 2018), or thinking about moving even further north where winters are colder and longer (Eimermann & Singleton, 2021). Lifestyle migrants are privileged in the sense that they can develop flexible attitudes towards migration, meaning that they can live in one place for some years before they decide to return or move on to the next adventure (ibid.).

Such flexible attitudes towards mobility have consequences for the extent to which migrants integrate in local society. Some research has already highlighted various ways in which lifestyle migrants integrate locally, for example through their engagement in local labour markets, contributions to volunteer groups, or community interactions as part of their children's school activities (O'Reilly, 2007; Benson, 2010; Eimermann et al., 2020). At the same time, a range of challenges hinder local integration of these lifestyle migrants, not least in relation to language barriers, institutional differences and intercultural communication problems, and conflicts of interest around the use of local resources and development priorities (Eimermann et al., 2019; van Rooij & Margaryan, 2019). While lifestyle migrants have enormous potential to stimulate new tourism development in SPAs through their micro-business ventures and international marketing activities, their potential to contribute to community tourism, as well as the obstacles they may face, is an under-researched topic.

NARRATIVE ANALYSIS AS CRITICAL INQUIRY

Critical inquiry in tourism and hospitality research can be summarized through six characteristics, three of which are the basis of this chapter's narrative analysis. These are 'asking critical questions', 'researchers' role in awareness-raising' and 'there is no Truth' (Mair, 2018). The critical questions reflect tourism development, asking 'who benefits and who loses, why are there winners and losers, why does this situation continue to exist, and what can be done to change this situation?' (ibid.). We used a *Foresight* approach to situate this study within latent material and political community structures (Eimermann, 2021).

Foresight is a methodology to advance local economic and social development strategies based on a structured dialogue between relevant local, regional and national stakeholders (Karlsdottir et al., 2017). We employed this methodology in 2018 and 2019 through five workshops in Åsele municipality (Sweden) with 16 participants in total. This rendered insights into international migrants' and longer-term residents' wishes for local future developments, based on broad questions regarding challenges and opportunities linked with quality of life, entrepreneurship, and migration. As such, the workshops were a tool to ask critical questions in a public forum, using the researcher's role in awareness-raising and stimulating more critical thinking among participants.

More concrete interviews were conducted after the workshops, and this chapter's analysis focuses on four interviewees' critical reflections regarding the Gafsele Open dogsledding event as a potential community-based tourism development. Interviews were conducted with workshop participants and other residents some weeks later to gain deeper understanding of the meaning of local socio-economic structures. The main questions regarded the organization of the annual dogsledding event (the Gafsele Open's 2019 edition), its links with local population developments, quality of life, and entrepreneurship, as well as realistic prospects for the event, the local village and the municipality.

Analysis of the transcribed interviews is based on understandings of narratives as expressions of identity through which stories are constructed about one's own history and experiences. Rather than objective reflections, such stories are limited and selective representations of activities put into context and given meaning by a narrator (Riessman, 2004). SPAs are unique geographical contexts for community studies, and narratives provide useful tools to account for this underexplored context. Narrative analysis identifies more nuances and detailed dimensions than thematic analysis, which makes it possible to connect layers of meaning in interviews to social relations in SPAs (Wiles et al., 2005; Sandberg & Tollefsen, 2010).

More concretely, this chapter's narrative analysis followed a socio-lingual model of structural categories (clauses) focusing on reference and temporal order (Labov, 1982; Labov & Waletzky, 1997). The model depends on relations between a sequence of real events and their ordering in the narrative account by organizing the interview text into six components of a narrative: Abstract, Orientation, Complicating Action, Evaluation, Resolution and Coda (Table 28.1). In interview transcripts, clauses that advance the action and describe the succession of events usually constitute a small part of the text that does not always consist of all six components. However, a focus on the Evaluation – which may take up a large part of the text – is central to the subsequent interpretation of the narrative (Sandberg & Tollefsen, 2010).

Labov's model is valuable in clarifying how narratives differ in their structural components and completeness, although it does not give any rules on the interpretation of the point of the story (Mishler, 1995). Therefore, it needs to be combined with previous studies to understand relations with developments and changes in wider society (Sandberg & Tollefsen, 2010). Thus, this chapter's previous section offers an understanding of the socio-economic context in which the analysis of the

Table 28.1 Labov's six interview components for narrative analysis (Mishler, 1995)

Component (abbreviation)	Explanation
Abstract (A)	A short summary introduction to the narrative
Orientation (O)	Statements that locate the story in place, time and behavioural situation
Complicating Action (CA)	The body of narrative clauses that advance the action in the story, the series of events that happen
Evaluation ©	A group of statements that express the significance of the story, its point
©da (C)	Statements that mark a return to the present mo ment

interviews was situated; in particular, how the individual's narrative is situated in social, cultural and institutional perspectives on Gafsele's changing role in developing tourism opportunities for Åsele. Using Labov's model implied reducing stories to a core structure to look at how people say what they do and who they are, as expressed through narratives. To better understand various views on the dogsledding community in this geographical context, we used a 'kind of radical surgery' to render the 'whole story' into a form that allows for comparison between the main messages in the individual narratives (Riessman, 1993, p. 43).

Interviews were audio-recorded and transcribed and then brief, bounded segments of interview text were identified as stories of experiences regarding opportunities and obstacles to contributing to local community tourism. These segments were analysed to identify structures in the narratives with a focus on how they are organized, what they say, how they are told, what this way of telling can mean, why the narrator tells the story in this way, and what the narrator wants to achieve with this choice of words (Riessman, 1993; 2004). We use the same pseudonyms for interviewees as previously (Eimermann & Singleton, 2021).

DIFFERING VIEWS ON COMMUNITY TRANSFORMATIONS

We identify the local dogsledding club as GSDC, which organized the Gafsele Open as a set of short distance races on the unique trails near Gafsele village, about 20 km from Åsele's municipal centre (Eimermann & Singleton, 2021; Figure 28.1). Dogsledders are called mushers and their assistants, who take care of the dogs, are handlers. Below, we link ideas from the literature review with each interviewee's core narrative in a separate subsection.

Old Deep Community Structures (Anne)

Local insights are derived from Anne (born 1951), a key person chairing the local branch of the Swedish outdoor association, who is also engaged as a volunteer in a very important meeting place; the local community centre. Anne emphasizes good relations between GSDC and local associations, saying that she engages in these activities to attract tourists and new residents. She considers the local club as firmly rooted in a traditional deep community where different members and associations have strong bonds and shared values and goals that have co-evolved over time. The contributions of long-term members (like herself and her family) who have been there from the beginning are particularly emphasized, while she also laments a lack of new contributors to continue the same work and values as the older generations retire.

BOX 28.1 TRANSCRIPT 1: ANNE (BORN 1951), A LOCAL KEY PERSON

(A): GSDC collaborates with the community centre and local organizations like the outdoor association to arrange the Gafsele Open. There are different people working in these associations, and we share the incomes equally. We sell hamburgers, hotdogs, buns and coffee. It makes good money to sell hamburgers. We [are trying] to find a combined hotdog and hamburger cart to develop this as soon as possible.

(O): We take care of things around the event, and we will continue doing that. During the event, we also serve meals on Thursday when it all starts, on Saturday evening and on Sunday. We are about 25–30 people in three working groups. We collaborate in the same way as for other events.

(CA): But there is now a different constellation in GSDC: we have decreased the number of board members from seven to three. Karl (my husband) has resigned as chairperson. Our daughter Selma, who was vice chair and a board member from the beginning, is now chairperson.

(E): But Selma now lives a 1½ hour drive away, she has young children and she works. She is very busy but she usually doesn't work on Fridays. Also, Karl and me have moved from Gafsele to a care home in Åsele town, but I am still chairing the outdoor association. We collaborate with the municipality and with Chris, who prepares the trail. And Peter's dogsledding company contributes too. They were co-organizers in the 2019 Gafsele Open. We had a contract making sure GSDC would get a share of the event's income.

(R): The mushers are free to do what they want between the races, but GSDC will keep on being in charge of the races. The club members are busy during the events so it's good that local volunteers help out.

(C): But the older generations are moving away (or dying). There are no new generations who continue our work in GSDC. That is not a good development, but we will see what happens.

The following stories develop Anne's indication of a generation shift in parallel with changes from a more traditional and deep community towards an enlarged community, developing from an early functional stage to a more mature conscious stage (Sharpe et al., 2016, p. 4). However, the transcripts below show that different people propose different directions in which the community should develop. They indicate social tensions at the individual level and political issues at the municipal level.

Towards an Enlarged Sense of the Dogsledding Community (Chris)

Chris moved from Switzerland to Gafsele in 2007 and has since become an established community member. Shortly after the interview, Chris became GSDC's new chair. His narrative puts the Gafsele Open in a historic, geographic and social context, indicating how the event has changed over the years from being run by locals to including more international migrants. Chris also indicates how sponsorship from another migrant entrepreneur (Peter) has made the event more commercial, taking realistic and small steps. He emphasizes the importance of slow and gradual transitions towards a more enlarged community and also points towards potential conflicts of interest between long-term residents and (transient) newcomers, especially those who want to change old community structures and practices too radically or too quickly. We link this with the role of population composition and age structure (many older people) for community development in this remarkably sparsely populated setting (D.B. Carson & Carson, 2014).

BOX 28.2 TRANSCRIPT 2: CHRIS (BORN 1975), FROM SWITZERLAND

(A): It all started with local dogsledding enthusiasts who managed to inspire people to collaborate and create this unique trail in the 1980s.

(O): The biggest event here so far was the dogsledding world cup in 2007. That is also when people from Europe started to migrate here.

(CA): But since 2010 it has not gone so well. People left the area, including the local enthusiasts who had started it all, and the dogsledding migrants who had arrived around 2007.

(E): Currently, almost all dogsled racers here are immigrants, many Swedes have started working in dogsledding tourism, which offers a better income. Peter, an immigrant running a business in dogsledding equipment, is sponsoring the Gafsele Open, with dogsledding harnesses as prizes. Without this sponsorship, prizes would have been minimal.

(R): In the beginning, GSDC wanted to keep the organization of the event in their own hands. They were afraid that foreign mushers would take over, and that they would have unrealistic goals or that they want to change too much at the same time.

(C): I believe people should be content and accept their situation, not continuously strive for more. Enjoy life because it is short anyway. I take small steps, one step at a time. We now have a contract with Peter's equipment company with the option of evaluating the collaboration after three years. GSDC trusts me with that, but Peter and I cannot take care of the contacts with other local clubs.

Chris also wants the Gafsele Open to be more prominent in the municipality's tourism development strategies. He would like municipal officials to make it possible to plan the event in the same week as other events (like ice fishing and a car rally on ice), taking place on a more centrally located lake. According to him, this would offer spill-over effects that would benefit all these events. But he feels the municipality prioritizes events in the central town over the more peripherally located Gafsele Open. Thus, Chris's story also illustrates how the migrants' views on latent local development potential may at times collide with those of local government and political stakeholders (Eimermann et al., 2019).

A More Conscious Dogsledding Community (Anders and Ellen)

This story reflects the views of more recent migrants (migrated in 2017) who want to develop the Gafsele Open into a more professional event and are frustrated by local attitudes of 'lagom'. This Swedish term signifies 'just the right effort', while Anders and Ellen (like many others) rather interpret this as 'not doing more than absolutely necessary'. They add that some people originally moving to the community to practise dogsledding stay in the area, even when they stop their activities due to high dogsledding costs, or when their priorities change (e.g., when they are having children). According to them, this shows the importance of dogsledding and outdoor activities in attracting new residents to this shrinking area. Anders and Ellen are conscious of the event's socio-economic potential for local rural development, and of GSDC members' interdependencies, while pointing at challenges for collaboration between members. They specify interdependencies with others beyond volunteers, such as a live band, entrepreneurs, and municipal officials, showing that the dogsledding community is becoming more conscious of its role beyond local village society (Sharpe et al., 2016, p. 4).

BOX 28.3 TRANSCRIPT 3: ANDERS (BORN 1957) AND ELLEN (BORN 1960), FROM THE NETHERLANDS

(A): A reason for us to become members of GSDC was to become part of the community. But there is no communication from the club, so we feel not at all engaged in the activities.

(O): GSDC and Chris are happy with the current situation but we think the Gafsele Open can be organized as a much bigger event. The event already involves a veterinarian, trail preparers, starters, several clubs and volunteers. You might as well promote the event more widely and invite more racers. That would mean earning more money for the club through racing fees. Other lifestyle migrants agree with us on this. But Chris is content the way it is – more racers also means more work and more volunteers.

(CA): We see many potentials for those offering accommodation and other entrepreneurs. As it is now, you can buy a hamburger but you cannot sit anywhere to eat it. Why not put up a tent with a heater? The event's stake-out area also offers space to put up small tents for entrepreneurs, to present themselves. And there could be a speaker to inform the audience about teams, racing times, results, etc. Also, a live band could be playing music in the local community centre in the evening.

(E): Our dream is to make this more of a circus-like event. Both the dogs' medical check and an opening ceremony could take place in Åsele's central square instead, which is only 20 km away from Gafsele. On a Friday afternoon, at the start of the weekend. Make people feel that something is happening.

(R): To make this dream come true, the municipal communication officer should be involved more. They have the knowledge and the tools, but do they promote the event enough?

(C): We came to Sweden because we like the more laid-back attitude. But the way it is now, the event is not alive, people are not really willing to invest more than necessary. If they continue like this, it will soon be over. GSDC only has a few members at the moment.

A Deeper Enlarged Dogsledding Community (Mona)

BOX 28.4 TRANSCRIPT 4: MONA (BORN 1993), FROM THE NETHERLANDS

(A): Anders has been a dog race referee elsewhere, deciding about procedures etc. But his age group is often quite stubborn. He doesn't want to get his green card [for racing], because he has taught others to race. He knows what he is doing. But this card is needed to take part in local training and races.

(O): I'd like to see more solidarity. I miss the cosiness of sitting around a campfire with other mushers.

(CA): But that is because I've experienced that elsewhere, I know it can be like that.

(E): What you need is good caravans for those camping on the trail's stake out area. But because of the cold, there can be less togetherness. People are freezing outside and retreat inside in their caravans.

(R): But you can sit in each other's caravans. I've had a handler who was a stranger to me. No problem.

(C): I realize that this has to do with money. The costs of taking part in a race will not decrease, and organizing more races to earn more income for the club is probably not feasible either. But I think GSDC could organize more training races, that would increase the sense of community. That would be doable.

Mona (migrated in 2015) estimates that dogsledding races in Gafsele usually attract 50 teams, and 150 teams when it is organized as the open Swedish championships. Teams come from Finland, France, Germany, Italy, the Netherlands, Norway, southern Sweden and Switzerland. Mona loves being outdoors, but she doesn't think the event will attract many tourists or migrants of her age (mid 20s): 'the dogsledding community is a world of its own, and it costs a lot of money'. She criticizes GSDC and the Gafsele Open's participants for not really being open to new people, but also older lifestyle migrants (like Anders) for relying on their previous experiences from elsewhere. Mona envisages more regular casual events and more efforts to foster a greater sense of solidarity and camaraderie among members of the dogsledding community, including mobile and temporary members such as visiting racers or seasonal dog handlers. If put into effect, these proposals would deepen the sense of an enlarged community.

CONCLUDING DISCUSSION

This chapter problematizes and complicates two dichotomous views on communities as either fully or not at all characterized by traditional industries and land use, strong place attachment, local social 'bonding' capital, high degrees of civic engagement and volunteering, and common values based on trust, reciprocity and reliance (Tuan, 1986; Sharpe et al., 2016). As younger people like Anne's daughter are leaving the area, GSDC has become a deep community based on trust, reciprocity and reliance between (extended) family members in the composition of this SPA's socio-economic population (which has many older people). Anne exemplifies a person rooted in a deep traditional sense of community, while Chris personifies a stranger who became a soulmate in the dogsledding community (Tuan, 1986). Over time, trust and new interdependencies between previously non-related people contributed to the GSDC's ongoing transformation into an enlarged community of practice with increased heterogeneity and dynamism (Shaffer & Anundsen, 1993).

Chris advocates small realistic steps to make the Gafsele Open a little more commercial without overwhelming local (older) stakeholders, yet with no intention of prolonging Anne's contacts with volunteers in other local associations. Other lifestyle migrants indicate preferences for different kinds of enlarged communities, including community constellations with more commercial and entrepreneurial orientation, as well as communities accommodating temporary and mobile members. This unveils interpersonal tensions that can negatively affect local community-based tourism development, particularly as different community members have different development aspirations regarding the scale, professionalism and commercial orientation of the club and tourism more broadly.

The interviewed lifestyle migrants are not a homogeneous group, and it is therefore unclear into what kind of enlarged community the GSDC is evolving (Sharpe et al., 2016). Anders and Ellen recognize and problematize the GSDC members' interdependence with each other, which suggests that a conscious community may be emerging that functions and interacts beyond the organizing of the Gafsele Open event, towards broader and longer-term prospects of tourism and community development. Mona strives for a deeper sense of enlarged community, with enhanced solidarity through internalized behaviour like organizing more training races and enjoying time around the campfire together after racing. Chris says that the new lifestyle migrants' goals are often unrealistic. Being a more established migrant, Chris is happy with Peter's efforts (another established migrant) but disagrees with other migrants' intentions to rapidly realize growth potentials. Anders, Ellen and Mona indicate that other individuals and the municipality want to continue doing things the way they have always been done. Together, their stories indicate tensions between individuals in the same community of practice, and experiences of social and physical distance between the more peripherally located Gafsele dogsledding trail and Åsele's main square and lakes. Heterogeneity is, thus, increased through people from abroad relating their experiences in northern Sweden with dogsledding in other countries.

Studying mobile people as an in-community subgroup (Skerrat & Steiner, 2013) within the initially locally based dogsledder community further adds to researchers', municipal officials' and community members' awareness of various transformations towards more mature stages of enlarged communities. As such, the chapter's critical narrative analysis drawing on Labov's (1982) model is a valuable tool to unpack and expose different stakeholder views, development aspirations and conflicts of interest when it comes to understanding the potentials and obstacles for lifestyle migrants to contribute to local community tourism. Hence, a question for researchers and policymakers concerns what can be done to help these (potentially) dysfunctional communities work out how to get along together (O'Reilly, 2012).

Communities in SPAs are highly transient places where different groups of mobile people from different geographic and socio-economic backgrounds and with different lifestyle aspirations meet and interact, often on a temporary basis. These community dynamics may differ quite substantially between even relatively proximate settlements (such as the case of Gafsele and Åsele). Future studies on the 'temporariness' of communities should be appreciative of different lived experiences, which can provide more in-depth insights into the localized effects of migration and other forms of mobility on community development, serious leisure and (tourism) entrepreneurship (Stebbins, 2009; Moscardo et al., 2013; Steiner et al., 2018). Such community-level studies should pay more attention to remote SPAs' geographic and socio-economic conditions, along with their mobility experiences (D.B. Carson & Carson, 2014), particularly as the longer-term consequences for in-migration and tourism mobilities in these outer peripheries are not yet understood. This would advance our understanding of the unique challenges and opportunities for community development in SPAs.

NOTE

1. Also see the Lifestyle Migration hub blog: https://lifestylemigrationhub.org/.

REFERENCES

Agnew, J.A., Livingstone, D.N., & Rogers, A. (eds) (2005 [1996]). *Human Geography: An Essential Anthology*. Oxford: Blackwell Publishing Ltd.

Benson, M. (2010). The context and trajectory of lifestyle migration: The case of the British residents of southwest France. *European Societies*, *12*(1), 45–64.

Benson, M., & O'Reilly, K. (eds) (2009). *Lifestyle Migration: Expectations, Aspirations and Experiences*. Farnham: Ashgate.

Benson, M., & O'Reilly, K. (2016). From lifestyle migration to lifestyle *in* migration: Categories, concepts and ways of thinking. *Migration Studies*, *4*(1), 20–37.

Bhattacharyya, J. (2004). Theorizing community development. *Community Development*, *34*(2), 5–34.

Carson, D.A., Carson, D.B., & Eimermann, M. (2018). International winter tourism entrepreneurs in northern Sweden: Understanding migration, lifestyle, and business motivations. *Scandinavian Journal of Hospitality and Tourism*, *18*(2), 183–198.

Carson, D.B., & Carson, D.A. (2014). Local economies of mobility in sparsely populated areas: Cases from Australia's Spine. *Journal of Rural Studies*, *36*, 340–349.

Carson, D.B., Carson, D.A., Eimermann, M., Thompson, M., & Hayes, M. (2020). Small villages and socio-economic change in resource peripheries: A view from northern Sweden. In L. Lundmark, D.B. Carson, & M. Eimermann (eds), *Dipping in to the North* (pp. 27–53). Singapore: Palgrave Macmillan.

Eimermann, M. (2021). Studying locals and lifestyle migrants in sparsely populated northern Sweden using foresight 'light'. *SAGE Research Methods Online*. https://dx.doi.org/10.4135/9781529763935.

Eimermann, M., & Singleton, B. (2021). Trails and tribulations. Lifestyle migration as nature-based integration in northern Sweden. In D. Friedrich, M. Hirnsperger, & S. Bauer (eds), *More than 'Nature'*. Vienna: LIT.

Eimermann, M., Tillberg Mattsson, K., & Carson, D.A. (2019). International tourism entrepreneurs in Swedish peripheries: Compliance and collision with public tourism strategies. *Regional Science Policy & Practice*, *11*(3), 479–492.

Eimermann, M., Tomozeiu, D., & Carson, D.A. (2020). Lifestyle migrants and intercultural communication in Swedish villages. In L. Lundmark, D.B. Carson, & M. Eimermann (eds), *Dipping in to the North* (pp. 107–132). Singapore: Palgrave Macmillan.

Eimermann, M., Adjei, E.K., Bjarnason, T., & Lundmark, L. (2022). Exploring population redistribution at sub-municipal levels: Microurbanisation and messy migration in Sweden's High North. *Journal of Rural Studies*, *90*, 93–103.

Granås, B. (2018) Destinizing Finnmark: Placemaking through dogsledding. *Annals of Tourism Research*, *72*, 48–57.

Huskey, L. (2005) Challenges to economic development: Dimensions of 'remoteness' in the North. *Polar Geography*, *29*(2), 119–125.

Karlsdottir, A., Smed Olsen, L., Greve Harbo, L., Jungsberg, L., & Rasmussen, R.O. (2017). *Future Regional Development Policy for the Nordic Arctic: Foresight Analysis 2013–2016*, Nordregio Report 2017:1. Stockholm: Nordregio.

Koster, R.L., & Carson, D.A. (2019). Considerations for differentiating among rural tourism geographies. In R.L. Koster & D.A. Carson (eds), *Perspectives on Rural Tourism Geographies* (pp. 253–271). Cham: Springer.

Labov, W. (1982). Speech actions and reactions in personal narrative. In D. Tannen (ed.), *Analysing Discourses: Text and Talk* (pp. 219–247). Washington, DC: Georgetown University Press.

Labov, W., & Waletzky, J. (1997). Narrative analysis: Oral versions of personal experience. *Journal of Narrative and Life History*, *7*(1–4), 3–38.

Mair, H. (2018). Critical inquiry in tourism and hospitality research. In R. Nunkoo (ed.), *Handbook of Research Methods for Tourism and Hospitality Management* (pp. 53–61). Cheltenham, UK & Northampton, MA, USA: Edward Elgar Publishing.

Mishler, E. (1995). Models of narrative analysis: A typology. *Journal of Narrative and Life History*, *5*, 87–123.

Moscardo, G., Konovalov, E., Murphy, L., & McGehee, N. (2013). Mobilities, community well-being and sustainable tourism. *Journal of Sustainable Tourism*, *21*(4), 532–556.

O'Reilly, K. (2007). Intra-European migration and the mobility-enclosure dialectic. *Sociology*, *41*(2), 277–293.

O'Reilly, K. (2012). *International Migration and Social Theory*. Basingstoke: Palgrave MacMillan.

Riessman, C.K. (1993). *Narrative Analysis*. Qualitative Research Methods Series No. 30. Newbury Park, CA: Sage.

Riessman, C.K. (2004). Narrative analysis. In M.S. Lewis-Beck, A. Bryman, & T.F. Liao (eds), *The SAGE Encyclopedia of Social Science Research Methods* (Vol. 1, pp. 706–709). Thousand Oaks, CA: Sage.

Robertsson, L., & Marjavaara, R. (2015). The seasonal buzz: Knowledge transfer in a temporary setting. *Tourism Planning & Development*, *12*(3), 251–265.

Sandberg, L., & Tollefsen, A. (2010). Talking about fear of violence in public space: Female and male narratives about threatening situations in Umeå, Sweden. *Social & Cultural Geography*, *11*(1), 1–15.

Shaffer, C., & Anundsen, K. (1993). *Creating Community Anywhere: Finding Support and Connection in a Fragmented World*. New York: G.P. Putnam's Sons.

Sharpe, E., Yuen, F., & Mair, H. (2016). Community development in leisure: Laying the foundations. In E. Sharpe, H. Mair, & F. Yuen (eds), *Community Development: Applications for Leisure, Sport, and Tourism* (pp. 3–13). State College, PA: Venture.

Skerrat, S., & Steiner, A. (2013). Working with communities-of-place: Complexities of empowerment. *Local Economy*, *28*(3), 320–338.

Stebbins, R. (2009). Serious leisure and work. *Sociology Compass*, *3*(5), 764–774.

Steiner, A., Woolvin, M., & Skerratt, S. (2018). Measuring community resilience: Developing and applying a 'hybrid evaluation' approach. *Community Development Journal*, *53*(1), 99–118.

Tillberg Mattsson, K., & Heldt Cassel, S. (2020). Immigrant entrepreneurs and potentials for path creating tourism development in rural Sweden. *Tourism Planning & Development*, *17*(4), 384–403.

Tuan, Y.-F. (1976). Humanistic geography. *Annals of the Association of American Geographers*, *66*(2), 266–276.

Tuan, Y.-F. (1986). *The Good Life*. Madison, WI: University of Wisconsin Press.

Van Rooij, N., & Margaryan, L. (2019). Integration of 'ideal migrants': Dutch lifestyle expat-preneurs in Swedish campgrounds. *Rural Society*, *28*(3), 183–197.

Wenger, E. (1998). *Communities of Practice: Learning, Meaning and Identity*. Cambridge: Cambridge University Press.

Wiles, J.L., Rosenberg, M.W., & Kearns, R.A. (2005). Narrative analysis as a strategy for understanding interview talk in geographic research. *Area*, *37*(1), 89–99.

29. Traveling workers as rural tourists: extending understandings from Tibetan travelers

Xiaotao Yang

THE LOCAL TURN IN TOURISM

Higgins-Desbiolles and her collaborators (2019) have urged the shift of attention from either profits or international tourism to local communities in order to (re)build their social capital. Tourism has long been viewed as an industry for profits and growth (Higgins-Desbiolles, 2006). Under such views, international tourism has attracted the most attention due to its lucrative profits. Multinational companies have reaped large amounts of profit from this business model. Tourism under this lens is merely about tourists and how the industry fulfills their travel needs. Local residents may have few employment options other than providing services to tourists (Nash, 1989; Higgins-Desbiolles & Bigby, 2022).

Higgins-Desbiolles et al. (2019) have challenged this economically focused view of tourism and argued for redefining tourism as a way of acknowledging and prioritizing the rights of members of local communities. The rights of local communities comprise various facets, such as the power to invite and receive visitors. In this way, domestic and local travel can be viewed as offering opportunities to improve the well-being of local residents, and for seeing them as the rightful owners of domestic resources, for either recreational or educational use.

Various benefits and functions of local tourism have also been identified by Higgins-Desbiolles et al. (2019) and Higgins-Desbiolles & Bigby (2022). Although domestic—or local—tourism may not have been seen as being as profitable as international tourism, it could boost "gross national happiness" (GNH, see Higgins-Desbiolles & Bigby, 2022). Ideally, local tourism can also cultivate people's passion towards their environment; such passion could be further expanded to include other communities and ultimately humankind. Moreover, local tourism has the potential to extend the positive role of tourism to wider society and perhaps even to ecological sustainability. Lastly, various local understandings of tourism could enrich the diverse worldviews and philosophy towards or beyond tourism.

As a response to Higgins-Desbiolles and her collaborators' call to prioritize the rights of local communities, this chapter reports on the local travel experience in a Tibetan context. This study was also motivated by the call of Chambers and Buzinde (2015) to delink from Western-centered knowledge production. Thus, I center on the Tibetans' experience and add their knowledge to the tourism and recreation field.

Peters and Higgins-Desbiolles (2012) pointed out a notable gap in tourism literature, arguing that few studies recognize Indigenous peoples as tourists. As a response to these claims, this chapter aims to privilege and center the Tibetan people's knowledge as tourists, rather than as objects for tourists to gaze upon. This chapter is guided by the question: how do Tibetan travelers narrate their travel experiences?

METHODOLOGICAL APPROACH

Research Design

In addition to considering whose travel knowledge should be centered, addressing how to construct their knowledge in a more equitable way is also important. Scholars from Indigenous studies and decolonial studies emphasized that Indigenous knowledge, epistemologies, and ontologies may be significantly different from Western-centered knowledge. Accordingly, scholars should avoid interpreting Indigenous knowledge with the existing dominant (Western-centered) framework (Escobar, 2010; Little Bear, 2000; Victor et al., 2016; Wilson, 2008; Winter, 2009).

Thus, three actions were taken during the research reported on in this chapter: First, instead of applying the existing definition of tourism, I explored the Tibetan participants' definition of tourism, which will be briefly summarized in next section. Second, I adopted the non-conventional approach of providing a literature review *after* fieldwork. A conventional structure in academic writing and representation normally requires researchers to review existing literature to establish a predetermined framework with which to interpret the participants' experience. Such a practice ensures that researchers move along the "right track" by leveraging pre-established analytical framework/tools/theories/terms from previous researchers. To depart from this convention, I attempted to put participants' voices at the center, and arranged other actions, which were based on their voices and perspectives. Indeed, I introduced participants' voices *before* reviewing related literature in order to ensure my reading was fully based on the issues brought up by participants. As Hollinshead (2012) has argued, non-Indigenous groups should not only "speak on all salient Indigenous matters of culture and nature (or 'culture-nature')", but also open up "the debate on such matters" (p. 49). Such a debate/dialogue could inform both Indigenous and non-Indigenous people. However, it is important to note that the role of the discussion section, which follows the participants' narrative in this chapter, is to extend our understanding, rather than devalue existing literature.

Third, I used visual narrative inquiry to collect participants' narratives, because of its flexibility in centering on the participants' experience. Bach (2007) referred it to the "use [of] photographs within a narrative inquiry framework" (p. 290). One main argument in support of narrative inquiry is that it does not devalue the participants' stories, but considers them as sophisticated outputs, which have been organized by the participants themselves. Accordingly, large numbers of verbatim quotations were presented in the analysis of the narratives. During the research, I invited participants to select three photos of their trips, describe the photos, and then explain why they chose those three photos. Compared with texts, photographs remove language constraints, which are major barriers for cross-cultural studies and people who "do not read or write in the dominant language" (Wang & Burris, 1997, p. 370). Most of my participants did not receive a university education, but they found this method interesting. Furthermore, I invited participants to decide whether I should use pseudonyms or their real names because as Wilson (2008), an Indigenous scholar suggests, to refer to real names is to acknowledge participants' contributions. Most of my participants preferred to go by their real names and were also comfortable sharing photographs in which their faces were identifiable.

The Fieldwork

To explore Tibetan people's experience and knowledge as tourists, four months of fieldwork was conducted in the Tibetan Autonomous Region (TAR), China in 2019. With the help of several key local informants, 35 Tibetan participants generously shared 93 travel stories that were self-identified as travel experience. According to the latest census of China, 11% of residents in TAR hold university degrees or higher qualifications and 65% of residents live in rural areas (Information Office of Tibetan Autonomous Region, 2021). Consequently, I tended to include those who did not hold a university degree and were rural residents. Eight out of 35 (22%) of my participants held a university degree and only these eight participants were urban residents. The rest were rural residents, although they were working or even running small businesses in the city. These 27 participants from rural areas comprised five construction workers, five small business owners, one self-educated poet, two painters, five temporary workers, three hotel and restaurant workers, four university students, and two farmers. Clearly, the profiles of my participants were roughly consistent with TAR's population structure.

Notably, nine of the shared work/leisure trips included participants taking advantage of the location of their work by arranging personal travel to nearby sites during their time off, and 17 trips were naturally arising travel opportunities throughout the normal course of their work, consisting of scattered moments of relaxation or sightseeing, which overlapped with their work responsibilities. In one example of the former, Drolma trained in Lhasa for one month, during which she visited nearby monasteries during the weekends. That means 28% of the shared work/leisure trips were work-related trips. Therefore, considerations of work-related travel can provide a preliminary understanding about travel experience of Tibetans because most participants were from rural areas—the main population of TAR. Next, I will discuss why most participants considered these trips as a form of travel.

It should be noted that re-storying participants' narratives was a challenging task because some of those stories were short and simple. Three reasons might contribute to the short length of their stories. First, some participants were not fluent in Mandarin, the main interview language. Although I proposed the option of conducting the interview with a professional translator, some participants who were not fluent in Mandarin still preferred to use Mandarin or to have a friend translate for them. Second, travel was not a familiar topic for my participants. For example, the transcript of my conversation with Pasang was only around 3500 words and that covered a description of three trips in total. I kept asking him if he had anything to add. He responded that that was everything he wanted to say. But while we drove around Lhasa City, he became talkative and gave many details about his construction work and related topics. His hard work and sweat had marked many corners of the city. I was impressed by his familiarity with construction work, rather than travel.

Third, some of the participants were good at communication through actions rather than language. Most participants worked in construction projects or farming, and while working, they did not communicate that much. For instance, Tsering Drolma, a local farmer in her 60s, asked why I was seeking to learn about Tibetan culture merely via verbal communication (i.e., the interviews). Based on her experience in farming, she thought knowledge and understanding came from practice, rather than language. Thus, she suggested that I should participate in local activities to supplement the oral interviews. This phenomenon is related to two academic debates. On the one hand, Spivak (1988) critiqued that Western intellectuals, even leftists, dominated the expression of subalterns in the Third World countries, as if the latter could speak for themselves. She argued for attentive listening to subalterns' voices. Indeed, some of my participants used plain language and I read these plain words many times, appreciating their original words. On the other hand, Tsering Drolma's words reminded me of Chilisa's (2012, p. 100) view that Indigenous knowledge includes embodied knowledge beyond the written literature. The two stories presented next in the findings section illustrate both the long and short original versions to enable readers to begin to appreciate Tibetan participants' narrative style.

FINDINGS

Through my research, two stories were constructed to shed light on experiences while traveling for work, and Tibetans' favorable feelings toward nature. In a way, these two stories can represent the 17 trips that were naturally arising travel opportunities throughout the normal course of participants' work and can enhance our understanding of Tibetans' travel. Moreover, participants expressed intense affect in response to natural beauty. Among the 22 tourism (*yukour*) trips within Tibet, 14 are about enjoying a *linka* or lush environment in the valleys, such as Nyingchi City. The second story is about Motok County, in Nyingchi City. The words in italics were participants' words that have been translated by the author from Chinese to English. As noted above, I tried to maintain the plain language style of the participants.

Story 1: Pasang's Work Trip to Nagqu— "I Work and Take Myself to '*Lvyou*'"

Whose voices?

The following story was developed from the interview with Pasang, and I have changed the original dialogue style of the interview into a solo-voiced story. Pasang was a construction worker in Lhasa during the time of the interview. He worked at various temporary jobs at various locations across TAR; therefore, he had traveled across Tibet for work duties. He went back to his hometown after completing a temporary job or to farm at home. He was in his 40s or 50s. He lacked education. His friend, Tsering, first introduced my study to several friends in a tea house. He was interested in attending; thus, we scheduled a meeting for a few days later. After he finished work, we had dinner together before the interview. The interview was conducted in Tibetan, with one of his Tibetan friends who was fluent in Chinese and Tibetan and who served as the translator.

"I work and take myself to '*lvyou*'"

> I went to Nagqu with another 17 workers to look after other people's herds. We selected Nagqu because the elevation is higher; thus, the salary is higher than that of other regions. The grasslands are extensive and endless. We also worked with some local workers who spoke different dialects. We slowed down our speaking speed, and gradually, we could communicate with basic words. The work was not physically demanding, as we merely followed the herds, milking, and picking up yak droppings. My work in Lhasa was really physically demanding and tiring. By contrast, herding was easy work. The most pleasant moment was when I got my salary. The compensation was higher than that of other regions. Moreover, I was able to drink more milk, eat more yak meat, and enjoy the endless grasslands. In my hometown, we only had cattle-yak, which are smaller in size than yak. Considering the high compensation, low work intensity, abundant food, and beautiful nature, I was free from any stress. I actually gained weight during my three months' work.
>
> I occasionally get such temporary work. Next week, I will leave for Nagqu again, to wash yaks. Yak racing will be held soon; thus, yak owners will wash their yaks before attending the race. I could earn some money and take this opportunity to watch the yak racing. My hometown does not have such events. I work and take myself to *lvyou* [his original word is "*lvyou*"—the Chinese word for tourism. In his elaboration in Tibetan, he suddenly used this Chinese word].

Chompel and Tsering's Work and Tourism Trip—A Lush Garden of Tibet

Whose voices?

The following story is directly based on the travel experience of Chompel's work trip to Motok and Tsering's trip to Mount Karlashi. Chompel shared another work trip with me, in which he again turned a break in his work into a holiday. Tsering shared three work-related trips; he worked driving trucks to Mount Karlashi for a travel agency. According to Chompel, the change of scenery and relaxing activities that accompanied this work greatly rejuvenated him and his ability to continue his work. Chompel and Tsering were from rural areas of Shigatse City. They both started work

Figure 29.1 Pasang enjoying the scenery in the grasslands while herding yaks

as construction workers and truck drivers, and in recent years started a small business. The mobility of their work also enabled them to travel across the TAR. They tried to obtain some small construction projects. Chompel was in his 40s, and he had graduated from middle school. A total of three rounds of interviews were conducted with him, in Mandarin Chinese. Tsering was in his 30s, and he lacked an education. Two rounds of interviews were conducted with him, half in Chinese and half in Tibetan, with his friend serving as the translator.

A lush garden of Tibet

> I went to Motok for a construction job in 2012. Motok County was the last county without car access in China, a situation which was improved in 2016. In 2012, it was seasonally accessible by car. I drove from Tingri County to Nyingchi City over three days, and another three days from Nyingchi City to Motok County. The latter was extremely dangerous. My truck was 2.2 meters wide, and the road was 2.3 meters wide. Sometimes, part of my tires was over the road edge. I had never driven on such a risky road before, but I had to

keep going. Another stress came from the condition of my truck; it was an old truck, so I was nervous about it breaking down. I prayed. Buddhism teaching repetitively teaches us believers to be kind to all living beings. I had always tried to be a kind person in my life to that point. I was reassured that Buddha would bless me to pass through that challenging situation when I reviewed my past behaviors. Praying filled me with the courage to keep driving. Praying turned out to be helpful. My old truck did not break down on the road.

My speed was slow as the truck was fully loaded with machines and steel. I and another two workers slept under the truck bed at night. I was worried about rain while lying down there. We would get damp with any drop of rain. Honestly, I was exhausted from driving along such a dangerous road. I regretted that I had found such a difficult job. Our destination, Beibeng Village, was beyond the county township. But we had to stay in Motok County for one week because the road ahead was blocked by a mudflow. We had no other choice but made the one-week waiting period our holidays.

I went out to enjoy and explore the scenery in Motok County. My hometown—Tingri—has almost no trees. It was my first time seeing in person crops of corn, rice, tea; large bamboos; and stone pots. I did not think any other place within Tibet was growing rice crops. There were many snakes on our way. It was also the first time in my life to see snakes in person. Most of them were dark, about 1 meter in length. I was nervous, but also excited. It had been so tough making my way to that place. As I saw these new plants and animals, I felt the difficulties were rewarded. Contrasting with my anxiety on the road, I now appreciated this work opportunity. Without this job, I would have been less likely to visit a place like this just for pure relaxation.

After one week's rest, we felt rejuvenated and continued on to the Beibeng Village. The road there was even worse than the one to Motok County. It was raining most of the days. The road was slippery and hard to drive on. After the rain, many animals came out, especially leeches. Even with our bodies fully covered by clothes, the leeches still found a way to reach our skin and suck blood. When I got up, I was scared because one leech was sucking my blood. I removed it immediately. However, I was wrong. Some local workers advised me to wait for the leech to fall off automatically after it sucked enough blood; otherwise, its lips would stay in my body. They were right. Since I had pulled it off by force, my leg swelled and stayed that way for three days. It was a novel place, but had too much novelty for me. I became stressed, particularly at the very beginning.

Due to the rain, we had to arrange the construction flexibly. For instance, if it was raining in the morning and stopped at noon, we then worked in the afternoon and night. But there was too much rain, we were not actually able to work too much and had some relaxing times socializing with the local people who were from the Luoba ethnic group. We did not speak the same language, so we could only communicate via Mandarin. But their customs were similar to our Tibetan ones. As time went by, I got used to the life in Beibeng.

Finally, after three months' work, we finished our construction in November. The snow was so heavy, I had to leave my truck around the place designated 80K (the name of a place). Other workers and I walked for around five hours to reach a place that was accessible by car. I was very happy because I got my wages. Another co-worker from Nyingchi City suggested that I visit this region. I agreed because I could afford a short tour with my pay, and I did not think I would have any non-work-related opportunities to visit this region later.

I spent three days there and visited a glacier, a lake, and a temple by the lake. The local people there were Tibetans, but our dialects were different. We could not communicate via the Lhasa-Shigatse Tibetan dialect that I spoke. I felt we were connected, especially when I saw the temple, which was built on the cliff of a mountain, similar to some temples in my hometown. Our traditional Tibetan construction skills were no joke! Many temples were able to be built in such challenging locations. It required lots of hard work and wisdom. It was small. Three nuns were dedicated to practicing Buddhism there. For them, not too

Figure 29.2 Chompel sightseeing in Motok

much space and materials were needed. I prayed in front of Buddha. I was grateful for Buddha's blessing during such a dangerous trip.

I do not think that travel for pure relaxation has been a choice in my life. I do make myself available for *neikour*, rather than *yukour*. Looking at my travel experience at Motok, I don't think my experience was significantly different from that of other tourists.

It was a relaxing experience for me, and I had many opportunities to communicate with the local people there. The Motok experience was similar to my other work experiences at various construction sites. I do not have the same travel conditions, either financial or time, as other tourists, but the unexpected challenges and opportunities can sometimes be used for sightseeing and relaxation as well. So *yukour* is more about the status of mood.

The trip to Motok was the most dangerous and hardest trip in my life.

TRAVELING WORKERS AS TOURISTS

Tourism is occasionally defined as "not work" (Graburn, 1989). Although work-related and tourism-oriented activities are common, only a few studies focused on the interaction between work and tourism because "tourism and work are usually perceived as two contradicting fields of human activity" (Uriely, 2001, p. 1). Winter (2009) argued that the typology of tourists was singularly understood as the product of industrialized society, in which tourists reside in the modern society. Tourists chase after nostalgia or authentic experience to find themselves in an otherized space/culture; or they can be post-modern tourists who know the roles of different types of tourists and perform accordingly (Wang et al., 2008). Winter (2009) underscored that these typologies of tourists lack cultural variations of ethnicity and nationality and further suggested embracing new ways of understanding tourism. This chapter extends the tourism knowledge in a Tibetan context.

Those who can afford travel with regard to finance or time are the frequently studied tourists. For middle and upper classes, participating in tourism activities can indicate the affordability of time and finance (Veblen, 1994 [1899]). The working classes are considered to be the main participants in institutionalized mass tourism as they tend to be involved in tourism to release their stress, which improves their productivity. From the social tourism perspective, Qiao et al. (2021) focused on the traveling experience, especially the perceived value, for Chinese rural migrant workers. They proposed a six-dimension scale of perceived values, namely, "life value, work value, social value, self-realization value, emotional value and epistemic value" (p. 6). Their findings are largely positive, claiming participation in tourism significantly benefits the workers' well-being and happiness, strengthens social connections, and brings new meanings to their current life. Moreover, and important in terms of the purposes of this Handbook, Qiao et al. (2021) highlighted that rural migrant workers have been largely ignored in tourism studies.

In my search of "farmers' travel" or "peasants' travel" on Google Scholar, only a few studies showed up. China has the largest rural population in the world (World Bank, 2020). Qiu et al. (2011) reviewed the literature on peasant tourism over three decades and concluded that only a few studies explored this market segment, despite its large population ratio. They also suggested that peasants' travel was closely related to a region's economic development; thus, farmers in eastern China travel more than those in western China. Evidently, rural areas, despite migrant workers or farmers who travel, are seldom considered as a market source of tourists, particularly under the industry-driven perception of tourism.

Uriely (2001) is one of the few scholars who made a closer examination of the mixture of work and travel. He classified four different types of travelers who combined work with tourism purposes: "travelling professional workers"; "migrant tourism workers"; "non-institutionalised working tourists"; and "working-holiday tourists". The first two were referred to as "travelling workers" because of their work orientation; and the latter two as "working tourists" for their tourist-orientation. The first two are similar to my study participants and are thus the focus of the following discussion. Traveling professional workers, a similar concept to Cohen's (1974) term "work-oriented travelers", refers to those who combine business and pleasure, where work is the primary purpose of the trip and tourism is a by-product. They are normally "career oriented, highly skilled and well rewarded" in professions such as "members of academic and diplomatic staff, air attendants, tourist guides, mercenaries and professional athletes" (Uriely, 2001, p. 3). The second work-oriented group, "migrant tourism workers", are mainly seasonal workers at popular tourist destinations (Uriely, 2001). Migrant tourism workers differ from traveling professional workers in their lower skill level and accordingly seek lower-paid and less secure work. Migrant tourism workers differ from conventional migrants in the way that traveling opportunities function as incentives for them to choose the work, with some of them even purposely working at various destinations to fit their tourist-related pursuits.

Business travelers are the most studied combination of work and travel, and fit into the "traveling professional workers" in Uriely's (2001) classification. Geerts (2014) made a comparison between business travel and leisure travel. Geerts argued that a certain activity of business travelers can be referred to as hypermobile, frequently undertaking short trips and thereby failing to recognize cultural differences between destinations. This mode of travel seems to be restricted to "corridors" (Lassen, 2006), a vivid description of the fact that the business travelers can only travel to limited sites without exploring the destination as much as possible. However, leisure travelers can visit a wide variety of destinations, which are not limited to urban areas.

Another element distinguishing the two types of travel is decision-making. Business travelers do not have the autonomy to decide where to travel; their employers make these decisions and pay for the trip, whereas leisure travelers have the freedom to select a desirable place, and pay for themselves. At present, business travelers cannot balance "the obligations to work and obligations to the family" (Geerts, 2014, p. 190). In addition, their travel attitudes (positive or negative) are heavily influenced by external and internal factors. External factors include "the frequency and length of travel, family situation, locations visited, and workload" (Geerts, 2014, p. 196), whereas internal factors involve whether he or she feels lonely and whether personal security is guaranteed (Ivancevic et al., 2003).

The boundary between work and leisure/travel has been described as blurred. In Uriely's (2001) classification, migrant tourism workers normally work at popular tourist destinations. Bianchi (2000) reflects upon the characteristics of migrant tourist-workers in the resort enclaves of southern Europe under the changing social and cultural circumstances shaped by capitalist restructuring. These workers regard their periodic participation in work as part of their touristic experience. The combination of tourism and work dissolves the boundaries between the two. Lichy and McLeay (2018) use photo elicitation to identify the types of business–leisure travelers in an education environment. They found that key motivations that shape typologies of what they refer to as "bleisure" travelers include: obtaining new skills and enlarging learning experience; escaping from daily routine; searching for novelty and leisure; sharing knowledge; and enhancing research abilities (e.g., additional work or funding). Their research integrating leisure into business trips successfully demonstrated the blurring boundary between work and leisure activities.

Sun et al. (2020) explored a new type of migrant tourism workers or owners in China, who lie in the continuum between work and tourism. Compared with the majority of migrant tourism workers moving from undeveloped countries to developed countries, these workers move from urban cities to rural tourist destinations to operate a business. They are usually well educated, and have obtained financial resources and social capital. They are therefore more powerful than local residents during host–guest interactions.

Dialog between Tibetan and Existing Literature's Understanding of Tourism and Traveling Workers

Travel takes two major forms in contemporary Tibetan society: *yukour* (tourism, ཡུལ་སྐོར།) and *neikour* (pilgrimage, གནས་སྐོར།). The word *yu* means other places and *kour* means walk around so that *yukour* indicates walk around other places. It is a comparable concept with *neikour* that means circulating around sacred places. *Yukour* suggests that non-religious reasons are involved. All participants but one believed that the Tibetan word *yukour* (ཡུལ་སྐོར།, which is similar to sightseeing, tourism, or leisure), was literally translated from the Chinese word *lvyou* because there was almost no nonreligious travel in traditional Tibetan society. However, Tibetans are familiar with the Chinese word *lvyou*, as Pasang suddenly used this Chinese word in his communication in Tibetan. The Tibetan word *yukour* can be referred to as leisure and tourism.

At least four aspects from the two abovementioned narratives could be used to extend our understanding.

First, although Uriely's (2001) classification is the closest in the literature on the combination of travel and work, it cannot speak to the above narratives. Neither type of "travelling workers", i.e., traveling professional workers and migrant tourism workers, is directly aligned with the two narratives above. Pasang and Chompel did not choose the work because of potential traveling opportunities. These occasional tourism opportunities were obtained after starting work or they took a touristic attitude during work. Two reasons might contribute to the difference between the above story and traveling workers in the literature: participants' financial status and their relaxed mood. They drove to workplaces that could provide work opportunities, as reflected in Chompel's story, when he had to continue driving even when the road conditions proved to be dangerous. In addition, the relaxation refreshed them and helped them discover the beauty of the area. Pasang's job included following the herds, milking, and picking up yak droppings, which does sound hard; but as is shown in the picture, he managed to find pockets of free time during his workday to enjoy the view of endless grasslands and sunshine with his wife, who was also his co-worker. Uriely (2001) only discussed migrant workers at tourist destinations; however, the relaxed mood could happen at non-tourist destinations. Therefore, the above two narratives supplement Uriely's (2001) categories of "travelling workers" by adding the people who were not motivated by traveling opportunities and yet adopted "touristic" moods alongside their work.

Second, business travelers are a frequently studied form that combines work and travel (Geerts, 2014; Lichy & McLeay, 2018; Uriely, 2001). The two abovementioned narratives reported "bleisure" literature, when contact with local culture may be brief, and they are not able to decide where to go. Their characteristics, summarized in the existing literature, indicate the advantage of their social and economic status, such as "highly skilled and well rewarded" (Uriely, 2001, p. 3) or escaping from daily routine and sharing knowledge (Lichy & McLeay, 2018). The narratives of Tibetan participants do not feature any of these characteristics. Thus, the opportunity to combine work and travel in this way may not be a priority for those who have better work; as shown in Pasang and Chompel's narratives, they were merely temporary workers with disadvantaged social and economic status. They combined their work with travel and as such it contributed to their subjective relaxed state and impacted their work.

Third, rural peasants' travel is an understudied area in tourism. Some participants were farmers and also worked as temporary workers in the cities, so that their narratives could advance the understanding of farmers' and workers' travel. Qiao et al. (2021) focused on Chinese rural migrant workers who participated in purposive forms of travel, such as trips organized by travel agencies. If we extend the definition of travel to the scattered leisure time as shown in Pasang and Chompel's stories, then peasants' travel will be comprehensively understood. In our case, Pasang and Chompel compared their work sites with their hometowns and identified novel encounters with animals, plants, scenery, or cultural practices as another major reason for them to consider their trips as tourism. Meanwhile, Sun et al. (2020) reported that migrants from urban to rural regions normally held the advantage in interaction with local hosts. In Pasang and Chompel's stories, they quickly adjusted to the local communities and felt like equals.

Fourth, the two stories highlight how the boundaries of work and travel are blurring. Postmodernist scholars argued for the blurring of boundaries between everyday life and travel experience (Uriely, 2005), noting that the enjoyment and pleasure of tourism are available in daily life. A person could visually "travel around" the world via online browsing, such as with maps or mass media. In addition, the increase in available attractions in a broader range of locations, such as theme parks and shopping malls, and the accessibility of global products through local retailers or online stores, allow people to experience the benefits of taking a trip without having to travel far from home. The boundaries between work and tourism could become even more blurred if researchers considered the aforementioned psychological factors. Engaging in part of their work in a relaxed mood and considering themselves tourists is a special characteristic that becomes evident from looking at participants' experiences.

CONCLUSION

This chapter reports two narrative accounts of Tibetan people's work-related travel, which consists of 28 travel stories shared by 35 participants, among whom 27 were from rural areas. It serves as an introduction to understanding Tibetan farmers' travel.

This work is also consistent with Higgins-Desbiolles et al.'s (2019) call to focus on local tourism, through which locals' well-being can be improved. This chapter's focus on Tibetans' understanding of tourism can enrich the existing literature in two important ways: (1) the notion of "traveling workers" could be extended to those who were not motivated by traveling opportunities and yet adopted "touristic" moods alongside their work; (2) the boundaries between work and travel are increasingly blurred. Thus, more studies should focus on the local turn of tourism as well as the experiences of rural peasants.

REFERENCES

Bach, H. (2007). Composing a visual narrative inquiry. In D.J. Clandinin (ed.), *Handbook of Narrative Inquiry: Mapping a Methodology* (pp. 280–307). Thousand Oaks, CA: Sage.

Bianchi, R.V. (2000). Migrant tourist-workers: Exploring the "contact zones" of post-industrial tourism. *Current Issues in Tourism*, *3*(2), 107–137.

Chambers, D., & Buzinde, C. (2015). Tourism and decolonisation: locating research and self. *Annals of Tourism Research*, *51*, 1–16.

Chilisa, B. (2012). *Indigenous Research Methodologies*. Thousand Oaks, CA: Sage Publications.

Cohen, E. (1974). Who is a tourist? A conceptual clarification. *Sociology*, *22*(4), 527–555.

Escobar, A. (2010). Latin America at a crossroads: Alternative modernizations, post-liberalism, or post-development? *Cultural Studies*, *24*(1), 1–48.

Geerts, W. (2014). Business travel and the environment: The strains of travelling for work and the impact on travellers' pro-environmental in situ behaviour. In C. Weeden & K. Boluk (eds), *Managing Ethical Consumption in Tourism* (pp. 188–206). Abingdon: Routledge.

Graburn, N. (1989). Tourism: The sacred journey. In V. Smith (ed.), *Hosts and Guests: The Anthropology of Tourism* (2nd edn, pp. 17–32). Philadelphia, PA: University of Pennsylvania Press.

Higgins-Desbiolles, F. (2006). More than an "industry": The forgotten power of tourism as a social force. *Tourism Management*, *27*(6), 1192–1208.

Higgins-Desbiolles, F., & Bigby, B. (2022). A local turn in tourism studies. *Annals of Tourism Research*, *92*(4), 103291.

Higgins-Desbiolles, F., Carnicelli, S., Krolikowski, C., Wijesinghe, G., & Boluk, K. (2019). Degrowing tourism: Rethinking tourism. *Journal of Sustainable Tourism*, *27*(12), 1926–1944.

Hollinshead, K. (2012). Tourism and Indigenous reverence: The possibilities for recovery of land and revitalisation of life. In A. Holden & D. Fennell (eds), *The Routledge Handbook of Tourism and the Environment* (pp. 43–54). Abingdon: Routledge.

Information Office of Tibetan Autonomous Region (TAR) (2021). The News Release Conference of the Seventh National Census was held on May 19 in TAR [5月19日西藏举行自治区第七次全国人口普查主要数据情况新闻发布会]. Retrieved June 17, 2022, from http://www.scio.gov.cn/m/xwfbh/gssxwfbh/xwfbh/xizang/Document/1704460/1704460.htm.

Ivancevic, J.M., Konopaske, R., & DeFrank, R.S. (2003). Business travel stress: A model, propositions and managerial implications. *Work and Stress*, *17*(2), 138–157.

Lassen, C. (2006). Aeromobility and work. *Environment and Planning A*, *38*(2), 301–312.

Lichy, J., & McLeay, F. (2018). Bleisure: Motivations and typologies. *Journal of Travel & Tourism Marketing*, *35*(4), 517–530.

Little Bear, L. (2000). Jagged worldviews colliding. In M. Battiste (ed.), *Reclaiming Indigenous Voice and Vision* (pp. 77–85). Vancouver: UBC Press.

Nash, D. (1989). Tourism as a form of imperialism. In V.L. Smith (ed.), *Hosts and Guest: The Anthropology of Tourism* (pp. 37–52). Philadelphia, PA: University of Pennsylvania Press.

Peters, A., & Higgins-Desbiolles, F. (2012). De-marginalising tourism research: Indigenous Australians as tourists. *Journal of Hospitality and Tourism Management, 19*, 76–84.

Qiao, G., Li, F., Xiao, X., & Prideaux, B. (2021). What does tourism mean for Chinese rural migrant workers? Perspectives of perceived value, *International Journal of Tourism Research, 24*(2), 227–239.

Qiu, J., Cha, A., & He, C. (2011). Peasant tourism: Concept, basic problems and study framework. *Tourism Tribune, 26*(10), 13–20.

Spivak, G.C. (1988). Can the subaltern speak? In C. Nelson & L. Grossberg (eds), *Marxism and the Interpretation of Culture* (pp. 271–313). Basingstoke: Macmillan Education.

Sun, J., Ling, L., & Huang, Z. (2020). Tourism migrant workers: The internal integration from urban to rural destinations. *Annals of Tourism Research, 84*, 102972.

Uriely, N. (2001). "Travelling workers" and "working tourists": Variations across the interaction between work and tourism. *International Journal of Tourism Research, 3*(1), 1–8.

Veblen, T. (1994 [1899]). The theory of the leisure class: an economic study of institutions. New York: Random House.

Victor, J.M., Goulet, L.M., Schmidt, K., Linds, W., Episkenew, J.A., & Goulet, K. (2016). Like braiding sweetgrass: Nurturing relationships and alliances in Indigenous community-based Research. *International Review of Qualitative Research, 9*(4), 423–445.

Wang, C., & Burris, M.A. (1997). Photovoice: Concept, methodology, and use for participatory needs assessment. *Health Education & Behavior, 24*(3), 369–387.

Wang, N., Liu, D., & Ma, N. (2008). *The Sociology of Tourism*. Tianjin, China: Nankai University Press.

Wilson, S. (2008). *Research Is Ceremony: Indigenous Research Methods*. Black Point, Nova Scotia: Fernwood Publishing.

Winter, T. (2009). Asian tourism and the retreat of Anglo-Western centrism in tourism theory. *Current Issues in Tourism, 12*(1), 21–31.

World Bank (2020). *Rural Population*. Retrieved June 17, 2022, from https://data.worldbank.org.cn/indicator/SP.RUR.TOTL?view=chart.

30. Conclusion: inspiring a radical politics of community-controlled rural tourism

Heather Mair

At the end of this volume, it should be crystal clear that the relationship between tourism and rural community development is complex, convoluted, contested and by any measure not the panacea it is sometimes made out to be. This collection brings together 43 scholars from around the world and at varying career stages, and their contributions focus in and on places that span the globe. The topics covered are many and wide-ranging and yet, in my view, a handful of primary arguments provide the glue that binds them all together: (1) tourism and rural community development are about far more than the economy; (2) there is opportunity for rethinking the economic dimensions of rural tourism; (3) the potential of 'new' voices and (non-Western) knowledges; and (4) the centrality of community. After engaging each of these primary arguments in light of the chapters and the work showcased in this collection, I will conclude by drawing on these and other ideas to contemplate what could (or should) be next for rural community development and tourism.

TOURISM AND RURAL COMMUNITY DEVELOPMENT ARE ABOUT FAR MORE THAN THE ECONOMY

While in some ways, most of the chapters in this volume attend to matters of economics, the authors quickly make the shift to talk about other dimensions of the tourism system and their roles in rural community development. In Chapter 2 (Part I, 'Conceptual foundations'), Michael Woods, for example, begins with a story about a community worrying about tourists returning to their area as pandemic restrictions are easing. This story allows the reader to understand the tension community members often feel between the pressure to embrace tourism to support economic diversification and recoup lost income, and the desire to avoid the social, cultural and even health impacts tourism can create (e.g., fear of overcrowding, exposure to COVID-19, etc.). However, Woods' real point is that it is the notion of community itself that is often left unpacked, and he draws on relational theory to make the case for a more complex and multifaceted way of understanding community in rural tourism development. In Chapter 8 (in Part II, 'Epistemological foundations, paradigms, and worldviews'), Michael Mackay, Nick Taylor, Harvey C. Perkins, and Tracy Nelson describe the use of social impact analysis on cycle trails, while also seeking to draw attention to early efforts by Mathieson and Wall (1982) to expand the analysis of tourism beyond economic considerations and to argue that we need to understand the 'non-economic and intangible effects [of tourism], such as social and cultural ones' (Pizam & Milman, 1986, p. 29).

Other examples of moving beyond economics include Chapter 19 (in Part IV, 'Processes and practices'), by Tazim Jamal and Blanca Carmago. These authors directly challenge the dominance of an economic lens as 'old ways of seeing and "measuring" success predominantly in terms of economic growth (quantifiable visitor numbers and revenues), [which] are clearly insufficient for a sustainable planet and the well-being of hundred of millions of rural inhabitants' (p. 254). Their focus on ethical considerations takes us away from economics and instead into essential discussions of values, morals, and principles as they are (or could be) made manifest in tourism. Chapter 22, by Dominic Lapointe (in Part IV) also provides much fodder for challenging the economic (capitalist) foundation of tourism development, which typically encourages a vision of tourism as the producer of primarily economic gain. Lapointe instead argues for engendering a way of thinking about the potential of tourism as a way to enhance well-being and quality of life in community places.

The last examples of contributions in this volume, which I see as modelling efforts to move beyond the economic gaze, can be found in Part V, 'Emerging matters and new directions' and include Chapter 24 by Kyle Rich, Chapter 25 by Carol Kline, Chapter 26 by Can-Seng Ooi and Becky Shelley, and Chapter 28, by Marco Eimermann, Doris A. Carson, and Linda Lundmark. First, Kyle Rich, by demonstrating the ways sport tourism events in rural areas are manifestations of rural–urban interdependencies, also supports a way of seeing tourism that is relational, grounded in notions of identity, and not just about economics. Next, with the focus on animals in tourism, Kline's work illustrates the danger of putting economic considerations first as it presents a direct threat to animal welfare. Following Kline's lead, an animal-centric position would represent an important step towards challenging the dominance of economic thinking in rural tourism.

Ooi and Shelley begin their chapter by noting it is, 'uncontroversial to consider tourism an economic resource' (p. 358) and yet go on to disengage from the idea of tourism's economic benefits by building on the notion of regenerative tourism, which risks being couched in economic terms, in order to foreground the development of cultural capital and intergenerational learning. Last, Eimermann et al. devote much attention to the social and relational impacts generated by primarily civic engagement activities of lifestyle migrants in sparsely populated areas. By focusing on the social and civic capacity of these intra-European migrants, our attention is shifted to, 'community tourism that is not necessarily based on supporting market hegemony … but rather on less commercial activities that are meant to benefit members of the community from a social, cultural and emotional well-being perspective' (p. 387). These chapters all make insightful cases for looking beyond the economics of tourism.

RETHINKING THE ECONOMIC DIMENSIONS OF RURAL TOURISM

Related to the ways in which the chapters touched on above are effectively steering rural tourism scholarship away from a path previously dominated by economics, much of the work included in this collection is also helpful in terms of identifying new ways to rethink or reframe tourism's economic dimensions. Chapter 4 by Kelly Vodden, Mery Angeles Perez, and Brady Reid (in Part I) set the stage by exploring the tension between the economic orientation supporting much rural development research and practice (see above) and meeting other community needs related to other sectors as well as social–cultural and ecological sustainability. Their chapter points to the need for new governance models, which can better balance community needs and values with economic considerations. These ideas are explored in more depth in the following subsections.

Of course, the notion of degrowth may come immediately to mind as the first conceptual arrow in our quiver as it is based on the idea that it is critical to not only challenge the economic foundations of tourism, but also the very notion of growth (and the credo that growth – or capitalist expansion – is unquestionably always a good and essential part of tourism development). First, these ideas arise most directly in Part III, 'Key conceptualizations' and specifically in Chapter 14, by Mark C.J. Stoddart and Yixi Yang, as a result of their focus on climate change in tourism. As Stoddart and Yang argue, degrowth is not anti-tourism, but 'asks host communities and governments to engage in the work of envisioning what kinds of tourism they want to cultivate for long-term social, environmental, and economic wellbeing' (p. 189). Given the role of tourism in climate change, the notion of degrowing tourism may be our only hope.

In Chapter 16, Alexander Safonov and C. Michael Hall deal head on with the notion of degrowth and this allows them to confront the economic incentive structure (i.e., government and industry policies) that shapes the growth strategy underlying most tourism development. They argue that 'despite having only one planet and supposedly recognising the dangers of biodiversity loss and climate change, the tourism industry, along with many others, remains committed to growth trajectories with respect to visitors, numbers of travellers, and their expenditure' (p. 208). The authors then position the notion of degrowth (*décroissance*) as not a strategy directly opposed to growth but a consideration of growth within system limits. This, of course, brings us to the local or rural community context, where although Safonov and Hall are sceptical, the scale of tourism activity and development might be sufficient for actually measuring and achieving degrowth.

While not referring to the idea of degrowth directly, Jennifer Sumner (Chapter 12) pulls on the same ideas and explores the potential for rethinking economics in rural tourism. The author describes the 'depredations of the neoliberal economy' (p. 154) and considers how specific manifestations of food in tourism (e.g., farmers' markets, 100 mile stores, etc.) may provide new opportunities for rural development by, 'eschewing the unsustainable growth imperative and committing to development focused on fairness, ecological limits, human benefit and sustainable futures' (p. 162).

In their focus on regenerative tourism, Tawsif Dowla and Karla A. Boluk (Chapter 18) also see the potential of degrowth as they open the door to new approaches to developing tourism (i.e., co-creative and regenerative) that reframe what was heretofore an overly economically driven orientation in rural tourism development. Finally, references to the notion of degrowth also surface in Part IV, 'Processes and practices', where Jamal and Camargo (Chapter 19) describe it as a 'paradigm' (p. 259) that could support the ethical principals and values they deem essential for tourism and rural community development in the future.

THE POTENTIAL OF 'NEW' VOICES (AND NON-WESTERN) KNOWLEDGES

There is an unquestionably important and growing movement to invite, accept, support, and possibly even adopt 'new' ways of knowing about tourism and rural areas and, broadly speaking, this includes Indigenous knowledges. The clearest example of this movement is the chapter by *tebrakunna* country, Emma Lee, and Bryan S.R. Grimwood (Chapter 10, in Part II, 'Epistemological foundations, paradigms, and worldviews'). These authors illuminate and challenge the tendency to accept tourism as a way to address colonial legacies and they then use the essential ideas of refusal and resistance to reshape our thinking about tourism and the role of Indigenous Peoples, knowledges, and power.

Additionally, there are many authors in this volume who, while not necessarily directly focused on Indigenous knowledges, are nonetheless seeking to encourage scholarship that makes room for ways of knowing about rural areas and tourism that have not traditionally been taken up. For example, Xiaotao Yang (Chapter 29, in Part V) draws on research with young Tibetans who use opportunities created by their work to become travellers. This allows the author to illustrate the ways Tibetan language and ways of knowing can be brought to bear on our field, and it broadens and deepens our understanding, not just of the work–travel relationship, but also our knowledge about, and conceptualisations of, travel itself. Indeed, Yang's contribution introduces us to two important Tibetan words, which capture the forms travel takes in contemporary Tibetan society: *yukour* (tourism, ཡུལ་སྐོར།) and *neikour* (pilgrimage, གནས་སྐོར།) (p. 412) and paints a moving picture of the role of both in the lives of her study participants.

In terms of approaches to research, Girish Prayag and Meghan Muldoon (Chapter 5 and Chapter 6, respectively, in Part II) both provide discussions that describe the opportunities and challenges involved with the use of post-positivist and interpretivist approaches in studies of tourism and rural community development. Prayag begins with a reflection on the role of assumptions in shaping research (and one can include here assumptions about whose voices and knowledge experiences are 'worthy' of inclusion in research) and encourages readers to be aware of, 'multiple theoretical lenses and multiple methods to understand how rural tourism can be developed and marketed for greater individual and community well-being and resilience' (p. 64). Muldoon demonstrates for the need for tourism scholars to be open, not just to creating room for participants' experiences, emotions, and knowledge in their work but also to be prepared to grapple with their own onto-epistemological positionality and related assumptions, which may be challenged when one is genuinely and earnestly open to hearing 'new' ways of thinking and knowing about tourism.

Relatedly, the chapter by Hollinshead and Mair (Chapter 9) makes the case for attending to one's own assumptions and sense of judgement about the often taken-for-granted-ness of rural tourism development. These considerations are dramatically extended in Chapter 7 by Christine N. Buzinde, who focuses on taking the broader political economic context of rural development and tourism into account, as well as building critical consciousness through dialogue with students and community members. Buzinde's argument for transformational learning in rural community and tourism development is built on a commitment to create room to hear and attend to the experiences of individuals who are marginalised and oppressed.

Chapter 17 (in Part III) by Sinead Francis-Coan and Annæ Buchmann illustrates how, by listening closely to local community members, one's very notion of the rural tourism landscape becomes complicated and deeply contested. Chapter 20, by Roslizawati Che Aziz (in Part IV, 'Processes and practices') draws our attention to appreciative inquiry (AI), an approach to community development, used frequently but not exclusively in rural community development circles. The author draws on extant research (e.g., Koster & Lemelin, 2009) to make the argument that, 'Efforts to understand the experiences of people living in poverty, Indigenous people, and women is becoming ever more important, which compels a need for non-traditional methods of data gathering and verification' (p. 277). Aziz illustrates how AI can be an approach for putting these 'non-traditional' knowledges and ways of knowing at the centre of tourism and rural community development.

Through their focus on gender in rural tourism, chapters by both Katherine Dashper (Chapter 11, in Part III) and Karla A. Boluk, Gaurav Panse, and Sung Eun Jeon (Chapter 27, in Part V) underscore the importance of bringing the oft-undervalued voices, perspectives and knowledges of women in rural tourism to the surface. Dashper makes a compelling case for realising that taken-for-granted notions of 'the community' are often homogeneous and miss the extent to which women are often marginalised (or excluded altogether). Her contribution points to the need to put women's experiences and knowledges at the centre of tourism and rural community development. Boluk et al. reinforce this argument and provide a theoretical opportunity to put gender and women's experiences in the frame with their use of the feminist ethic of care. This allows the authors to first highlight women's experiences and knowledge as tourism social entrepreneurs and then to provide opportunities for seeing new ways to support and promote care in rural tourism.

THE CENTRALITY OF COMMUNITY

Undergirding the three previous sections, which present my view of many of this volume's contributions to understanding issues of economics and knowledges, is an unwavering commitment to community. Perhaps it has become so obvious that it almost goes without saying, at least since Murphy made the argument in 1985, that members of communities must be an integral part of tourism development decision-making. The scholars whose work is featured in this volume would likely all tend to agree that this is a common refrain. Indeed, my own journey into the world of tourism research began with a doctoral project designed to broaden and support community participation in rural communities in south-western Ontario, Canada (e.g., Reid et al., 2000; Reid et al., 2004; Mair & Reid, 2007). Further, my work (Mair, 2006; 2009) as well as that of many of this volume's contributors, has long reflected on the complex question of who is, and who can speak for, the community, and yet, the essential, overriding question of power to control tourism development decision-making seems to come through especially clearly. Certainly, the sub-area of scholarship in rural tourism development has grown dramatically in these last few decades and its vibrancy and fluidity are profoundly evident in this collection.

For instance, Chapter 3 (in Part I) by Donald G. Reid goes directly to the question of who controls tourism development in rural communities by focusing on how tourism and rural development unfold and why. In advocating community control, Reid tackles the challenge of power hierarchies in communities head on. He also notes that the best form of tourism planning happens before the development takes place (rather than retroactively, and reactively, as is often the case). As Reid writes, 'All the players in the system, particularly those that will be directly affected by the plan, need to be identified and their role and involvement in the control of the project determined a priori' (p. 35).

Bernard Lane contributes two chapters (Chapter 13, in Part III, and Chapter 23, in Part IV), and both, arguably, contend with the question of who controls tourism development in rural communities. First, in describing the often-contentious process of interpreting heritage through tourism in rural communities, Lane sees a clear role for community members to play in deciding first what heritage is and second how it should be part of the tourism offering. In Chapter 23, Lane also clearly delineates the role of tourism academics in providing not only substantive expertise in support of the rural tourism planning process, but also the need for academics to utilise process (facilitation) skills to ensure community members have the tools and capacity to participate meaningfully in tourism development decision-making. Relatedly, David W. Marcouiller (Chapter 15, in Part III) discusses the importance of understanding the role of risk and resiliency for communities engaging in tourism, particularly in regard to his broader observation that the least advantaged members of communities reliant upon tourism, often found in less economically diverse regions, can also be especially vulnerable to its whims and external shocks. The author sees much opportunity in collective action and what he calls a 'progressive approach to decision-making' (p. 195) to ensure community engagement and preparedness.

Perhaps most radically, chapters by Jamal and Carmago (Chapter 19) and by Gianna Moscardo (Chapter 21), both in Part IV, present perspectives that hold even more tightly to ideas of power and control because their focus is on what is, or what could be, the very purpose of tourism development in rural communities. Jamal and Carmago's focus on principles of justice and ethics, are, of course, grounded in a concern for community. Further, they develop a salient description of what they call common principles of 'good' rural tourism (p. 260), which include, among other essential facets: local ownership; direct resident participation; local control; local economic and social benefits, which enable gender equity and women's empowerment; and individual and communal agency and empowerment. Moreover, Moscardo's notion of destination community well-being (DCW), by putting at the centre the question of how tourism can benefit community well-being first and foremost, opens up opportunities for rethinking the very ideas of decision-making and governance as we can ask who decides (and how) what well-being is and the role tourism can or cannot play therein. Building on Flora's Community Capitals Framework (e.g., Flora et al., 2016), Moscardo describes a DCW tourism planning process, which is circular and centred on a destination community's needs and well-being rather than the 'objectives of tourism' (see Figure 21.1, p. 289).

Taken together, the chapters in this collection provide a solid foundation for understanding that the trajectory of tourism and its relationship to rural community development should be in the hands of the community. This trajectory must also move beyond traditional, economic orientation, justification, and goals if it is going to be of benefit socially, culturally, politically, and ecologically and help communities thrive.

WHERE TO NEXT?

One of the great opportunities of editing a collection such as this is the chance to build on what has been said and to contemplate what might come next. In many ways, I feel most strongly about the need to continue to deepen research, practice and writing in response to each of the four primary arguments described above. Rural tourism scholarship that is concerned with community development needs *more* efforts to break away from the dominance of economic-centrism or at least to continue to rework, rethink and reframe how we research and write about the contemporary political economic context in which all tourism systems operate. We need *more* efforts to bring 'new' knowledges to tourism development and scholarship, especially Indigenous and other non-Western ways of knowing into our field, and to continue to follow the paths lit by some of this volume's contributors as they argue for what likely amounts to a great (even revolutionary) upheaval in regard to how we know what we know about tourism and community development, and why. And we need to keep pushing our thinking to develop *more* radical approaches to community-controlled tourism and to find ways to keep building trust in and with communities so our scholarship can take some risks and experiment with alternative forms of engagement, empowerment, and community control.

Nonetheless, other areas of scholarship come quickly to mind as needing *more* attention too and, given the space, this volume likely should have included additional investigations into these areas. Climate change, of course, is the defining challenge of our time and much more work is crucial here. We need to continue to work to understand how rural communities can use, adapt, and also resist tourism (at least the typical manifestations of tourism, which merely exacerbate our problems) in their efforts to respond to and prepare for dramatic weather events, and what are clearly going to be catastrophic climate events going forward. Rural tourism and its role in just transitions, for example, is a rich area for future exploration and taps into concerns raised throughout this collection about power, justice, and equity in rural communities.

The relationship between tourism, rural community development and the more-than-human world is also in need of additional and especially critical investigations. Challenging ideas related to so-called 'nature-based' tourism is an important first step, as are emerging onto-epistemological orientations such as post-humanism. As this area of scholarship develops, incorporating Indigenous knowledges and ways of knowing will be essential.

There is also a need to expand scholarship to investigate the changing political economic context, especially the growing threat of right-wing, autocratic populist movements around the world, which are often disproportionately present and powerful in rural communities. Research linking efforts to formulate and implement community development processes (tourism-related and otherwise) that are working to comprehend and perhaps even respond to the foundations of these movements (disinformation, alienation, isolation, precarity, etc.) would serve the field and rural communities well.

This volume ends where it began, reflecting on an extended, if uneven, global pandemic, which at least for some, continues to offer a chance for a deep reckoning. In 2021, countless papers and pieces were published; with scholars taking advantage of a chance to rethink the way tourism has developed around the world (see for example, Lew et al.'s 2020 special issue in *Tourism Geographies*). Without taking away from these innovations and the quality of these offerings, I remain sceptical and deeply concerned about the opportunity to reform tourism. As I observed more than 15 years ago (Mair, 2006), the economic (capitalist) foundations of tourism are deep rooted and hard to shake, and while this period of pandemic-inspired reflection was inspiring, I agree with Development Studies scholars, Büscher et al.: 'In other words: any apparent short-term gains from this pandemic or similar interruption of business-as usual will be ineffective without concerted efforts for broader and deeper political-economic change' (p. 2).

Of course, the seeds of inspiring and activating this deeper political economic change can be found in many of the pieces included in this contribution and I am profoundly inspired by the work of all the scholars featured here. Their collective effort to focus their lenses sharply on the 'community' aspect of tourism and rural community development, is hopeful, inspiring, and motivating. I hope readers agree.

REFERENCES

Büscher, B., Feola, G., Fischer, A., Fletcher, R., Gerber, J.F., Harcourt, W., et al. (2021). Planning for a world beyond COVID-19: Five pillars for post-neoliberal development. *World Development, 140*, 105357.

Flora, C.B., Flora, J.L., & Gasteyer, S. (2016). *Rural Communities* (5th edn). Boulder, CO: Westview Press.

Koster, R.L.P., & Lemelin, R.H. (2009). Appreciative inquiry and rural tourism: A case study from Canada. *Tourism Geographies, 11*(2), 256–269.

Lew, A.A., Cheer, J.M., Haywood, M., Brouder, P., & Salazar, N.B. (2020). Visions of travel and tourism after the global COVID-19 transformation of 2020. *Tourism Geographies, 22*(3), 455–466.

Mair, H. (2006). Global restructuring and local responses: Investigating rural tourism policy in two Canadian communities. *Current Issues in Tourism, 9*(1), 1–45.

Mair, H. (2009). Searching for a new enterprise: Themed tourism and the re-making of one small Canadian community. *Tourism Geographies, 11*(4), 462–483.

Mair, H., & Reid, D.G. (2007). Tourism and community development vs. tourism for community development: Conceptualizing planning as power, knowledge, and control. *Leisure/Loisir, 31*(2), 403–425.

Mathieson, A., & Wall, G. (1982). *Tourism, Economic, Physical and Social Impacts*. London: Longman.

Murphy, P. (1985). *Tourism: A Community Approach*. Abingdon: Routledge.

Reid, D.G., Mair, H., & Taylor, J. (2000). Community participation in rural tourism development. *World Leisure Journal, 42*(2), 20–27.

Reid, D.G., Mair, H., & George, W. (2004). Community tourism planning: A self-assessment instrument. *Annals of Tourism Research, 31*(3), 623–639.

Pizam, A., & Milman, A. (1986). The social impacts of tourism. *Tourism Recreation Research, 11*(1), 29–33.

Index